REPRODUCTION OF MARINE INVERTEBRATES

Volume I

Acoelomate and Pseudocoelomate Metazoans

ADVISORY BOARD

REPRODUCTION OF MARINE INVERTEBRATES

Volume I

Acoelomate and Pseudocoelomate Metazoans

Edited by

Arthur C. Giese

Department of Biological Sciences and Hopkins Marine Station Stanford University Stanford, California

John S. Pearse

Division of Natural Sciences University of California Santa Cruz, California

ACADEMIC PRESS New York and London 1974

A Subsidiary of Harcourt Brace Jovanovich, Publishers

ACADEMIC PRESS, INC.
111 Fifth Avenue, New York, New York 10003

United Kingdom Edition published by
ACADEMIC PRESS, INC. (LONDON) LTD.
24/28 Oval Road, London NW1

LIBRARY OF CONGRESS CATALOG CARD NUMBER: 72-84365

PRINTED IN THE UNITED STATES OF AMERICA

CONTENTS

LIST OF CONTRIBUTORS

Numbers in parentheses indicate the pages on which the authors' contributions begin.

Richard D. Campbell (133), *Department of Developmental and Cell Biology, University of California, Irvine, California*

Paul E. Fell (51), *Department of Zoology, Connecticut College, New London, Connecticut*

Arthur C. Giese (1), *Department of Biological Sciences and Hopkins Marine Station, Stanford University, Stanford, California*

Catherine Henley (267), *Department of Zoology and Laboratories for Reproductive Biology, University of North Carolina, Chapel Hill, North Carolina*

Robert P. Higgins (507), *Office of Environmental Sciences, Smithsonian Institution, Washington, D. C.*

W. D. Hope (391), *Department of Invertebrate Zoology, Smithsonian Institution, Washington, D. C.*

William D. Hummon (485), *Department of Zoology, Ohio University, Athens, Ohio*

John S. Pearse (1), *Division of Natural Sciences, University of California, Santa Cruz, California*

Helen Dunlap Pianka (201),* *Friday Harbor Laboratories and the Department of Zoology, University of Washington, Seattle, Washington*

Nathan W. Riser (359), *Marine Science Institute, Northeastern University, East Point, Nahant, Massachusetts*

Wolfgang Sterrer (345), *Bermuda Biological Station, St. George's West, Bermuda*

Anne Thane (471), *Zoological Institute, University of Aarhus, Aarhus C, Denmark*

*Present address: 3221 Clearview Drive, Austin, Texas.

PREFACE

The extensive and widely scattered information on reproduction in marine invertebrates has never been summarized, and most books covering comparative physiology and ecology of marine invertebrates omit consideration of reproduction for lack of such a summary. Yet reproduction is one of the fundamental activities of organisms. This was even pointed out by Aristotle in ancient times. This treatise should be of particular interest because some of the marine invertebrates have remained relatively undifferentiated and even archaic in their reproductive processes, while others have become specialized and have changed considerably from their prototypes in this respect. Thus they provide a unique perspective on the evolution of reproductive mechanisms and behavior in the animal kingdom. During the course of reproduction many invertebrates undergo profound changes in anatomy, physiology, and behavior. Moreover, reproductive success is closely linked to environment so that reproductive activities are often sensitive to, and synchronized by, environmental changes. This treatise should therefore fill a need and be of use to students of marine biology, ecology, and reproduction in general.

The work will consist of seven volumes. In the first six volumes, specialists deal with particular marine invertebrate groups from a similar vantage point to provide up-to-date coverage. All groups of free-living marine invertebrates are considered rather than only groups which have already received considerable attention. In Volume VII the authors will deal with general aspects of reproductive physiology and ecology.

We are indebted to our Advisory Board for suggestions on the scope and organization of the treatise, to the Board and a larger community of biologists for encouragement and suggestion of prospective authors, and to all the authors who enthusiastically assumed responsibility for chapters which required of them much effort and time. Finally, we are indebted to Dr. Vicki Buchsbaum Pearse for her painstaking editorial assistance and to the staff of Academic Press for their advice and help with the development of the manuscript.

ARTHUR C. GIESE
JOHN S. PEARSE

Chapter 1

INTRODUCTION: GENERAL PRINCIPLES

Arthur C. Giese

and

John S. Pearse

On the basis of the number of phyla and classes, the most numerous kinds of animals are invertebrate and marine. Marine invertebrates thus provide the widest scope of diversity in animal life, some relatively simple in organization (e.g., sponges, cnidarians, and flatworms), and others much more complex (e.g., echinoderms, molluscs, and arthropods). Most invertebrates probably originated and became diversified in the sea; the sea is considered to have changed little in physical and chemical properties since life began (Rubey, 1951; Robertson, 1953; Pearse and Gunter, 1957). In general, however, studies of reproduction in animals have been concerned mainly with vertebrates, especially with particular classes of vertebrates.* Reviews on invertebrates deal only with specialized aspects of reproduction.†

Coverage in this treatise of all groups of free-living marine metazoans will focus attention on those groups of invertebrates for which there is presently little or no information as well as on those possessing unique features which might point to fundamental problems in reproductive biology and might be especially revealing for an understanding of the regulation of reproduction. It is the intention of the editors that this treatise may serve as a framework from which further, more coordinated studies on reproductive biology will proceed.

Protozoans have been excluded because there has been little work on reproduction of marine forms, and because many aspects of their reproduction seem very different from those in metazoans (see Fenchel, 1969). Work on parasites and nonmarine invertebrates** is also excluded since these groups seem to mainly illustrate modifications and specializations of general patterns seen in related free-living marine forms.

Because reproduction is of such universal biological importance, it is probably governed by general principles. Recognition, elucidation, and formulation of such principles is therefore a central objective of this treatise.

1.1 Basic Events and Terminology of Reproduction

Reproduction occurs by sexual or asexual means (Fig. 1). In *asexual* reproduction, which occurs by budding, fission, fragmenta-

*See, for example, general: Frazer, 1959; Bullough, 1961; Van Tienhoven, 1968; the journal *Biology of Reproduction;* the series *Advances in Reproductive Physiology;* the yearly chapter in the *Annual Review of Physiology;* fish: Götting, 1961; Breder and Rosen, 1966; Nakano, 1968; Hoar, 1969; reptiles: Tinkle, 1969; Licht *et al.*, 1969; birds: Farner and Follett, 1966; Lack, 1967; Lofts and Murton, 1968; mammals: Marshall and Parkes, 1950-1956; Asdel, 1964; Harrison, 1969; Sadlier, 1969.

†For example, breeding cycles: Giese, 1959; sex determination: Bacci, 1965; Crew, 1965; genital morphology: Beklemishev, 1969; hermaphroditism: Ghiselin, 1969; oogenesis: Raven, 1961; Schuetz, 1969a; Davidson, 1968; Busson-Mabillot, 1969; spermatogenesis: Roosen-Runge, 1969; fertilization: Austin, 1968; Franklin, 1970; development: Kumé and Dan, 1968.

**See, for example, the recent reviews on insects: de Wilde, 1964; Davey, 1965; Engelmann, 1970.

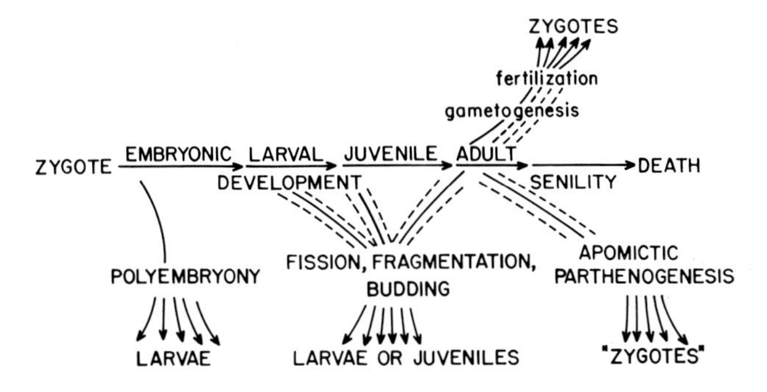

FIG. 1. Diagram of the types of reproduction that can occur during the life cycle of an animal.

tion, formation of reproductive bodies (e.g., gemmules, frustules, statoblasts), polyembryony, or ameiotic (apomictic) parthenogenesis, no recombination of genetic materials takes place (Vorontosova and Liosner, 1960; Herlant-Meewis, 1965). Asexual reproduction may be cyclic and may occur only at certain times of the year or during certain stages of the life cycle. Asexual reproduction enables organisms to reproduce rapidly during favorable conditions, especially in unstable environments, or, as in some freshwater forms, it provides a means of surviving adverse conditions. These topics will be discussed in the chapter on asexual reproduction in a future volume.

Genetic recombination occurs in *sexual* reproduction, almost always as a result of meiosis and the subsequent fusion (amphimixis) of genetically different gametes. Genetic recombination also occurs in meiotic or automictic parthenogenesis, in which meiosis takes place and the haploid products recombine to form new diploid cells (parthenogamy); in a strict sense, this should be considered a form of sexual reproduction. Moreover, ameiotic diploid and meiotic haploid forms of parthenogenesis are sometimes considered to be types of sexual reproduction because they obviously are derived from sexual reproductive processes (i.e., the process of oogenesis is modified). The production of males by parthenogenesis is termed arrhenotoky, while parthenogenetic production of females is termed thelytoky. See White (1954), Bacci (1965), Crew (1965), and Engelmann (1970) for more complete discussions on parthenogenesis.

Gametes are formed through *gametogenesis*, a process which is markedly similar among most animals (Fig. 2). Generally, primordial

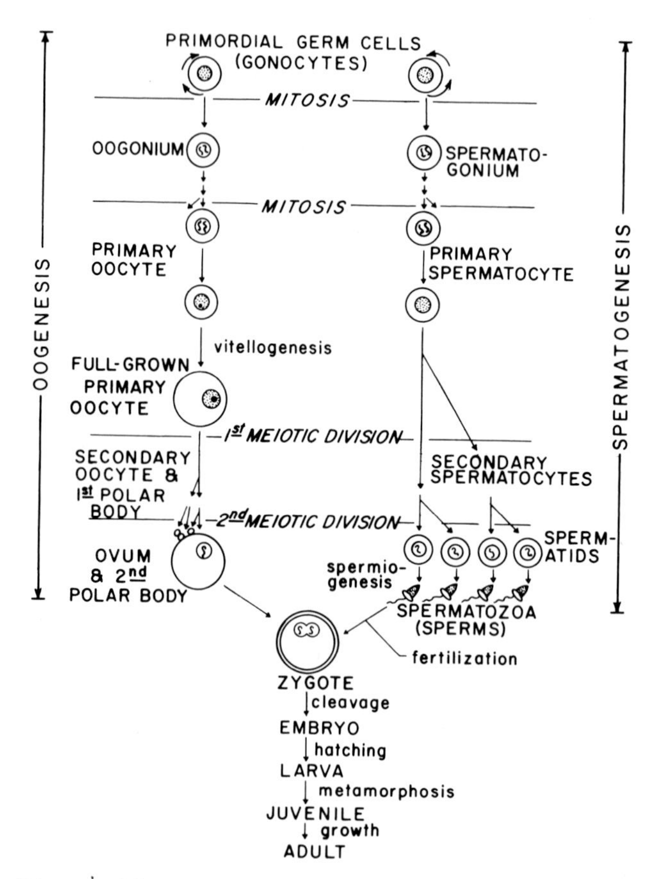

FIG. 2. Diagram of the general cellular changes that occur during gametogenesis.

germ cells (gonocytes) appear early in the development of an animal and either migrate to, or form the locus of, the gonads. Primordial germ cells multiply by mitosis and produce gonial cells. Gonial cells (spermatogonia in males, oogonia in females) can sometimes be separated into multiplying (primary) and terminal (secondary) stages. Primary gonial cells usually go through a limited number of mitotic divisions (usually 2-6), and the resulting secondary gonial cells transform into primary spermatocytes (in males) or primary oocytes (in females). In practice, primordial germ cells, primary and secondary gonial cells, and young primary spermatocytes and primary oocytes are difficult to distinguish from each other; all are usually relatively large cells, each containing a conspicuous large nucleus.

Both primary spermatocytes and primary oocytes undergo a number of characteristic nuclear changes during prophase of the first meiotic division. These changes involve the duplication, pairing, condensation, and partial separation of the homologous chromosomes, and the more or less discrete stages are termed leptotene, zygotene, pachytene, diplotene, and diakinesis (see White, 1954; Raven, 1961; Bacci, 1965). In spermatogenesis, these changes are followed by the first meiotic division of the primary spermatocytes to form the secondary spermatocytes. The second meiotic division occurs soon after the secondary spermatocytes are formed, and produces haploid spermatids (four spermatids for each primary spermatocyte). Differentiation of the spermatids into mature spermatozoa (sperms) is termed spermiogenesis or spermateleosis. Spermiogenesis is always completed, and usually large numbers of sperms are accumulated, before the sperms are shed.

Nuclear changes are arrested when the diplotene stage is reached in the primary oocytes, and the paired, partially separated, chromosomes are dispersed in the nuclear sap of the enlarged nucleus (termed the germinal vesicle). Usually the nucleus contains one conspicuous nucleolus at this stage. Subsequent primary oocyte development goes through two distinct phases (Raven, 1961; Schuetz, 1969a). First there is a period of intense RNA synthesis and the large number of ribosomes accumulated typically cause the cytoplasm of the cell to be strongly basophilic. This period is followed by growth of the primary oocyte through the uptake and accumulation of nutrients (vitellogenesis) (Schjeide *et al.*, 1970), and often the cytoplasm becomes weakly basophilic or even acidophilic. Primary oocyte growth is usually assisted by accessory cells in the ovary called nutrient, follicle or nurse cells (see Davidson, 1968, for distinction of these cell types). Full-grown primary oocytes vary in size among different species, but usually they range from about 75 μm to several millimeters in diameter. After reaching full size, the primary oocytes may undergo the two meiotic divisions (maturation divisions) to form haploid ova. Secondary oocytes, like secondary spermatocytes, are usually very transitory and are not often encountered. Some invertebrates (e.g., many echinoids) accumulate ova in the ovaries and these are shed during spawning. In many other invertebrates, however, "eggs" are shed as full-grown, diploid, primary oocytes, and the meiotic divisions are completed after spawning (e.g., asteroids) or fertilization (e.g., platyhelminths). These differences in the timing of the completion of meiosis with respect to fertilization have led to considerable confusion in terminology (Schuetz, 1969a). In this treatise, the term *ova* refers to haploid female gametes only; the more general term, *eggs*, is used for female gametes (primary oocytes or ova), zygotes, or embryos which are held or shed by the female.

Most individuals of many species of marine invertebrates have separate sexes, and are either males or females which produce sperms or eggs, respectively; such species are said to be *unisexual, dioecious,* or *gonochoric.* Most individuals of many other species are *bisexual, monoecious,* or *hermaphroditic* and produce both sperms and eggs either simultaneously or sequentially. Usually, but not always, simultaneous hermaphrodites are not self-fertile. Sequential hermaphrodites initially may produce sperms or eggs, processes referred to as *protandry* or *protogyny,* respectively. Sometimes hermaphroditic individuals are found in normally gonochoric species; when such hermaphrodites are rare the species is said to have stable gonochorism, but when they are relatively common it is said to have labile gonochorism. For recent discussion of sex determination and hermaphroditism, see Crew (1965), Bacci (1965), and Ghiselin (1969).

Usually, if not always, sexual reproductive activities are cyclic within individuals, that is, they proceed through a repetitive series of events, the distinctive feature being the order of occurrence rather than the duration of each event (Kleitman, 1949). Central to these cyclic events is the *gametogenic cycle* within an individual (Fig. 3).

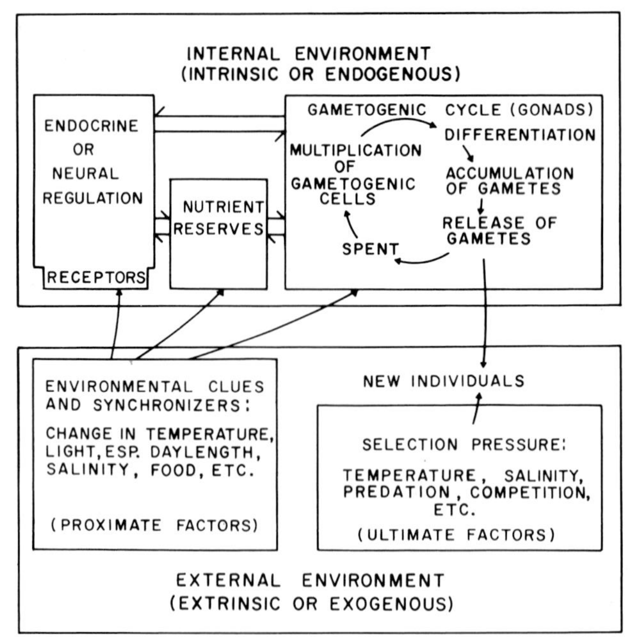

FIG. 3. Diagram of a gametogenic cycle in relation to internal and external regulation.

A gametogenic cycle usually includes (1) an accumulation of nutrients to be utilized during gametogenesis, (2) proliferation of gonial cells and their differentiation into gametes (gametogenesis), (3) accumulation of ripe gametes, (4) release of the gametes, and (5) a reproductively quiescent or spent period when relict gametes are removed or the adult dies. Often, little activity seems to occur in the gonads following the release of the gametes, and this period is sometimes called the "rest period." However, intense biochemical activity such as RNA synthesis in young sea urchin oocytes can occur during this period (Davidson, 1968). Moreover, periods with little synthetic activity can occur during other stages of a gametogenic cycle (as in the storage of gametes in echinoids).

Gametogenic cycles within individuals are perhaps *intrinsic,* their timing being regulated by *endogenous* (internal) factors. These factors may include the accumulation of nutrients and an interplay of hormones between the controlling centers and the gonads. Repression of gametogenic cycles by endogenous control may result in quiescent periods or sometimes, refractory periods. *Extrinsic* or *exogenous* (external) factors may initiate and synchronize the timing of gametogenic events with environmental changes. Whenever individual gametogenic cycles are synchronized within a species population so that the individuals breed simultaneously during a breeding season, exogenous regulation is probably involved.

The *breeding season* of a species is the period of the year when most individuals in a population have numerous ripe gametes available to be released either by spawning in the sea *(spawning season)* or by copulatory transfer during mating *(mating season).* Breeding seasons are usually delineated in terms of months. In some species there is only one gametogenic cycle in each individual during each breeding season. The gametes may be released all at once or intermittently during the season. Individuals of other species, especially those in the tropics, may go through several gametogenic cycles during a single breeding season. Individuals of species which lack well-defined breeding seasons (i.e., those in which the population apparently reproduces continuously) probably undergo successive gametogenic cycles throughout the year. In such cases there is usually little synchronization between different individuals.

In many species, sperms, zygotes, or embryos are retained by the female for various periods of time after spawning or mating. For example, in opisthobranchs and many other gastropods, sperms are stored after copulation, and fertilization and egg-laying may continue for weeks, or even months, after mating (Fretter and Graham, 1964;

Beeman, 1970). In males of the European oyster *Ostrea edulis*, on the other hand, sperms are spawned directly into the sea, fertilizing the eggs held in the mantle cavity of females, and larvae are released several weeks later (Yonge, 1960). In the barnacle *Balanus balanoides* copulation and fertilization occur in the fall or winter, the resulting embryos are brooded in the mantle cavity, and the larvae are hatched and released in the spring (Barnes, 1957). Sometimes the periods of brooding, egg-laying, or larval release are termed breeding or spawning seasons; to avoid confusion they should be referred to simply as *brooding, egg-laying,* or *larval release periods,* respectively. The period of larval release also has been called the "swarming period" but this term is better reserved for swarming by polychaete epitokes and other animals during spawning.

Precise breeding seasons may be important to provide larvae or juveniles with favorable environmental conditions. It is essential to recognize and distinguish between (1) exogenous factors which serve as clues to synchronize the cycles and (2) environmental conditions that exert selective pressure on the timing of the breeding season of the species. Baker (1938) distinguished these as the *proximate* and *ultimate* causes of breeding cycles, respectively. Knowledge of proximate control mechanisms of reproductive cycles, both endogenous and exogenous, is still fragmentary for most marine invertebrates. Seasonal fluctuations of sea temperatures appear to be among the most important proximate control factors ("Orton's Rule," Thorson, 1946). Even less is known about ultimate adaptive significance but most cycles are timed so that the production of larvae or juveniles is synchronized with periods favorable for feeding ("Crisp's Rule," Qasim, 1956).

Often the sperms or eggs are packaged in spermatophores or egg cases, respectively, for shedding. Sperms must be released from the spermatophores before they can fertilize the eggs and various degrees of embryonic development usually proceed in the egg cases before the larvae or juveniles are released. In this treatise, the term *embryo* refers to all developmental stages which occur within the parent, egg case, or egg membrane. *Larvae* are free swimming stages which pass through a metamorphosis to form the immature juvenile individual. Larvae may be demersal (bottom dwelling) or planktonic (floating), and they may be nourished by yolk reserves (lecithotrophic), or feed on plankton (planktotrophic) or on bottom material (benthotrophic). *Juveniles* are miniature adults, resulting from larval metamorphosis, or posthatching stages which grow in size to attain adulthood without metamorphosis.

1.2 Methods of Estimating Sexual Reproductive Activity

Methods used to determine reproductive conditions in marine invertebrates vary among different groups studied and among different researchers (Giese, 1959a). Because the methods are so varied and often not quantitative, much information in the literature on reproduction has limited value. The values and limitations of several of the more common methods are discussed in the following sections.

1.2.1 The Gonad Index

The most widely used quantitative method for estimating reproductive activity is the gonad index. The gonad index method has been used with several groups of animals, including marine invertebrates (Moore, 1934; Giese, 1967a). The method, while not precise, is especially suitable for handling large numbers of individuals. The gonad index is calculated in several ways, but usually it is the ratio of the gonad wet weight (or volume) to the wet weight (or volume) of the whole animal, expressed as a percentage.

Sometimes the body cavity contains variable amounts of water resulting in meaningless fluctuations of the gonad index; this limitation may be avoided by determining ratios for eviscerated body wet weight (Pearse, 1965) or dry weight. Other measurements have also been used, such as cross-sectional area of gonads (Boolootian *et al.*, 1962), or linear measurements of the body and gonad such as diameter or length. Care should be taken to equate dimensions when volumetric and linear measurements are used together, because weight and volume values increase by approximately the cube of linear measurements (see, e.g., Lewis, 1966).

Determination of the gonad index is an attempt to measure the relative reproductive condition of animals of variable sizes so that changes in their gonads at different times can be compared (Giese, 1967a). It rests upon the assumption that the ratio of body parts varies little with change in size of the animal. Such an assumption is not always valid for exceptionally large or small animals; possible correlation between gonad indices and animal sizes should be checked by regression analyses (e.g., Pearse, 1965, 1970). The relation of gonad size to animal size is not proportional between juveniles and adults (of sexually mature size) as shown for sea urchins by Fuji (1960, 1967). In animals with large, fluid-filled body cavities, such as sea urchins, the proportion of the body mass that is body wall tissue decreases with increase in body size. Such a change introduces some error in gonad index calculations. Nevertheless, when similar-sized, sexually mature animals are used in a sample, as is usually the case, their gonadal and other body component indices are generally independent of animal size (Fuji, 1967; Giese, 1967b, 1969; Giese *et al.*, 1967).

A major limitation of the gonad index method is that unless accompanied by microscopic examination of the gonads, it indicates little as to what is occurring within the gonads. In species possessing little nutritive tissue in the gonads an increase in gonad index may be interpreted as a buildup of gametogenic cells and gametes, while a decrease, as spawning (Giese, 1959b). In species possessing considerable nutritive tissue in the gonads (e.g., most sea urchins), both an increase and a decrease in gonad index may be a consequence of changes in the number or mass of nutritive cells without a corresponding change in gametogenic tissue (Moore, 1937; Pearse, 1969a,b). Moreover, in such forms gametogenesis may occur at the expense of nutritive tissue, without a corresponding change in gonad index. Finally, even in some species with little or no nutritive tissue in the gonads, a fraction of the oocytes may provide nutrients to others, and only minor fluctuations in the gonad index may occur during oogenic growth (e.g., in the antarctic sea star *Odontaster validus*, Pearse, 1965). However, if the cellular composition of the gonads, and their changes, are known, the gonad index method can be most useful in showing changes over a reproductive cycle.

1.2.2 Observations of Spawning Animals

Perhaps the simplest method of estimating reproductive activity is the observation of spawning animals in the field or laboratory. While such observations are inexact, they provide important data on the presence of mature gametes. Sperms and fertilizable eggs obtained through dissection or induced spawning (shaking, standing out of sea water, temperature shock, electrical stimulation, injection of potassium chloride or spawning "hormones") (Costello *et al.*, 1957; Kumé and Dan, 1968) add to such information. However, measurement of spawning by any of these methods is not quantitative and estimates the degree of "ripeness" only grossly. Moreover, spawning determinations can be misleading since in many species in which quantitative methods (gonad index and histology) demonstrate cyclic seasonal reproduction, occasional individuals containing ripe gametes may be found at almost any time of the year.

1.2.3 Gonadal Smears and Sections

Microscopic examination of gonadal smears or histologic sections is often used to estimate reproductive activity. Information from such observations can define the state of reproductive activity within the gonads. Usually such observations are not quantitative because the gonads are graded into a series of "maturity stages" which are not the same for different investigators. Attempts have therefore been made to quantify analyses of histologic sections of gonads, e.g., by estimating oocyte diameters. When the size distribution frequencies of the oocytes are plotted against time, it is possible to ascertain the progressive development of successive oocyte crops, and by this means, the reproductive cycle. This method, introduced with fish oocytes (Clark, 1934), has more recently been used with many invertebrates: asteroids (Pearse, 1965), echinoids (Pearse and Giese, 1966), sipunculids (Towle and Giese, 1967), sponges (Fell, 1969), molluscs (Webber and Giese, 1969), polychaetes (Olive, 1970), and ophiuroids (Fenaux, 1970). Holland (1967) and Pearse (1969a,b, 1970) have measured the thickness of the layers of nutritive tissue, spermatogenic tissue and spermatozoan mass in sea urchin testes and followed the change in thickness of each with time, and Holland and Holland (1969) have followed changes in area (in section) of different gonadal cell populations.

When quantitative histological methods are coupled with gonad index measurements, a fairly clear estimate of reproductive changes in a population can be obtained. Analysis of seasonal reproductive changes in the sea urchin *Echinometra mathaei* is shown in Fig. 4 as an example; it can be seen that: (1) gonad growth in the autumn, as indicated by the gonad index, was due to increase in nutritive tissue, (2) gametogenesis involving oocyte growth and proliferation of spermatogenic cells occurred in the spring and early summer, and (3) spawning, as indicated by decreases in gonad indices, percent animals with numerous gametes, thickness of spermatozoa, and relative proportion of ova, occurred in the summer and autumn.

1.2.4 Following Changes in Individual Animals

Most analyses of reproductive activities are static in the sense that they are based on samples taken periodically through the year. Yet when analyzed as shown in Fig. 4, such data show dynamic changes, indicating the initiation of gametogenesis and the span of various phases of the reproductive cycle. Mass rearing of animals in the laboratory would also provide much information on reproductive

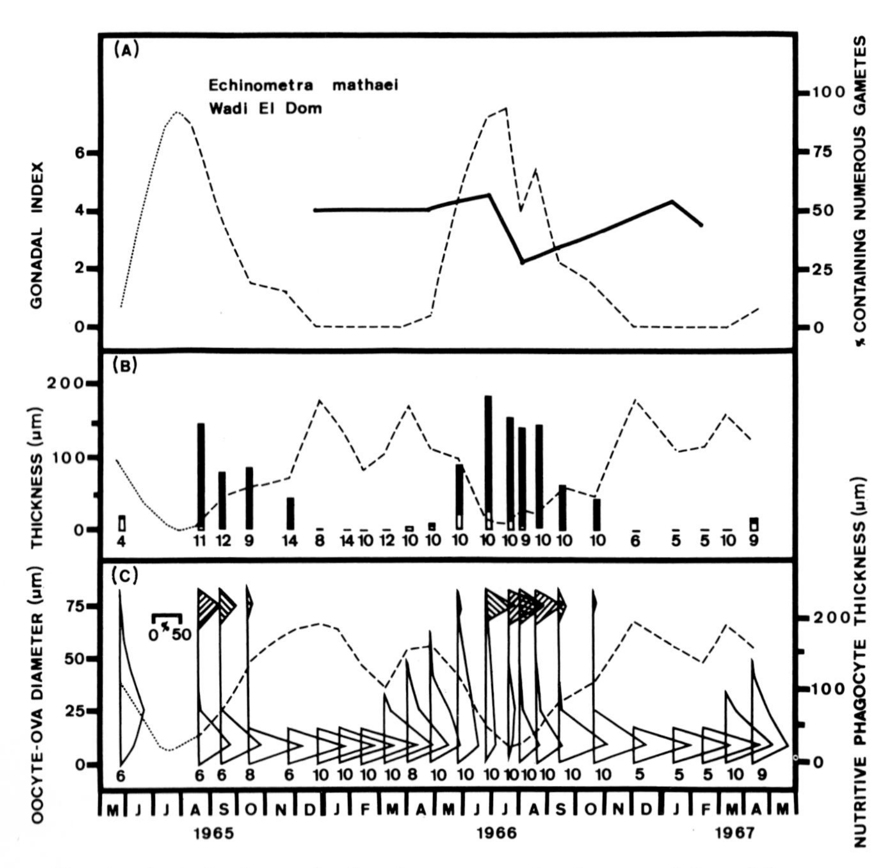

FIG. 4. Analysis of reproductive changes occurring in a population of the sea urchin *Echinometra mathaei* in the Gulf of Suez. (A) Mean gonad index (solid line) and percentage of animals containing numerous gametes (broken line). (B) Testes, mean thickness of the spermatogenic cell layer (open bars), spermatozoa (solid bars), and nutritive tissue (broken line). (C) Ovaries, mean size distributions of the oocytes (open polygons) and ova (cross hatched polygons), and mean thickness of the nutritive tissue (broken line). [From Pearse (1969b).]

changes, but this has been achieved only infrequently (e.g., Roosen-Runge, 1970) and the results obtained may be difficult to relate to nature.

In many studies it is desirable to follow reproductive changes that occur within the same individual. Crisp and Davies (1955) were able to do this in the barnacle *Elminius modestus* by watching gonadal development through the transparent bases of the animals after they settled on glass slides. Periodic biopsies of the gonads of the same animal have been done in some cases (e.g., on sea urchins, Boolootian, 1963; and sea stars, Delavault and Buslé, 1970), but the results may be difficult to interpret because of possible damage done to the animals. Tritiated thymidine has been used to label gonial cells during periods of DNA synthesis in order to estimate time requirements to reach subsequent stages of spermatogenesis in sea urchins (Holland and Giese, 1965) and gastropods (Beeman, 1970). Culture of gonads *in vitro* also is a promising method for studying factors which regulate gametogenesis, especially endocrine (see Lutz, 1970; Ziller-Sengal, 1970).

1.2.5 Sampling Eggs, Embryos, Larvae, and Juveniles

The reproductive activity of a species may be estimated by sampling its eggs, embryos, larvae, or juveniles during and after its breeding period. Eggs, larvae, and juveniles have been sampled from the plankton in the vicinity of the breeding population to estimate the breeding period and larval release time, for example, with brittle stars (Thorson, 1936; Semenova *et al.*, 1964), oysters (Korringa, 1947), euphausiids (Marr, 1962), prawns (George, 1962), and lobsters (Johnson, 1960). The spawning patterns of many species in the White Sea have been analyzed by the use of plankton samples (Mileikovsky, 1970a). Gonadal maturation of adult sea urchins also has been correlated with the presence of the larvae in the plankton (Fenaux, 1968). Species which brood their embryos are especially suitable for this type of analysis because the fraction of adults which are brooding at any time and the stage of the embryos can be readily estimated (e.g., Boolootian *et al.*, 1959a; Knudsen, 1964; Reese, 1968; Steele and Steele, 1969; Chia and Rostron, 1970).

It is also desirable to have information on the settling of larvae and the success of the juveniles to estimate the ecological conditions which make one period of the year more favorable for reproduction than others. Attempts to gain such information on a quantitative basis over a suitable period of time have been rare, however, and the information so gained is difficult to interpret. Loosanoff (1964), for example, estimated sea star settlement over a period of 25 years and could find little correlation between the success of the settlement and environmental conditions.

We may summarize the types of data useful for understanding the changes that occur during reproduction as follows: estimates of (1) the amount of material that is expended during reproduction as indicated by changes in gonad indices, (2) the changes that occur within the gonadal cells during a gametogenic cycle as shown by light and electron microscopy, (3) the time course of gametogenesis, as revealed by successive samples taken from a population of animals or, when feasible, from the same animal, (4) the time and conditions under which spawning occurs, as revealed by field observations, and (5) the reproductive success as shown by analysis of the plankton and newly settled young. To our knowledge, no one has assembled all these types of data in the study of an annual reproductive cycle for any species.

1.3 The Timing and Patterns of Reproduction

The timing of reproduction in populations of different species of marine invertebrates takes many patterns. Reproduction may occur rhythmically or sporadically during part or all of the year, or it may occur more or less continuously throughout the year. When rhythmic, the period of breeding may be daily, semimonthly, monthly, semiannual, annual, or biennial (Fig. 5).

Continuous production of gametes throughout the year is probably rare in an individual, inasmuch as successive gametogenic cycles usually have at least some pause between them. Gametogenesis may be staggered among different individuals during the year, however, so that reproduction appears to occur continuously throughout the year for the entire population of the species. The term continuous is used in this sense in this account. Such "continuous" reproduction may be expected in areas of little seasonal change, such as in the deep sea and parts of the tropics. In such areas there would be little if any selective pressure favoring reproduction at any one time of the year. Data on deep-sea animals are still fragmentary, but at least a few species are reported to reproduce throughout the year (Madsen, 1961; Sanders and Hessler, 1969).

Continuous reproduction has been reported in many populations of tropical species (e.g., Stephenson, 1934; Panikkar and Aiyar, 1939; Paul, 1942; Antony Raja, 1963; Goodbody, 1965; Rahaman, 1967; Warner, 1967). In tropical populations of the Indo-Pacific sea urchin *Echinometra mathaei*, different individuals are in different phases of their gametogenic cycles and about 50-60% of the individuals are usually ripe at any one time (Pearse, 1968; Pearse and Phillips, 1968). The gametogenic cycle in individuals of this species takes about 6 months (Pearse, 1969b), so most individuals probably undergo two gametogenic cycles each year. Continuous or nonrhythmic

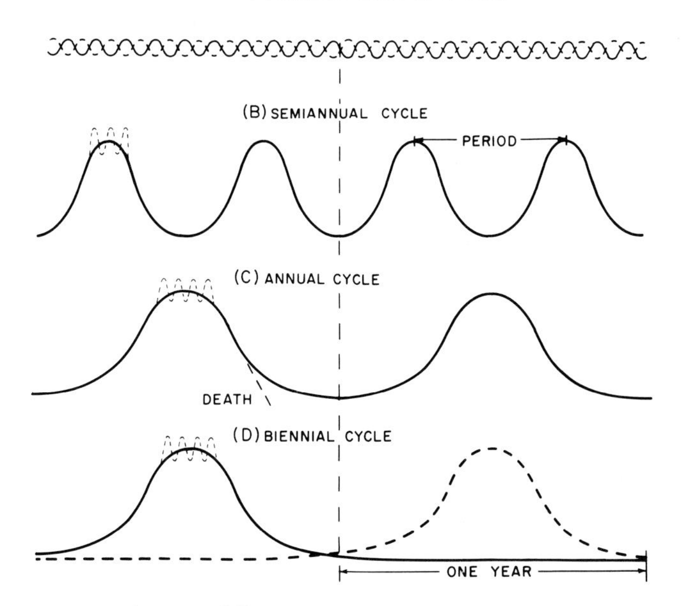

FIG. 5. Diagram showing different patterns of breeding of marine invertebrates. Each wave crest represents the height of the breeding season when the gonad is largest and gravid. The dotted waves superimposed on the large waves indicate successive gametogenic cycles of which there may be only one or many in each breeding season. Some species die after breeding; this is shown in pattern (C) but it could occur at the conclusion of breeding in any of the patterns illustrated.

reproduction also occurs in many terrestrial animals in the tropics where seasonal fluctuations are slight (e.g., frogs: Berry and Varughese, 1968; reptiles: Inger and Greenberg, 1966; Tinkle, 1969; birds: Lofts and Murton, 1968; snails: Berry, 1965). Reproduction in continuously breeding tropical species is not likely to be of the same intensity throughout the year, however; when closely examined, many of these populations show periods of more intense reproduction (e.g., Stephenson, 1934; Giese *et al.*, 1964; Pearse and Phillips, 1968) and these fluctuations probably reflect minor environmental fluctuations. Variations in reproductive intensities also are found in continuously breeding species in other areas, e.g., an arctic-boreal pteropod (Mileikovsky, 1970b).

Some species inhabiting shallow-water temperate areas where definite environmental fluctuations occur, also reproduce throughout the year (e.g., Crisp and Davies, 1955; Giese, 1959a; Boolootian *et al.*, 1959a; Olive, 1970). As pointed out long ago by Orton (1920), these species are often wide-ranging or cosmopolitan in distribution, and may not be very sensitive to the environmental fluctuations that occur in any one locality. Continuous or nearly continuous reproduction also is common among species in the interstitial fauna (Swedmark, 1964; Pollock, 1970). Species of this fauna are all microscopic, or nearly so, and they cannot produce many gametes at any one time; perhaps production of gametes throughout the year by successive generations is needed to increase their reproductive capacity. In this regard, Williams (1966) makes the interesting suggestion that short-lived species (such as interstitial animals) need to take more reproductive "risk" to survive than long-lived animals.

Discontinuous patterns of reproduction are almost always rhythmic. So-called rhythmic reproductive patterns with a circadian period (e.g., Yoshida, 1959) are mainly spawning rhythms; the ripe animals release their gametes at a specific time of day during the breeding season (see Section 1.6 on spawning). Semimonthly or monthly reproductive rhythms ("lunar rhythms") also may be spawning rhythms superimposed on a seasonal reproductive rhythm. Oysters, for example, which release their larvae more or less semimonthly (Korringa, 1947) and polychaetes which swarm monthly (Hauenschild, 1960), do so only during the summer breeding season. However, in some species gametogenic cycles in individuals may be completed with each month's cycle and monthly breeding periods can occur throughout most or all of the year (e.g., Marshall and Stephenson, 1933; Pearse, 1972).

A breeding season usually lasts for a period of more than one month, and it is followed by a period of a month or more when reproduction is quiescent in most or all the individuals. A good number of tropical species have semiannual breeding seasons. Semiannual breeding seasons are especially characteristic of areas influenced by semiannual monsoons, including southern India (e.g., Paul, 1942; George, 1962; Antony Raja, 1963) and Singapore (Wickstead, 1960), but they also occur in animals on the Great Barrier Reef (Stephenson, 1934) and at Hawaii (Reese, 1968) and may be of widespread occurrence in the tropics. Semiannual breeding seasons also occur in some temperate species (e.g., Pfitzenmeyer, 1962; Fell, 1969; Fish and Preece, 1970), but they have not been reported for polar species.

The most common and best studied breeding pattern is the annual breeding season characteristic of most animals of shallow, temperate seas. The timing of the breeding season varies from species to species and between localities for the same species, but generally it is in the summer in the Mediterranean (Lo Bianco, 1909), in the North Atlantic off western Europe (Orton, 1920; Thorson, 1946) and eastern North America (Costello *et al.*, 1957); and in the winter or spring in the East Pacific off western North America (Giese, 1959b). With respect to the difference in timing of reproduction on the east and west coasts of North America, it is interesting that oysters from Long Island normally spawn in the summer, but when raised off California they spawn in the spring (Berg, 1969).

Many species in tropical or subtropical areas also have single annual breeding seasons (e.g., Stephenson, 1934; Lewis, 1966; Reese, 1968; Pearse, 1969a). However, in cases where the same species has been studied over a wide latitudinal range, reproduction seems to occur throughout the year in populations near the equator, but seasonally some distance from the equator (Rao, 1937; Pearse, 1968, 1969b, 1970).

Annual breeding seasons are also common in both the arctic (Thorson, 1936; MacGinitie, 1955; Grainger, 1959) and the antarctic (Pearse, 1965, 1966; Pearse and Giese, 1966). Some polar species breed in the middle of the winter and their slowly developing larvae or juveniles probably reach a feeding stage in phase with the brief summer plankton bloom (Pearse, 1965, 1969c; Thurston, 1970). Others, with even more slowly developing embryos or larvae (e.g., arctic barnacles: Feyling-Hanssen, 1953; antarctic copepods: Littlepage, 1964) spawn in the fall, while those with relatively rapid rates of development spawn in the late spring and summer (Thorson, 1936; Marr, 1962; Mileikovsky, 1970b). A large proportion of polar species brood their embryos and larvae (Thorson, 1950; MacGinitie, 1955; Bullivant and Dearborn, 1967); presumably the brooded embryos and larvae are not so dependent on the brief plankton bloom nor subject to the hazards of polar pelagic life. While at least one antarctic isopod, which broods, breeds throughout the year (White, 1970), it is not known whether brooding species generally breed continuously or periodically.

Although continuous reproduction seems to occur in some deep-sea species (Sanders and Hessler, 1969), annual breeding seasons are known for others. The annual breeding season of the bathyl echinoid *Allocentrotus fragilis* is probably related to seasonal plankton production in overlying waters (Boolootian *et al.*, 1959b; Giese, 1961). Breeding seasons of abyssal ophiuroids in an essentially constant environment are more difficult to explain (Schoener, 1968). Wolff (1962) and George and Menzies (1967, 1968) also report that abyssal isopods have seasonal breeding, but their data are incomplete and inconclusive. Like many polar species, a high proportion of deep-sea species brood their embryos and young (Thorson, 1950; Madsen, 1961).

Biennial breeding, in which two full years are required for individuals to mature and spawn, is apparently unique to some species in polar latitudes (MacGinitie, 1955; Dunbar, 1957, 1962, 1968). Individuals of polar species with biennial cycles also tend to be larger than those of comparable forms (or even the same species) in warmer latitudes. McLaren (1966) has shown that larger animals grown over a 2-year period produce many more gametes than annuals, and he suggested that this increased fecundity might be selectively advantageous in polar seas where food is markedly seasonal.

1.4 Endogenous Regulation of Gametogenesis

Reproduction in most animals, including marine invertebrates, is most likely under endogenous regulation by nervous, endocrine and/or neuroendocrine control systems. The sequential changes of a well-defined gametogenic cycle need to be coordinated with other activities in the organism. Moreover, gametogenic synchrony in the multiple gonads of many species suggests coordination by endogenous control (e.g., in asteroids and echinoids; Pearse, 1965, 1969a). Furthermore, as discussed above, most species reproduce during discrete breeding seasons. Such periodicity implies that changes in environmental factors serve as clues to regulate reproduction, most likely through an endogenous system. Feedback from the gonads and nutrient reserves, in turn, is likely to modulate endogenous control (see Fig. 3).

Despite the probability of such endogenous control of reproduction, little information is available for marine invertebrates in comparison with the immense amount of information on vertebrate and insect endocrine systems. Charniaux-Cotton and Kleinholz (1964), Highnam and Hill (1969), and Tombes (1970) summarize much of the scattered information on invertebrate endocrinology, including some on marine invertebrate reproduction. Some of this information is diagrammed in Fig. 6. Endocrine control of reproduction in marine invertebrates is perhaps best known in polychaetes, especially those that undergo sexual metamorphosis to form mature epitokous "swarmers" (e.g., Clark, 1965; Durchon, 1967). Elimination of neurosecretory cells in the brain by decapitation or microsurgery modifies gamete development and metamorphosis, indicating the presence of both inhibiting and stimulating hormones. Moreover, these neurosecretory cells seem to be under photoperiodic control, linking reproductive and environmental changes (Hauenschild, 1964).

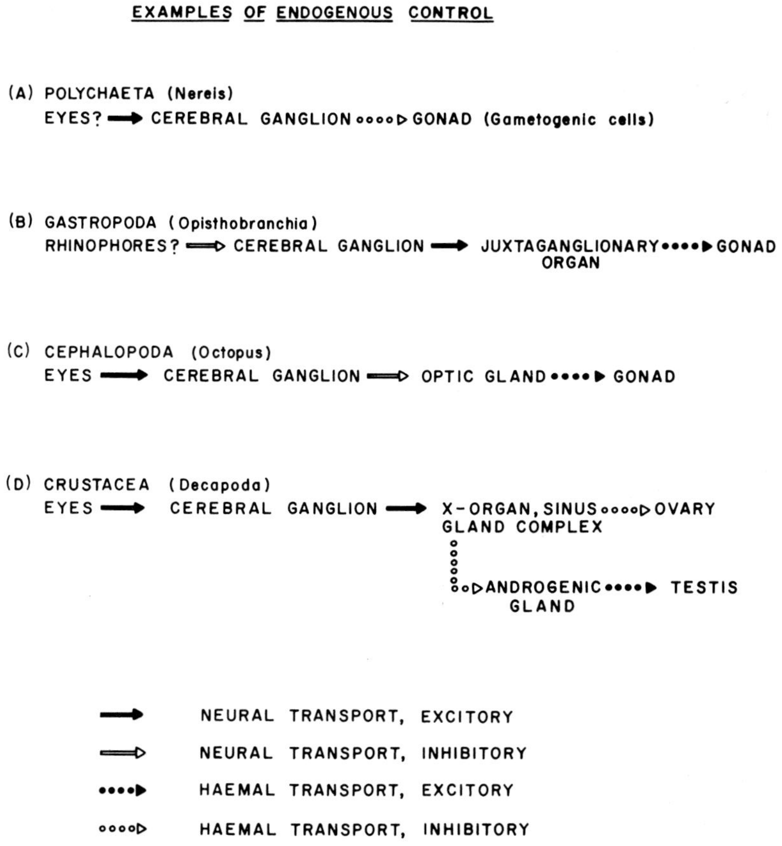

EXAMPLES OF ENDOGENOUS CONTROL

(A) POLYCHAETA (*Nereis*)
EYES? ➡ CEREBRAL GANGLION oooo▷ GONAD (Gametogenic cells)

(B) GASTROPODA (Opisthobranchia)
RHINOPHORES? ⇒ CEREBRAL GANGLION ➡ JUXTAGANGLIONARY ORGAN ••••▶ GONAD

(C) CEPHALOPODA (*Octopus*)
EYES ➡ CEREBRAL GANGLION ⇒ OPTIC GLAND ••••▶ GONAD

(D) CRUSTACEA (Decapoda)
EYES ➡ CEREBRAL GANGLION ➡ X-ORGAN, SINUS GLAND COMPLEX oooo▷ OVARY
X-ORGAN, SINUS GLAND COMPLEX oooo▷ ANDROGENIC GLAND ••••▶ TESTIS

➡ NEURAL TRANSPORT, EXCITORY
⇒ NEURAL TRANSPORT, INHIBITORY
••••▶ HAEMAL TRANSPORT, EXCITORY
oooo▷ HAEMAL TRANSPORT, INHIBITORY

FIG. 6. Diagrammatic representation of endogenous control systems in several marine invertebrates. [(A) Data from Clark (1965), (B) data from Vicente (1966), (C) data from Wells (1960), (D) data from Adiyodi and Adiyodi (1970).]

Endocrine control of reproduction is also well documented in molluscs and crustaceans. Evidence has been presented for gonadotropic (cerebral) and inhibitory (tentacular) hormones in 12 species of opisthobranch gastropods (Vicente, 1966). Also notable are studies on the endocrine control of sexual development in amphipods and other higher crustaceans (reviewed by Charniaux-Cotton and Kleinholz,

1964). Male primary and secondary characters are under the control of androgenic glands located on the vas deferens, while female characters seem to be under the control of endocrine cells in the ovaries. These endocrine organs, in turn, are under the inhibitory control of the x-organ sinus gland complex located in the eyestalks, perhaps linking reproductive activities to photoperiodic control (see also, Highnam and Hill, 1969; Adiyodi and Adiyodi, 1970; Berreur-Bonnenfant, 1970). Regulatory systems strikingly similar to the above are present in cephalopods; nonneural optic glands produce gonadotropins and the optic glands are under inhibitory control of neurosecretory cells of the brain (Wells, 1960).

Endogenous control of reproduction is by no means restricted to more complex invertebrates. Brien (1965), for example, has shown that some diffusible substance regulates sexuality in *Hydra*; in male-female grafts, the female part gradually becomes male. A similar dominance of maleness was found in male-female grafts of nemertineans (Bierne, 1970). Such indications of endocrine control in these simpler animals support our contention that endogenous control systems regulate reproduction in some way in most or all metazoans. However, these systems have been demonstrated in only a relatively few species, and much more work could profitably be done to elucidate them in both simple and complex marine invertebrates.

1.5 Exogenous Regulation of Gametogenesis

It is important that marine invertebrates reproduce at a time when the young have a good chance of surviving. Thus if the young are produced when temperatures, salinities, or food conditions are unfavorable and they do not develop, the objective of reproduction will not have been achieved. Other factors might also be mentioned, but the main point is that the organism must synchronize its reproduction with the environmental conditions that will be most favorable for the success of the young. Often reproductive activities, such as gametogenesis, are begun much in advance of such favorable conditions, and some environmental changes are detected to act as clues or "zeitgebers" which synchronize reproduction with the subsequent favorable conditions. These changes could then exert proximate exogenous control on reproduction. Such environmental changes need not be restricted to only one factor, but may consist of several factors which interact in synchronizing reproductive activities.

One factor may, however, be dominant, at least under one set of circumstances.

Much effort has been expended searching for the environmental factors which exert proximate control over discrete breeding periods in marine invertebrates. In spite of such effort, possible environmental factors controlling reproduction are poorly understood and even their identification is largely speculative. Often, correlation of environmental and reproductive fluctuations in nature cannot be substantiated by experimental manipulation of reproduction in the laboratory. On the other hand, data from experimental manipulation of reproductive activities may be misleading by emphasizing only one of many interacting time clues, and perhaps not those which are usually important in nature.

Little seems to have been done to clarify the relative importance of different environmental factors in reproduction of marine invertebrates since the subject was reviewed by Giese (1959a). The factors which may be important in this regard are discussed below.

1.5.1 Temperature

Seasonally changing sea temperatures may influence the reproductive activities of marine animals. Because of the high heat capacity of water and the large volume of the ocean, sea temperatures often vary rhythmically through the year, not capriciously as on land. Changes in sea temperature therefore, might provide marine animals with reliable clues to seasonal changes and may serve to synchronize their reproduction. Much accumulated evidence indeed supports the idea that changes in sea temperatures synchronize reproduction of marine animals, and many reviews on temperature emphasize this relationship (e.g., Gunter, 1957; Kinne, 1963, 1970).

Apellöf (1912) was among the first to correlate temperature changes in the sea with reproduction, and Orton (1920), after documenting many more correlations, proposed that sea temperature change is the most important exogenous factor regulating reproduction of marine invertebrates. Orton (1920) suggested that each species has a critical breeding temperature which must be reached before reproductive activities proceed. Correlation of temperature with the reproduction of various species of invertebrates has continued since Orton's review (e.g., Nelson, 1928; see reviews by Giese, 1959a; Wilson and Hodgkin, 1967; Pearse, 1968; Mileikovsky, 1970a,b). The correlative evidence now seems overwhelming. Moreover, Runnström (1927, 1929) and others (e.g., Harvey, 1956; Fenaux, 1968, 1970; Mileikovsky, 1970a,b; Sastry, 1970) have shown that the breeding periods of many North Atlantic species at any one locality are related to their zoogeographic distribution. Mediterranean-boreal forms, for example, spawn during a much shorter period in the summer in the northern parts of their distributions than in the southern parts. Arctic-boreal forms, on the other hand, spawn during a shorter winter period in the southern parts of their distributions than in more northerly regions. As shown in Fig. 7, temperatures favorable for development correlate with the temperature regimes near the centers of distribution for a species, suggesting that temperature is of major importance to the timing of reproduction. In 1946, Thorson proposed that the relationship between sea temperature and reproduction be termed "Orton's Rule." He further distinguished the effects of gradual temperature changes on gametogenesis from the effects of sharp temperature changes which induce spawning.

Support for Orton's Rule has also come from laboratory manipulation of reproduction by temperature changes (Fig. 8). Townsend (1940) induced summer-breeding sea urchins to begin gametogenesis in midwinter by maintaining them at simulated summer temperatures. Induction of midwinter reproduction in summer-breeding animals has also been successful with oysters (Loosanoff, 1945; Loosanoff and Davis, 1950, 1952; Aboul-Ela, 1960), polychaetes and scallops (Turner and Hanks, 1960), scallops (Sastry, 1963; Sastry and Blake, 1971), snails (Hanks, 1963), sea hares (Smith and Carefoot, 1967), shrimp (Little, 1968), and archiannelids (Schmidt and Westheide, 1971). Crisp (1957) reported that winter-breeding barnacles could be induced to develop gametes and spawn in midsummer by lowering the temperature to winter levels. These experiments provide strong evidence that sea temperature fluctuations are indeed of paramount importance to the synchronization of reproductive activities in shallow water, temperate species.

However, it can be contended that Orton's Rule does not strictly apply to all or even most marine species. Galtsoff (1961), for example, pointed out that "physiological races" of the same species spawn at quite different temperatures (e.g., Loosanoff and Nomejko, 1951). It may be said, in response, that Orton's Rule may be valid for each physiological race of a species rather than for the species as a whole.

Seemingly more damaging to the generality of Orton's Rule is the fact that many species are known to have discrete breeding seasons in areas where temperature fluctuations are slight, such as parts of the polar, tropical, and deep seas. However, many areas which are generally considered to be stenothermal are subject to slight, sea-

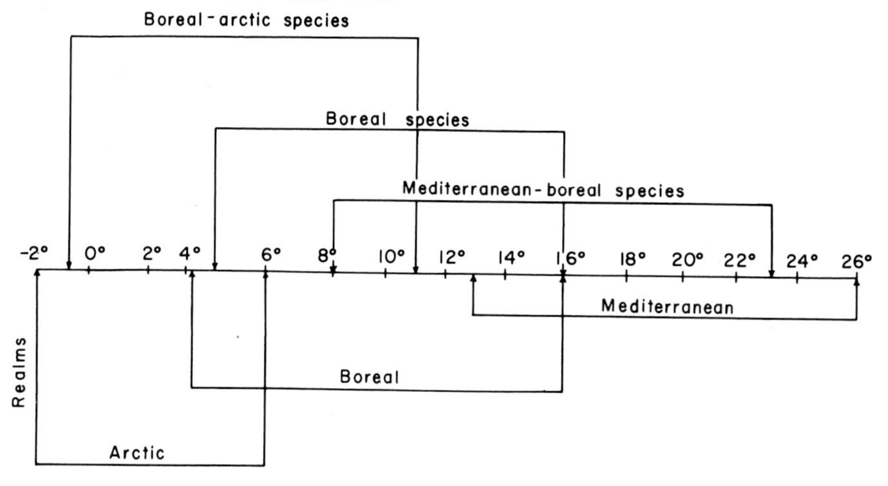

FIG. 7. Annual temperature ranges of three European marine realms, arctic, boreal and mediterranean, and the thermal zones within which boreal-arctic, boreal and mediterranean-boreal species are able to develop. [From Fenaux (1968), data from Runnström (1927).]

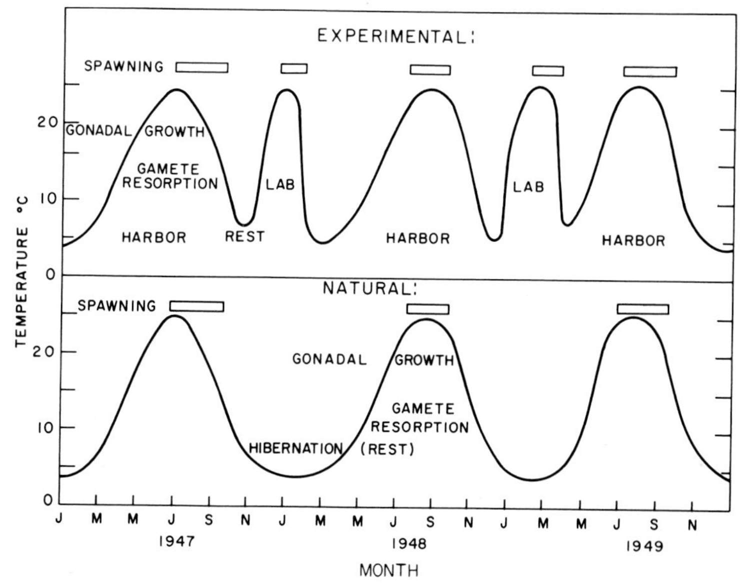

FIG. 8. Schematic representation of breeding in the oyster (*Crassostrea virginica*) correlated with temperature under natural (bottom) and under experimental conditions (top); during the winter the experimental animals were taken into the laboratory and held at temperatures equivalent to those of the summer; these animals underwent a complete gametogenic cycle. [Data from Loosanoff and Davis (1952).]

sonal temperature changes, and rigorous experiments are required to exclude the possibility that such temperature fluctuations influence reproduction. At McMurdo Sound in the antarctic the temperature varies seasonally by only half a degree centigrade (Pearse, 1965), and in the arctic it varies by several degrees (Thorson, 1936; Grainger, 1959; Mileikovsky, 1970b). Temperature also varies seasonally by several degrees in the tropics and the breeding seasons of different populations of the same tropical species appear to be related to temperature fluctuations (Yonge, 1940; Pearse, 1968). Slight but seasonal temperature changes may occur at abyssal depths as well. Therefore, without more data, it cannot be said that periodic breeding in any of these regions offers conclusive evidence contrary to Orton's Rule.

We do not know how a change in temperature brings about the complex series of events that lead to gametogenesis and spawning in species in which these activities are known to be regulated by temperature changes. Presumably, receptors initiate these events, but none have been identified. Because temperature change influences both feeding activity and general metabolic levels, it may also influence reproduction by its effect on the utilization of available nutrients. Sastry and Blake (1971), for example, have shown that the incorporation in scallop gonads of carbon-14, injected into the animals as labeled leucine, is higher in animals kept at 15°C than in those kept at 5°C. Clearly, more experimental studies are needed to elucidate the relation of temperature changes to reproduction.

1.5.2 Photoperiod

It is often stated that daylength could serve as an ideal synchronizer for biological events because it is so invariant year after year for the same latitude, at the same time of year. The farther the position from the equator the greater the seasonal difference in daylength and the more readily could daylength, if sensed, serve as a cue of the season. Daylength probably could not serve this purpose at the equator because day and night are of nearly equal length all year.

It is well known that in many terrestrial and freshwater animals (and plants) well removed from the equator, daylength is very effective in synchronizing seasonal activities, for example in birds and insects. In the males of some species of birds, it is possible by manipulation of daylength, to induce breeding five times in a single year (Wolfson, 1964). One might also expect photoperiodic control of reproduction in some marine invertebrates because there are well-defined links between the photoreceptors and endogenous control centers in annelids, molluscs and crustaceans, for example. It is therefore surprising that only a little evidence for photoperiodic control has been found. Much of this evidence is reviewed by Segal (1970).

Correlations between spring-summer breeding and increasing daylength or fall-winter breeding and decreasing daylength suggest to some investigators the possibility of photoperiodic control. Without experimental data, such correlations are not convincing. Moreover, correlations may be observed for many years yet fail in one year. For example, the chiton *Katharina tunicata* breeds in the summer, suggesting that increasing daylength is important but in one year out of ten it showed three breeding seasons, indicating that factors other than daylength are involved (Giese, 1969). Similarly the sea star *Pisaster ochraceus* breeds in spring-summer, suggesting correlation with increasing daylength, but it breeds at the same time when kept in the dark (Greenfield, 1959).

To demonstrate that daylength is a regulatory cue for reproduction, it is necessary to show that by manipulation of daylength, the breeding season can be displaced in correlation with the lighting regime. A few such experiments on marine invertebrates are considered below.

According to Boolootian (1963), male sea urchins (*Strongylocentrotus purpuratus*) from California showed proliferation of gonial cells with few later stages when exposed to 14 hours of light and 10 hours of darkness daily for 12-15 weeks, but after a change in the daily lighting regime to 6 hours of light and 18 hours of dark, large numbers of spermatocytes and spermatids appeared along with copious spermatozoa. These experiments are difficult to evaluate, however, because the time of year the experiments were done was not given.

Perhaps the most completely analyzed example of photoperiodic control in a marine invertebrate is the study of the polychaete *Platynereis dumerilii* in which 20 days after full moon laboratory-grown individuals come out of their burrows to swarm and breed. It is probable that moonlight is added to daylight, making a long day with which the breeding is correlated. If, during the dark of the moon, light at moonlight intensities is applied, the entire cycle can be shifted. If the worms are continuously illuminated at moonlight intensities, synchrony of swarming and breeding is lost (Hauenschild, 1960). Nevertheless, Hauenschild's experiments are difficult to apply to nature; he manipulated the worms' breeding rhythm so they spawned at times corresponding to the new moon, while at Naples, where he collected them, they reportedly spawn near the full moon, and at Roscoff they spawn near both phases of the moon (Korringa, 1957).

Gonad development in female cuttlefish *Sepia officianalis* is inhibited if the animals are illuminated for more than 12 hours per day, and occurs more rapidly as the light period is reduced to 6 hours or even to 1 hour per day (Richard, 1967). It is known that blinding *Octopus* by cutting optic nerves or destroying the connections in the optic lobes between the eyes and the brain hastens onset of sexual maturity (Wells, 1960), suggesting that light delays maturity, but no studies have been made on *Octopus* with different light regimes.

In the annelid-mollusc line, cerebral inhibition of sexual maturation is general and presumably induced by illumination although experiments are few (Durchon, 1967). Few studies have been made with crustaceans. In the barnacle *Balanus balanoides* light of more than 18 of each 24 hours inhibits formation of egg masses, which form well even in the complete absence of light. It is interesting that when breeding takes place in darkness, a higher temperature is required than when some illumination is provided each day, suggesting a stimulatory influence of light upon gametogenesis (Barnes, 1963). Prospects for further studies in the field of photoperiodism in marine invertebrates are enticing.

1.5.3 Salinity

Fluctuation in salinity has been suggested as a possible factor regulating reproduction in some marine invertebrates. However, in most areas of the sea, fluctuations in salinity are slight, irregular, and of short duration, and they have rarely been shown to influence reproduction (Kinne, 1964; Green, 1968). On the other hand, shallow marine areas and estuaries may experience great and prolonged changes in salinity during periods of heavy rainfall. Along the coast of India reproduction of some marine invertebrates correlates with salinity changes (Panikkar and Aiyar, 1939; Paul, 1942). Similar observations have been made in the harbor at Lagos, Nigeria (Sandison, 1966a, b; Hill, 1967). Wilson (1969) has shown that gametogenesis and spawning in a estuarine mussel *Xenostrobus securus* are related to salinity fluctuations, and spawning occurs at different times in several populations, in each case correlated with a critical salinity level. Even more convincing is the finding that when the rains occur at different times in successive years, the onset of gametogenesis, maturation of gametes and spawning occur at different times each year. Moreover, these events are always correlated with the change in salinity, irrespective of the prevailing temperature or daylength, further attesting to the controlling influence of salinity upon reproduction. Another interesting case is the sea star *Asterias rubens* in the North Sea and the brackish waters of the western Baltic; the periods of gametogenesis and spawning are different depending on the salinity (Schlieper, 1957) (Fig. 9). The influence of salinities higher than seawater on reproduction has been studied in the Caspian Sea (Zenkevitch, 1963).

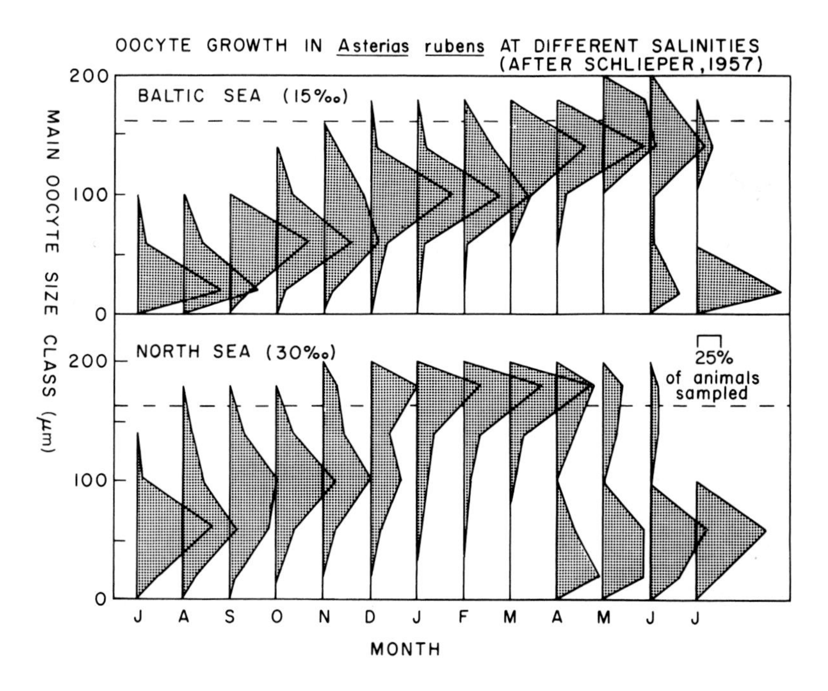

FIG. 9. Diagram of oocyte growth in *Asterias rubens* in areas of different salinities. [Data from Schlieper (1957).]

Salinity requirements for reproduction may reflect salinity requirements for spawning and development of the larvae, not the tolerance of the adults which can often endure a wide range of salinity. Spawning generally occurs only at a salinity in which the larvae developing from zygotes have a chance of surviving (Wilson, 1969). This illustrates the adaptive nature of the reproductive response to salinity.

1.5.4 Abundance of Food

The relative abundance of food may also be a factor in the regulation of gametogenesis. Seasonal plankton blooms might regulate gametogenesis by making available adequate supplies of food at specific times only. This may be true in temperate and, especially, in polar seas where plant production occurs mainly in the summer. At least spring and summer breeding periods of many temperate species are correlated with periods of increased plant production. Pearse (1966) also suggested that gametogenesis in an antarctic sea star might be somehow regulated by the brief period of summer plant production.

Differences in food supply may account for differences in reproduction of adjacent populations. Adjacent populations of sea stars, for example, produce more gametes in areas of greater food production, although the timing of reproduction seems little affected (Pearse, 1965; Crump, 1971). Moreover, Sutherland (1970) found that high intertidal limpets with a seasonal food supply reproduced seasonally, while subtidal limpets with a more constant food supply reproduced throughout the year. These limpet populations were within a few meters of each other.

If food is required for the induction of gametogenesis in marine invertebrates, it should be possible to prevent gametogenesis by starvation at the time when the animals would be breeding in nature, and conversely, to induce breeding by feeding off season. Gonadal growth has been manipulated in such a way in the sea urchin *Strongylocentrotus purpuratus* (Holland, 1964) (Fig. 10). Somewhat similar results have been obtained with the barnacle *Balanus balanoides*. Prolonged starvation, starting early in the season, prevents gametogenesis and fertilization, although starvation begun later in the season does not (Barnes and Barnes, 1967). Continuous reproduction occurs with continuous feeding in the barnacle *Elminius modestus* (Crisp and Davies, 1955).

Manipulation of feeding has been shown to have very little effect on the timing of breeding in several species which possess large storage organs (digestive organs) that can nourish the animal. For example, starvation of the chiton *Katharina tunicata*, beginning when the gonads are just starting to enlarge and continuing until growth of the gonads and gametogenesis have been completed in the controls, has no effect on the rate of development and differentiation of the gametes. Although the gametes produced in starved individuals were normal in appearance, they were present in reduced numbers (Nimitz and Giese, 1964). Somewhat similar results were ob-

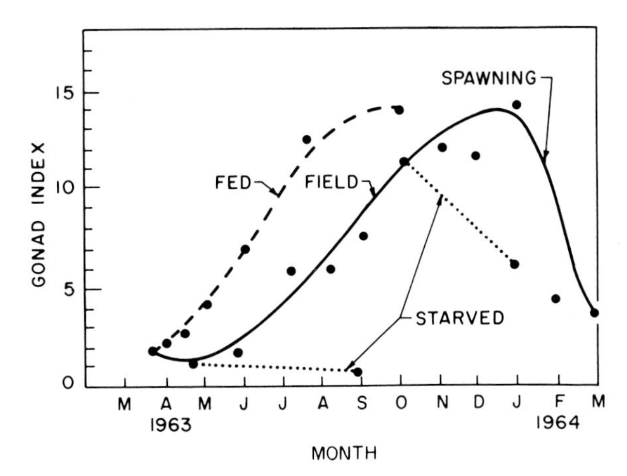

FIG. 10. Off season gonadal growth in the sea urchin *Strongylocentrotus purpuratus* when fed unlimited amounts of the kelp *Macrocystis* in the laboratory. Note the decrease in the gonadal size with starvation. [From Holland (1964).]

tained with *Cryptochiton* (Vasu and Giese, 1966) and the sea star *Pisaster ochraceus* (Giese, unpublished). In both species the size of the storage organs varies inversely with the size of the gonads, as it does in many other species which have been investigated (asteroids: Delaunay, 1926, Farmanfarmaian *et al.*, 1958, Crump, 1971; echinoids: Lawrence *et al.*, 1965a, 1966; Pearse, 1969a; chitons: Lawrence *et al.*, 1965b; abalone: Boolootian *et al.*, 1959b; Webber and Giese, 1969) (Fig. 11). In some marine invertebrates, however, size changes of the storage organs parallel size changes of the gonads (asteroids: Pearse, 1965; cuttlefish: Takahashi, 1961). Sometimes the food supply available may not vary much but feeding activity occurs in seasonal cycles (asteroids: Mauzey, 1966; Christensen, 1970; echinoids: Fuji, 1967).

Seasonal fluctuations of specific reserve materials (lipids, carbohydrates) appear to be correlated with reproductive fluctuations (Giese, 1966, 1967a, 1969). Oysters, for example, accumulate large stores of glycogen prior to and during gametogenesis (Yonge, 1960; Galtsoff, 1964). Blackmore (1969) found a seasonal build-up of reserves in adult limpets before and during gametogenesis, but such a build-up did not occur in juveniles. It is tempting to speculate that gametogenesis is sensitive to levels of reserve nutrients which are dependent upon seasonal fluctuations in the quantity and the quality of food supplies. Environmental factors may affect breeding by their action upon the development of the right kind and abundance of food required by a species. Even daylength might act by affecting the production of food required by a given species, to the extent that it affects the growth of plants used as food.

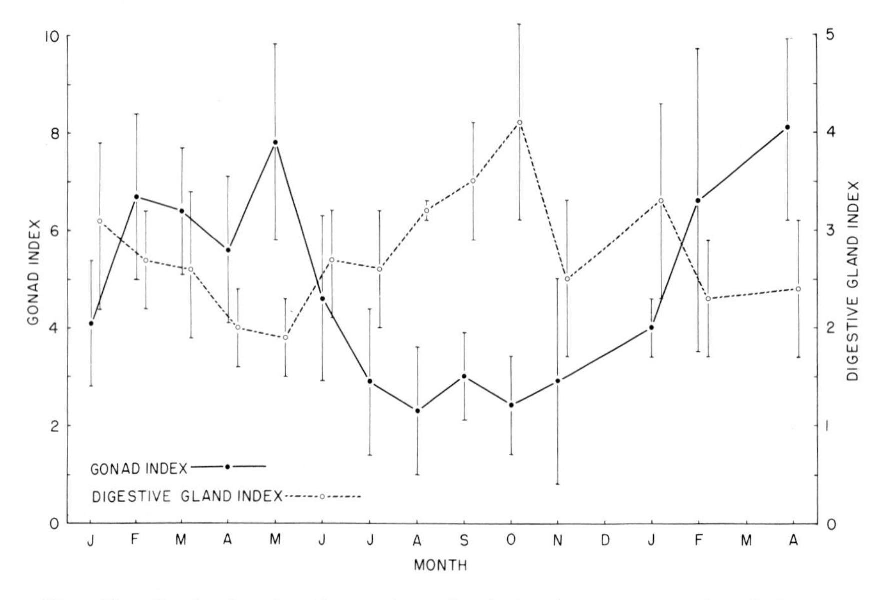

FIG. 11. Graph showing the reciprocal relation between gonad and digestive gland indices during the year in the chiton *Katharina tunicata*. [From Lawrence *et al.* (1965b).]

1.5.5 Chemical factors

Qualitative changes in nutrient supplies, such as trace metabolites in the food, might serve as synchronizers of reproduction. The chemical composition of many planktonic organisms and seaweeds varies seasonally (Wort, 1955; Jensen, 1969; Rao, 1969) and qualitative nutrient changes may be detected by animals feeding upon them. Such changes could serve as proximate exogenous regulators of reproduction. Moreover, metabolites may leak from plants and animals in a seasonal pattern and be detected by other animals as exogenous time clues. Such regulation has been demonstrated with freshwater rotifers (Gilbert, 1971), but not yet with any marine invertebrate.

It has been shown, however, that metabolites produced by an animal can influence the reproductive activities of another member of the same species. The most notable examples are sex determination in the echiuroid *Bonellia* (Baltzer, 1925) and the slipper shell *Crepidula* (Gould, 1952). In both cases, an individual develops into a female when settling alone but into a male when settling on or near a female; presumably, a chemical (pheromone) diffuses from the female and influences the newly settled young. Wilczynski (1960) disputes Baltzer's report, however, and gives evidence suggesting sex is genetically determined in *Bonellia*.

As discussed below, chemicals released during spawning of one individual often stimulate other members of the same species to spawn, leading to synchronous, epidemic spawning. It is also possible, though not demonstrated, that diffusible chemicals may synchronize gametogenic cycles in a population (Lucas, 1961). It should be stressed, however, that even if pheromones synchronize gametogenesis in different individuals, such a system alone could not synchronize reproduction so that it would occur every year at the same time; some exogenous time clues must be detected by at least some individuals in the population.

1.6 Spawning

At present there exist few critical studies on spawning although there is a vast literature in which spawning is mentioned. This lack of study is no doubt due to the difficulties encountered. Spawning is not easy to induce when the gonad is not gravid, but when gravid, almost any stimulus—even disturbing the specimen in the field—may cause spawning. Since in most marine invertebrates, the gonads are enclosed by connective tissue and musculature, any factor which causes contraction of the musculature may cause spawning in gravid animals. To obtain ripe gametes, for example, embryologists often use electrical shock or injection of potassium chloride (Costello *et al.*, 1957; Kumé and Dan, 1968). In most cases, these procedures probably stimulate muscle contraction and thus spawning. It remains to be determined which factors operate to induce spawning in nature.

It may be argued that a "natural" spawning stimulus is not required and that spawning may be spontaneous after the gonads reach ripeness. However, the fact that some laboratory-held ripe echinoids (Harvey, 1956) and asteroids (Pearse, 1965) do not spawn when the individuals in nature do, indicates that some specific exogenous stimulus starts spawning and insures synchronized mass spawning in a population of the species.

Many investigators cite a rapid increase or decrease in temperature such as might occur during tidal exchanges as a stimulus to spawning, these being different from the gradual temperature changes which induce gametogenesis (e.g., Thorson, 1946; Loosanoff and Davis, 1952; Galtsoff, 1964; Pearse, 1965). Conceivably, sharp temperature changes could induce contraction of the gonadal muscles.

Rapid salinity changes during heavy rains also may stimulate spawning in ripe animals; this has been suggested to explain spawning along the Indian coast during the monsoons (Panikkar and Aiyar, 1939; Antony Raja, 1963), in estuaries (e.g., Wilson, 1969) and in the intertidal zone (Giese, 1959b).

Wave action, or its absence, also has been reported to stimulate spawning in some species. Young (1945) suggested that mussels spawn after the physical shock of strong waves, while Heath (1905) found that some chitons spawn in isolated tide pools when the waters are undisturbed.

Light changes have been reported to stimulate spawning in a number of species, as reviewed by Segal (1970). Thus a polychaete released larvae when exposed to light (Knight-Jones, 1951); presumably spawning normally occurs in the early morning. Some cnidarians (Yoshida, 1959) and other animals (Kumé and Dan, 1968) spawn when placed in darkness after an exposure to light; these species presumably spawn in nature in the early evening.

Many species have been reported to spawn in synchrony with lunar phases; this suggests, but does not prove, that the spawning stimulant is related to the changing lunar or tidal phases. Fluctuation in moonlight photoperiod may account for lunar spawning periodicities, especially in the palolo worm; but evidence is lacking that moonlight changes are a direct spawning stimulant (Hauenschild, 1969). Tidal pressure changes may also account for the twice-monthly "lunar" larval release cycle in the European oyster, but experimental evidence also is still lacking (Korringa, 1947, 1957).

Chemicals, especially those exuding from potential larval food, are possible spawning stimuli although there is little direct evidence. Barnacles release brooded larvae (which might be considered analogous to spawning) during spring diatom blooms; increased cirral feeding activity stimulates dissolution of the lamellar material holding the brooded larvae (Barnes, 1957) (Fig. 12). An exudate from green algae appears to induce spawning in oysters (Miyazaki, 1938). However, many other materials, often of questionable significance in nature (such as thyroidin and thyroxin), also stimulate oysters to spawn (Galtsoff, 1964). Even pollutants in paper mill wastes can induce spawning in mussels (Breese *et al.*, 1963).

Chemical exudates from males or females of the same species have been shown to be important for the synchronization of spawning in many species, including cnidarians, echinoderms, ascidians, annelids, and especially molluscs (Giese, 1959a; Galtsoff, 1964). Usually a substance in the "sperm water" of the male stimulates females and other males to spawn, and substances in the "egg water" can stimulate males and other females to spawn in some cases. The adaptive value of spawning stimulants in the spawn material may be very high for species with epidemic spawning. Presumably, several of the ripest individuals are induced to spawn by exogenous stimulants (e.g., a sharp temperature change) and the spawning stimulant in their spawn stimulates other less ripe individuals to spawn until the synchronized spawn reaches epidemic proportions and provides for maximum fertilization.

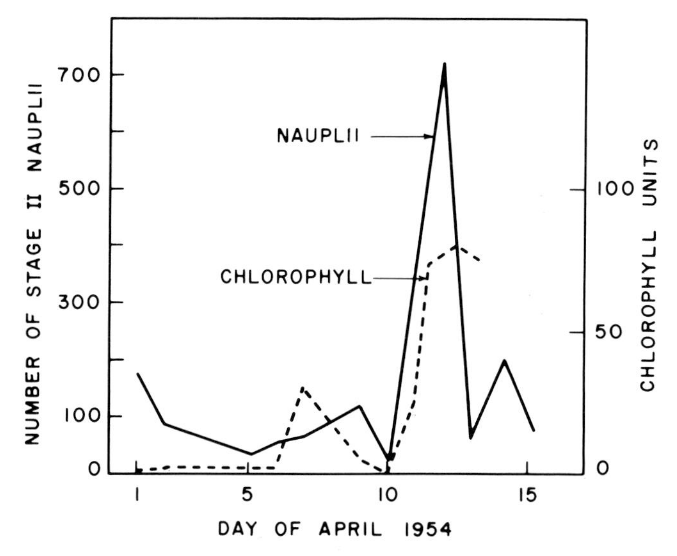

FIG. 12. Diagram showing the correlation between barnacle nauplii caught in vertical plankton hauls at Millport, Scotland, and chlorophyll extracted from the water (in arbitrary units). [Data from Barnes (1957).]

Endogenous control recently has been demonstrated for spawning in asteroids (Chaet, 1967; Kanatani, 1969; Schuetz, 1969b), for spawning in echinoids (Cochran and Engelmann, 1972), and egg-laying in opisthobranchs (Kupfermann, 1967; Strumwasser *et al.*, 1969; Coggeshall, 1970). Kanatani and his co-workers have identified several of the hormonal substances regulating spawning in asteroids, including a 23-amino acid polypeptide gonad stimulator formed in the radial nerves (Kanatani, 1969) and 1-methyladenine which stimulates meiosis and spawning (Kanatani *et al.*, 1969). Presumably, some external factor acts on the radial nerves (perhaps through the sensory complex in the terminal tube feet), inducing secretion of the polypeptide and subsequent production of 1-methyladenine (Kanatani and Shirai, 1970). Highnam and Hill's (1969) suggestion that the radial nerve polypeptide acts as a spawning pheromone has not been demonstrated, however.

1.7 Ecological Considerations of Reproductive Cycles

Much work on reproductive cycles of marine invertebrates has been directed toward elucidation of the exogenous proximate factors synchronizing the cycles. However, it should be remembered that the timing of various modes of reproduction is a response to selection pressures which reflect environmental conditions favorable for successful reproduction. This and related aspects of reproduction have received considerable attention from people working with reptiles and birds (e.g., Williams, 1966; Lack, 1967; Tinkle, 1969). The "ultimate" cause (Baker, 1938) of reproductive cycles is usually assumed to be associated with conditions favorable for the developing larvae or young. For the marine environment this has been termed "Crisp's Rule" by Qasim (1956) who wrote: "The breeding cycles may be so regulated that the larvae hatch during the season which is most favorable for finding planktonic food, under conditions that prevail over the greater part of their range." Although the statement sounds reasonable and is probably true in general, there are few data which directly support it. Such data are difficult to obtain; however, in temperate seas, most species do breed in the spring and summer when the necessary planktonic food is available for the larvae (Thorson, 1950). Barnes' (1957) demonstration that diatom blooms induce barnacles to release their larvae illustrates close adaptive synchrony (see Fig. 12). Moreover, Olive (1970) explained breeding of a polychaete throughout the year in England in terms of ample food for both larvae and adults all year.

As pointed out by Thorson (1950), conditions in polar seas are decidedly unfavorable for planktonic larvae. Not only do low temperatures slow development, but the period when phytoplanktonic food is available is brief. Many polar species apparently have responded to these conditions by brooding their young and thus presumably avoiding unfavorable food conditions (Thorson, 1950). Other polar species, which are often dominant in terms of numbers of individuals, breed in the winter or early spring, and their slow-developing larvae reach a feeding stage during the summer period of phytoproduction (Pearse, 1969c).

Conditions other than food also may be important for the success of the larvae. Most obvious is a temperature favorable for larval development. Also, as shown by Wilson (1969), salinity changes in estuaries may be significant; spawning times in an estuarine mussel are so phased that the salinities are suitable for larval development. Another factor which may be influential in determining breeding times is predation on larvae. For example, the spawning of some clams in the North Sea apparently corresponds to the time when predacious brittle-stars are not feeding (Thorson, 1958).

Several authors have tried to explain the discrete breeding seasons of species in areas of little seasonal change in terms of reduction of interspecific competition. Thus, in the tropics, where food and other conditions are supposedly equable throughout the year, competition may be less keen because of staggered breeding times (Thorson, 1950). Grainger (1959) explained the different breeding times of various carnivorous polar zooplankters in the same terms. However, this hypothesis does not adequately explain why different species (or different populations of one species) breed at particular times, nor how they are informed of their turns. It seems more likely that some environmental conditions (e. g., food, temperature, or light) are more favorable at one time than another for the reproduction of at least some of the species. Reproduction of other species may be attuned somehow to that of the species more directly regulated, but this has never been demonstrated.

Selection pressures may not always be at the level of the larvae, and conditions for juveniles or even adults may instead be critical for determining breeding times. Reese (1968), for example, suggested that the spring and autumn breeding seasons in two species of hermit crabs are related to the summer spawning season of a third species which dominates the other two in shell selection (Fig. 13).

The foregoing indicates how inexact is our knowledge of the ultimate causes of breeding cycles in marine animals. Now, when human activities are drastically changing many areas of the world, including the sea, it is important to ascertain the environmental factors critical to the successful reproduction of different marine animals. Mileikovsky (1968, 1970c) suggests that "human influence" should be taken into account in discussions of the reproduction of various species and indeed human influence has become important in many regions. Although precise data will always be difficult to obtain, workers are encouraged to consider the adaptive aspects of reproductive cycles when evaluating their work.

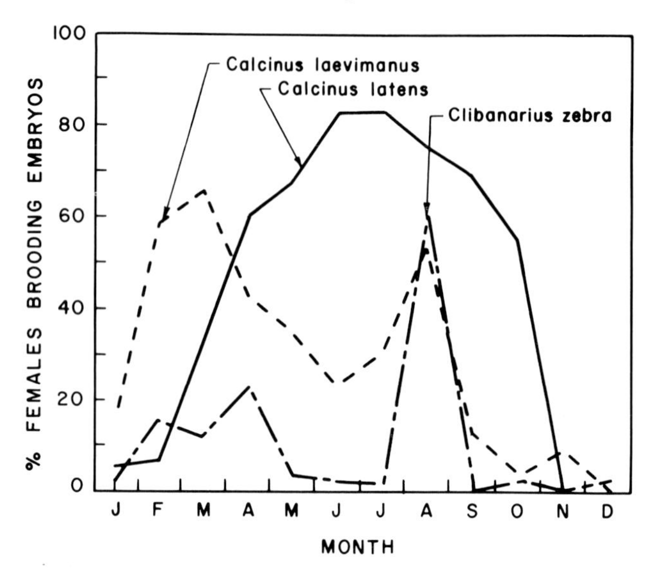

FIG. 13. Graph showing times when three species of hermit crabs are brooding embryos in Kaneohe Bay, Oahu, Hawaiian Islands. Note that two species brood mainly before or after the peak brooding time of *Calcinus latens* which dominates as an adult in shell selection. [Modified from data of Reese (1968).]

1.8 Conclusion

It is evident that some progress has been made on various aspects of reproduction in marine invertebrates. Terms have been defined and ways of assessing reproductive activity have been developed which, together, make possible a relatively quantitative approach by which much more meaningful information can be obtained about reproduction. From present studies it is seen that the timing of reproductive rhythms is variable—in some species, reproduction appears to be continuous for the population, though probably not for individuals, in others it is semiannual, annual, or even biennial. The length of the breeding season varies also and in some species this depends upon latitude.

Endogenous control mechanisms operate in regulating reproduction in a wide variety of marine invertebrates although the information on these is incomplete and sometimes only suggestive. Endogenous control mechanisms are, however, quite susceptible to experimental analysis. Feedback and interplay between the hormones is also probable though little investigated.

It is also evident that not one but a number of exogenous factors may serve to synchronize breeding in different marine invertebrates. In species at or near the equator, neither change in daylength nor temperature would seem to be useful cues, but change in food and other factors might be. In polar regions changes in daylength and food would appear to be of greater importance than change in temperature. However, seasonal temperature changes in both polar and tropical seas and their importance to reproduction should not be discounted anywhere without rigorous experimentation. In temperate regions it is likely that change in temperature plays a major role as a synchronizer of breeding, though there is some evidence for changes in daylength and food as cues for some species. Change in salinity serves as a synchronizer of breeding in estuaries and harbors and possibly influences reproduction elsewhere, perhaps not directly but by influencing food supply. However, in all cases the information is incomplete and the mechanism only suggestive.

Although some information is available about provocation of spawning, analysis of its mechanism awaits further study. It is evident that spawning is a climax event occurring during a reproductive season when ripe gametes are available. At that time many factors provoke it experimentally but these have been little related to spawning in nature.

Finally, it is evident that all the mechanisms involved in reproduction are aimed ultimately at survival of the young produced, for example, their appearance when food is available and when temperature, salinity, etc., are favorable. An analysis of mechanisms regulating reproduction in terms of proximate and ultimate causes has only begun.

Thus while some progress has been made, much more information is needed for a more meaningful analysis of reproductive processes in marine animals. The increasing interest in these problems, as evidenced by the larger number of individuals initiating research in this field, promises continued advancement of knowledge about reproduction in marine invertebrates.

Acknowledgments

Much of this chapter was written while the second author was on the faculties of the W. M. Keck Laboratories, California Institute of Technology, and the Hopkins Marine Station, Stanford University; we thank both of these institutions for providing aid and facilities. We also are indebted to Drs. Peter W. Glynn, Stephen W. Arch, and Vicki B. Pearse for criticizing earlier versions of the manuscript, and to Mrs. Juliana Spranza for drawing the figures.

1.9 References

Aboul-Ela, I. A. (1960). Conditioning *Ostrea edulis* L. from the Limfjord for reproduction out of season. *Medd. Danmarks Fiskeri-og Havundersøg* (*Ny Ser.*) No. 25 2, 1-15.

Adiyodi, K. G., and Adiyodi, R. G. (1970). Endocrine control of reproduction in decapod crustacea. *Biol. Rev. Cambridge Phil. Soc.* **45**, 121-165.

Antony Raja, B. T. (1963). Observations on the rate of growth, sexual maturity and breeding of four sedentary organisms from Madras Harbour. *J. Mar. Biol. Ass. India* **5**, 113-132.

Apellöf, A. (1912). Invertebrate bottom fauna of the Norwegian Sea and North Atlantic. *In* "The Depths of the Ocean" (Sir J. Murray and J. Hjort, eds.), pp. 457-560. Macmillan, New York.

Asdell, S. A. (1964). "Patterns of Mammalian Reproduction," 2nd ed., viii+670 pp. Cornell Univ. Press, Ithaca, New York.

Austin, C. R. (1968). "Ultrastructure of Fertilization," viii+196 pp. Holt, New York.

Bacci, G. (1965). "Sex Determination," vii+306 pp. Pergamon, Oxford.

Baker, J. R. (1938). The Evolution of breeding seasons. *In* "Evolution, Essays on Aspects of Evolutionary Biology Presented to Professor E. S. Goodrich on his Seventieth Birthday" (G. R. DeBeer, ed.), pp. 161-177. Oxford Univ. Press, London and New York.

Baltzer, F. (1925). Untersuchungen über die Entwicklung und Geschlechtsbesimmung der *Bonellia*. *Pubbl. Sta. Zool. Napoli* **6**, 223-285.

Barnes, H. (1957). Processes of restoration of synchronization in marine ecology. The spring diatom increases and the "spawning" of the common barnacle, *Balanus balanoides* (L.). *Année Biol.* **33**, 67-85.

Barnes, H. (1963). Light, temperature and the breeding of *Balanus balanoides*. *J. Mar. Biol. Ass. U.K.* **43**, 717-727.

Barnes, H., and Barnes, M. (1967). The effect of starvation and feeding on the time of production of egg masses in the boreo-arctic cirripede *Balanus balanoides* (L.). *J. Exp. Mar. Biol. Ecol.* 1, 1-6.

Beeman, R. D. (1970). An autoradiographic and phase contrast study of spermatogenesis in the anaspidean opisthobranch *Phylloplysia taylori* Dall 1900. *Arch. Zool. Exp. Gen.* 111, 5-22.

Beklemishev, W. N. (1969). "Principles of Comparative Anatomy of Invertebrates" (J. M. MacLenna, transl., Z. Kabata, ed.), Vol. 1, Promorphology, xxx+490 pp. Univ. Chicago Press, Chicago, Illinois.

Berg, C. J., Jr. (1969). Seasonal gonadal changes of adult oviparous oysters in Tomales Bay, California. *Veliger* **12**, 27-36.

Berreur-Bonnenfant, J. (1970). Organotypic culture *in vitro* and endocrine mechanisms in crustaceans. *In* "Invertebrate Organ Cultures" (H.Lutz, organizer), pp. 211-227. Gordon and Breach, New York.

Berry, A. J. (1965). Reproduction and breeding fluctuations in *Hydrocena monterosatiana* a Malayan limestone archaeogastropod. *Proc. Zool. Soc. London* **144**, 219-227.

Berry, P. Y., and Varughese, G. (1968). Reproductive variation in the Torrent frog *Amolops larutensis* (Boulenger). *J. Linn. Soc. London, Zool.* **47**, 547-561.

Bierne, J. (1970). Influence des facteurs hormonaux gonadoinhibiteur et androgène sur la différenciation sexuelle des parabionts hétérosexués chez un némertien. *Année Biol.* (*Ser. 4*) **9**, 395-400.

Blackmore, D. T. (1969). Studies of *Patella vulgata* L. II. Seasonal variation in biochemical composition. *J. Exp. Mar. Biol. Ecol.* **3**, 231-245.

Boolootian, R. A. (1963). Response of the testes of purple sea urchins to variations in temperature and light. *Nature (London)* **197**, 403.

Boolootian, R. A., Giese, A. C., Farmanfarmaian, A., and Tucker, J. S. (1959a). Reproductive cycles of five west coast crabs. *Physiol. Zool.* **32**, 213-220.

Boolootian, R. A., Giese, A. C., Tucker, J. S., and Farmanfarmaian, A. (1959b). A contribution to the biology of a deep sea echinoid, *Allocentrotus fragilis* (Jackson). *Biol. Bull.* **116**, 362-372.

Boolootian, R. A., Farmanfarmaian, A., and Giese, A. C. (1962). On the reproductive cycle and breeding habits of two western species of *Haliotis*. *Biol. Bull.* **122**, 183-193.

Breder, C. M., and Rosen, D. E. (1966). "Modes of Reproduction in Fishes," xv+941 pp. Natural History Press, New York.

Breese, W. P., Millemann, R. E., and Dimick, R. E. (1963). Stimulation of spawning in the mussels, *Mytilus edulis* Linnaeus and *Mytilus californianus* Conrad, by kraft-mill effluent. *Biol. Bull.* **125**, 197-206.

Brien, P. (1965). Considérations à propos de la reproduction sexuée des invertébrés. *Année Biol.* **4**, 329-365.

Bullivant, J. S., and Dearborn, J. H. (1967). The fauna of the Ross Sea. Part 5. General accounts, station lists, and benthic ecology. *Bull. N. Z. Dept. Sci. Ind. Res.* **176**, 1-77.

Bullough, W. S. (1961). "Vertebrate Sexual Cycles," 2nd ed., viii+123 pp. Methuen, London.

Busson-Mabillot, S. (1969). Données récentes sur la vitellogenèse. *Année Biol.* **8**, 199-228.

Chaet, A. (1967). Gamete release and shedding substance of sea stars. *Symp. Zool. Soc. London* **20**, 13-24.

Charniaux-Cotton, H., and Kleinholz, L. H. (1964). Hormones in invertebrates other than insects. *In* "The Hormones; Physiology, Chemistry, and Applications." (G. Pincus, K. V. Thimann, and E. B. Astwood, eds.) Vol. IV, pp. 135-198, Academic Press, New York.

Chia, F.-S., and Rostron, M. A. (1970). Some aspects of the reproductive biology of *Actinia equina* (Cnidaria, Anthozoa). *J. Mar. Biol. Ass. U. K.* **50**, 253-264.

Christensen, A. M. (1970). Feeding biology of the sea-star *Astropecten irregularis* Pennant. *Ophelia* **8**, 1-134.

Clark, F. N. (1934). Maturity of the California sardine *(Sardina caerulea)* determined by ova diameter measurements. *Calif. Fish Game, Fish Bull.* No. 42, 1-49.

Clark, R. B. (1965). Endocrinology and the reproductive biology of the polychaetes. *Oceanogr. Mar. Biol. Ann. Rev.* **3**, 211-255.

Cochran, R. C., and Engelmann, F. (1972). Echinoid spawning induced by a radial nerve factor. *Science* **178**, 423-424.

Coggeshall, R. E. (1970). A cytologic analysis of bag cell control of egg laying in *Aplysia*. *J. Morphol.* **132**, 461-486.

Costello, D. P., Davidson, M. E., Eggers, A., Fox, M. H. and Henley, C. (1957). "Methods for Obtaining and Handling Marine Eggs and Embryos," xv+247 pp. Lancaster Press, Lancaster, Pennsylvania.

Crew, F. A. E. (1965). "Sex-determination," 4th ed., viii+188 pp. Methuen, London.

Crisp, D. J. (1957). Effect of low temperature on the breeding of marine animals. *Nature* (London) **179**, 1138-1139.

Crisp, D. J., and Davies, P. A. (1955) Observations *in vivo* on the breeding of *Elminius modestus* grown on glass slides. *J. Mar. Biol. Ass. U. K.* **34**, 357-380.

Crump, R. G. (1971). Annual reproductive cycles in three geographically separated populations of *Patiriella regularis* (Verrill), a common New Zealand asteroid. *J. Exp. Mar. Biol. Ecol.* **7**, 137-162.

Davey, K. G. (1965). "Reproduction in Insects," 168 pp. Freeman, San Francisco, California.

Davidson, E. H. (1968). "Gene Activity in Early Development," xi+375 pp. Academic Press, New York.

Delaunay, H. (1926). Sur l'excrétion azotée des astéries (*Asterias rubens* Lin.) *C. R. Soc. Biol.* **94**, 1289-1290.

Delavault, R., and Truslé, J. (1970). Experimental analysis of sexuality in *Asterina gibbosa* with particular use of organotypic culture. *In* "Invertebrate Organ Cultures" (H. Lutz, organizer), pp. 229-240. Gordon and Breach, New York.

Dunbar, M. J. (1957). The determinants of production in northern seas: A study of the biology of *Themisto libellula* Mandt. *Can. J. Zool.* **35**, 797-819.

Dunbar, M. J. (1962). The life cycle of *Sagitta elegans* in Arctic and Subarctic seas, and the modifying effects of hydrographic differences in the environment. *J. Mar. Res.* **20**, 76-91.

Dunbar, M. J. (1968). "Ecological Development in Polar Regions, A Study in Evolution," viii+119 pp. Prentice-Hall, Englewood Cliffs, New Jersey.

Durchon, M. (1967). "L'Endocrinologie des Vers et des Mollusques," 241 pp. Masson, Paris.

Engelmann, F. (1970). "The Physiology of Insect Reproduction," ix+307 pp. Pergamon, Oxford.

Farmanfarmaian, A., Giese, A. C., Boolootian, R. A., and Bennett, J. (1958). Annual reproductive cycles in four species of west coast starfishes. *J. Exp. Zool.* **138**, 355-367.

Farner, D. S., and Follett, B. K. (1966). Light and other environmental factors affecting avian reproduction. *J. Anim. Sci.* **25**, 90-115.

Fell, P. E. (1969). The involvement of nurse cells in oogenesis and embryonic development in the marine sponge, *Haliclona ecbasis*. *J. Morphol.* **127**, 133-150.

Fenaux, L. (1968). Maturation des gonades et cycle saisonnier des larves chez *A. lixula, P. lividus* et *P. microtuberculatus* (Echinides) à Villefranche-sur-Mer. *Vie Milieu* **19A**, 1-52.

Fenaux, L. (1970). Maturation of the gonads and seasonal cycle of planktonic larvae of the ophiuroid *Amphiura chirajei* Forbes. *Biol. Bull.* **138**, 262-271.

Fenchel, T. (1969). The ecology of marine microbenthos. IV. Structure and function of the benthic ecosystem, its chemical and physical factors and the microfauna communities with special reference to the ciliated Protozoa. *Ophelia* **6**, 1-182.

Feyling-Hanssen, R. W. (1953). The barnacle *Balanus balanoides* (Linné, 1766) in Spitzbergen. *Nor. Polarinst. Skr.* **98**, 1-64.

Fish, J. D., and Preece, G. S. (1970). The annual reproductive patterns of *Bathyporeia pilosa* and *Bathyporeia pelagica* (Crustacea: Amphipoda). *J. Mar. Biol. Ass. U. K.* **50**, 475-488.

Franklin, L. E. (1970). Fertilization and the role of the acrosomal region in non-mammals. *Biol. Reprod. Suppl.* **2**, 159-176.

Frazer, J. F. D. (1959). "The Sexual Cycles of Vertebrates," 168 pp. Hutchinson Univ. Library, London.

Fretter, V. and Graham, A. (1964). Reproduction. *In* "Physiology of Mollusca" (K. M. Wilbur and C. M. Yonge, eds.), Vol. 1, pp. 127-164. Academic Press, New York.

Fuji, A. (1960). Studies of the biology of the sea urchin. I. Superficial and histological changes in gametogenic process of two sea urchins, *Strongylocentrotus nudus* and *S. intermedius*. *Bull. Fac. Fish. Hokkaido Univ.* **11**, 1-14.

Fuji, A. (1967). Ecological studies on the growth and food consumption of Japanese common littoral sea urchin, *Strongylocentrotus intermedius* (A. Agassiz). *Mem. Fac. Fish. Hokkaido Univ.* **15**, 83-160.

Galtsoff, P. S. (1961). Physiology of reproduction in molluscs. *Amer. Zool.* **1**, 273-289.

Galtsoff, P. S. (1964). The American Oyster *Crassostrea virginica* Gmelin. *U. S. Fish Wildl. Serv. Fish Bull.* **64**, 1-480.

George, M. J. (1962). On the breeding of Penaeids and the recruitment of their post larvae into the backwaters of Cochin. *Indian J. Fish.* **9**, 100-116.

George, R. Y. and Menzies, R. J. (1967). Indications of cyclic reproductive activity in abyssal organisms. *Nature* (London) **215**, 878.

George, R. Y., and Menzies, R. J. (1968). Further evidence for seasonal breeding cycles in the deep sea. *Nature* (London) **220**, 80-81.

Ghiselin, M. T. (1969). The evolution of hermaphroditism among animals. *Quart. Rev. Biol.* **44**, 189-208.

Giese, A. C. (1959a). Comparative physiology: annual reproductive cycles of marine invertebrates. *Ann Rev. Physiol.* **21**, 547-576.

Giese, A. C. (1959b). Reproductive cycles of some west coast invertebrates. *In* "Photoperiodism and Related Phenomena in Plants and Animals" (R. Withrow, ed.). *Amer. Ass. Advan. Sci. Publ. No.* 55, 625-638.

Giese, A. C. (1961). Further studies on *Allocentrotus fragilis*, a deep sea echinoid. *Biol. Bull.* **121**, 141-150.

Giese, A. C. (1966). On the biochemical constitution of some echinoderms. *In* "Physiology of Echinodermata" (R. A. Boolootian, ed.), pp. 757-796. Wiley (Interscience), New York.

Giese, A. C. (1967a). Some methods for study of the biochemical constitution of marine invertebrates. *Oceanogr. Mar. Biol. Ann. Rev.* **5**, 159-186.

Giese, A. C. (1967b). Changes in body component indices and respiration in the purple sea urchin *Strongylocentrotus purpuratus*. *Physiol. Zool.* **40**, 194-200.

Giese, A. C. (1969). A new approach to the biochemical composition of the mollusc body. *Oceanogr. Mar. Biol. Ann. Rev.* **7**, 175-229.

Giese, A. C., Krishnaswamy, S., Vasu, B. S., and Lawrence, J. (1964). Reproductive and biochemical studies on a sea urchin, *Stomopneustes variolaris*, from Madras Harbor. *Comp. Biochem. Physiol.* **13**, 367-380.

Giese, A. C., Hart, M. A., Smith, A. M., and Cheung, M. A. (1967). Seasonal changes in body component indices and chemical composition in the Pismo clam *Tivela stultorum*. *Comp. Biochem. Physiol.* **22**, 549-561.

Gilbert, J. J. (1971). Some notes on the control of sexuality in the rotifer *Asplanchna sieboldi*. *Limnol. Oceanog.* **16**, 309-319.

Goodbody, I. (1965). Continuous breeding in populations of tropical crustaceans, *Mysidium columbiae* (Zimmer) and *Emerita portoricensis* Schmidt. *Ecology* **46**, 195-197.

Götting, K.-J. (1961). Beitrage zur Kenntnis der Grundlagen der Fortpfdanzung und Frucht barkeitsbestimmung bei marinen Teleosteern. *Helgoländer Wiss. Meeresunters.* **8**, 1-41.

Gould, H. N. (1952). Studies on sex in the hermaphrodite mollusk *Crepidula plana*. IV. Internal and external factors influencing growth and sex development. *J. Exp. Zool.* **119**, 93-160.

Grainger, E. H. (1959). The annual oceanographic cycle at Igloolik in the Canadian Arctic. 1. The zooplankton and physical and chemical observations. *J. Fish. Res. Bd. Can.* **16**, 453-501.

Green, J. (1968). "The Biology of Estuarine Animals," x+401 pp. Univ. Washington Press, Seattle, Washington.

Greenfield, L. J. (1959). Biochemical and environmental factors involved in the reproductive cycle of the sea star *Pisaster ochraceus* (Brandt). Doctoral dissertation, Stanford Univ., 143 pp.

Gunter, G. (1957). Temperature. *In* "Treatise on Marine Ecology and Paleoecology" (J. W. Hedgpeth, ed.). Vol. 1, pp. 159-184. *Geol. Soc. Amer., Mem. 67.*

Hanks, J. E. (1963). Reproduction and larval development of the New England clam drill, *Polinices duplicatus* (Say) (Naticidae: Gastropoda). *Proc. 16th Int. Congr. Zool.* **1**, 227.

Harrison, R. J. (1969). Reproduction and reproductive organs. *In* "The Biology of Marine Mammals" (M. T. Anderson, ed.), pp. 253-348. Academic Press, New York.

Harvey, E. B. (1956). "The American Arbacia and Other Sea Urchins," xiv+298 pp. Princeton Univ. Press, Princeton, New Jersey.

Hauenschild, C. (1960). Lunar periodicity. *Cold Spring Harbor Symp. Quant. Biol.* **25**, 491-497.

Hauenschild, C. (1964). Postembryonale Entwicklungssteuerung durch ein Gehim-Hormon bei *Platynereis dumerilii*. *Zool. Anz. Suppl.* **27**, 111-120.

Hauenschild, C. (1969). Untersuchunger am pazifischen Palolowurm *Eunice viridis* (Polychaeta) in Samoa. *Helgoländer Wiss. Meeresuntersuch.* **18**, 254-295.

Heath, H. (1905). The breeding habits of chitons of the California coast. *Zool. Anz.* **29**, 390-392.

Herlant-Meewis, H. (1965). Les facteurs qui conditionnent la reproduction asexuée chez les invertébrés. *In* "Regeneration in Animals and Related Problems" (V. Kiortsis and H. A. L. Trampusch, eds.), pp. 3-19. North-Holland Publ., Amsterdam.

Highnam, K. C., and Hill, L. (1969). "The Comparative Endocrinology of the Invertebrates," ix+270 pp. Arnold, London.

Hill, M. B. (1967). The life cycles and salinity tolerance of the serpulids *Mercierella enigmatica* Fauvel and *Hydroides uncinata* Phillipi at Lagos, Nigeria. *J. Anim. Ecol.* **36**, 303-321.

Hoar, W. S. (1969). Reproduction. *In* "Fish Physiology" (W. S. Hoar and P. J. Randall, eds.), Vol. 3, Reproduction and Growth. Academic Press, New York.

Holland, N. D. (1964). Cell proliferation in post-embryonic specimens of the purple sea urchin *(Strongylocentrotus purpuratus)* an autoradiographic investigation employing tritiated thymidine. Doctoral dissertation, Stanford Univ., 224 pp.

Holland, N. D. (1967). Gametogenesis during the annual reproductive cycle in a cidaroid sea urchin *(Stylocidaris affinis). Biol. Bull.* 133, 578-590.

Holland, N. D., and Giese, A. C. (1965). An autoradiographic investigation of the gonads of the purple sea urchin *(Strongylocentrotus purpuratus). Biol. Bull.* **128**, 241-258.

Holland, N. D., and Holland, L. Z. (1969). Annual cycles in germinal and non-germinal cell populations in the gonads of the sea urchin *Psammechinus microtuberculatus. Pubbl. Staz. Zool., Napoli* 37, 394-404.

Inger, R. F., and Greenberg, B. (1966). Annual reproductive patterns of lizards from a Bornean rain forest. *Ecology* **47**, 1007-1020.

Jensen, A. (1969). Seasonal variations in the content of individual tocopherols in *Ascophyllum nodosum, Pelvetia canaliculata* and *Fucus serratus* (Phaeophyceae). *Proc. 5th Int. Seaweed Symp.* (R. Margalef, ed.), pp. 493-500. Direccion General de Pesca Marina, Madrid.

Johnson, M. W. (1960). Production and distribution of larvae of the spiny lobster, *Panulirus interruptus* (Randall) with records on *P. gracilis* Streets. *Bull. Scripps Inst. Oceanogr.* **7**, 413-462.

Kanatani, H. (1969). Mechanism of starfish spawning: action of neural substance on the isolated ovary. *Gen. Comp. Endocrinol. Suppl.* **2**, 582-589.

Kanatani, H. and Shirai, H. (1970). Mechanism of starfish spawning. III. Properties and action of meiosis-inducing substance produced in gonad under influence of gonad-stimulating substance. *Develop. Growth Differentiation* **12**, 119-140.

Kanatani, H., Shirai, H., Nakanishi, K., and Kurokawa, T. (1969). Isolation and identification of meiosis-inducing substance in starfish *Asterias amurensis. Nature (London)* **221**, 273-274.

Kinne, O. (1963). The effects of temperature and salinity on marine and brackish water animals. I. Temperature. *Oceanogr. Mar. Biol. Ann. Rev.* **1**, 301-340.

Kinne, O. (1964). The effects of temperature and salinity on marine and brackish water animals. II. Salinity and temperature salinity combinations. *Oceanogr. Mar. Biol. Ann. Rev.* **2**, 281-339.

Kinne, O. (1970). Temperature. Animals. Invertebrates. *In* "Marine Ecology" (O. Kinne, ed.), Vol. 1, Part 1, pp. 407-514. Wiley (Interscience), New York.

Kleitman, N. (1949). Biological rhythms and cycles. *Physiol. Rev.* **29**, 1-30.

Knight-Jones, E. W. (1951). Gregariousness and some other aspects of settling behavior in *Spirorbis. J. Mar. Biol. Ass. U. K.* **30**, 201-222.

Knudsen, J. W. (1964). Observations on the reproductive cycles and ecology of the common Brachyura and crab-like Anomura of Puget Sound, Washington. *Pac. Sci.* **18**, 3-33.

Korringa, P. (1947). Relations between the moon and periodicity in the breeding of marine animals. *Ecol. Monogr.* **17**, 347-381.

Korringa, P. (1957). Lunar Periodicity. *In* "Treatise on Marine Ecology and Paleoecology" (J. W. Hedgpeth, ed.), Vol. 1, pp. 917-934. *Geol. Soc. Amer., Mem.* 67.

Kumé, M. and Dan, K. (1968). Introduction. *In* "Invertebrate Embryology" (M. Kumé and K. Dan, eds., J. C. Dan, transl.), pp. 1-70. NOLIT Publ. House, Belgrade (TT 67-5805D, Clearing-house for Fed. Sci. and Tech. Info., Springfield, Virginia).

Kupfermann, I. (1967). Stimulation of egg laying: possible neuroendocrine function of bag cells of abdominal ganglion of *Aplysia californica. Nature (London)* **216**, 814-815.

Lack, D. (1967). "The Natural Regulation of Animal Numbers," viii+343 pp. Oxford Univ. Press, London and New York.

Lawrence, J. M., and Giese, A. C. (1969). Changes in the lipid composition of the chiton, *Katharina tunicata,* with the reproductive and nutritional state. *Physiol. Zool.* **42,** 353-360.

Lawrence, J. M., Lawrence, A. L., and Holland, N. D. (1965a). Annual cycle in the size of the gut of the purple sea urchin, *Strongylocentrotus purpuratus* (Stimpson). *Nature (London)* **205,** 1238-1239.

Lawrence, A. L., Lawrence, J. M., and Giese, A. C. (1965b). Cyclic variations in the digestive gland and glandular oviduct of chitons (Mollusca). *Science* **147,** 508-510.

Lawrence, J. M., Lawrence, A. L., and Giese, A. C. (1966). Role of the gut as a nutrient-storage organ in the purple sea urchin (*Strongylocentrotus purpuratus*). *Physiol. Zool.* **39,** 281-290.

Lewis, J. B. (1966). Growth and breeding in the tropical echinoid, *Diadema antillarum* Philippi. *Bull. Mar. Sci.* **16,** 151-158.

Licht, P., Hoyer, H. E., and van Oordt, P. G. W. J. (1969). Influence of photoperiod and temperature on testicular recrudescence and body growth in the lizards, *Lacerta sicula* and *Lacerta muralis. J. Zool.* **157,** 469-501.

Little, G. (1968). Induced winter breeding and larval development in the shrimp *Palaemonetes pugio* Holthius (Caridea, Palaemonidae). *Crustaceana Suppl.* **2,** 19-26.

Littlepage, J. L. (1964). Seasonal variation in liquid content of two Antarctic marine crustacea. *In* "Biologie Antarctique" (R. Carrick and M. Holgate, eds.), pp. 463-470. Herman, Paris.

Lo Bianco, S. (1909). Notizie biologiche reguardanti specialmente il periodo di maturita sessuale degli animali del golfo di Napoli. *Mitt. Zool. Stat. Neapel* **19,** 513-761.

Lofts, B., and Murton, R. K. (1968). Photoperiodic and physiological adaptations regulating avian breeding cycles and their ecological significance. *J. Zool.* **155,** 327-394.

Loosanoff, V. L. (1945). Precocious gonad development in oysters induced in midwinter by high temperature. *Science* **102,** 124-125.

Loosanoff, V. L. (1964). Variation in time and intensity of settling of the starfish, *Asterias forbesi,* during a twenty-five year period. *Biol. Bull.* **126,** 423-439.

Loosanoff, V. L., and Davis, H. C. (1950). Conditioning *V. mercenaria* for spawning in winter and breeding its larvae in the laboratory. *Biol. Bull.* **98,** 60-65.

Loosanoff, V. L., and Davis, H. C. (1952). Repeated semiannual spawning of northern oysters. *Science* **115,** 675-676.

Loosanoff, V. L., and Nomejko, C. A. (1951). Existence of physiologically different races of oysters, *Crassostrea virginica. Biol. Bull.* **101,** 151-156.

Lucas, C. E. (1961). External metabolites in the sea. *In* "Papers in Marine Biology and Oceanography", pp. 139-148. Pergamon, New York.

Lutz, H., organizer (1970). "Invertebrate Organ Cultures", xiv+252 pp. Gordon and Breach, New York.

MacGinitie, G. (1955). Distribution and ecology of the marine invertebrates of Point Barrow, Alaska. *Smithson. Misc. Collect. (Publ. 4221)* **128,** 1-201.

Madsen, F. J. (1961). On the zoogeography and origin of the abyssal fauna in view of the knowledge of the Porcellanasteridae. *Galathea Rep.* **4,** 177-218.

Marr, J. W. S. (1962). The natural history and geography of the antarctic Krill (*Euphausia superba* Dana). *Discovery Rep.* **32,** 33-463.

Marshall, A. J., and Parkes, A. S. (1950-56). "Physiology of Reproduction", Vols. 1 (2 parts) to 3, 3rd ed. (I xix+688, xx+877; II xx+880; III xv+1168 pp). Little Brown, Boston, Massachusetts.

Marshall, S. M., and Stephenson, T. A. (1933). The breeding of reef animals, Part 1. The Corals. *Great Barrier Reef Exped. 1928-29, Sci. Rep.* **3(8)**, 219-245.

Mauzey, K. P. (1966). Feeding behavior and reproductive cycles in *Pisaster ochraceus. Biol. Bull.* **131**, 127-144.

McLaren, I. A. (1966). Adaptive significance of large size and long life of the chaetognath *Sagitta elegans* in the arctic. *Ecology* **47**, 852-855.

Mileikovsky, S. A. (1968). The influence of human activities on breeding and spawning of littoral marine bottom invertebrates. *Helgoländer Wiss. Meeresunters.* **17**, 200-208.

Mileikovsky, S. A. (1970a). Seasonal and daily dynamics in pelagic larvae of marine shelf bottom invertebrates in nearshore waters of Kandalaksha Bay (White Sea). *Mar. Biol.* **5**, 180-194.

Mileikovsky, S. A. (1970b). Breeding and larval distribution of the pteropod *Clione limacina* in the North Atlantic, Subarctic and North Pacific Oceans. *Mar. Biol.* **6**, 317-334.

Mileikovsky, S. A. (1970c). The influence of pollution on pelagic larvae of bottom invertebrates in marine nearshore and estuarine waters. *Mar. Biol.* **6**, 350-356.

Miyazaki, I. (1938). On a substance which is contained in green algae and induces spawning action of the male oyster. (Preliminary note.) *Bull. Jap. Soc. Sci. Fish.* **7**, 137-138.

Moore, H. B. (1934). A comparison of the biology of *Echinus esculentus* in different habitats. Part 1. *J. Mar. Biol. Ass. U. K.* **19**, 869-885.

Moore, H. B. (1937). A comparison of the biology of *Echinus esculentus* in different habitats. Part III. *J. Mar. Biol. Ass. U. K.* **21**, 711-719.

Nakano, E. (1968). Fishes. *In* "Fertilization" (C. B. Metz and A. Monroy, eds.), Vol. 2, pp. 295-324. Academic Press, New York.

Nelson, T. C. (1928). On the distribution of critical temperatures for spawning and for ciliary activity in bivalve molluscs. *Science* **67**, 220-221.

Nimitz, M. A., and Giese, A. C. (1964). Histochemical changes correlated with reproductive activity and nutrition in the chiton *Katharina tunicata. Quart. J. Microsc. Sci.* **105**, 481-495.

Nordenskiöld, E. (1935). "The History of Biology", p. 34. Tudor Publ., New York.

Olive, P. J. W. (1970). Reproduction of a Northumberland population of the polychaete *Cirratulus cirratus. Mar. Biol.* **5**, 259-273.

Orton, J. H. (1920). Sea temperature, breeding and distribution in marine animals. *J. Mar. Biol. Ass. U. K.* **12**, 339-366.

Panikkar, N. K., and Aiyar, R. G. (1939). Observations on breeding in brackish water animals of Madras. *Proc. Indian Acad. Sci.* **9B**, 343-364.

Paul, M. D. (1942). Studies on the growth and breeding of certain sedentary organisms in the Madras Harbor. *Proc. Indian Acad. Sci.* **15B**, 1-42.

Pearse, A. S., and Gunter, G. (1957). Salinity. *In* "Treatise on Marine Ecology and Paleoecology" (J. W. Hedgpeth, ed.), Vol. 1, pp. 129-157. *Geol. Soc. Amer. Mem.* 67.

Pearse, J. S. (1965). Reproductive periodicities in several contrasting populations of *Odontaster validus* Koehler, a common antarctic asteriod. *Antarct. Res. Ser.* **5**, 39-85.

Pearse, J. S. (1966). Antarctic asteroid *Odontaster validus:* constancy of reproductive periodicities. *Science* **152,** 1763-1764.

Pearse, J. S. (1968). Patterns of reproductive periodicities in four species of Indo-Pacific echinoderms. *Proc. Indian Acad. Sci.* **68B,** 247-279.

Pearse, J. S. (1969a). Reproductive periodicities of Indo-Pacific invertebrates in the Gulf of Suez. I. The echinoids *Prionocidaris baculosa* (Lamarck) and *Lovenia elongata* (Gray). *Bull. Mar. Sci.* **19,** 323-350.

Pearse, J. S. (1969b). Reproductive periodicities of Indo-Pacific invertebrates in the Gulf of Suez. II. The echinoid *Echinometra mathaei* (de Blainville). *Bull. Mar. Sci.* **19,** 580-613.

Pearse, J. S. (1969c). Slow developing embryos and larvae of the antarctic sea star *Odontaster validus. Mar. Biol.* **3,** 110-116.

Pearse, J. S. (1970). Reproductive periodicities of Indo-Pacific invertebrates in the Gulf of Suez. III. The echinoid *Diadema setosum* (Leske). *Bull. Mar. Sci.* **20,** 697-720.

Pearse, J. S. (1972). A monthly reproductive rhythm in the diadematid sea urchin *Centrostephanus coronatus* Verrill. *J. Exp. Mar. Biol. Ecol.* **8,** 167-186.

Pearse, J. S., and Giese, A. C. (1966). Food, reproduction and organic constitution of the common Antarctic echinoid *Sterechinus neumayeri* (Meissner). *Biol. Bull.* **130,** 387-401.

Pearse, J. S., and Phillips, B. F. (1968). Continuous reproduction in the Indo-Pacific sea urchin *Echinometra mathaei* at Rottnest Island, Western Australia. *Aust. J. Mar. Freshwater Res.* **19,** 161-172.

Pfitzenmeyer, H. T. (1962). Periods of spawning and settling of the softshelled clam, *Mya arenaria,* at Solomons, Maryland. *Chesapeake Sci.* **3,** 114-120.

Pollock, L. W. (1970). Distribution and dynamics of interstitial Tardigrada at Woods Hole, Massachusetts, U.S.A. *Ophelia* **7,** 145-165.

Qasim, S. Z. (1956). Time and duration of the spawning season in some marine teleosts in relation to their distribution. *J. Cons., Cons. Perma. Int. Explor. Mer* **21,** 144-155.

Rahaman, A. A. (1967). Reproductive and nutritional cycles of the crab *Portunus pelagicus* (Linnaeus) (Decapoda; Brachyura) of the Madras Coast. *Proc. Indian Acad. Sci.* **65B,** 76-82.

Rao, H. S. (1937). On the habitat and habits of *Trochus niloticus* Linn. in the Andaman Seas. *Rec. Indian Mus.* **39,** 47-82.

Rao, M. U. (1969). Seasonal variations in growth, alginic acid and mannitol contents of *Sargassum wightii* and *Turbinaria conpides* from the Gulf of Mannar, India. *Proc. 5th Int. Seaweed Symp.* (R. Margalef, ed.), pp. 579-584. Dirección General de Pesca Marina, Madrid.

Raven, C. P. (1961). "Oogenesis: The Storage of Developmental Information," viii-274 pp. Pergamon, New York.

Reese, E. S. (1968). Annual breeding seasons of three sympatric species of tropical intertidal hermit crabs, with a discussion of factors controlling breeding. *J. Exp. Mar. Biol. Ecol.* **2,** 307-318.

Richard, A. (1967). Rôle de la photopériode dans le déterminisme de la maturation génitale femelle du Céphalopode *Sepia officinalis* L. *C. R. Acad. Sci. Paris Ser. D.* **264,** 1315-1318.

Robertson, J. D. (1953). Further studies on ionic regulation in marine invertebrates. *J. Exp. Biol.* **30,** 277-296.

Roosen-Runge, E. C. (1969). Comparative aspects of spermatogenesis. *Biol. Reprod. Suppl.* **1,** 24-39.

Roosen-Runge, E. C. (1970). Life cycle of the hydromedusa *Phialidium gregarium* (A. Agassiz, 1862) in the laboratory. *Biol. Bull.* **139**, 203-221.

Rubey, W. W. (1951). Geologic history of sea water. *Bull. Geol. Soc. Amer.* **62**, 1111-1147; reissue 1964. *In* "The Origin and Evolution of Atmospheres and Oceans" (P. J. Brancasio and A. G. W. Camberon, eds.), pp. 1-63. Wiley, New York.

Runnström, S. (1927). Über die Thermopathie der Fortpflanzung und Entwicklung mariner Tiere in Beziehung zu ihrer geographischen Verbreitung. *Bergens Museum Årbok, Naturv. rekke* **2**. 1-67.

Runnström, S. (1929). Weiters Studien über die Temperaturanpassung der Fortpflanzung und Entwicklung mariner Tiere. *Bergens Museum Årbok, Naturv. rekke* **10**, 1-46.

Sadleir, R. M. F. S. (1969). "The Ecology of Reproduction in Wild and Domestic Mammals," xii+321 pp. Methuen, London.

Sanders, H. L., and Hessler, R. R. (1969). Ecology of the deep-sea benthos. *Science* **163**, 1419-1424.

Sandison, E. E. (1966a). The effect of salinity fluctuations in the life cycle of *Balanus pallidus stutsburi* Darwin in Lagos Harbor, Nigeria. *J. Anim. Ecol.* **35**, 363-378.

Sandison, E. E. (1966b). The effect of salinity fluctuations on the life cycle of *Gryphaea gasar* (Adamson) Dautzenberg in Lagos Harbor, Nigeria. *J. Anim. Ecol.* **35**, 379-390.

Sastry, A. N. (1963). Reproduction of the bay scallop, *Aequipecten irradians* Lamarck, influence of temperature on maturation and spawning. *Biol. Bull.* **125**, 146-153.

Sastry, A. N. (1970). Reproductive physiological variation in latitudinally separated populations of the bay scallop *Aequipecten irradians* Lamarck. *Biol. Bull.* **138**, 56-65.

Sastry, A. N., and Blake, N. J. (1971). Regulation of gonad development in the bay scallop, *Aequipecten irradians* Lamarck. *Biol. Bull.* **140**, 274-283.

Schjeide, O. A., Galey, F., Grellert, E. A., San Lim, R. F., de Vellis, J. J., and Mead, J. P. (1970). Macromolecules in oocyte maturation. *Biol. Reprod. Suppl.* **2**, 14-43.

Schlieper, C. (1957). Comparative study of *Asterias rubens* and *Mytilus edulis* from the North Sea and the western Baltic Sea. *Année Biol.* **33**, 117-127.

Schmidt, P. and Westheide, W. (1971). Der Einfluss der Temperatur auf Oocytenentwicklung und Eiablage bei dem interstitiellen Archianneliden *Trilobodrilus axi. Mar. Biol.* **10**, 94-100.

Schoener, A. (1968). Evidence for reproductive periodicity in the deep sea. *Ecology* **49**, 81-86.

Schuetz, A. W. (1969a). Oogenesis: processes and their regulation. *Advan. Reprod. Physiol.* **4**, 99-148.

Schuetz, A. W. (1969b). Chemical properties and physiological actions of a starfish radial nerve factor and ovarian factor. *Gen. Comp. Endocrinol.* **12**, 209-221.

Segal, E. (1970). Light. Animals. Invertebrates. *In* "Marine Ecology" (O. Kinne, ed.), Vol. 1, Part 1, pp. 159-211. Wiley (Interscience), New York.

Semenova, T. N., Mileikovsky, S. A., and Nesis, K. N. (1964). Morphology, distribution and seasonal occurrence of larval brittle stars *Ophiocten sericeum* (Forbes) S. L. B. in the plankton of the north-western Atlantic, Norwegian and Barents Seas. *Okeanologiya* **4**, 660-683.

Smith, S. T., and Carefoot, T. H. (1967). Induced maturation of gonads of *Aplysia punctata. Nature (London)* **215**, 652-653.

Steele, D. H. and Steele, V. J. (1969). The biology of *Gammarus* (Crustacea, Amphipoda) in the northwestern Atlantic. I. *Gammarus duebeni* Lillj. *Can. J. Zool.* **47**, 235-244.

Stephenson, A. (1934). The breeding of reef animals. Part II. Invertebrates other than corals. *Great Barrier Reef Exped. 1928-29, Sci. Rep.* 3, 219-245.

Strumwasser, F., Jacklet, J. W., and Alvarez, R. (1969). A seasonal rhythm in the neural extract induction of behavioral egg-laying in *Aplysia. Comp. Biochem. Physiol.* **29**, 197-206.

Sutherland, J. P. (1970). Dynamics of high and low populations of the limpet *Acmaea scabra* (Gould). *Ecol. Monogr.* **40**, 169-188.

Swedmark, B. (1964). The interstitial fauna of marine sand. *Biol. Rev. Cambridge Phil. Soc.* **39**, 1-42.

Takahashi, T. S. (1961). Studies on the utilization of cuttlefish *Ommastrephes sloani pacificus*. III. The seasonal variations in the gravimetric constitution and chemical composition of the various parts of the body. *Bull. Jap. Soc. Sci. Fish.* **26**, 95-98.

Thorson, G. (1936). The larval development, growth and metabolism of arctic marine bottom invertebrates. *Medd. Groenland* **100**(**6**), 1-155.

Thorson, G. (1946). Reproduction and larval development of Danish marine bottom invertebrates. *Medd. Komm. Danmarks Fiskeri-og Havunders., Serie: Plankton* **4**, 1-523.

Thorson, G. (1950). Reproductive and larval ecology of marine bottom invertebrates. *Biol. Rev. Cambridge Phil. Soc.* **25**, 1-45.

Thorson, G. (1958). Parallel level-bottom communities, their temperature adaptation, and their "balance" between predators and food animals. *In* "Perspectives in Marine Biology" (A. A. Buzzati-Traverso, ed.), pp. 67-86, Univ. of California Press, Berkeley, California.

Thurston, M. H. (1970). Growth in *Bovallia gigantea* Pfeffer (Crustacea:Amphipoda). *In* "Antarctic Ecology" (M. W. Holgate, ed.), Vol. 1, pp. 269-278. Academic Press, New York.

Tinkle, D. W. (1969). The concept of reproductive effort and its relation to the evolution of life histories of lizards. *Amer. Natur.* **103**, 501-516.

Tombes, A. S. (1970). "An Introduction to Invertebrate Endocrinology," xiii+217 pp. Academic Press, New York.

Towle, A. and Giese, A. C. (1967). The annual reproductive cycle of the sipunculid *Phascolosoma agassizii. Physiol. Zool.* **40**, 229-237.

Townsend, G. (1940). Laboratory ripening of *Arbacia* in winter. *Biol. Bull.* **79**, 363.

Turner, H. J., and Hanks, J. E. (1960). Experimental stimulation of gametogenesis in *Hydroides dianthus* and *Pecten irradians* during the winter. *Biol. Bull.* **119**, 145-152.

Van Tienhoven, A. (1968). "Reproductive Physiology of Vertebrates," xiii+468 pp. Saunders, Philadelphia, Pennsylvania.

Vasu, B. S., and Giese, A. C. (1966). Variations in the body fluid nitrogenous constituents of *Cryptochiton stelleri* (Mollusca) in relation to nutrition and reproduction. *Comp. Biochem. Physiol.* **19**, 737-744.

Vicente, N. (1966). Sur les phénomènes neurosécrétoires chez les Gastéropodes Opisthobranches. *C. R. Acad. Sci. Paris Ser. D* **263**, 382-385.

Vorontosova, M. A., and Liosner, L. D. (1960). "Asexual propagation and regeneration" (F. Billett, ed.; P. M. Allen, transl.), xxiv+489 pp. Pergamon, New York.

Warner, G. F. (1967). The life history of the mangrove tree crabs *Aratus pisoni. J. Zool.* **153**, 321-335.

Webber, H. H., and Giese, A. C. (1969). Reproductive cycle and gametogenesis in the black abalone *Haliotis cracheroidii* (Gastropoda: Prosobranchiata). *Mar. Biol.* **4**, 152-159.

Wells, M. J. (1960). Optic glands and the ovary of *Octopus*. *Symp. Zool. Soc. London* **2**, 87-107.

White, M. G. (1970). Aspects of the breeding biology of *Glyptonotus antarcticus* (Eights) (Crustacea, Isopoda) at Signy Island, South Orkney Islands. *In* "Antarctic Ecology" (M. W. Holgate, ed.), Vol. 1, pp. 279-285. Academic Press, New York.

White, M. J. D. (1954). "Animal Cytology and Evolution," 2nd ed., Cambridge Univ. Press, London and New York.

Wickstead, J. H. (1960). A quantitative and qualitative study of some Indo-west-Pacific plankton. *Colonial Office, Fish. Publ. London No.* **16**, 1-200.

Wilczynski, J. Z. (1960). On egg dimorphism and sex determination in *Bonellia viridis* R. *J. Exp. Zool.* **143**, 61-75.

Wilde, J. de (1964). Reproduction. *In* "The Physiology of Insecta" (M. Rockstein, ed.), pp. 9-58. Academic Press, New York.

Williams, G. C. (1966). Natural selection, the costs of reproduction, and a refinement of Lack's principle. *Amer. Natur.* **100**, 687-692.

Wilson, B. R. (1969). Survival and reproduction of the mussel *Xenostrobus securis* (Lamarck) (Mollusca; Bivalvia; Mytilidae) in a Western Australian estuary. Part II: Reproduction, growth and longevity. *J. Natur. Hist.* 3, 93-120.

Wilson, B. R., and Hodgkin, E. P. (1967). A comparative account of the reproductive cycles of five marine mussels (Mollusca; Bivalvia; Mytilidae) in the vicinity of Fremantle, Western Australia. *Aust. J. Mar. Freshwater Res.* **18**, 175-203.

Wolff, T. (1962). The systematics and biology of bathyl and abyssal Isopoda Asellota. *Galathea Rep.* **6**, 1-320.

Wolfson, A. (1964). Animal Photoperiodism. *In* "Photophysiology" (A. C. Giese, ed.), Vol. 2, pp. 1-49. Academic Press, New York.

Wort, D. J. (1955). The seasonal variation in chemical composition of *Macrocystis integrifolia* and *Nereocystis luetkeana* in British Columbia waters. *Can. J. Bot.* 33, 323-340.

Yonge, C. M. (1940). The biology of reef-building corals. *Great Barrier Reef Exped. 1928-29, Sci. Rep.* **1**, 353-391.

Yonge, C. M. (1960). "Oysters," xiv+209 pp. Collins, London.

Young, R. T. (1945). Stimulation of spawning in the mussel *(Mytilus californianus)*. *Ecology* **26**, 58-69.

Yoshida, M. (1959). Spawning in coelenterates. *Experientia* **15**, 11-12.

Zenkevitch, L. (1963). "Biology of the Seas of the U.S.S.R.,, (S. Botcharskaya, transl.), 955 pp. Wiley (Interscience), New York.

Ziller-Sengal, C. (1970). Organ culture in the invertebrates. *In* "Organ Culture" (J. A. Thomas, ed.). Academic Press, New York.

Chapter 2

PORIFERA

Paul E. Fell

2.1 Introduction

There is relatively little information available on the life histories of marine sponges and on the ecological and physiological factors which regulate the included sequences of events. Although there have been numerous studies on various aspects of sponge development, the complete life history of few sponges is known in any detail. Even where certain developmental events are well described morphologically, little is known of regulatory mechanisms. This is true, at least to a large extent, because there are at the present time almost no procedures for maintaining sponges under controlled conditions for extended periods of time. Precise analyses of regulatory factors will depend heavily upon the development of such procedures (see Fell, 1967; Simpson, 1968).

This chapter is not intended to be an exhaustive review of the literature on sponge reproduction. Rather it is an attempt to define the state of our knowledge in this area of biology.

2.2 Asexual Reproduction

2.2.1 Occurrence and Types

Asexual reproduction in sponges may take a variety of forms and may serve a number of biological functions. Beyond reproduction in the narrowest sense, it may provide mechanisms for dispersal and/or for survival under extreme environmental conditions.

2.2.1.1 Gemmules that Develop Directly into Adult Sponges

Although a number of marine sponges are believed to produce gemmules which develop directly into adult sponges, members of only two genera, *Haliclona* and *Suberites,* have been studied in detail (see Topsent, 1888; Müller, 1914; Prell, 1915; Annandale, 1915; Herlant-Meewis, 1948; Hartman, 1958; Wells *et al.*, 1964). Typically, the gemmules are irregular hemispherical bodies flattened against the substratum to which they continue to adhere following the degeneration of the parent sponge. In many cases nearly all of the substratum occupied by the sponge is covered by gemmules which are closely associated in one or more pavementlike layers (see Fig. 1). The substratum may consist of shells, rocks, eel grass, and other similar objects; and in the case of *Suberites domuncula,* the sponge invariably encrusts a hermit crab or dromiid crab (Müller, 1914; Herlant-Meewis, 1948; Hartman, 1958). Masses of gemmules may also occur within the body of the sponge, usually associated with some foreign object (Müller, 1914; Annandale, 1915).

The gemmules differ in size, those of a particular species falling within a characteristic size range. The gemmules of the different species which have been studied vary in size from about 200 μm to approximately 1000 μm in diameter. Each gemmule consists of a mass of cells enclosed within a protective capsule composed of spongin. The capsule of *Haliclona* is fortified with spicules, but that of *Suberites* is aspiculous. In *Haliclona loosanoffi* the spicules occur primarily in the upper convex portion of the capsule; and in *Suberites* the convex portion of the capsule is usually thicker than the portion in contact with the substratum. The capsule of *Suberites domuncula*

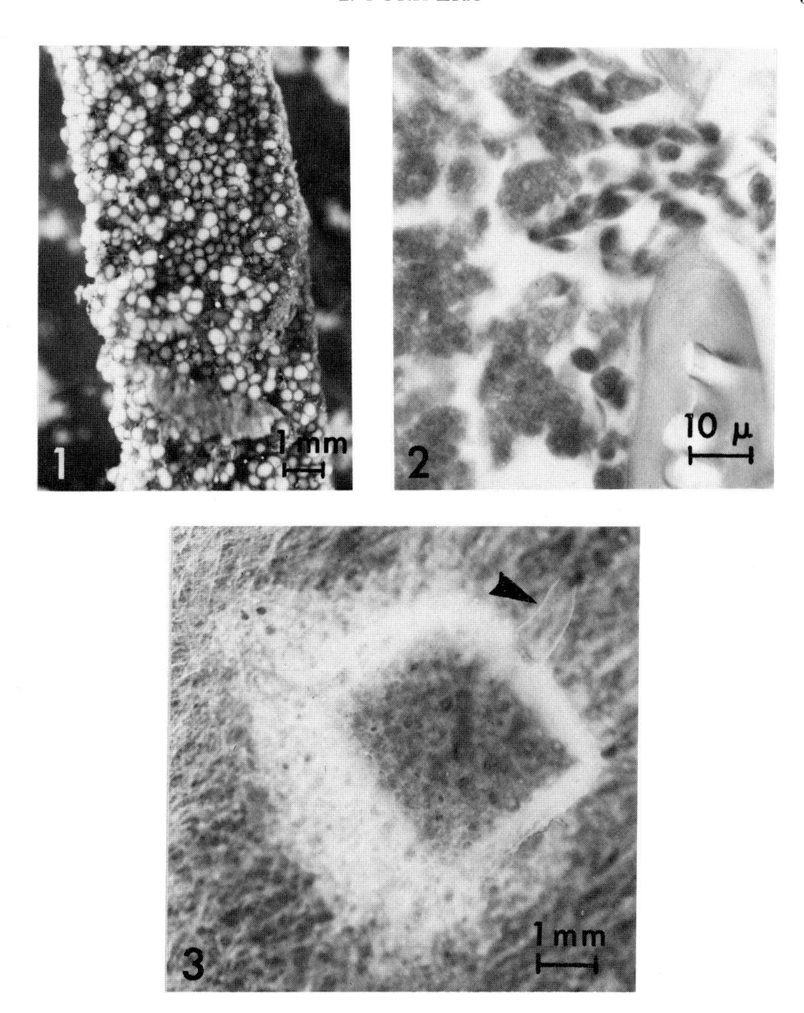

FIG. 1. Living gemmules of *Haliclona loosanoffi* covering a blade of eel grass.

FIG. 2. A small portion of a developing gemmule of *Haliclona ecbasis* showing the central cells (left) filled with granules, the spongin capsule with spicules (right), and small, darkly staining cells associated with the forming capsule. (Hematoxylin and eosin.)

FIG. 3. A living sponge *(Haliclona loosanoffi)* germinated from a mass of gemmules in the laboratory. Note the oscular tube (arrow).

consists of two distinct layers, and that of *Laxosuberites lacustris* is multilayered with spaces between the layers. There is no micropyle

(Müller, 1914; Annandale, 1915; Herlant-Meewis, 1948; Hartman, 1958). The gemmules of most *Suberites* may occasionally contain a few skeletal spicules, but in *Laxosuberites lacustris* the gemmules are regularly supported by large numbers of such spicules (Annandale, 1915).

The cells comprising a gemmule are of relatively uniform size and appearance and are closely packed together. They are filled with numerous reserve granules (Herlant-Meewis, 1948; Hartman, 1958). In *Suberites domuncula* the capsule is lined on the inside by a layer of flattened epithelial cells which are distinct from the other cells of the gemmule (Herlant-Meewis, 1948).

The gemmules of *Cliona vastifica* are attached to the walls of the galleries which the sponge bores into mollusc shells, and an incomplete capsule separates each gemmule from the body of the sponge. The outside of the capsule is usually covered by skeletal spicules situated tangential to its surface (Topsent, 1888).

The formation of gemmules may take a variety of forms. In *Suberites domuncula* the basal portion of the sponge becomes disorganized, and the cells lining the canals and flagellated chambers mix with the cells of the mesenchyme. The cells aggregate on the substratum (or previously formed gemmules); and the amoebocytes (archaeocytes?) and spherulous cells displace the collencytes, pinacocytes, and choanocytes toward the periphery of the mass. Then each condensation divides into a definitive gemmule rudiment and a sheath of peripheral cells, and a thin acellular membrane forms between the two parts (Herlant-Meewis, 1948).

Within the enclosed region the amoebocytes grow and accumulate numberous granules in their cytoplasm. On the other hand, the spherulous cells cytolyze and apparently supply the amoebocytes with reserve substances. They are a special type of nurse cell; but some of the other cells of the sponge, including some of the amoebocytes, may also be consumed (Herlant-Meewis, 1948). Although corresponding information is lacking for *Suberites*, it is of some interest that in the Spongillids, the gemmular nurse cells are identical to those involved in oogenesis (Leveaux, 1939, 1941).

The cells of the sheath and the peripheral cells of the gemmule rudiment elongate perpendicular to the surface of the acellular membrane on each side, forming simple columnar epithelia. The cells of these epithelia then secrete a two-layered capsule around the gemmule. When the capsule is completed, the internal secretory cells form the flattened epithelial lining; and the cells of the outer epithelium degenerate. All of the processes of capsule formation proceed from the base towards the apex of the gemmule (Herlant-Meewis, 1948).

In *Suberites domuncula* it appears that gemmules may also form by the enclosure of portions of the endosome within capsular material followed by the development of the amoebocytes at the expense of the other types of cells. Finally, very small gemmules may form by the unequal subdivision of large ones (Herlant-Meewis, 1948).

It appears that in *Suberites (Ficulina) ficus* (Lévi, 1956), as well as in *Laxosuberites lacustris* (Annandale, 1915), *Prosuberites microsclerus* (Wells *et. al.*, 1964), and *Cliona vastifica* (Topsent, 1888), gemmule formation may occur together with sexual reproduction.

In *Haliclona loosanoffi* the gemmules are formed by archaeocytes which accumulate at the base of the sponge from all parts of the endosome. As in *Suberites,* the cytoplasm of the gemmular cells becomes filled with reserve granules, but an involvement of special nurse cells in the process does not appear to exist. The spongin capsule evidently is produced by an internal layer of granule-free cells (Hartman, 1958; see also Fig. 2 of this chapter).

For discussions of gemmule formation in the freshwater sponges, see Pourbaix (1935, 1936), Leveaux (1939), Rasmont (1956), Brien (1967a), and De Vos (1971).

Gemmules that are attached to eel grass or other plant material may be dispersed by currents when the plants become fragmented. This is apparently the case for *Haliclona loosanoffi* (Fell, in preparation). They may also be detached from firm substrata, such as shells (Hartman, 1958). The gemmules of *Laxosuberites lacustris* are generally attached to rocks which are exposed during the dry season. As the gemmules dry, many groups of them peel away from the substratum and apparently are carried into new locations by wind and water currents. The air spaces in the capsule render the gemmules especially light (Annandale, 1915). In both cases there is evidently a large mortality. Many gemmules are washed up on the beach, and others are presumably buried in the mud or subjected to other unfavorable conditions.

There have been no detailed studies of gemmule germination in marine sponges. Presumably the capsule is ruptured by an enzymatic process; and the cells emerge, divide, and progressively form a small adult (see Fig. 3). Apparently in most cases many gemmules give rise to a single sponge. For a description of gemmule germination in the freshwater sponges (Spongillids) see Brien (1932), Brøndsted and Carlsen (1951), Bröndsted (1943, 1953), Rasmont (1962, 1963, 1968), Ruthmann (1965), and Schmidt (1970).

2.2.1.2 Gemmules that Develop into Larvae

Gemmules that give rise to larvae have been described in a number of marine sponges, including *Mycale (Esperella) fibrexilis* (Wilson, 1891, 1894, 1902); *Euplectella marshalli* and *Leucopsacus orthodocus* (Ijima, 1901, 1903); *Stylocordyla borealis* (Burton, 1928); *Callyspongia diffusa* (Sivaramakrishnan, 1951); and *Spongia reticulata* and *Hymeniacidon perleve* (Bergquist *et al.*, 1970). (See Table I.) In most cases the gemmules consist of a compact mass of cells enclosed only by a single layer of flattened follicle cells and/or by a thin layer of spongin. They apparently are formed by the aggregation of archaeocytes which contain varying amounts of reserve substances. These aggregates may increase in size by growth of their component cells, cell division, and/or fusion with other small aggregates. In *Spongia reticulata* the developing gemmules are surrounded by nurse cells (Bergquist *et al.*, 1970). In the gemmules, cellular boundaries may be obscured by the close packing of the cells and by the abundance of reserve granules in their cytoplasm. However, as a gemmule developes, it beomes subdivided into smaller masses and then into more or less readily distinguishable cells. In many of the species mentioned above, the gemmules ultimately develop into free-swimming parenchymula larvae (see Section 2.4.2).

The formation of gemmules of this sort is frequently associated with the disorganization of the endosome of the parent sponge (Wilson, 1894; Sivaramakrishnan, 1951).

In a number of cases the disappearance of cell boundaries in the gemmules has been assumed to be the result of the formation of syncytia. For example, according to Sivaramakrishnan (1951), the gemmules of *Callyspongia diffusa* are formed by the aggregation of syncytial masses which are produced by the fusion of a single amoebocyte with a number of choanocytes. These masses subsequently fuse together, forming a single large syncytium. Then during the development of the gemmule into a larva, the syncytium breaks up into its component cells. In this and other cases where syncytia are believed to occur, it is desirable that their existence be verified by electron microscopy.

In some cases the asexual origin of these larvae has been disputed (Maas, 1896; Minchin, 1897). For example, Maas questioned Wilson's interpretation of the development of *Mycale (Esperella) fibrexilis* and was convinced that the forming gemmules described by him were in fact growing oocytes surrounded by nurse cells and that the division of the gemmule into smaller masses was simply cleavage of the fertilized egg. Wilson (1894) himself had noted the occurrence of spermatic cysts and of small oocytes, surrounded by masses of mesenchymal cells, scattered about among the gemmules. However, he believed that these were an indication of a later sexual reproductive season. The occasional occurrence of what appear to be oocytes has also been reported in *Euplectella marshalli* and *Leucopsacus orthodocus* (Ijima, 1901).

The sexual and asexual production of larvae by the same specimen may be relatively common among certain intertidal sponges (Bergquist *et al.*, 1970). In seven species which were sampled on the northern coast of New Zealand at monthly intervals for a period of 2 years, it was found that oocytes occur only at certain times of the year but that gemmules are present almost continuously. In *Spongia reticulata*, but apparently not in the other species studied, the development of the gemmules into larvae occurs only at the time that the sexually produced embryos develop into larvae. The larvae develop within the course of a month, and large numbers of oocytes are found only during the two preceding months. In all of the sponges, the gemmules have the same structure as the advanced embryos; and larval development is identical in the two situations. (Bergquist *et al.*, 1970)

In other cases in which samples have been examined throughout the year, it appears that asexual reproduction occurs to the near or complete exclusion of sexual reproduction. This seems to be true for *Callyspongia diffusa* at Madras, India (Sivaramakrishnan, 1951) and *Halichondria moorei* on the northern coast of New Zealand (Bergquist *et al.*, 1970).

The origin of the armored larvae described by Topsent (1920, 1948) and Trégouboff (1942) in *Thoosa* and *Alectona* remains obscure, but it has been suggested that they also may be asexually produced (Borojević, 1967).

At the present time this form of asexual reproduction remains poorly understood, and there is a clear need for further studies.

2.2.1.3 Buds

Some marine sponges produce stalked external buds. Among these are *Donatia (Tethya) deformis* (Edmondson, 1946); *Tethya aurantium (lyncurium)* (Burton, 1948; Connes, 1967; Bergquist *et al.*, 1970); *Aaptos aaptos* (Bergquist *et al.*, 1970); and apparently *Cliona vastifica* (Hartman, 1958).

In *Donatia deformis* the buds, which arise as clubshaped extensions of the sponge surface, consist of bundles of spicules surrounded by masses of archaeocytes. The mature buds possess a long

TABLE I
THE PRODUCTION OF GEMMULES BY MARINE SPONGES

Species	Time of occurrence	Locality	Author
A. Gemmules That Develop Directly			
Suberites domuncula	Present throughout the year	Banyuls-sur-Mer, France	Herlant-Meewis, 1948
Suberites (Ficulina)	—	North Sea and Barents Sea	Müller, 1914
Suberites ficus	Form at the end of the summer and germinate in the spring	Channel Coast of France	Topsent, 1888
	Apparently present all year	Block Island Sound, Rhode Island	Hartman, 1958
Suberites sericeus	—	Chilka Lake on the Bay of Bengal	Annandale, 1915
Prosuberites microsclerus	November and December	Hatteras Harbor, North Carolina	Wells *et al.*, 1964
Laxosuberites lacustris	Produced throughout the year but especially at the approach of the dry season; germinate in November and early December	Chilka Lake on the Bay of Bengal	Annandale, 1915
Haliclona loosanoffi	Form beginning in late August; germinate in the spring	Long Island Sound	Hartman, 1958
	Throughout most of the year	Hatteras Harbor, North Carolina	Wells *et al.*, 1964
Haliclona ecbasis	Apparently rare; found in September	San Francisco Bay	Fell, 1970

Haliclona permollis	Apparently overwintering structures	Oregon Coast	Elvin, personal communication
Haliclona (Chalina) oculata	—	Channel Coast of France	Topsent, 1888
	Apparently present all year	Fishers Island Sound, Connecticut	Fell in preparation
Haliclona (Chalina) gracilenta	—	Channel Coast of France	Topsent, 1888
Cliona vastifica	Reported in September and March	Chilka Lake on the Bay of Bengal	Annandale, 1915
	Apparently present all year	Channel Coast of France	Topsent, 1888
Cliona truitti	September through December	Hatteras Harbor, North Carolina	Wells *et al.*, 1964
B. Gemmules That Give Rise to Larvae			
Mycale (Esperella) fibrexilis	July to August	Woods Hole, Massachusetts	Wilson, 1937
Mycale macilenta	Throughout most of the year	Northern coast of New Zealand	Bergquist *et al.*, 1970
Tedania brucei	August to October	Green Turtle Cay, the Bahamas	Wilson, 1894
Tedania nigrescens	Throughout most of the year	Northern coast of New Zealand	Bergquist *et al.*, 1970
	—	Madras Harbor, India	Sivaramakrishnan, 1951
Callyspongia diffusa	Larvae released throughout the year; peak is during the rainy season (September to November)	Madras Harbor, India	Sivaramakrishnan, 1951

continued

TABLE I (*continued*)
THE PRODUCTION OF GEMMULES BY MARINE SPONGES

Species	Time of occurrence	Locality	Author
	Throughout most of the year	Northern coast of New Zealand	Bergquist *et al.*, 1970
Ircinia (Hircinia) sp.	—	Madras Harbor, India	Sivaramakrishnan, 1951
Ircinia noveazealandiae	Throughout most of the year	Northern coast of New Zealand	Bergquist *et al.*, 1970
Spongia reticulata	Throughout most of the year	Northern coast of New Zealand	Bergquist *et al.*, 1970
Haliclona heterofibrosa	Throughout most of the year	Northern coast of New Zealand	Bergquist *et al.*, 1970
Halichondria moorei	Throughout most of the year	Northern coast of New Zealand	Bergquist *et al.*, 1970
Halichondria panicea	Throughout most of the year	Northern coast of New Zealand	Bergquist *et al.*, 1970
Hymeniacidon perleve	Throughout most of the year	Northern coast of New Zealand	Bergquist *et al.*, 1970
Tethya aurantium	Sporadic	Northern coast of New Zealand	Bergquist *et al.*, 1970
Stylocordyla borealis	Apparently throughout the year	Widespread in the Atlantic Ocean	Burton, 1928
Euplectella marshalli	Apparently throughout the year	Sagami Sea, Japan	Ijima, 1901
Leucopsacus orthodocus	—	Sagami Sea, Japan	Ijima, 1901, 1903
Vitrollula fertile		Sagami Sea, Japan	Ijima, 1901 1903

stalk, and their distal knob supports a varying number of long, slender rays. Under laboratory conditions the buds are released from the parent sponge within one or two days following their initial appearance. The freed buds develop into small sponges under suitable circumstances; and during this process, the rays are withdrawn. When large numbers of buds are released, they may fuse with one another. Although a single bud is capable of producing a small sponge, aggregates of them do so more rapidly (Edmondson, 1946).

The production of buds in *Donatia* is usually accompanied by the development of long, retractile filaments. The filaments have essentially the same internal structure as the buds, but they apparently do not separate from the parent sponge. Although excised filaments are capable of transforming into small sponges, filaments seem to have at best a minor role in asexual reproduction.

Mycale contarenii produces spheroidal buds without stalks (Devos, 1965). Each bud is situated in a small depression in the external surface of the parent sponge. The bud apparently forms by the pinching off of a small portion of the sponge containing all of the various types of cells. The further development of the bud involves the complete reorganization of the cells, beginning at the periphery and progressing towards the center; and results in the formation of a small sponge (Devos, 1965).

Little information is available on the periods of bud production. Many specimens of *Tethya aurantium*, collected off Norway from June through September, showed signs of budding (Burton, 1948). Budding specimens of *Mycale contarenii* were found in May, August, and December at Roscoff, France (Devos, 1965).

On the New Zealand coast *Aaptos aaptos* produces stalked buds only in intertidal regions, while *Tethya aurantium* and *Tethya ingalli* produce such buds at all depths. Intertidal specimens of *Tethya aurantium* and *Aaptos aaptos* also reproduce by the formation of stolonic buds which remain attached to the parent sponge (Bergquist *et al.*, 1970).

2.2.1.4 Fragmentation

Sponges, especially those of branching habit, may be fragmented by water turbulence, predators, and other causes (Burton, 1949). Probably in most cases the portions torn away from the original sponge are carried into situations where they perish; however, such fragments may occasionally give rise to new sponges. Sponges are well known for their regenerative capacities (see Korotkova, 1970).

In its most extreme form fragmentation results in the dissociation of the cells of an organism. Wilson (1907) discovered that the cells of sponges can be readily dissociated by squeezing pieces of sponge through fine silk cloth and that the dissociated cells of some species can reaggregate and re-form functional sponges. Among the sponges of which the reconstitution has been studied are *Microciona prolifera* (Galtsoff, 1925; Wilson and Penney, 1930), *Suberites (Ficulina) ficus* (Fauré-Fremiet, 1932), *Callyspongia diffusa* (Sivaramakrishnan, 1951), *Ophlitaspongia seriata* (Borojevic and Lévi, 1964), *Ephydatia fluviatilis* and *Spongilla lacustris* (Müller, 1911; Brien, 1937; Efremova, 1970), and *Sycon raphanus* (Huxley, 1912; Tuzet and Connes, 1962).

The ease with which sponge cell suspensions can be obtained and manipulated has made them good material for experimental studies of cellular adhesion. Especially in recent years, many studies on the specificity and the mechanism of sponge cell adhesion have been performed (see Margoliash *et al.*, 1965; Gasic and Galanti, 1966; Sarà *et al.*, 1966; Moscona, 1968; Curtis, 1970; Humphreys, 1970; Mac Lennan, 1970; John *et al.*, 1971; McClay, 1971).

2.2.2 Factors Influencing Asexual Reproduction

At the present time little is known of the factors influencing asexual reproduction in marine sponges. The following discussion considers only the factors which influence the production and germination of gemmules that develop directly into adult sponges.

2.2.2.1 Water Temperature

In temperate regions there may be marked seasonal changes in water temperature. For example, in Long Island Sound the water temperature in shallow coastal areas is generally 20°C or greater from June through September, but it is close to 0°C from December through February (see Hartman, 1958; Simpson, 1968; and Table II, this chapter). Small, sheltered portions of the Sound may even be covered by ice for a substantial part of the cold period.

In Long Island Sound *Haliclona loosanoffi* survives the winter in the form of gemmules which germinate when the water temperature rises in the spring. According to Hartman (1958), gemmules develop during the late summer and early autumn when the water temperature is still high; and then in late October and early November, as the water temperature drops below approximately 16°C, the parent sponges begin to degenerate. In the Mystic Estuary (on Fishers Island

TABLE II

THE LIFE HISTORIES OF THREE HIBERNATING SPONGES FROM LONG ISLAND SOUND

Month	Water temperature (°C) at low tide[a]		*Haliclona loosanoffi*[a] (occurrence of)			*Halichondria bowerbanki*[a] (occurrence of)		*Microciona prolifera*[b] (occurrence of)	
			Adult sponges	Large oocytes and/or embryos	Gemmules	Large oocytes and/or embryos	Hibernating form of the sponge	Large oocytes and/or embryos	Hibernating form of the sponge
March	—	6.0	—	—	+	—	+	—	+
April	12.0	15.0	—	—	+	—	±	—	+
May	14.5	19.5	+	+	+	+	—	+	—
June	22.5	23.0	+	+	±	+	—	+	—
July	23.0	23.0	+	+	±	+	—	+	—
August	26.0	23.0	+	—	+	+	—	+	—
September	22.0	19.0	+	—	+	—	—		
October	19.0	8.5	±	—	+	—	—	—	—
November	13.0	4.5	±	—	+	—	—	—	+
December	1.5	1.0	—	—	+	—	+		
January	0	−1.5	—	—	+			—	+
February	0.5	0	—	—	+	—	+	—	+

[a]From the Mystic Estuary, Connecticut. Water temperature was measured once during the first half and once during the second half of each month from 1969 to 1970. (Fell, in preparation.)

[b]From the Milford-New Haven area of Connecticut [after Simpson (1968)].

Sound) specimens of this sponge begin to form gemmules during the last part of July; and although many gemmules apparently germinate in May, some ungerminated gemmules persist through early June (see Table II). Thus the gemmules are produced approximately 7-10 months before they normally germinate and up to 3 or more months before the onset of winter conditions.

During the period of gemmule formation, the parent sponges are rapidly growing. This fact suggests that gemmulation may be coupled in some way with the growth of the sponge. Laboratory studies employing the freshwater sponge, *Ephydatia fluviatilis,* indicate the importance of size and nutrition to gemmulation. With very small sponges developed from gemmules, it was found that up to a point gemmules form earlier and the size of the gemmules produced is larger with a greater initial size of the parent sponge and/or with feeding (Rasmont, 1961, 1963). It has been proposed on the basis of a variety of experiments that in *Ephydatia* gemmulation is initiated by a diffusible factor(s) that is rapidly produced by growing sponges. Presumably this factor can achieve a critical concentration only in sponges above a certain minimal size and usually only in the deeper portions of the sponges where few flagellated chambers occur. At high concentrations this factor may also inhibit the germination of gemmules (see below) (Rasmont, 1963).

Light and temperature also appear to be important interacting factors. Continuous illumination completely inhibits gemmule formation (Rasmont, 1970).

In *Haliclona loosanoffi* germination of the gemmules may be inhibited initially by the surrounding parental tissue and later by low temperature, but the situation apparently is not a simple one. At least in some localities, the parent sponges begin to degenerate when the water temperature is still relatively high and comparable to that occurring at the time of year when the gemmules usually germinate.

An inhibitory influence of parental tissue on the germination of gemmules has been clearly demonstrated in *Ephydatia fluviatilis* (Rasmont, 1963, 1968). In this species the gemmules are capable of germination as soon as they are formed if they are isolated from the parent sponge. However, under natural conditions the gemmules remain embedded within living tissue until late winter when the water temperature is low (Rasmont, 1962). In growing sponges the inhibition of germination is virtually complete; but in starved sponges kept at summer temperatures, many gemmules are found to germinate within the parental mesenchyme. That low temperature prevents the germination of gemmules is shown by the fact that gemmules may be stored at 4°C for up to 4 years, and they readily germinate when placed at 20°C (Rasmont, 1963).

Ephydatia mülleri, like *Haliclona loosanoffi,* degenerates early in the year when the water temperature is still high. In this case the gemmules undergo diapause, and the isolated gemmules are incapable of immediate germination. The diapause is slowly broken at summer temperatures, but low temperature (3°C) markedly accelerates the process. It is of interest that *Ephydatia mülleri* and *Ephydatia fluviatilis* may occur together under apparently identical environmental conditions (Rasmont, 1962).

At Hatteras Harbor, North Carolina the water temperature is about 29°-30°C from July through September; and during this period many of the sponges which are characteristic of more northern regions appear less active or degenerate. *Haliclona loosanoffi* exists only in the form of gemmules during most of the summer. Although some specimens of *Haliclona* degenerate during the winter, the sponge is abundant at this period when the water temperature drops to at least 5°C. Gemmules exposed by the specimens that degenerate during the summer germinate in September, while those exposed by specimens that die during the winter germinate in April (Wells *et al.*, 1964). It therefore appears that high temperature, as well as low temperature, may inhibit the germination of the gemmules of this species. Precise studies on the conditions permitting germination of the gemmules would be very interesting.

Hibernation in the form of gemmules is similar in some respects to another form of hibernation that occurs in certain sponges which do not produce gemmules. During the winter in Long Island Sound, shallow-water speciments of *Microciona prolifera,* of *Haliclona canaliculata,* and of *Halichondria bowerbanki* degenerate to relatively thin encrustations consisting primarily of amoebocytes (Hartman, 1958; Simpson, 1968; and Table II, this chaper). Flagellated chambers, canals, and subdermal spaces disappear, and the surface of the sponge frequently becomes covered by a layer of detritus. The critical temperature for *Microciona* appears to be about 10°C. In November, as the water temperature falls below this level, hibernation sets in; and in May when the temperature rises above it, the sponge begins to grow and to reestablish a normal tissue organization (Simpson, 1968). In this case hibernation apparently is a more or less direct response to low temperature. It is strikingly similar to reduction in *Hymeniacidon (Stylotella) heliophila* induced by other "unfavorable" conditions (Wilson, 1907).

Specimens of *Microciona prolifera* inhabiting different regions apparently respond somewhat differently to low temperature. In the northern extent of its range, the water temperature varies between about 3° and 12°C, rising only slightly above the critical level for the Long Island Sound specimens (Simpson, 1968). At Beaufort, North Carolina where the monthly means in water temperature fluctuate between about 5.5° and 28°C, this sponge grows most rapidly during the winter (McDougall, 1943). This is also the case at Hatteras Harbor, North Carolina (Wells *et al.*, 1964).

2.2.2.2 Other Factors

Some marine sponges are subject to periodic desiccation. In Chilka Lake, a brackish-water lagoon that communicates by a narrow

TABLE III

The Apparent Form of Sexuality in Various Sponges

Form of sexuality	Species	Author
Gonochorism	*Tetilla (Tethya) serica* (o)[a]	Egami and Ishii, 1956
	Aaptos aaptos (o)	Sarà, 1961
	Hippospongia communis (v)	Tuzet and Pavans de Ceccatty 1958
	Stelletta grubii(o)	Liaci and Sciscioli, 1967
	Hymeniacidon sanguinea (v)	Sarà, 1961
	Axinella damicornis (o)	Siribelli, 1962
	Axinella verrucosa (o)	Siribelli, 1962
Successive hermaphrodism	*Polymastia mammillaris* (o)	Sarà, 1961
	Hymeniacidon (Stylotella) heliophila? (v)	Fincher, 1940
	Haliclona ecbasis ? (v)	Fell, 1970
	Octavella galangaui? (v)	Tuzet and Paris, 1964
	Halisarca dujardini ? (v)	Lévi, 1956
Contemporaneous hermaphrodism	*Microciona prolifera* (v)	Simpson, 1968
	Cliona viridis	Sarà, 1961
	Plakortis nigra	Lévi, 1953
	Reniera simulans	Tuzet, 1932
	Reniera elegans	Tuzet, 1932
	Halichondria panicea (v)	Fell, unpublished
	Oscarella (Halisarca) lobularis (v)	Meewis, 1938
	Halisarca metschnikovi (v)	Lévi, 1956
	Tylodesma annexa	Lévi, 1956
	Ephydatia fluviatilis (v)	Leveaux, 1941
	Spongilla lacustris (v)	Leveaux, 1941

[a](o) Oviparous; (v) viviparous.

connection with the Bay of Bengal, the water level varies 5-6 feet between the dry and flood seasons. In late autumn the water level begins to fall and reaches a low in early summer immediately before the monsoon floods begin (Annandale and Kemp, 1915). *Laxosuberites lacustris* survives the dry season, during which time the rocks that it encrusts are exposed, in the form of gemmules. The gemmules apparently are produced most abundantly just before the dry season begins, and they germinate in November and early December when the salinity is relatively high (Annandale, 1915).

Topsent (1888) has reported that along the Channel coast of France the gemmules of *Suberites ficus* form at the end of the summer and germinate during the following spring. However, gemmules appear to be present throughout the year in specimens of *Suberites ficus* and *Suberites domuncula* in Long Island Sound (Hartman, 1958) and in those of the latter species at Banyuls-sur-Mer, France (Herlant-Meewis, 1948). It has been suggested that such gemmules give rise to new sponges when the parental sponges perish at the end of their growth. As the sponges grow to large sizes, the basal portions degenerate, and eventually the entire sponge apparently dies. The objects encrusted by a sponge are then released; and if those bearing gemmules are dispersed, several new sponges can potentially be produced from the original one. It has not yet been possible to germinate the gemmules of *Suberites* in the laboratory, although many conditions apparently have been tested (Herlant-Meewis, 1948).

2.3 Sexual Reproduction

2.3.1 Sexual Differentiation

Among the various marine sponges there are found examples of gonochorism, successive hermaphrodism, and contemporaneous hermaphrodism (see Table III). No sexual dimorphism exists where the sexes are separated either in space or in time. In contemporaneous hermaphrodites the male and female gametes generally are not restricted to different regions of the sponge, but rather oocytes and spermatic cysts are intermingled. However, in *Oscarella (Halisarca) lobularis* the spermatic cysts tend to be situated superficial to the oocytes (Meewis, 1938).

Contemporaneous hermaphrodism can be readily detected in most cases, but the distinction between gonochorism and successive hermaphrodism is not always easy to make. In *Polymastia mammillaris* it appears that a period during which most of the specimens contain developing oocytes is followed by one in which most of them contain developing spermatocytes (there is only one synchronous cycle each of oocyte and spermatocyte differentiation during the reproductive season). Furthermore, some specimens contain cysts of primary spermatocytes in proximity to empty oocyte follicles. The collective evidence indicates the sequential production of different kinds of gametes within individual specimens. The fact that specimens may contain in different regions both mature oocytes and developing or mature spermatozoa suggests that the transition does not occur synchronously throughout the body of the sponge (Sarà, 1961).

In certain other sponges, in which there is an asynchronous differentiation of oocytes and spermatocytes in separate specimens throughout the reproductive season, a relatively small number of specimens are found that contain both kinds of gametes in close association. It is not clear whether such specimens are to be considered abnormal or whether they represent a transitional phase between producing one type of gamete and the other. *Hymeniacidon (Stylotella) heliophila* (Fincher, 1940), *Octavella galangaui* (Tuzet and Paris, 1964), and *Haliclona ecbasis* (Fell, 1970) are among the sponges in which successive hermaphrodism may also occur.

In *Hymeniacidon sanguinea*, degenerating small and medium-sized oocytes are found both in specimens with normal oocytes and those with normal spermatozoa. However, this sponge seems to be functionally gonochoric (Sarà, 1961).

The occasional occurrence of male and female gametes in distant parts of the same specimen could possibly result from the fusion of larvae, the fusion of adults, or, in oviparous forms, the fusion of zygotes (Sarà, 1961; see also Section 2.4.3.2, this chapter). Egami and Ishii (1956) experimentally fused sexually undifferentiated specimens of the gonochoric sponge, *Tetilla (Tethya) serica*; and when the heterosexual combinations become sexually mature, there was no mixing of the two kinds of gametes and apparently no influence of one sex upon the other.

Oscarella (Halisarca) lobularis appears to be a contemporaneous hermaphrodite; however, not all specimens possess both kinds of gametes. Some specimens contain oocytes and spermatic cysts, and others contain only oocytes. Although many specimens apparently have been examined, none containing only male gametes have been found (Meewis, 1938). A similar situation may possibly exist in *Grantia compressa* (Gatenby, 1920). Although gonochorism in oviparous sponges and hermaphrodism among viviparous species appear to be prevalent conditions, it is evident from Table III that these are not rigid relationships (see Sarà, 1961).

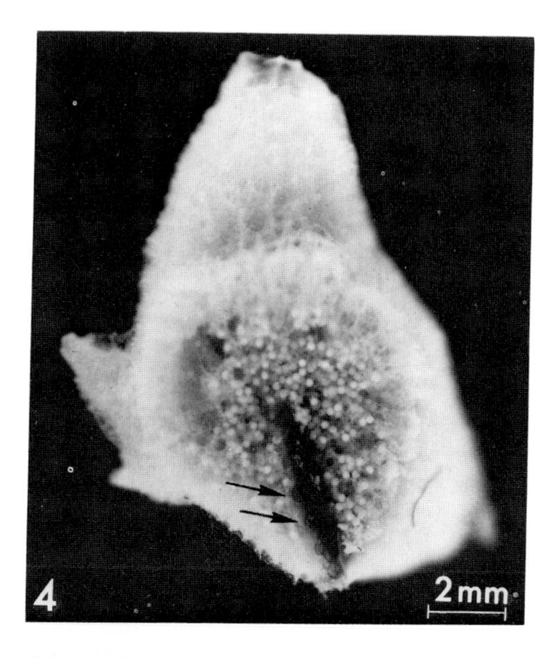

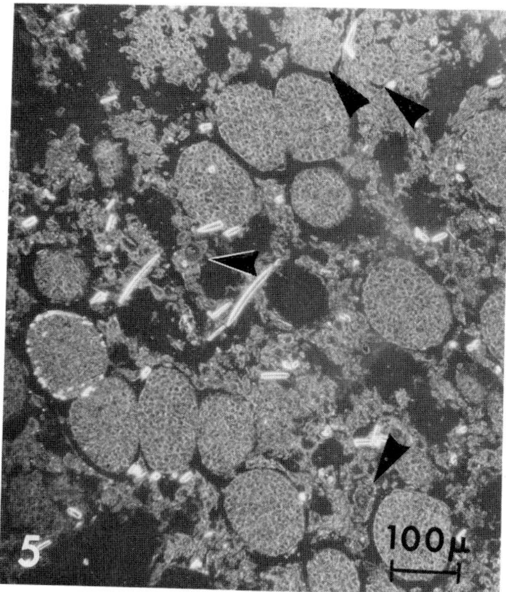

FIG. 4. A section of a preserved specimen of *Haliclona loosanoffi* showing the endosome filled with large oocytes and embryos. Note the empty gemmular capsules (arrows) attached to a blade of eel grass.

FIG. 5. An 8 μm section of *Haliclona ecbasis* showing oocytes (arrows) and embryos within the endosome. (Dark field optics.)

The number of various sponges examined is too small to establish sex ratios for those in which the sexes are separate. However, there are indications that in some sponges this may not be a simple 1:1 ratio. For example, for 80 specimens of *Hymeniacidon sanguinea* the ratio of males to females was nearly 6:1 (Sarà, 1961), and for 85 specimens of *Stelletta grubii* it was 1:4 (Liaci and Sciscioli, 1967).

Nothing is known about the factors which determine sex type, and studies on all aspects of sexual differentiation are greatly needed.

2.3.2 Anatomy of the Reproductive System

Sponges do not possess true gonads or specific reproductive ducts, and in most cases a major portion of the sponge body is involved in reproduction (see Fig. 4 and 5). In a number of sponges, including *Octavella galangaui* (Tuzet and Paris, 1964). *Stelletta grubii* (Liaci and Sciscioli, 1967), *Suberites (Ficulina) ficus* and *Polymastia mammillaris* (Lévi, 1956), *Haliclona ecbasis* (Fell, unpublished), and many calcareous sponges (Tuzet, 1964; Vacelet, 1964), the gametes are distributed more or less completely throughout the endosome and are surrounded by relatively normal somatic tissue containing flagellated chambers. In other sponges, including *Hippiospongia lachne* (Storr, 1964), *Microciona prolifera* (Simpson, 1968), *Oscarella (Halisarca) lobularis* (Meewis, 1938), *Clathrina coriacea* (Sarà, 1955), and the freshwater sponges, *Spongilla lacustris* and *Ephydatia fluviatilis* (Brien and Meewis, 1938; Leveaux, 1941, 1942), the endosome is differentiated into two more or less distinct zones, a superficial one with canals and flagellated chambers and a basal or central zone, in which the gametes and embryos develop, with no or relatively few flagellated chambers. The first situation appears to occur in both oviparous and viviparous forms which produce relatively small eggs, while the second seems to exist largely in viviparous sponges which produce large, yolky eggs (see Lévi, 1951a). In *Halisarca dujardini* the spermatic cysts occur in an intermediate zone displaced somewhat towards the external surface of the sponge (Lévi, 1956).

In most sponges the oocytes are relatively independent, showing no particular association. However, in *Halisarca dujardini* and *Oscarella lobularis* the oogonia form temporary aggregates at various points in the endosome. In the latter species, in which the differentiation of female gametes is asynchronous, the oogonia may associate with mature oocytes and embryos (Lévi, 1951b). The oocytes and larvae of some sponges, including some species of *Haliclona* (Lévi, 1956; Bergquist and Sinclair, 1968) and *Hippiospongia lachne* (Storr,

1964), are grouped in small clusters; and those of *Adocia simulans* and *Phyllospongia foliascens* (Lévi, 1956) are clustered inside of chambers enclosed by loose mesenchyme. These latter structures are essentially rudimentary ovaries. Similarly, the spermatic cysts (see Section 2.3.4.1) may be regarded as primitive testes.

In most instances it appears that the gametes or larvae leave the body of the parent sponge by way of the efferent canal system and the oscula.

2.3.3 Origin of the Germ Cells

Extensive studies based on the appearance of various types of cells in histological sections indicate that the germ cells may arise in various sponges from two different sources, from choanocytes and from archaeocytes (see Tuzet, 1964, for a review). However, such studies, no matter how carefully they are performed, are subject to a number of uncertainties, and in many cases the results obtained are at best tentative. Experimental analyses are very desirable, but they will be technically challenging.

Most of the more recent studies of calcareous sponges suggest that both the oocytes and the spermatocytes derive either directly or indirectly from choanocytes or flagellated larval cells (Tuzet, 1964; Vacelet, 1964). However, Borojević (1969) has presented evidence for the differentiation of oogonia from pinacocytes in *Ascandra minchini* and other Calcinea. On the other hand, the oocytes of most acalcareous sponges appear to derive from archaeocytes, while the spermatocytes seem to develop in some cases from archaeocytes and in other cases from choanocytes (see Tuzet, 1964).

Among the acalcareous sponges the origin of spermatocytes from choanocytes seems to be well established in certain cases. For example, in *Aplysilla rosea* the transformation of flagellated chambers into spermatic cysts has been studied with the electron microscope, and a continuity between typical choanocytes and primary spermatocytes was demonstrated (Tuzet *et al.*, 1970). Tuzet (1947) has suggested that in *Reniera elegans* the archaeocytes derive from choanocytes and that some of the archaeocytes become germ cells. If the transformation of choanocytes into archaeocytes were general among sponges, then the germ cells of acalcareous sponges would also appear to derive either directly or indirectly from choanocytes. However, experimental evidence for such transformation is lacking, and it is impossible to determine with certainty the direction of the change (archaeocyte to choanocyte or choanocyte to archaeocyte) from histological preparations. On the other hand, the differentiation of choanocytes from archaeocytes has been shown to occur in cultures of the internal mass of the larva of *Mycale contarenii* (Borojević, 1966); and choanocytes, as well as the gametes, apparently differentiate from archaeocytes following winter hibernation in *Microciona prolifera* (Simpson, 1968) and during the germination of gemmules. It therefore appears that the germ cells of siliceous sponges may potentially develop from archaeocytes either directly, as usually seems to be the case for oocytes, or indirectly from choanocytes. However, the situation is not simple. In many cases most of the choanocytes of the young sponge apparently develop from the flagellated cells of the larval epithelium (see Section 2.4.3.3), and such choanocytes are capable of propagating themselves by mitosis (see, for example, Simpson, 1963; Borojević, 1966; Shore, 1971). Possibly the only common source of all of the germ cells produced during the life history of a sponge is the blastomeres of the undifferentiated embryo.

At the present time nothing is known concerning the factors that determine which archaeocytes or choanocytes will differentiate as germ cells.

2.3.4 Cytodifferentiation of the Gametes

2.3.4.1 Spermatogenesis

Spermatogenesis occurs within spermatic cysts (see Fig. 6). The formation of such cysts has not been extensively studied in calcareous sponges. However, among the acalcareous sponges, about which there is more information, three patterns of cyst formation have been described. In *Hippospongia communis* (Tuzet and Pavans de Ceccatty, 1958), *Octavella galangaui* (Tuzet and Paris, 1964), and *Aplysilla rosea* (Tuzet *et al.*, 1970), all of the choanocytes of a flagellated chamber apparently lose their collar and flagellum and develop into spermatogonia. The spermatogonia constitute a more or less solid mass which is surrounded by a single layer of flattened follicle cells and/or by spongin. The follicle cells apparently are derived from mesenchymal amoebocytes. In other sponges, including *Halisarca dujardini* and *Halisarca metschnikovi* (Lévi, 1956), *Hymeniacidon (Stylotella) heliophila* (Fincher, 1940), and the freshwater sponges, *Spongilla lacustris* and *Ephydatia fluviatilis* (Leveaux, 1942), amoeboid cells aggregate at various points in the mesenchyme, forming the rudiments of the spermatic cysts. These masses frequently are subdivided as groups of cells are enclosed within follicles which are comparable to those formed around the transformed flagellated chambers mentioned above. Finally, in *Reniera simulans* (Tuzet, 1930b) a single sperm-mother cell divides asymmetrically, producing one daughter cell with a prominent nucleolus and another with no visible nucleolus. The former apparently becomes a follicle cell which progressively envelops the mass of spermatogonia produced by the division of the latter. In this case each spermatic cyst ultimately contains 32 developing spermatocytes which are surrounded by a single extended follicle cell. However, in many other sponges the cysts vary considerably in size.

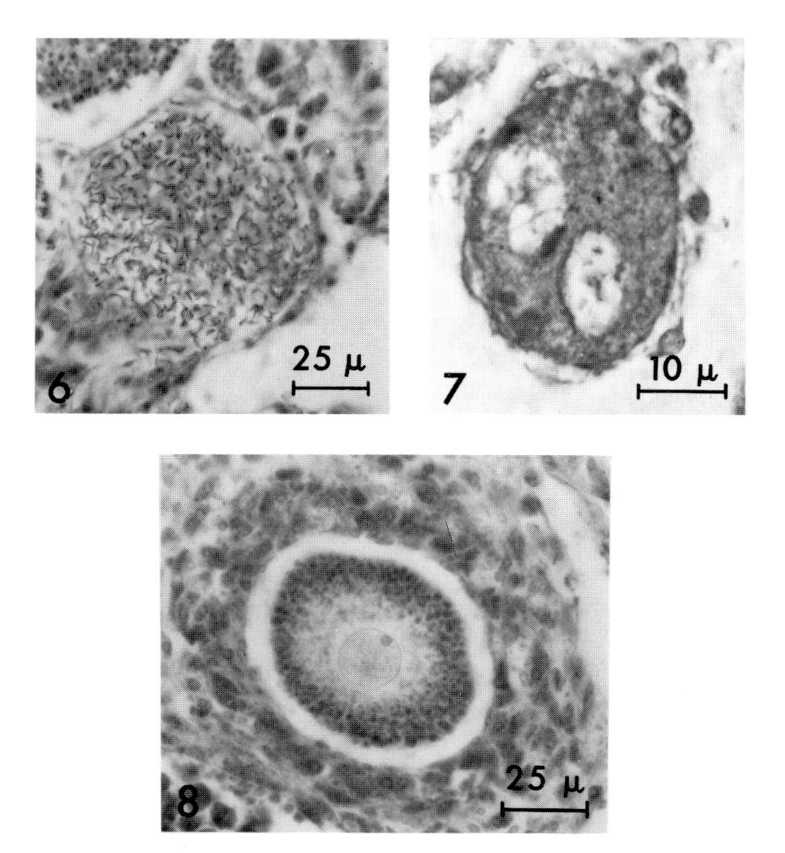

FIG. 6. A spermatic cyst of *Microciona microjoanna*. Note that the spermatozoa are oriented at random within the cyst. (Hematoxylin and eosin.)

FIG. 7. An oocyte of *Haliclona ecbasis* with 2 nuclei. (Phase contrast optics.)

FIG. 8. An oocyte of *Halichondria panicea* in the early stages of yolk formation. Note the mass of cells surrounding the oocyte outside of the follicle. (Hematoxylin and eosin.)

In some cases there are several spermatogonial generations; in other cases the germ cells apparently develop directly into primary spermatocytes. The nuclei of the primary spermatocytes pass through the characteristic stages of meiotic prophase; and in *Aplysilla rosea* they have been observed with the electron microscope to contain synaptinemal complexes (Tuzet *et al.*, 1970). The primary spermatocytes divide, producing the secondary spermatocytes; and the latter divide in turn, forming spermatids. As the spermatocytes divide, their size is reduced.

As the spermatids differentiate into spermatozoa, they undergo profound changes. The nucleus condenses into a compact chromatic structure. A flagellum extends from one of a pair of centrioles positioned close to the nucleus; and the mitochondria become grouped around the base of the flagellum. In some cases the Golgi complex, situated at the opposite pole of the nucleus, forms an acrosome. A definite acrosome is seen in the spermatozoa of *Grantia compressa* (Gatenby, 1927), *Reniera simulans* (Tuzet, 1930b), *Hippospongia communis* (Tuzet and Pavans de Ceccatty, 1958), and *Octavella galangaui* (Tuzet and Paris, 1964); but in many cases it is inconspicuous or apparently absent. Finally, some of the cytoplasm may be eliminated.

The head of the mature spermatozoan may be spherical, ovoid, or conical, depending upon the species. Similarly, the intermediate segment, containing the mitochondria, may assume a variety of forms and may exhibit different degrees of development. The length of the head is frequently about 0.5 -1.5 μm, and that of the entire spermatozoan is generally about 10 - 50 μm (Fincher, 1940; Lévi, 1956; Tuzet and Pavans de Ceccatty, 1958; Tuzet and Paris, 1964; Reiswig, 1970).

In some sponges, including *Hippospongia communis* (Tuzet and Pavans de Ceccatty, 1958) and *Haliclona ecbasis* (Fell, unpublished), there is a tendency for the spermatozoa to be oriented with their flagella directed towards one pole of the cyst, while in other sponges, including *Halisarca dujardini* (Lévi, 1956) and *Microciona microjoanna* (see Fig. 6), the spermatozoa are randomly oriented. The factors that determine the orientation of the spermatozoa are unknown. In all of the species mentioned above, the differentiation of the spermatozoa in any one cyst is highly synchronous (see below). In the freshwater sponges, *Spongilla lacustris* and *Ephydatia fluviatilis*, the spermatozoa may be oriented either with their heads central and their tails extended towards the periphery or at random, depending upon whether the differentiation of the cells in the cyst is synchronous or asynchronous, respectively (Leveaux, 1942).

There is little information available on the temporal aspects of spermatogenesis. The time consumed by various steps in the differentiation process can be most readily estimated by successive sampling either of individual specimens, where there is a high degree of synchrony within such individuals, or of populations where synchrony exists both within and among different specimens. In *Axinella damicornis* and *Axinella verrucosa* (Siribelli, 1962) and in *Polymastia mammillaris* (Sarà, 1961), it appears that the differentiation of spermatozoa from primary spermatocytes may occur within approximately two weeks (water temperature was not recorded).

The differentiation of spermatozoa within individual specimens may follow one of three patterns. All of the cells in all of the cysts of the sponge may be synchronized in their development. This situation is found in *Polymastia mammillaris* (Sarà, 1961), *Axinella damicornis* and *Axinella verrucosa* (Siribelli, 1962) and *Tylodesma annexa* (Lévi, 1956). In many sponges, including *Cliona viridis* and *Hymeniacidon sanguinea* (Sarà, 1961), *Pronax plumosa* and *Halisarca dujardini* (Lévi, 1956), *Hippospongia communis* (Tuzet and Pavans de Ceccatty, 1958), *Aplysilla rosea* (Tuzet *et al.*, 1970) and *Haliclona ecbasis* and *Microciona microjoanna* (Fell, unpublished), the differentiation of the cells of a cyst is synchronous, but different cysts are at different stages of development. In such cases cysts at various stages may be clustered together. Finally, in *Oscarella lobularis* and *Halisarca metschnikovi* (Lévi, 1956), *Plakortis nigra* (Lévi, 1953), *Octavella galangaui* (Tuzet and Paris, 1964), and *Aaptos aaptos* (Sarà, 1961), the development of the cells in any particular cyst is asynchronous.

The factors that determine these patterns are not at all understood. Most of the sponges, in which all of the spermatozoa differentiate at about the same time, are oviparous, while many of the sponges, in which maturation of spermatozoa is consecutive, are viviparous. However, there are exceptions (see Section 2.3.4.2).

2.3.4.2 Oogenesis

According to several authors, the oogonia of calcareous sponges move from beneath the choanoderm into the flagellated chambers where they undergo two successive divisions that result in the production of oocytes. The oocytes then leave the chambers and assume positions under the choanocytes. This sequence of events has been described in *Sycon raphanus, Sycon elegans, Grantia compressa, Petrobiona massiliana, Clathrina coriacea, Leucosolenia botryoides,* and several species of *Leucandra* (see Dendy, 1914; Vacelet, 1964; Tuzet, 1964). However, Sarà (1955) questions the existence of oogonial generations in *Clathrina* and *Leucosolenia*, believing that the germ cells develop directly into oocytes. Furthermore, Borojević (1969) found no evidence of two generations of oogonia in *Ascandra*. The division of oogonia within the flagellated chambers has also been described in two species of acalcareous sponge, *Reniera elegans* (Tuzet, 1947) and *Octavella galangaui* (Tuzet and Paris, 1964), but there appears to be only one oogonial division. By contrast, the oogonia of *Halisarca dujardini* form aggregates in the mesenchyme, and within these masses they divide and produce the oocytes which later disperse (Lévi, 1956). Finally, oogonial generations have not been discerned in many of the acalcareous sponges.

In *Reniera elegans, Hippospongia communis, Octavella galangaui,* and many of the calcareous sponges, it appears that the chromosomes pair precociously, during the oogonial divisions (see Tuzet, 1964; Tuzet and Paris, 1964; Vacelet, 1964).

The oocytes at first undergo a moderate increase in size which does not involve the accumulation of yolk.* During this period of growth, the early stages of meiotic prophase and the formation of the germinal vesicle may occur. This is true in *Reniera elegans* (Tuzet, 1947), the freshwater sponges, *Spongilla lacustris* and *Ephydatia fluviatilis* (Leveaux, 1941), and the calcareous sponges (Tuzet, 1964). In *Hippospongia communis* meiotic prophase begins at the end of this period and immediately precedes the maturational divisions (Tuzet and Pavans de Ceccatty, 1958). On the other hand, these phenomena have not yet been observed at any stage in many sponges.

In some cases the smaller oocytes have an irregular form which suggests that they may move about through the mesenchyme.

Small oocytes with two nuclei are sometimes observed in a number of sponges (see Fig. 7) These include *Grantia compressa* (Duboscq and Tuzet, 1937), *Petrobiona massiliana* (Vacelet, 1964), *Spongilla lacustris* and *Ephydatia fluviatilis* (Leveaux, 1941), *Halisarca dujardini* (Lévi, 1956), and *Haliclona ecbasis* (Fell, 1970). In *Halisarca* large numbers of oocytes with two nuclei occur, and their growth parallels that of the other oocytes. In *Spongilla* and *Ephy-*

*The term "yolk" is used here in the broadest sense, meaning reserve material. The existence of specific yolk substances, such as specific yolk proteins, has not been demonstrated. As will be discussed more fully below, yolk formation is frequently accompanied by the engulfment of cells; and in a few cases yolk spheres have been shown to have an ultrastructure resembling that of phagosomes (see Borojević, 1967).

datia large oocytes that have completed their growth may also possess two nuclei. In some cases the nuclei may be male and female pronuclei, but in other cases the oocytes with two nuclei are clearly abortive.

Following the period of initial growth, the oocytes begin a more extensive growth which is usually accompanied by the accumulation of yolk. In most cases this growth involves the activity of nurse cells, but in many instances the role of the nurse cells is not well understood.

Among the calcareous sponges the nurse cells usually seem to be either choanocytes or cells that derive from choanocytes. In *Grantia compressa* and *Sycon raphanus* the oocytes apparently migrate from their position under the choanocytes to the vicinity of the atrium where the nurse cells are situated. The nurse cells of *Sycon* develop from choanocytes at the proximal ends of the radial tubes. The differentiation of these cells has not been followed in *Grantia.* Each oocyte becomes associated with 2 or 3 cells which form a nurse complex. In *Grantia* the oocytes apparently engulf at least one of the cells of the complex; however, in *Sycon* it is not clear whether the oocytes ingest whole nurse cells or simply absorb materials from them. After a period of growth the oocytes move back under the choanocytes (Duboscq and Tuzet, 1937).

In *Sycon elegans* (Duboscq and Tuzet, 1944), *Petrobiona massiliana* (Vacelet, 1964), *Leucosolenia botryoides*, *Clathrina coriacea*, and several other species (Duboscq and Tuzet, 1942; Tuzet, 1947; Sarà, 1955), the oocytes remain under the choanoderm during their major growth while in *Leucandra gossei* the oocytes apparently may either remain under the choanoderm or migrate to the atrium as in *Grantia compressa* and *Sycon raphanus* (Duboscq and Tuzet, 1942). In the course of its growth each oocyte of *Leucosolenia* incorporates a number of choanocytes (Duboscq and Tuzet, 1942) and/or a few (frequently one) abortive oocytes (Sarà, 1955). In *Leucandra gossei* the oocytes also engulf a small number of cells that resemble germ cells (Duboscq and Tuzet, 1942; Tuzet, 1964).

There exist two very different opinions concerning the nurse cells of *Clathrina.* According to Tuzet (1947), the nurse cells are special amoebocytes which are numerous only in specimens that contain reproductive elements. These cells are filled with characteristic granules which they transfer to the growing oocytes. There is apparently no engulfment of nurse cells. The granules derived from the nurse cells are degraded within the oocytes. On the other hand, Sarà (1955) has found that the oocytes of *Clathrina* are surrounded by numerous choanocytes and occasional smaller oocytes. The oocytes engulf these cells, and the nuclei of the ingested cells are visible in their cytoplasm. According to Sarà (1955), granular porocytes may envelop the masses of cells that surround the oocytes, but they do not interact with the latter. In *Clathrina blanca* many abortive oocytes appear to be engulfed by the normal oocytes in advanced stages of growth (Sarà, 1955).

A more complex situation appears to exist in *Ascandra minchini* (Borojević, 1969). In this species the oocytes undergo substantial growth by the phagocytosis of choanocytes, and then they move into nests of nurse cells to complete their development. As the oocytes and embryos develop within a nest, they are apparently displaced through it by the influx of new oocytes; and the larvae are released individually into the atrium. The nurse cells of the nests derive from pinacocytes which grow by engulfing eosinophils and choanocytes. These nurse cells appear to establish cytoplasmic continuity with the oocytes and to transfer phagosomes to them. When the transfer is complete the nurse cells apparently separate from the oocytes and subsequently degenerate.* Thus the major growth of the oocytes and *Ascandra minchini* is similar to that observed by Sarà (1955) in *Clathrina coriacea*, while the final stages of oocyte growth resemble the process described by Tuzet (1947). In *Ascandra falcata* and *Clathrina contorta* the nests of nurse cells are less well developed, and the oocytes may engulf eosinophils and hyaline amoebocytes while they are resident in the nests (Borojević, 1969).

Finally, in *Sycon elegans* (Duboscq and Tuzet, 1944) and *Petrobiona massiliana* (Vacelet, 1964) modified choanocytes become associated with the cell that transports the spermatozoon to the oocyte (see below), and they may supply the oocyte with materials through this intermediary. However, this type of nurse cell relationship is not well understood.

The nurse cells of acalcareous sponges are generally amoebocytes which are frequently characterized by numerous granules in their cytoplasm (see insert in Fig. 9). These cells may be extremely abundant during the reproductive season. In many sponges the nurse cells aggregate around the oocytes that have attained a characteristic

*The transfer of materials to oocytes through cytoplasmic bridges has been clearly demonstrated in certain animals. However, in the case considered here it would seem difficult to determine whether nurse cells transfer materials to the oocytes and then separate from them or whether they simply fuse with the oocytes. The occurrence of degenerating nurse cells devoid of phagosomes does not necessarily imply that such nurse cells transferred their phagosomes to oocytes.

size, forming more or less extensive masses around them. Presumably this aggregation is a chemotactic response to a substance(s) produced by the oocytes.

In most cases the nurse cells are ingested by the oocytes and their substance is converted into yolk spheres. This process is well exemplified by *Spongilla lacustris* and *Ephydatia fluviatilis* (Leveaux, 1941) and by *Hymeniacidon (Stylotella) heliophila* (Fincher, 1940). Although in the freshwater sponges the nurse cells involved in oogenesis and in gemmulation appear to be identical, the nature of the reserve granules formed in the two situations is very different (Leveaux, 1941).

In *Haliclona ecbasis*, as well as in several other haliclonids including *H. permollis* and *H. loosanoffi*, the nurse cells are engulfed by the oocytes, but are not immediately degraded (see Fell, 1969; and Figs. 9 and 10, this chapter). Although the nucleus of the engulfed nurse cells disappears, the cytoplasm appears to remain essentially unchanged throughout oogenesis and the early stages of embryogenesis. The nurse cells accumulate within the oocytes, and the fully developed oocytes contain about 1300 of them. During cleavage of the zygote the engulfed nurse cells are progressively fragmented and their contents are incorporated into the cytoplasm of the blastomeres. It is not clear whether this occurs by fusion or by phagocytosis. Nothing is known of the physiological state of the nurse cells inside of the oocyte and within the blastomeres, but there exists the interesting possibility that these cells may represent more than simply nutritive stores.

In some of the siliceous sponges mentioned above, the growing oocytes are surrounded by an incomplete layer of flattened follicle cells and the nurse cells are scattered among the follicle cells at the surface of the oocyte. Only after an oocyte has completed the ingestion of nurse cells is it surrounded by a continous follicular epithelium. This epithelium resembles that enclosing the spermatic cysts and it apparently has the same origin (see Leveaux, 1941; Fell, 1969).

In other sponges, including *Tedanione foetida* (Wilson, 1894) and *Halichondria panicea* (Fell, unpublished), oocytes of relatively small size are enclosed by what appears to be a continuous follicular epithelium; and a dense mass of granular nurse cells accumulates around it (see Fig. 8). These nurse cells apparently secrete material which moves across the epithelium and is incorporated into yolk spheres. However, more extensive studies, including ultrastructural ones, will be required to definitely establish the existence of such a

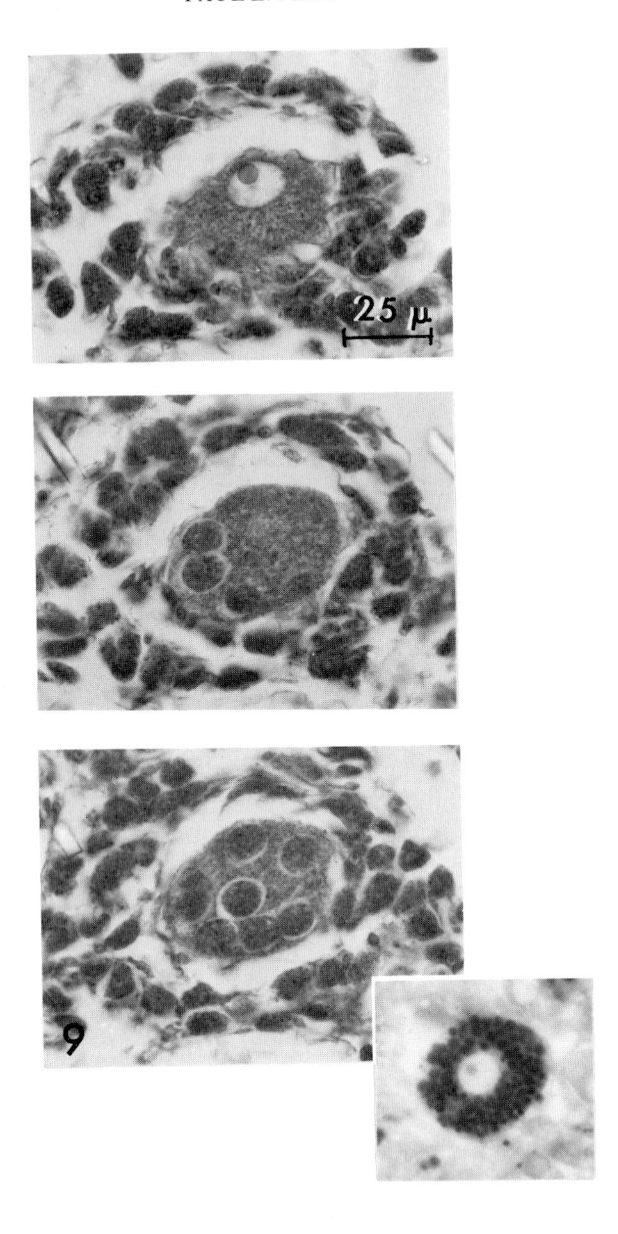

FIG. 9. Consecutive 8 μm sections through an oocyte of *Haliclona permollis* containing approximately 10 nurse cells. (Hematoxylin and eosin.) The insert shows a nurse cell filled with numerous granules. (Osmium tetroxide, embedded in Epon, Azure II methylene blue.)

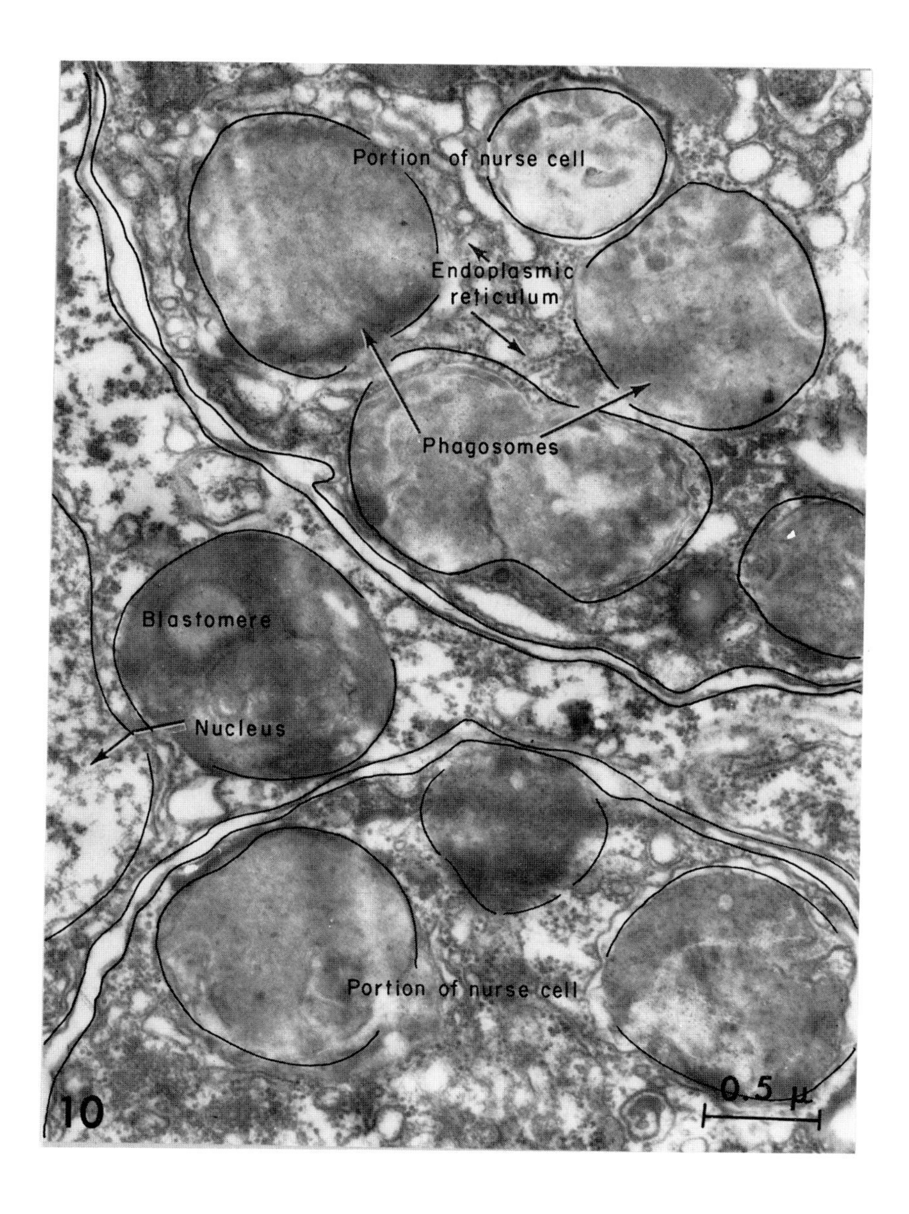

FIG. 10. An electron micrograph showing a blastomere of *Haliclona ecbasis* with fragments of engulfed nurse cells. (Osmium tetroxide, lead citrate.) (Kallman and Fell, unpublished.)

phenomenon. In *Halichondria* it appears that the larger oocytes ingest some nurse cells.

Apparently also in in *Hippospongia communis* nurse cells both secrete materials in the vicinity of the oocytes (and zygotes) and are engulfed by them; however, there is no follicular epithelium separating the former from the latter. During the later stages of growth the nurse cells immediately surrounding each oocyte are arranged in a columnar epithelium (Tuzet and Pavans de Ceccatty, 1958). In *Octavella galangaui* the oocytes are surrounded by a multilayered follicle during the period of their major growth, but there is no indication that the follicle cells make material contributions to the oocytes (Tuzet and Paris, 1964).

Finally, in a number of sponges, including *Halisarca dujardini* (Lévi, 1956), *Cliona viridis* (Tuzet, 1930), and *Reniera elegans* (Tuzet, 1947), the growing oocytes are amoeboid and apparently move about in the mesenchyme. Those of the first two species ingest some of the surrounding cells, but those of *Reniera elegans* do not. In some cases choanocytes, as well as amoebocytes, are engulfed.

Nothing is known concerning the synthetic activities of developing oocytes or, when they are present, of special nurse cells. As the occytes engulf, absorb, and/or synthesize materials, yolk granules accumulate in their cytoplasm. In many cases these granules first appear in the peripheral cytoplasm and increase in size as the oocyte grows. The fully developed oocytes of a number of sponges, including *Hymeniacidon (Stylotella) heliophila* (Fincher, 1940), *Halichondria panicea* (Fell, unpublished), and the freshwater sponges, *Spongilla lacustris* and *Ephydatia fluviatilis* (Leveaux, 1941), contain dense yolk spheres of various sizes, ranging up to about 15-20 μm in diameter. On the other hand, only very small yolk granules are found in the oocyte of *Haliclona limbata, Adocia cinerea,* and *Halichondria coalita* (Meewis, 1941). There is frequently a region relatively free of yolk around the nucleus.

Evidently all calcareous sponges are viviparous; however, both oviparity and viviparity exist among the acalcareous sponges. Within the latter group the oocytes of oviparous species tend to be smaller and more numerous than those of viviparous forms (Lévi, 1951a, 1956). There is no information available concerning the relationship between egg production and the nutrition of the parent sponge.

According to Colussi (1958) many of the oocytes of *Sycon raphanus* degenerate at the end of the period of major growth, but this phenomenon has not been noted by other authors.

In the growing oocytes of a number of sponges, including *Leucosolenia botryoides* (Duboscq and Tuzet, 1942), *Petrobiona massiliana*

(Vacelet, 1964), *Sycon elegans* (Dubosq and Tuzet, 1944), *Grantia compressa* (Duboscq and Tuzet, 1937), *Reniera elegans* (Tuzet, 1947), and *Stelletta grubii* (Liaci and Sciscioli, 1967), there is an extrusion of nuclear material into the cytoplasm. Little is known concerning the nature of this material, but it would not be surprising if it consisted of ribosomes or ribosomal precursors.

The maturational divisions with the formation of polar bodies have been observed in a number of calcareous sponges, but in only relatively few acalcareous sponges (see Tuzet, 1964). These latter species include *Reniera elegans* (Tuzet, 1947), *Halisarca dujardini* and *Halisarca metschnikovi* (Lévi, 1956), *Hippospongia communis* (Tuzet and Pavans de Ceccatty, 1958), and *Octavella galangaui* (Tuzet and Paris, 1964). In most cases these divisions occur at the end of the growth of the oocyte, but in *Hippospongia communis* they occur before the period of major growth.

Duboscq and Tuzet (1937) have described the elimination of chromatin, which persists at the equator of the spindle, during the maturational divisions in *Grantia compressa* and *Sycon raphanus*. However, the significance of this phenomenon remains unclear.

Little is known concerning the time required for the different stages of oocyte differentiation. However, it appears that oogenesis is frequently a much more lengthy process than spermatogenesis (see Section 2.3.5). In *Petrobiona massiliana* the time required for small oocytes in meiotic prophase to complete their development has been found to be approximately 6 weeks (Vacelet, 1964). Although no precise data are available, it appears that a somewhat shorter period of oocyte development may exist in *Microciona prolifera* (Simpson, 1968). On the other hand, in *Stelletta grubii* (Liaci and Sciscioli, 1967) and in *Axinella damicornis* and *Axinella verrucosa* (Siribelli, 1962) differentiation of the oocytes apparently require several months.

In a number of sponges, including *Polymastia mammillaris* and *Aaptos aaptos* (Sarà, 1961), *Stelletta grubii* (Liaci and Sciscioli, 1967), *Halisarca dujardini* (Lévi, 1956), *Petrobiona massiliana* (Vacelet, 1964), *Sycon raphanus* (Duboscq and Tuzet, 1937), *Sycon elegans* (Duboscq and Tezet, 1944), and several species of *Leucandra* (Duboscq and Tuzet, 1942), the differentiation of the oocytes is synchronous. On the other hand, in *Grantia compressa* (Duboscq and Tuzet, 1937), *Oscarella lobularis* (Lévi, 1951b), *Adocia simulans* (Lévi, 1956), *Hymeniacidon sanguinea* (Sarà, 1961), *Haliclona ecbasis* (Fell, 1969), and many other sponges, the oocytes develop asynchronously. Among the acalcareous sponges, many of the species with synchronous differentiation of the oocytes are oviparous. The pattern of oocyte development and of spermatocyte development are not necessarily the same for a particular species. For example, in *Aaptos aaptos* (Sarà, 1961) and *Stelletta grubii* (Liaci and Sciscioli, 1967) the oocytes develop synchronously but the spermatocytes do not.

2.3.5 Gametogenic Cycles within Populations

The gametogenic cycles of only a few sponges have been studied in any detail. Such cycles are most readily analyzed in large populations of sponges which permit extensive sampling. However, in many localities the sponge populations are relatively small. In such cases samples taken at any one time may not be truly representative; but if sampling is continued over a period of several years, a reasonably accurate picture of reproduction may be obtained. In some instances long-term studies may be hindered by fluctuations in the abundance of sponges at particular localities from one period to another.

Several different patterns of gamete production have been described. In the oviparous sponges, *Stelletta grubii* (Liaci and Sciscioli, 1967), *Axinella damicornis* and *Axinella verrucosa* (Siribelli, 1962), and *Polymastia mammillaris* (Sarà, 1961), there is a definite reproductive period with one cycle of oocyte production per year. The oocytes appear several months before the spermatocytes, and only after the former are fully developed does the differentiation of the latter begin. For example, in *Stelletta grubii* the oocytes develop from July to September (at Bari, Italy); and then developing spermatozoa appear. During the next three months mature gametes of both kinds are found in decreasing numbers (see Fig. 11A). In this case there are asynchronous cycles of spermatozoan differentiation in individual specimens; however, in *Polymastia* and the two species of *Axinella* the differentiation of spermatozoa is not only synchronous within individual specimens but also among the individuals of the pupulation. In *Stelletta grubii* and the species of *Axinella* spermatogenesis and oogenesis occur in separate specimens. *Polymastia* is apparently proterogynous. *Axinella damicornis* and *Axinella verrucosa* occur together in the Gulf of Naples, but are reproductively isolated since the gametes of *damicornis* mature in May while those of *verrucosa* mature in September (Siribelli, 1962).

In *Petrobiona massiliana* (Vacelet, 1964) and in *Sycon* and *Leucandra* (Duboscq and Tuzet, 1942, 1944), oogenesis is also synchronized within populations, most of the specimens of a species examined at a particular time and place having oocytes or embryos at the same stage of development. In *Petrobiona massiliana* (Vacelet,

1964), *Sycon raphanus* (Duboscq and Tuzet, 1937; Tuzet, 1964), and *Sycon elegans* (Duboscq and Tuzet, 1944), there may be a succession of reproductive cycles, and oocytes may coexist with embryos or larvae in the same specimen. When two stages occur simultaneously, they are frequently situated in different regions of the sponge. In *Sycon elegans* (Duboscq and Tuzet, 1944) the younger of two stages generally occurs in the newer portions of the specimen, but this relationship does not appear to exist in *Petrobiona* (Vacelet, 1964).

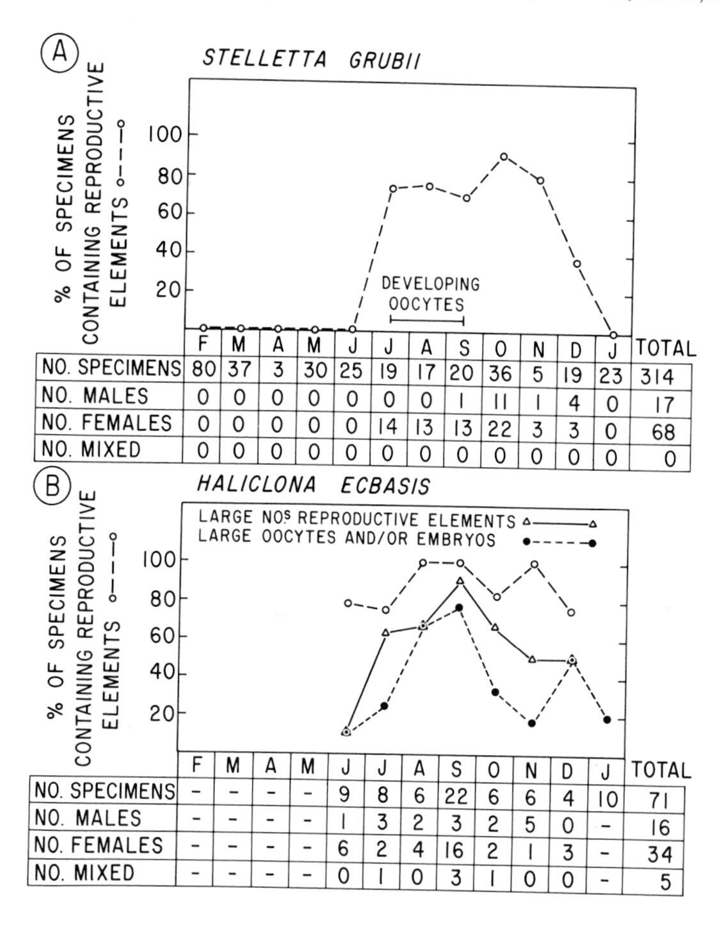

	F	M	A	M	J	J	A	S	O	N	D	J	TOTAL
NO. SPECIMENS	80	37	3	30	25	19	17	20	36	5	19	23	314
NO. MALES	0	0	0	0	0	0	0	1	11	1	4	0	17
NO. FEMALES	0	0	0	0	0	14	13	13	22	3	3	0	68
NO. MIXED	0	0	0	0	0	0	0	0	0	0	0	0	0

	F	M	A	M	J	J	A	S	O	N	D	J	TOTAL
NO. SPECIMENS	-	-	-	-	9	8	6	22	6	6	4	10	71
NO. MALES	-	-	-	-	1	3	2	3	2	5	0	-	16
NO. FEMALES	-	-	-	-	6	2	4	16	2	1	3	-	34
NO. MIXED	-	-	-	-	0	1	0	3	1	0	0	-	5

FIG. 11. (A) The occurrence of reproductive specimens of the oviparous sponge, *Stelletta grubii*, in the vicinity of Bari, Italy. [Data from Liaci and Sciscioli (1967) presented in a different form.] (B) The occurrence of reproductive specimens of the viviparous sponge, *Haliclona ecbasis*, in San Francisco Bay. The sponge was not abundant during the spring of the year represented; only the fall reproductive peak is shown. [Data from Fell (1970).]

Many other sponges apparently have a more or less definite reproductive period with repeated, asynchronous cycles of oocyte and spermatozoan differentiation. Included in this category are *Grantia compressa* (Tuzet, 1964), *Hippiospongia lachne* (Storr, 1964), *Microciona prolifera* (Simpson, 1968), *Hymeniacidon sanguinea* (Sarà, 1961), and *Haliclona ecbasis* (Fell, 1970). In some species and in certain localities, specimens with reproductive elements are found throughout a major portion of the year; but in such cases there are definite periods during which gamete and embryo production are maximal. For example, in particular regions of San Francisco Bay *Haliclona ecbasis* is of somewhat sporadic occurrence, but some specimens containing gametes and/or embryos can usually be found whenever the sponge is abundant (see Table IV). However, a high frequency of specimens containing large numbers of reproductive elements is found only in the spring and during the late summer and fall (see Fig. 11B). A similar pattern of reproduction is found in *Hippiospongia lachne* (Storr, 1964), but the number and time of occurrence of the reproductive peaks is different in different localities (see Section 2.3.6).

TABLE IV

THE OCCURRENCE OF REPRODUCTIVE ELEMENTS IN HALICLONA ECBASIS IN SAN FRANCISCO BAY

Year[a]	Month	Sperm	Small oocytes	Large oocytes	Embryos	Larvae
1964, 1966	January	?	+	+	+	?
1964	March	?	+	+	+	?
1964	April	?	+	+	+	?
1964, 1966	May	?	+	+	+	+
1965	June	+	+	+	?	—
1965	July	+	+	+	+	+
1965	August	+	+	+	+	+
1964, 1965	September	+	+	+	+	+
1964, 1965	October	+	+	+	+	+
1964, 1965	November	+	+	+	+	+
1965	December	—	+	+	+	+

[a]The sponge was absent (or extremely scarce) at the collecting stations during the summer of 1964 and the first several months of 1965, and few collections were made in 1966.

In the Mystic Estuary (on Fishers Island Sound) *Haliclona loosanoffi* and *Halichondria bowerbanki* begin to reproduce sexually soon after their emergence from winter hibernation (see Table II). During late June and July, large numbers of small sponges, developed from swimming larvae, are found on the new crop of eel grass (see Fig. 19). These young sponges grow rapidly; and in the case of *Haliclona*, soon begin to form gemmules. In both sponges the sexual reproductive season ends well in advance of the next period of dormancy. The specimens of *Halichondria* derived from larvae apparently do not reproduce sexually until the following summer; and in *Haliclona* it appears that only the specimens developed from gemmules engage in sexual reproduction (Fell, in preparation). In Milford Harbor, Connecticut these sponges reproduce sexually during the late summer and fall (Hartman, 1958). The later reproductive period at Milford is correlated with a later rise in water temperature there. A similar sequence of events was also found to exist for *Microciona prolifera* in the Milford-New Haven region of Long Island Sound (Hartman, 1958; Simpson, 1968).

The occurrence of reproduction in these sponges is strikingly different at Hatteras Harbor, North Carolina, where there are also large annual changes in water temperature but where the water is generally warmer (range from 4.7° to 30.2°C). At this locality *Haliclona loosanoffi* has two reproductive periods, one during June and July and another during October and November. The sponge occurs in a nonreproductive state during the winter and survives the warm summer months (from late July through early September) only in the form of gemmules. On the other hand, *Microciona prolifera* produces larvae from June through November with a major peak in June and a minor peak in September (Wells *et al.*, 1964).

At Plymouth, England, *Sycon coronatum* and *Grantia compressa* are annuals and may have two reproductive periods in their life history, one in the late spring and another in the later summer and fall. Large specimens reproduce in the late spring and then disintegrate during the summer. The small sponges produced in the spring reproduce in the late summer and fall, grow during the winter, and then reproduce again in the following spring. Specimens produced in the fall apparently reproduce only once (Orton, 1914, 1920).

In some cases occasional small oocytes may be found during most of the year, although larger oocytes and embryos are present only during a definite reproductive period. This is true for *Halisarca dujardini* at Roscoff, France (Lévi, 1956) and *Halichondria bowerbanki* in the Mystic Estuary (Fell, in preparation).

In *Hymeniacidon (Stylotella) heliophila* (Fincher, 1940) and *Octavella galangaui* (Tuzet and Paris, 1964), spermatocytes appear before oocytes. These sponges may be protandrous. On the other hand, in *Microciona prolifera* oocytes appear first (Simpson, 1968). As has been mentioned above, this sponge is a contemporaneous hermaphrodite.

2.3.6 Factors Influencing Gametogenesis

Very little is known concerning the factors that influence gametogenesis. In a number of cases it appears that water temperature may be an important factor. For example, the production of oocytes and embryos in *Hippiospongia lachne* was followed at monthly intervals in three different localities, and the occurrence of reproductive peaks was correlated with water temperature (Storr, 1964). The greatest proportion of reproductive specimens was found during March and April in British Honduras, from April through June in the Bahamas, and during June and July in the vicinity of Cedar Keys, Florida. These peaks occurred when the water temperature was approaching 29°C. In all three localities after the water temperature had risen above that temperature, there was a decline in the frequency of reproductive specimens. In British Honduras and the Bahamas some reproductive specimens were found throughout the year, but in the Cedar Keys area reproduction ceased in the fall when the water temperature dropped sharply. The lower critical temperature for reproduction appeared to be about 23°C. In the first two localities where the changes in water temperature were less extreme, the proportion of reproductive specimens began to rise again in the fall after the water temperature had fallen below 29°

Data obtained from studies of *Haliclona loosanoffi* in Hatteras Harbor, North Carolina and Long Island Sound suggest that the temperature range within which reproduction occurs is between roughly 20° and 27°C (see Hartman, 1958; Wells *et al.*, 1964). However, at least at certain localities, it seems very probable that factors other than water temperature are also important in determining the duration of the reproductive period (see Table II).

Lévi (1956) has recorded the reproductive seasons for a large number of sponge species on the northern coast of Finistère, France. Most of these sponges reproduce during the summer months when the water is warmest (see Table V). The occurrence of reproduction during periods when the water is relatively warm appears to be a widespread phenomenon (see Table VI); however, it is not known to what extent water temperature has a direct influence on reproduction. Although there is little precise information available on the subject, it seems clear from a number of studies that reproduction can occur only within a certain temperature range and that this range is

TABLE V

THE REPRODUCTIVE SEASONS OF 43 SPONGES IN THE VICINITY OF ROSCOFF, FRANCE[a]

Number of species	Reproductive period[b]											
	F	M	A	M	J	J	A	S	O	N	D	J
2			—	—								
1			—	—	—	—						
1				—	—							
1				—	—	—						
1					—							
2					—	—						
1					—	—	—	—				
9						—	—					
7							—					
3							—	—				
1							—	—	—	—	—	—
12								—				
1								—	—			
1									—			
Approximate water temperature (°C)	8.5	8.5	9.0	12.0	13.0	14.0	15.5	16.0	14.0	13.0	12.5	8.5

[a] Data of Lévi (1956) pp. 8 and 9, presented in a different form.

[b] The months during which specimens contain large oocytes and/or embryos.

TABLE VI
THE REPRODUCTIVE SEASONS OF SOME MARINE SPONGES

Species	Reproductive period	Locality	Author
Leucosolenia botryoides	Oogenesis, February to March	Bay of Naples	Sarà, 1955
	Late summer	Roscoff, France	Duboscq and Tuzet, 1942
Leucosolenia variabilis	Larvae, August and early September	Roscoff, France	Minchin, 1896
Clathrina (Leucosolenia) coriacea	Oogenesis, September to November	Bay of Naples	Sarà, 1955
	Larvae, September	Roscoff, France	Minchin, 1896
Sycon coronatum	Late spring and early summer, and again in late summer and fall	Plymouth, England	Orton, 1914 and 1920
Sycon raphanus	Oogenesis, February to November	Banyuls-sur-Mer, France	Duboscq and Tuzet, 1937
Sycon elegans	Early spring through early winter	Villefranche-sur-Mer, France	Duboscq and Tuzet, 1944
Grantia compressa	Late spring and early summer, and again in late summer and fall	Plymouth, England	Orton, 1914, 1920; Dendy, 1914
	June to early September	Northumberland coast, England	Jorgensen, 1917
Rhabdodermella nuttingi	Early embryos, May; larvae, August	Monterey Bay, California	Fell, unpublished
Petrobiona massiliana	April or May to December or January; larvae first appear in July	Marseille, France	Vacelet, 1964
Octavella galangaui	May to early September; larvae, August to early September	Banyuls-sur-Mer, France	Tuzet and Paris, 1964

Halisarca dujardini	Oocytes, April to August; sperm, May; larvae released, June to September	Roscoff, France	Lévi, 1956
Halisarca metschnikovi	Mature gametes, May; larvae released, early June	Roscoff, France	Lévi, 1956
Stelletta grubii	July to December; mature gametes, October to December	Bari, Italy	Liaci and Sciscioli, 1967
Tetilla (Tethya) serica	Breeding season, July to September; peak, August	Misaki, Japan	Egami and Ishii, 1956; and Watanabe, 1957
Tethya aurantium	Eggs, July; spawning, beginning of August	Roscoff, France	Lévi, 1956
Cliona celata	Larval settlement, early August to early October	Long Island Sound	Hartman, 1958
Cliona viridis	Gametogenesis, May	Bay of Naples	Sarà, 1961
Cliona stationis	Eggs, May to mid-June	Black Sea	Nassonow, 1883; 1924
Polymastia robusta	Shedding of eggs, September	Roscoff, France	Borojević, 1967
Polymastia mammillaris	Spermatogenesis and shedding of eggs, November to December	Bay of Naples	Sarà, 1961
Suberites (Ficulina) ficus	Shedding of eggs, early October	Roscoff, France	Lévi, 1956
Laxosuberites aquae-dulcioris	Mature larvae, July	Chilka Lake on the Bay of Bengal	Annandale, 1915
Laxosuberites lacustris	Mature larvae, April	Chilka Lake on the Bay of Bengal	Annandale, 1915
Prosuberites microsclerus	Larval settlement, May to July and September to December	Hatteras Harbor, North Carolina	Wells *et al.*, 1964
Aaptos aaptos	Gametogenesis, late November to early December	Bay of Naples	Sarà, 1961

continued

TABLE VI (*Continued*)
THE REPRODUCTIVE SEASONS OF SOME MARINE SPONGES

Species	Reproductive period	Locality	Author
Raspailia pumila	August and September	Rance and Morlaix Estuaries, France	Lévi, 1956
Axinella damicornis	Mature gametes, May	Gulf of Naples	Siribelli, 1962
	Mature eggs, end of September	Roscoff, France	Lévi, 1951c
Axinella verrucosa	Mature gametes, September	Gulf of Naples	Siribelli, 1962
Halichondria bowerbanki	Larval settlement, August to November	Long Island Sound	Hartman, 1958
Halichondria panicea	Late spring and summer	San Francisco Bay	Fell, unpublished data
Hymeniacidon sanguinea	June to September	Bay of Naples	Sarà, 1961
Hymeniacidon (Stylotella) heliophila	Late summer and early autumn	Beaufort, North Carolina	Fincher, 1940
Lissodendoryx similis	Larvae, August to February	Madras, India	Ali, 1956
Lissodendoryx isodictyalis (carolinensis)	July to August	Beaufort, North Carolina	Wilson, 1937
	Larval settlement, July to September; peak, August	Hatteras Harbor, North Carolina	Wells *et al.*, 1964
Microciona prolifera	July to August	Beaufort, North Carolina	Wilson, 1937
	Larval settlement, August and September	Beaufort, North Carolina	McDougall, 1943
	Larval settlement, June to November; major peak, June	Hatteras Harbor, North Carolina	Wells *et al.*, 1964
	May to August.	Long Island Sound	Simpson, 1968
	Larval settlement, August to mid-October	Long Island Sound	Hartman, 1958

Microciona microjoanna	Late spring and early summer; larvae, June	San Francisco Bay	Fell, 1967
Mycale syrinx	Larvae, October and November	Gulf of Naples	Wilson, 1935
Mycale contarenii	July and August	Roscoff, France	Borojević, 1966
Mycale cecilia	Embryos, August; larval settlement, June and July, and again in October and November; peak July	Hatteras Harbor, North Carolina	Wells *et al*, 1964
Tedania gurjanovae	July	San Juan Archipelago	Bakus, 1964a
Tedanione foetida	Oocytes, September and October	Green Turtle Cay, the Bahamas	Wilson, 1894
Adocia (Reniera) tubifera	Larval settlement, June and early July; peak, early June	Beaufort, North Carolina	McDougall, 1943
Haliclona loosanoffi	Larval settlement, August to Mid-October; peak, early October	Milford Harbor, Long Island Sound	Hartman, 1958
	Larval settlement, June and July, and again in October and November; major peak, July	Hatteras Harbor, North Carolina	Wells *et al*., 1964
Haliclona (Chalina) oculata	Oocytes and embryos, July	Block Island Sound, Rhode Island	Hartman, 1958
Haliclona ecbasis	Reproductive peaks, spring and again during the late summer and early fall	San Francisco Bay	Fell, 1970
Haliclona permollis	Late spring and summer	Monterey Bay, California	Fell, 1967
	Oocytes, March to October, most abundant April to June; embryos, April to October, most abundant May to July	Oregon Coast	Elvin, personal communication

continued

TABLE VI (*Continued*)

Species	Reproductive period	Locality	Author
Haliclona indistincta	May to July	Isle Verte, Brittany, France	Lévi, 1956
Hippiospongia lachne	Major reproductive peak, spring and early summer	British Honduras	Storr, 1964
	Major reproductive peak, April to June	the Bahamas	Storr, 1964
	Major reproductive peak, June and July	Cedar Keys, Florida	Storr, 1964
Hippospongia communis	Larvae released, end of March to third week of June; peak late May	Coast of Tunisia	Vaney and Allemand-Martin, 1918; Tuzet and Pavans de Ceccatty, 1958
Phyllospongia foliascens	December and January	Red Sea	Lévi, 1956
Farrea sollasii	Throughout the year	Sagami Sea, Japan	Okada, 1928

different for different species and perhaps for the same species in different localities (Simpson, 1968). Where the annual fluctuations in water temperature are extreme, temperature may be the dominant factor regulating reproduction, but light, the availability of nutrients, and other factors may also be important.

Size (age?) may be another important factor. For example, it appears that during the reproductive period of *Suberites* (*Ficulina*) *ficus*, usually only small specimens (2-5 cm) produce oocytes (Lévi, 1956). On the other hand, only the larger specimens of *Clathrina coriacea* (Sarà, 1955) and of *Hippiospongia lachne* (Storr, 1964) are found to contain oocytes and embryos. In the latter case the minimum size of reproductive specimens is smaller in warmer localities.

In some cases it appears that the reproductive cycles of intertidal and subtidal specimens of the same species may differ considerably. Along the New Zealand coast specimens of *Tethya aurantium* and *Aaptos aaptos* from subtidal regions (in 3 m or more of water) produce oocytes at regular intervals, but specimens of these sponges rarely produce oocytes in the intertidal zone. This latter region apparently is repopulated by larvae produced by offshore specimens and by asexual reproduction (Bergquist *et al.*, 1970).

Probably in most cases a number of exogenous and endogenous factors interact in the regulation of reproductive cycles. The relative importance of different factors will be learned only through experimental studies in which certain factors are varied independently of others.

2.3.7 Reproductive Behavior and Spawning

There is little behavior associated with reproduction in these sessile organisms. Spermatozoa are shed into the excurrent canals and are discharged through the oscula. The rupturing of the walls of the spermatic cysts is probably an enzymatic process. Many marine sponges are viviparous, and consequently the female gametes are not spawned. In some oviparous sponges, oocytes are released by the same route as the spermatozoa with the result that fertilization is external. However, in other oviparous species it appears that fertilization is internal and that zygotes are released into the sea. The situation is unknown in many cases because fertilization has not been observed. In some instances in which oocytes (or zygotes) were released in the laboratory but did not develop, it is not clear whether shedding was premature, precluding normal internal fertilization, or whether conditions for normal external fertilization and/or development were inadequate. *Tetilla (Tethya) serica** sheds oocytes (Kumé, 1952; Egami and Ishii, 1956), while *Cliona celata* apparently releases zygotes (Warburton, 1961). However, since many of the published accounts are unclear on this point, the term "egg" will be used in the discussion that follows.

In *Polymastia robusta* (Borojević, 1967), *Suberites (Ficulina) ficus, Tethya aurantium,* and *Raspailia pumila* (Lévi, 1956), the eggs are usually shed singly or in small groups, while in *Axinella damicornis* (Lévi, 1951c), there is a massive and rhythmic discharge of eggs. In *Polymastia* the eggs are embedded in a mucoid material which envelops the parent sponge.

Spawning occurs in the morning in *Suberites (Ficulina) ficus* and *Axinella damicornis,* following the cessation of water circulation. In the latter sponge, spawning occurred in the laboratory on 4 successive days, beginning about 2 hours after sunrise and lasting for approximately 30 minutes (Lévi, 1951c, 1956). These observations suggest that light may coordinate spawning in these sponges. In *Polymastia robusta* there was a simultaneous spawning of all the sponges in the laboratory, and all of the eggs were shed in less than 24 hours (Borojević, 1967). On the other hand, specimens of *Tetilla (Tethya) serica* release eggs (or spermatozoa) almost daily during the long breeding season (Watanabe, 1957).

The formation of polar bodies occurs either during or immediately following the release of the eggs in *Axinella damicornis* (Lévi, 1951c) and *Cliona stationis* (Nassonow, 1883).

The release of spermatozoa by three species of sponges was observed by Reiswig (1970) on the coral reefs along the northern coast of Jamaica (Discovery Bay). The spermatozoa were released either in a relatively dense column *(Verongia archeri)* or in diffuse turbid clouds *(Geodia* sp. and *Neofibularia nolitangere)* which extended 2 - 3 m above the specimens. Only single specimens of *Verongia* and *Geodia* were observed to release spermatozoa, but there was a simultaneous shedding of spermatozoa by many specimens of *Neofibularia*. In this latter case, the initiation of shedding spread from specimen to specimen down current as the cloud of spermatozoa advanced. The shedding specimen(s) of *Verongia, Geodia,* and *Neofibularia* were observed in mid-February, early July, and late October, respectively. In all three cases shedding occurred in the middle of the afternoon on a day near to a new or full moon.

*Insufficient details are given to be at all certain, but the internally developing larvae of *Tetilla cranium* (see Burton, 1931) may be asexually produced.

In many instances no distinction has been made in the literature between the breeding period (when spawning and fertilization occur) and the entire reproductive period. Furthermore, reference to the latter is frequently based on incidental observations. Table VI summarizes what is known concerning the breeding and reproductive periods of a number of marine sponges. Additional information may be found in the papers of Maas (1893, p.335), Topsent (1887, pp. 138-159; and 1895, p. 11), Lévi (1956, pp. 8 and 9), Hartman (1958, pp. 81-85), and Wells *et al.* (1964, pp. 760 and 761).

In many sponges, in which there is an asynchronous differentiation of gametes, breeding apparently occurs throughout most of the reproductive period. But obviously, as the length of the reproductive period is reduced, the breeding period becomes relatively shorter since breeding can not occur until at least some gametes have differentiated. On the other hand, in many oviparous sponges, breeding occurs only at the end of a period of synchronous oocyte development.

2.4 Development

2.4.1 Fertilization

Fertilization is internal in the vast majority of sponges and is associated with special problems. As mentioned above, the oocytes are scattered about in the mesenchyme. There are no reproductive ducts, and in viviparous sponges the embryos frequently develop within the follicles of the oocytes. Spermatozoa are released into the sea and are drawn into oocyte-containing specimens in the incurrent water stream. Once inside an oocyte-containing sponge, the spermatozoa must usually penetrate one or more barriers in order to reach the oocytes. There are therefore two basic problems related to fertilization: (1) locating the oocytes, and (2) passing the barriers that exist between the canal system and the oocytes.

In all cases in which internal fertilization has been observed, it appears that spermatozoa are transported to the oocytes by cellular intermediaries. The intermediate appears always to be a choanocyte. After a spermatozoan has entered a choanocyte, the latter loses its collar and flagellum and may increase in size. The spermatozoan is enclosed within a vesicle inside the choanocyte, and it sooner or later loses its flagellum. The choanocyte containing the spermatozoan, leaves the flagellated epithelium, applies itself against the surface of an oocyte, and transfers the spermatozoan to it (see Fig. 12). Thus the spermatozoan apparently is transported across the barriers that exist between the canal system and the oocytes by a modified choanocyte, the carrier cell. In those sponges in which the oocytes become enclosed within a follicle, the carrier cells make contact with the oocytes before the follicle forms. This sequence, or parts of it, has been observed in *Grantia compressa* (Gatenby, 1920, 1927; Duboscq and Tuzet, 1937), *Sycon raphanus* (Duboscq and Tuzet, 1937), *Sycon elegans* (Duboscq and Tuzet, 1944), *Leucosolenia botryoides* (Duboscq and Tuzet, 1942), *Clathrina (Leucosolenia) coriacea* (Tuzet, 1947; Sarà, 1955), *Reniera simulans* (Tuzet, 1932) and *Reniera elegans* (Tuzet, 1947), *Cliona viridis* (Tuzet, 1930a; Sarà, 1961), *Hippospongia communis* (Tuzet and Pavans de Ceccatty, 1958), *Petrobiona massiliana* (Vacelet, 1964), *Octavella galangaui* (Tuzet and Paris, 1964), and several other species.

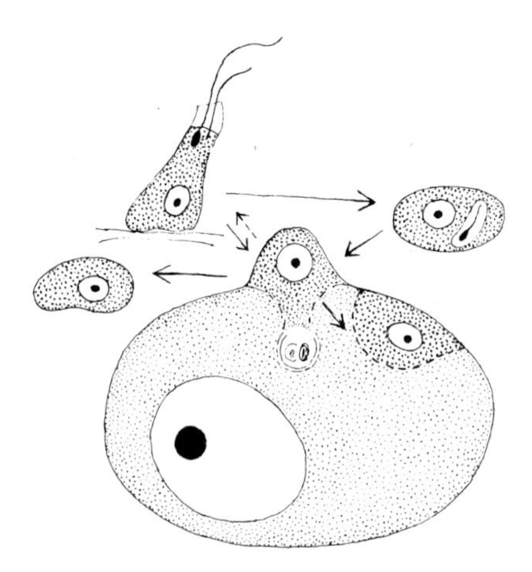

FIG. 12. Transport of the spermatozoan to the oocyte by a carrier cell. The spermatozoan enters a choanocyte, which, after losing collar and flagellum, may either transfer it to the oocyte directly or move through the mesenchyme to the oocyte. After the spermatozoan has been transferred to the oocyte, the carrier cell leaves the latter in some cases and is engulfed by it in others. [Modified from Duboscq and Tuzet (1937); Tuzet (1947); Tuzet and Paris (1964).]

In some sponges, including *Octavella galangaui*, *Hippospongia communis*, and many of the calcareous sponges, the oocytes are situated immediately below the choanoderm when the carrier cells establish contact with them. However, in other sponges, including *Sycon raphanus*, *Grantia compressa*, *Reniera elegans*, and *Cliona viridis*, they are dispersed in the mesenchyme. Among the mechanisms by which the spermatozoa may ultimately make contact with the oocytes are the following: (1) The spermatozoa may enter choanocytes at random and only the choanocytes, which are initially associated with oocytes and/or which subsequently make chance contact with them, may function in fertilization; (2) The spermatozoa may enter choanocytes at random, and the carrier cells may locate the oocytes by chemotaxis; and (3) The spermatozoa may enter choanocytes that are in proximity to oocytes, by chemotaxis. Although there is no conclusive evidence for chemotaxis, a number of observations are very suggestive. For example, in *Sycon raphanus* spermatozoa are seen only in choanocytes that are situated over oocytes (Duboscq and Tuzet, 1937). Chemotactic location of oocytes by spermatozoa appears to be rare in the animal kingdom, but it has been shown to occur in certain hydroids (Miller, 1966). Probably the mechanism that brings the gametes together is somewhat different in different species.

In some sponges, including *Reniera elegans* (Tuzet, 1947) and *Grantia compressa* and *Sycon raphanus* (Duboscq and Tuzet, 1937), choanocytes with spermatozoa are associated with very small oocytes, but fertilization does not occur until the oocytes have completed or nearly completed their growth. In these particular species the carrier cells apparently follow the oocytes into the mesenchyme. On the other hand, in *Clathrina (Leucosolenia) coriacea* (Tuzet, 1947), *Petrobiona massiliana* (Vacelet, 1964), *Octavella galangaui* (Tuzet and Paris, 1964), and *Hippospongia communis* (Tuzet and Pavans de Ceccatty, 1958), fertilization occurs either before or soon after the oocytes enter the period of major growth.

At the time that a spermatozoan is transferred to an oocyte, it usually appears as two masses within a clear vesicle. One mass is the head of the spermatozoan, and the other is the intermediate segment. Apparently cytoplasmic continuity is established between the carrier cell and the oocyte, and the vesicle containing the spermatozoan moves from one cell to the other. In *Grantia compressa* and *Sycon raphanus* the spermatic vesicle enters the oocyte through what appears to be a tubular invagination of the oocyte surface (Duboscq and Tuzet, 1937). In many sponges the carrier cell apparently separates from the oocyte after the transfer has been made, but in *Octavella galangaui* (Tuzet and Paris, 1964) and perhaps also in *Leucandra nivea* (Tuzet, 1964) it is incorporated into the oocyte. Although the carrier cell of *Sycon elegans* is not engulfed, its nucleus apparently degenerates before the spermatozoan is transferred to the oocyte (Duboscq and Tuzet, 1944). Since the actual union of gametes appears to be very different from what occurs in most other animals (see Colwin and Colwin, 1964, for a review), more extensive information on this process would be of considerable interest. Electron microscopic studies are very desirable, but will be extremely difficult because of the problems in finding the objects of study.

In nearly all cases the spermatozoan undergoes little change within the cytoplasm of the oocyte until the maturation divisions occur. Then the male and female pronuclei develop and fuse. In *Sycon raphanus* three pronuclei are sometimes observed, which indicates occasional polyspermy (Duboscq and Tuzet, 1937). In *Hippospongia communis* the maturation divisions occur before the major growth of the oocyte, and the fusion of pronuclei takes place during the early stages of growth. The diploid, anucleolate nucleus is situated at the periphery of the cell throughout the remainder of the growth period, during which the diameter of the cell more than triples (Tuzet and Pavans de Ceccatty, 1958).

The process of gamete fusion has not been studied in cases of external fertilization. In *Tetilla (Tethya) serica* the shed, unfertilized egg possesses numerous radiating processes that may be longer than the diameter of the egg itself. These processes extend through a clear vitelline membrane. Following fertilization, the vitelline membrane is elevated from the surface of the egg, beginning at one point (presumably the site of sperm penetration) and extending progressively around the egg within about one minute. At the same time the egg adheres to the substratum and the processes are withdrawn into the perivitelline space where they form a number of layers around the egg (Kumé, 1952).

Fertilization has been observed in only a few acalcareous sponges; and its occurrence is usually inferred from the presence of spermatozoa, as well as oocytes, during the reproductive period. This situation is understandable when one considers the nature of the material. In many instances, the larger oocytes are filled with yolk spheres which, if fertilization occurs at these later stages, would make observation of the process difficult. Furthermore, in many species the oocytes develop asynchronously so that the few oocytes that are fertilized at any one time might be easily missed. However, maturation divisions have been observed in the oocytes of relatively few sponges; and the possibility exists that in some species unfertilized, diploid oocytes may cleave and develop into larvae (Bergquist *et al.*, 1970). While this possibility should be kept in mind, maturation and fertilization of oocytes should be assumed to occur in all cases until direct evidence to the contrary is forthcoming.

2.4.2 Embryonic Development

The oviparous sponges shed oocytes or zygotes into the sea where embryonic development takes place. In the species in which early development has been observed, cleavage is total and nearly equal and leads to the formation of a morula. Development beyond the morula stage may follow one of four courses (see Fig. 13). In *Tetilla (Tethya) serica* the zygote adheres to the substratum, and the morula develops directly into an adult sponge. Pseudopods and later single cells and groups of cells penetrate the fertilization membrane and move out radially on the substratum. These processes progressively

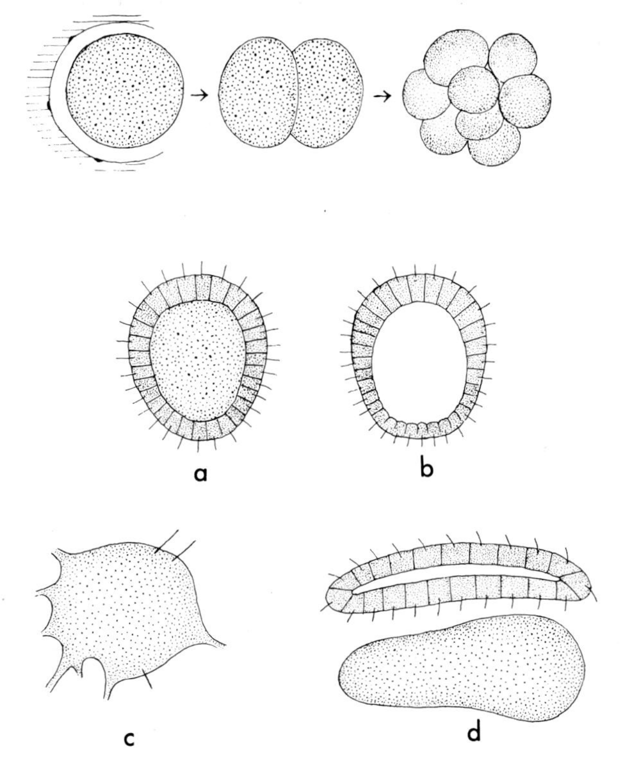

FIG. 13. The development of four oviparous, Tetractinomorph sponges. The morula develops: (a) in *Tethya aurantium*, into a swimming parenchymula larva [modified from Lévi, (1956); (b) in *Cliona stationis*, into a swimming amphiblastula larva [modified from Nassonow (1883)] (c) in *Tetilla (Tethya) serica*, directly without the formation of a larva (the dorsal aspect of an embryo flattened on the substratum is shown) [modified from Watanabe (1957)]; and (d) in *Polymastia robusta*, into a creeping benthic larva (a sagittal section and the dorsal aspect of the larva are shown) [modified from Borojević (1967)].

come together under the embryo and form the rudiment of the stalk. At the same time spicules, canals, and flagellated chambers develop within the body of the sponge (Watanabe, 1957, 1960). The cleaving zygotes of *Polymastia robusta* are also held to the substratum, but in a common mass of mucus produced by the maternal sponge. The morula develops into a simple creeping larva which has the form of a flattened blastula. This larva consists of a relatively small number of undifferentiated cells, each of which possesses a single flagellum. It moves out of the mass of mucus and changes very little during a long free-living period (Borojević, 1967). The development of *Raspailia pumila* appears to be very similar to that of *Polymastia* (Lévi, 1956). In *Cliona stationis* the young embryo apparently can move over the substratum, but it soon develops into a swimming amphiblastula larva.* The hollow larva consists of two types of cells, tall columnar ones at the anterior end and along the sides and oval ones at the posterior end. All of the cells of the larva possess a single flagellum (Nassonow, 1883). Finally, the morula of *Tethya aurantium* develops into a swimming parenchymula larva. The larva, which shows no definite anterior-posterior differentiation, consists of a central mass of yolk-filled cells enclosed within a columnar flagellated epithelium (Lévi, 1956).

The larva of *Polymastia robusta* develops within approximately 24 hours after the shedding of the eggs (Borojević, 1967), and those of *Cliona stationis* (Nassonow, 1924) and *Tethya aurantium* (Lévi, 1956) differentiate in about 2 - 3 days (no temperatures are given).

In *Raspailia pumila* the embryos may fuse beginning at the morula stage (Lévi, 1956; and see Section 2.4.3.2 this chapter).

The majority of the sponges are viviparous; and the embryos, which in many cases are surrounded by the oocyte follicle, develop within the maternal mesenchyme. In all cases cleavage of the zygote is total, but there is considerable variation in the pattern of cleavage among the different sponges that have been studied. In many species, including *Adocia cinerea* (Meewis, 1941), *Haliclona limbata* (Meewis, 1939), *Haliclona indistincta* (Lévi, 1956), *Octavella galangaui* (Tuzet and Paris, 1964), *Hippospongia communis* (Tuzet and Pavans de Ceccatty, 1958), *Oscarella (Halisarca) lobularis* (Meewis, 1938), *Farrea sollasii* (Okada, 1928), and many of the calcareous sponges (Duboscq and Tuzet, 1937; Tuzet, 1948; Borojević, 1969), cleavage is equal or nearly equal during the early stages of development (see Fig. 17). On the other hand, in *Adocia simulans* (Lévi,

*Warburton (1961) does not give a detailed description of the larva of *Cliona celata;* however, he implies that it is a parenchymula.

1956) and the freshwater sponge, *Ephydatia fluviatilis* (Brien and Meewis, 1938), cleavage is very unequal almost from the start. In *Ephydatia* cytokinesis is frequently retarded during the early cleavage stages, and binucleate and plurinucleate blastomeres are formed. The cleaving zygotes of *Microciona prolifera* (Simpson, 1968) and *Spongia reticulata* (Bergquist *et al.*, 1970) undergo a substantial growth; and in the latter species, cleavage is very irregular. In all sponges the yolk is progressively broken down and utilized during the later stages of cleavage (see Fig. 14).

In many sponges nucleoli are present in the nuclei of the embryonic cells from the beginning of development. In a number of animals the presence and size of nucleoli have been correlated with ribosomal RNA synthesis (Brown, 1966). Unfortunately, but understandably, there is presently no information available on macromolecular synthesis during sponge embryogenesis.

Most of the acalcareous sponges produce a parenchymula larva which develops at the end of the period of cleavage (see Lévi, 1956, for a review). At some point during cleavage, which varies according to the species, cells of different sizes and appearances emerge. In *Haliclona limbata* (Meewis, 1939) and *Adocia cinerea* (Meewis, 1941) small cells with granular nuclei appear initially interspersed among the larger cells with vesicular, nucleolate nuclei. The former subsequently migrate towards the periphery of the embryo where they form the larval epithelium, while the latter become concentrated in the interior. In other sponges, such as *Chalinula fertilis* (Maas, 1893), *Halichondria coalita* (Meewis, 1941), and *Ephydatia fluviatilis* (Brien and Meewis, 1938), the smaller proepithelial cells are already peripherally situated at their first appearance. In these sponges they appear first at the future anterior pole of the larva and then progressively at more posterior levels. There is little or no cellular migration, except for a limited epiboly.

Cell division continues to occur during the differentiation and segregation of the different kinds of cells; and since the smaller cells divide most rapidly, their number increases dramatically. In most cases the small peripheral cells align and form a pseudostratified columnar epithelium in which there may be approximately 2-8 nuclear layers and in which each cell possesses a single flagellum (see Fig. 15 and 16). The larva of *Hymeniacidon sanguinea* apparently possesses a simple columnar, flagellated epithelium (Lévi, 1956).

The larvae of many sponges are entirely covered by a flagellated epithelium; however, in a large number of cases the flagella at the posterior pole are of a different length and/or density than else-

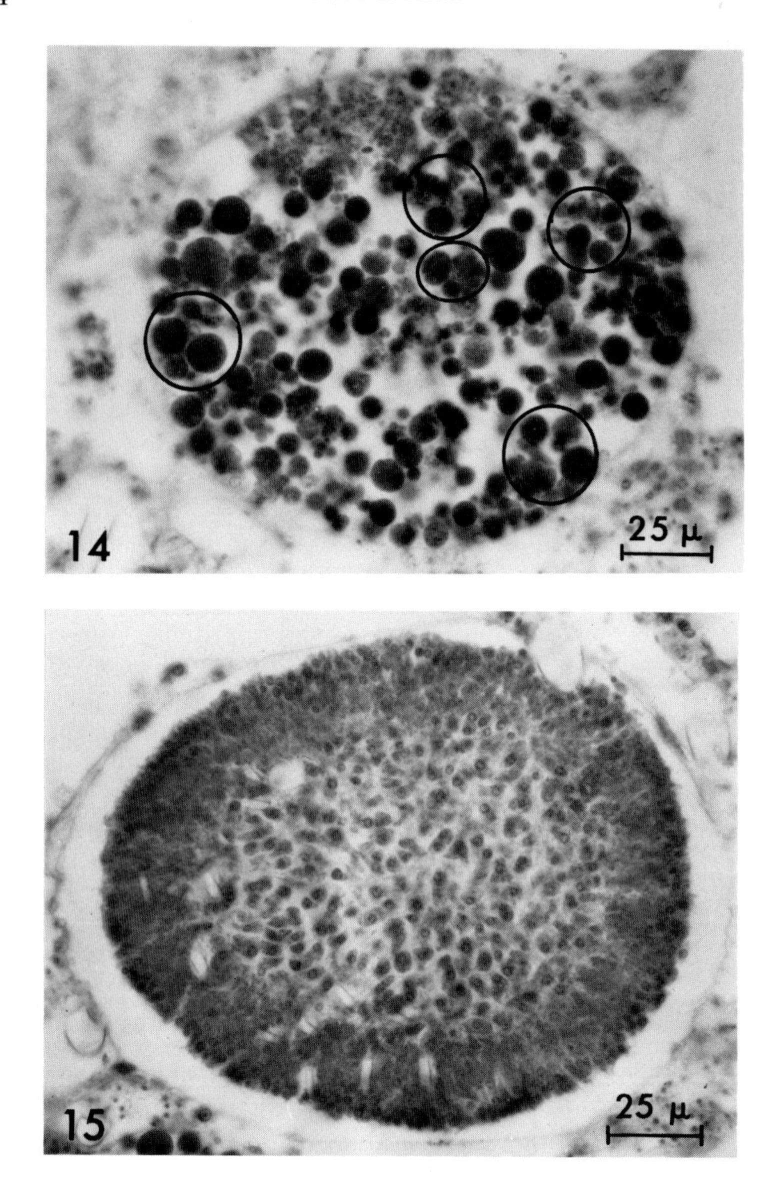

FIG. 14. An embryo of *Halichondria panicea* showing the blastomeres (circles) filled with large yolk spheres.

FIG. 15. A developing larva of *Halichondria panicea* following the breakdown of yolk spheres. Note the layer of epithelial cells at the periphery. (Both hematoxylin and eosin.)

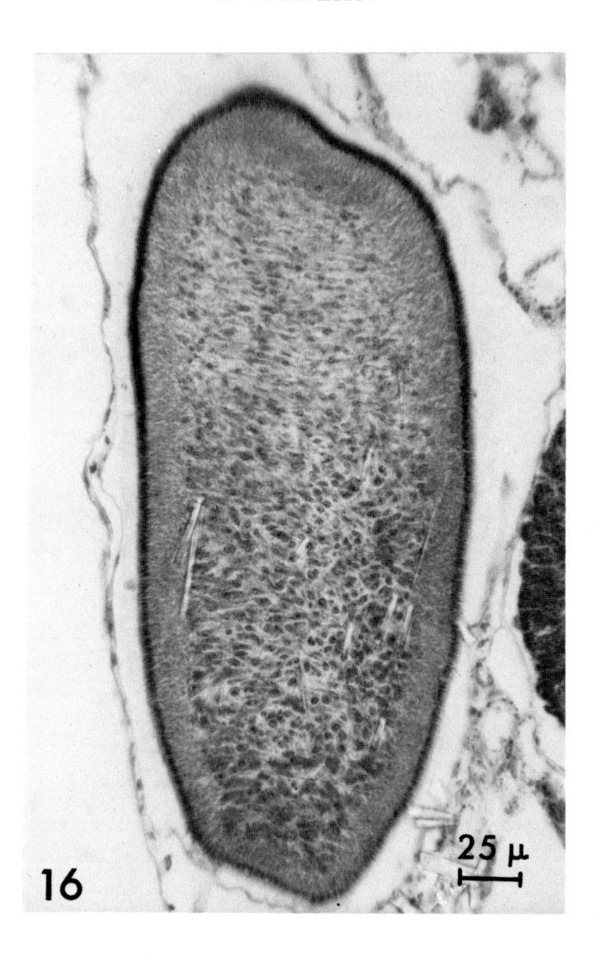

FIG. 16. An advanced larva of *Halichondria panicea*. Note the flagella extending from the surface and the central spicules. (Hematoxylin and eosin.)

where. On the other hand, the larvae of many other species lack posterior flagella (see Fig. 18). In the larva of *Haliclona limbata*, flagella are absent from the anterior pole, and they have only a transient existence at the posterior pole (Meewis, 1939). Finally, in a number of species flask-shaped secretory cells appear among the flagellated cells, especially at the anterior pole.

Although the parenchymula is typically solid, the larvae of *Ephydatia fluviatilis* (Delage, 1892; Brien and Meewis, 1938) and *Spongilla moorei* (Brien, 1967) have a large anterior cavity lined by pinacocytes. In certain other sponges, such as *Myxilla rosacea* (Maas,

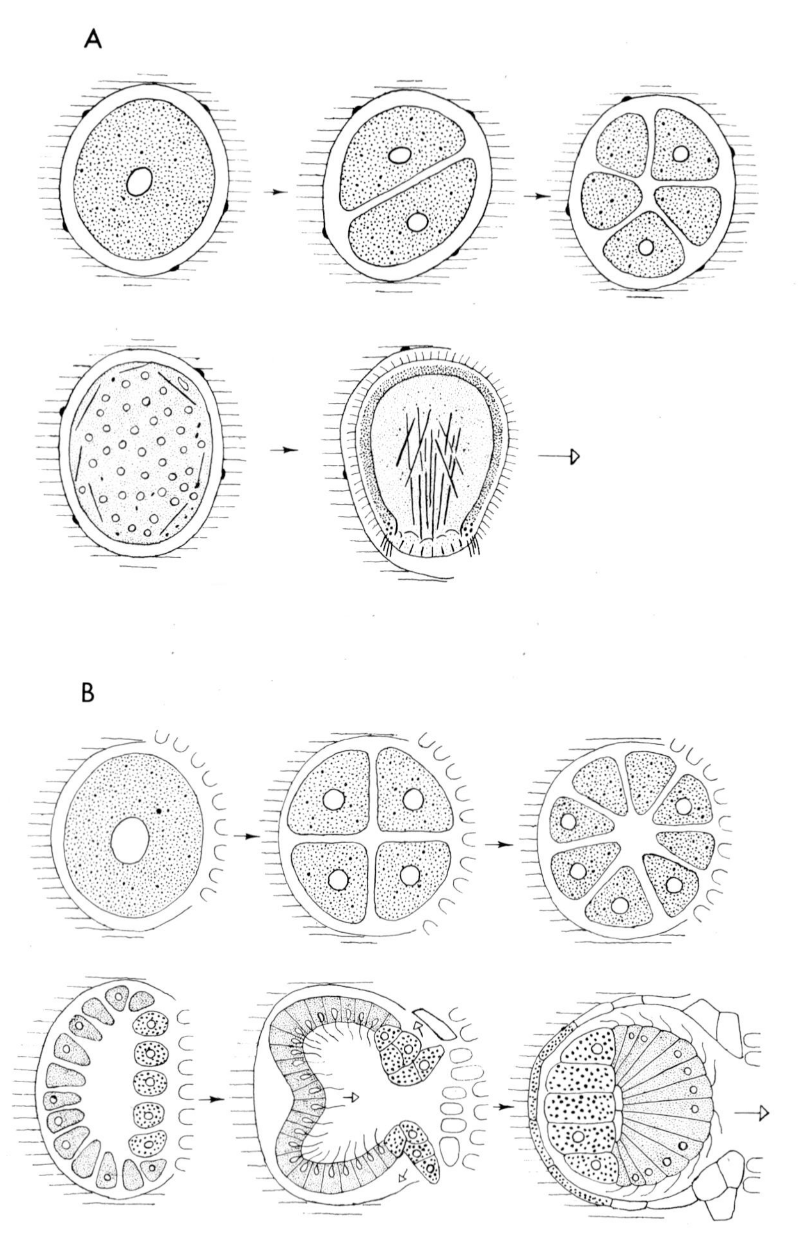

FIG. 17. Two patterns of larval development in viviparous sponges: (A) the formation of a parenchymula larva [modified from Maas (1893); Meewis (1939, 1941); Fell (1969)]. (B) The formation of an amphiblastula larva [modified from Duboscq and Tuzet (1937); Lutfy (1957b); Vacelet (1964); Fell, unpublished observations].

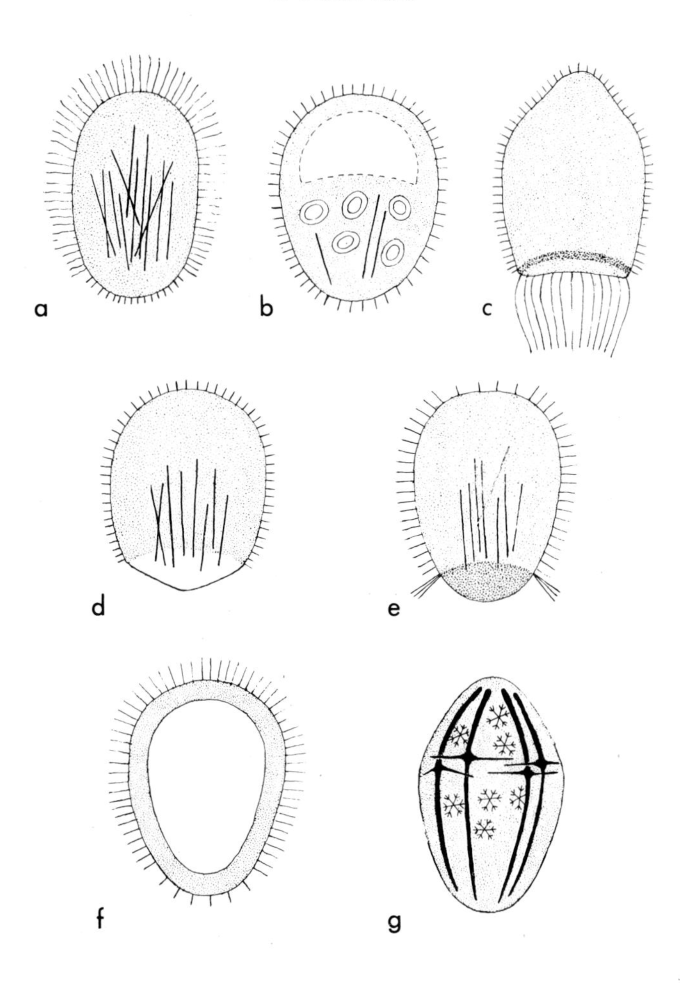

FIG. 18. Different types of acalcareous sponge larvae. (a) *Halichondria bowerbanki* [modified from Hartman (1958)]; (b) *Ephydatia fluviatilis* and *Spongilla moorei* [modified from Brien and Meewis (1938); Brien (1967b)]; (c) *Euspongia officinalis* [modified from Maas (1893)]; the larvae of *Aplysilla sulfurea* (Delage, 1892), *Hippospongia communis (equina)* (Vaney and Allemand-Martin, 1918), and *Phyllospongia foliascens* (Lévi, 1956) are similar; (d) *Mycale simularis (Esperella sordida)* [modified from Delage (1892)], the larvae of *Myxilla rosacea* and *Axinella crista-galli* (Maas, 1893), *Tedania gurjanovae* (Bakus, 1964a), *Lissodendoryx similis* (Ali, 1956), *Microciona coccinea* and *Ophlitaspongia seriata* (Bergquist and Sinclair, 1968), and *Hymeniacidon sanguinea* (Lévi, 1956) are similar; (e) *Adocia cinerea* [modified from Meewis (1941)]; the larvae of *Reniera densa* (Delage, 1892), *Gellius varius* (Maas, 1893), *Haliclona ecbasis* (Fell, 1969), and *Callyspongia diffusa* (Sivaramakrishnan, 1951) are similar; (f) *Oscarella (Halisarca) lobularis* [modified from Meewis (1938)]; and (g) *Farrea sollasii* [modified from Okada (1928)].

1893), a cavity appears during cleavage but does not persist in the larva. However, in these and other cases the cells at the anterior end of the larva form a very loose tissue (also see Wilson, 1935). Thus there is frequently a concentration of internal cells towards the posterior pole of the larva. But in other sponges, including *Haliclona limbata* (Meewis, 1939), *Halichondria coalita* (Meewis, 1941), and *Phyllospongia foliascens* (Lévi, 1956), this is not the case. In fact, the larva of *Halichondria* is very compact throughout (see Fig. 16). The cells of the internal mass usually develop into collencytes, archaeocytes, and other kinds of cells of the adult mesenchyme.

The parenchymulae of many sponges contain spicules which are frequently arranged in a bundle near the posterior pole. In *Adocia cinerea* (Meewis, 1941) and *Haliclona ecbasis* (Fell, 1969), the spicules appear at the periphery of the embryo during late cleavage. They are situated tangential to the surface of the embryo and form a loose meshwork about it. But as the larval epithelium begins to differentiate, the spicules are brought into a central position. In many other sponges, including *Halichondria coalita* (Meewis, 1941) and *Hymeniacidon perleve* (Bergquist *et al.*, 1970), the spicules appear during the formation of the larval epithelium and lie at first immediately beneath it. On the other hand, in *Ephydatia fluviatilis* the spicules appear relatively early but are situated in the center of the embryo from the start (Brien and Meewis, 1938). The larvae of a number of sponges, including *Tedanione foetida* (Wilson, 1894), *Haliclona limbata* (Meewis, 1939), and *Halichondria moorei* (Bergquist and Sinclair, 1968), possess no spicules at the time they are released from the parent sponge. However, in *Haliclona limbata* the spicules appear during the free-swimming phase of the larva.

From the preceding discussion it is clear that the parenchymulae of the majority of the sponges are highly differentiated. They contain most of the cell types of the adult, with the exception of choanocytes and reproductive elements (see Lévi, 1964, for an account of the ultrastructure of the larva of *Mycale contarenii).* The larvae of *Ephydatia fluviatilis* (Brien and Meewis, 1938) and *Spongilla moorei* (Brien, 1967b) are somewhat unique in that they contain flagellated chambers, as well as the rudiments of the excurrent canal system. However, the differentiation of choanocytes begins in the larvae of a few other sponges. In *Haliclona limbata* (Meewis, 1939) choanoblasts and occasionally choanocytes in flagellated chambers are observed in the free swimming larva; and in the Hexactinellid, *Farrea sollosii* (Okada, 1928), the rudiments of the flagellated chambers appear in the larva when it is still within the parent sponge.

By contrast, the larvae of a few of the acalcareous sponges are very simple. *Halisarca metschnikovi* and *Halisarca dujardini* produce relatively undifferentiated parenchymula larvae (Lévi, 1956). Apparently in both of these sponges the cells of the morula, which have peripherally situated nuclei, pinch off some of their yolk-filled cytoplasm into the center of the mass. In *Halisarca metschnikovi* the embryo very early consists of a simple epithelium surrounding a mass of yolk in which internal blastomeres are suspended; and cells move from the epithelium into the interior at later stages. On the other hand, in *Halisarca dujardini* all of the blastomeres are initially incorporated into the peripheral epithelium; and only at a relatively late stage of development is the center of the larva populated by cells from the peripheral layer. This displacement of cells into the interior of the larva is most pronounced at the future posterior pole. In both cases the cells of the peripheral epithelium divide, become more closely associated, and develop flagella, while the internal mass of cells develops very slowly. The flagellated epithelium is of the simple columnar type. The swimming larvae exhibit a polarity which results from the fact that the flagellated cells are larger and less crowded at the posterior pole.

Oscarella (Halisarca) lobularis (Meewis, 1938) and *Octavella galangaui* (Tuzet and Paris, 1964) produce blastular larvae. In *Oscarella* some of the blastomeres cytolyze during the later cleavage stages, and the early embryo consists of a peripheral epithelium surrounding masses of blastomeres bathed in an amorphous substance. The peripheral cells pinch off lobes of cytoplasm which together with most of the central cells progressively disintegrate and contribute materials to the interblastomeric substance. The other internal cells are incorporated into the peripheral epithelium. Eventually the peripheral cells align in a pseudostratified columnar epithelium and develop flagella. These processes result in the formation of a completely flagellated amphiblastula larva in which the cells of the posterior pole are larger than elsewhere (Meewis, 1938; see Lévi and Porte, 1962, for an account of the ultrastructure of this larva). In *Octavella* cleavage results in the formation of a compact morula. The peripheral cells form a simple columnar epithelium, and the central cells are progressively incorporated into it. There is no cytolysis. The hollow larva is composed entirely of flagellated cells with the exception of a few large cells at the posterior pole (Tuzet and Paris, 1964).

The calcareous sponges produce more or less differentiated blastular larvae. In all cases the initial blastula develops directly as the blastomeres become oriented around a central space. In various species of *Clathrina* (Minchin, 1896, 1900; Tuzet, 1948) and *Ascandra* (Borojevic, 1969) [the Calcinea], the larva is a simple flagellated blastula which, in some species, has a small number of large, nonflagellated cells at the posterior pole (see Borojević, 1969, for a description of the ultrastructure of the larva of *Ascandra falcata*). Sooner or later some of the cells of the flagellated epithelium lose their flagellum and migrate into the central cavity. In *Clathrina (Leucosolenia) coriacea* and *Clathrina cerebrum* individual cells from all regions of the epithelium, except for the extreme anterior pole, move into the interior of the larva, while in *Clathrina reticulum* the immigration of cells appears to be at first restricted to the posterior pole (Minchin, 1896; Tuzet, 1948; Borojević, 1969).

In *Grantia compressa* and *Sycon raphanus* (Duboscq and Tuzet, 1937), *Leucosolenia botryoides* (Tuzet, 1948), and *Petrobiona massiliana* (Vacelet, 1964) [the Calcaronea], the formation of the larva is more complex (see Fig. 17B). During cleavage two types of cells emerge: rounded granular cells with large nuclei, which are situated at the pole of the blastula that is directly beneath the choanoderm, and elongated cells with small nuclei. The latter form flagella which extend into the blastular cavity. Then an opening appears among the granular cells, and the blastula turns inside out as the opening enlarges and its edges fold down around the invaginating flagellated pole. The granular cells come together again at the opposite pole, and an amphiblastula larva with external flagella is formed. In the final stages of differentiation the cells of the larva increase in size and form a tall columnar epithelium with the result that the blastular cavity is very much reduced. The granular, nonflagellated cells occupy the posterior half of the larva.

As Lévi (1956) has emphasized, it is frequently very difficult to distinguish in the so-called viviparous sponges whether it is a case of true viviparity or of simple incubation. In many species the oocytes contain large quantities of stored nutrient materials, and there appears to be little or no transfer of materials from the parent to the developing embryos. The embryos, in many cases still surrounded by the oocyte follicle, evidently gain only protection from the parental tissue. Such protection could include a number of forms: simple mechanical protection, protection from silt, protection from small predators and scavengers, protection against displacement into unfavorable situations, etc.

On the other hand, in a number of sponges there are indications that cells and/or cellular products may be added to the developing embryos.* For example, in *Microciona prolifera* (Simpson, 1968) and *Spongia reticulata* (Bergquist *et al.*, 1970), the mature oocytes apparently have a diameter which is only about one fifth that of the advanced embryos. This means that during cleavage the embryos must incorporate materials from the parent sponge. In *Microciona* some of the embryos are surrounded by masses of nurse cells which are identical to those that surround the growing oocytes, and it has been suggested that some of these somatic cells may be added to the embryos (Simpson, 1968). In the oviparous sponge, *Cliona celata*, 20-35 granular amoeboid cells are attached to the surface of the shed zygote. During cleavage these cells move into the interior of the embryo and still persist in the mature larva. Their fate has not been determined, but it appears that the maternal granular cells may serve as a nutritive reserve for the larva (Warburton, 1961). In *Grantia compressa* and *Sycon raphanus* the embryos are situated immediately under the maternal choanoderm; and at all stages of development, even during the early cleavage stages, modified choanocytes appear to be incorporated into the embryo. Many of these are resorbed, but others are still present at the end of larval development (Duboscq and Tuzet, 1937; Lutfy, 1957a). Beginning at the early blastula stage, a placental membrane is formed around the embryo by invaginated choanocytes which have lost their collar and flagellum. The cells of this membrane, which lie adjacent to the granular cells of the amphiblastula, are filled with yolk (see Fig. 17B); and it appears that there may be a transfer of material from the former to the latter. Placental cells may also be ingested by the cells of the embryo (Gatenby and King, 1929; Duboscq and Tuzet, 1937; Lutfy, 1957b). In *Petrobiona massiliana,* in which a similar type of nurse capsule exists, a large number of the nurse cells move into the central cavity of the larva (Vacelet, 1964). In *Ascandra falcata* the embryos develop within nests of parental cells, and during the early stages of development some of these cells move between the blastomeres and enter the embryo. However, such nurse cells disappear before the larva is released (Borojević, 1969). Finally, the young larvae of *Clathrina coriacea* and *Leucosolenia botryoides* pass into the lumina of the radial tubes where they complete their development. The larva of *Clathrina* is enclosed within a placental membrane which is similar to that of *Sycon* and *Grantia.* On the other hand, the larva of *Leucosolenia* rests on a layer of yolk-filled nurse cells (transformed

*There is a need for caution in interpreting developmental stages for in some cases it is difficult to determine whether one is dealing with the growth of oocytes, the growth of early embryos, or the asexual development of larvae.

choanocytes), some of which penetrate into the interior of the larva and others of which are engulfed by the posterior granular cells (Tuzet, 1948).

Little precise information concerning the parent dependent nutrition of embryos is available, and more studies in this area are needed. In addition, almost nothing is known about other factors that influence embryonic development.

2.4.3 Larvae

Since asexually produced larvae are identical to those produced in sexual reproduction, they will also be considered in the discussion that follows.

2.4.3.1 Larval Release

Little is known concerning the mechanism by which larvae are released; however, it would not be surprising if it involves the production of an enzyme(s) by the larvae. In *Halisarca dujardini* the larvae rotate within their follicles, the walls of the follicles rupture, and the larvae leave the parent sponge by way of the excurrent canals and oscula (Lévi, 1956). Evidently a similar sequence is followed in the release of most other sponge larvae. In *Callyspongia diffusa* peripherally situated larvae appear to be released directly through small openings in the dermal membrane (Sivaramakrishnan, 1951). The larvae of *Microciona coccinea* and *Ophlitaspongia seriata*, which are immobile prior to their release, are expelled at a rate of 4-5 per minute and may be either eased out of the osculum or ejected 1.5-2.0 cm from it. Not all of the larvae are released; and those which are not are progressively resorbed (Bergquist and Sinclair, 1968).

In *Oscarella lobularis* and *Hymeniacidon sanguinea* (Lévi, 1951c) and *Halisarca metschnikovi* (Lévi, 1956), the larvae are released in the laboratory a short time after sunrise, about one hour following cessation of water circulation. It therefore appears that the release of larvae by these sponges may be regulated by light. On the other hand, specimens of *Halisarca dujardini* may release larvae all during the day which follows their collection (Lévi, 1956). In still other cases, large numbers of larvae are extruded within several hours after the specimens are brought into the laboratory; and in such instances the release of larvae is very likely induced by shock (Ali, 1956). There is a clear need for extensive field observations. It is not known whether the release of larvae by intertidal sponges is ever correlated with particular phases of the tide. Information on this subject might be obtained by using a tide-simulation aquarium system (Bergquist *et al.*, 1970).

In *Halisarca metschnikovi* tall oscular chimneys develop a short time prior to the release of the larvae, and the sponges have an inflated appearance during the period of larval emission. Specimens, that are filled with the white larvae, are themselves almost white instead of the usual light yellow (Lévi, 1956). Similarly, in *Microciona coccinea* and *Ophlitaspongia seriata* the superficial excurrent canals become greatly inflated during the period of larva production (Bergquist and Sinclair, 1968).

Careful determinations of the period of larval release have been made for few sponges. In many species, including *Halichondria bowerbanki* and *Haliclona loosanoffi* (Hartman, 1958), *Lissodendoryx similis* (Ali, 1956), *Lissodendoryx isodictyalis* (Wells *et al.*, 1964), *Microciona prolifera* (McDougall, 1943; Hartman, 1958; Wells *et al.*, 1964), *Halisarca dujardini* (Lévi, 1956), *Mycale syrinx* (Wilson, 1935), *Prosuberites miscosclerus* (Wells *et al.*, 1964), and *Hippospongia communis (equina)* (Vaney and Allemand-Martin, 1918; Tuzet and Pavans de Ceccatty, 1958), the release of larvae may occur for a period of two or more months in the localities where studies have been made (see Table VI). However, data on larval settlement and other information suggest that in a number of cases the release of large numbers of larvae may be restricted to a much shorter period. Other sponges, including *Halisarca metschnikovi* (Lévi, 1956) and *Adocia (Reniera) tubifera* (McDougall, 1943), apparently release larvae for a relatively brief period.

2.4.3.2 Larval Behavior and Settling

Although there are many references to larval behavior in the literature, in only a few instances has an attempt been made to relate larval behavior to the ecological distribution of the adult sponge (see Bergquist *et al.*, 1970, for a review). Larval behavior has been studied only in the laboratory for obvious reasons, and the possibility that the larvae may behave somewhat differently in their natural environment should be kept in mind.

The larvae of most sponge species swim for a certain period of time and then creep on the bottom before fixing to the substratum.* In some sponges, including *Microciona coccinea* and *Ophlita-*

*There are always some inactive larvae. These may have been prematurely released or perhaps damaged during their release (Bergquist *et al.*, 1970).

spongia seriata (Bergquist *et al.*, 1970) and *Callyspongia diffusa* (Sivaramakrishnan, 1951), the larvae swim for only 3 or 4 hours, while in others, including *Clathrina (Leucosolenia) cerebrum* and *Leucosolenia variabilis* (Minchin, 1896), *Microciona prolifera* and *Cliona celata* (Warburton, 1966), *Mycale macilenta* (Bergquist *et al.*, 1970), *Oscarella (Halisarca) lobularis* (Meewis, 1938), *Lissodendoryx similis* (Ali, 1956), and *Haliclona limbata* (Meewis, 1939), they may swim for as long as 1 or 2 days. On the other hand, the larvae of *Halichondria moorei* (Bergquist *et al.*, 1970) and *Polymastia robusta* (Borojević, 1967) are benthic. They creep along the substratum for 20 - 60 hours and for 18 - 20 days, respectively.

Phototaxis and geotaxis may be particularly important factors determining the ecological distribution of sponges (see Bergquist *et al.*, 1970; and Table VII, this chapter). The larvae of a number of sponges, including *Hippospongia communis (equina)* (Vaney and Allemand-Martin, 1918), *Mycale simularis (Esperella sordida), Reniera densa*, and *Aplysilla sulfurea* (Delage, 1892), *Callyspongia diffusa* (Sivaramakrishnan, 1951), *Lissodendoryx similis* (Ali, 1956) *Mycale syrinx* (Wilson, 1935), *Mycale macilenta* (Bergquist *et al.*, 1970), and *Haliclona ecbasis* (Fell, 1969), are negatively phototactic; those of others, including *Ophlitaspongia seriata, Microciona coccinea*, and *Halichondria moorei* (Bergquist *et al.*, 1970), *Microciona prolifera* (Warburton, 1966; Simpson, 1968), and *Haliclona indistincta* (Lévi, 1956), show no phototaxis; and the larvae of *Sycon (Sycandra) raphanus* (Maas, 1906) and *Haliclona* sp. (Bergquist *et al.*, 1970) are positively phototactic. The larvae of *Mycale syrinx* are less responsive to light during the creeping phase than they are during the period of swimming (Wilson, 1935; see also McDougall, 1943 and Hartman, 1958). There is less clear information on geotaxis. The larvae of many sponges, including *Lissodendoryx similis* (Ali, 1956), *Halichondria moorei* (Bergquist *et al.*, 1970), and *Microciona prolifera* (Warburton, 1966) apparently are negatively geotactic, however, the larvae of *Microciona* become positively geotactic during the latter part of their free existence.

It appears that the behavior of sponge larvae is relatively complex; and many factors, including currents, wave action, turbidity, temperature, and the nature of the substratum, probably also influence the movement of the larvae (Bakus, 1964a; Warburton, 1966; Simpson, 1968). Finally, although larval behavior is undoubtedly important in determining the distribution of adult sponges, differential survival probably also plays an important role. Almost no information on larval distribution and survival is available (see Hartman, 1958).

In certain cases, it appears that the larvae need a clean, firm surface for successful settling, but that beyond meeting these requirements, the nature of the substratum can vary considerably. For example, the larvae of *Haliclona loosanoffi* settle in approximately equal numbers on white marble, dark gray phyllite, and oyster shells (Hartman, 1958); and they also settle in abundance on eel grass (Fell, in preparation). Similarly, the larvae of *Hippiospongia lachne* attach to sea grass, sea whips, mollusc shells, and pieces of coral; however, many of the sponges that are poorly anchored do not survive to a large size (Storr, 1964). Sponge larvae may also attach to paraffin, glass, cellophane, and agar (see for example, Wilson, 1937; Brien and Meewis, 1938; Bergquist and Sinclair, 1968). Although the larvae of *Callyspongia diffusa* develop normally on the various substrata, they attach to *Ulva* more rapidly than they do to paraffin or glass (Sivaramakrishnan, 1951).

The larvae of various species of *Cliona* settle on and bore into the shells of living and dead molluscs, coral, coralline algae, and other calcareous matter. However, Warburton (1966) found that in the laboratory, the larvae of *Cliona celata* settle and metamorphose on glass and on calcite with about the same frequency. The direct development of larvae into free-living specimens is apparently rare under natural conditions, but large specimens eventually overgrow their shells which are later completely destroyed (Hartman, 1958). The mechanisms by which the sponges penetrate into calcareous substrata are discussed by Nassonow (1883, 1924), Hartman (1958), Warburton (1958), and Cobb (1969).

Under laboratory conditions, the larvae of some sponges, including *Haliclona limbata* (Meewis, 1939) and *Halisarca metschnikovi* (Lévi, 1956), may be held at the surface of the water by surface tension and there undergo essentially normal metamorphosis. However, this phenomenon is probably of rare occurrence under natural conditions.

Under unfavorable conditions settlement may be very much delayed or completely prevented (Meewis, 1938, 1939).

A number of laboratory and field observations suggest that sponge larvae may aggregate during settling and that larvae and/or postlarvae may fuse, forming heterogenomic specimens. The larvae (or postlarvae) of *Lissodendoryx* sp. (Wilson, 1907), *Mycale syrinx* (Wilson, 1935), *Tethya aurantium* (Lévi, 1956), *Polymastia robusta* (Borojević, 1967), *Ophlitaspongia seriata* (Fry, personal communication), *Ephydatia fluviatilis* (Van de Vyver, 1970), and certain calcareous sponges (Borojević, 1969) have been observed to fuse

TABLE VII

LARVAL BEHAVIOR AND ADULT HABITAT OF SEVERAL MARINE SPONGES

Species	Nature and duration of larval locomotion	Phototaxis	Geotaxis	Habitat of adult	Author
Haliclona sp.	Rapid (1 cm/second) directional swimming with constant rotation for 9-10 hours, then crawls on the bottom for about 2 hours	Positive during swimming phase, none or negative during creeping phase	Initially negative	Around and under boulders of middle littoral zone	Bergquist *et al.*, 1970
Haliclona ecbasis	Rapid directional swimming, turning about longitudinal axis, then spins on bottom (duration of phases not precisely determined)	Negative	Initially negative	Attached to mussels and other objects on the lower sides and bottom of floats	Fell, 1969, 1970
Microciona prolifera	Swims (several mm/second) continuously at the surface for about 20-30 hours, then creeps	None	Negative during swimming phase and	On undersides and occasionally on upper	Warburton, 1966; Simpson, 1968

	on bottom for a similar period of time		positive during creeping phase	surfaces of intertidal rocky substrata	
Microciona coccinea	Swims in an open spiral with constant rotation for 3-4 hours, then creeps on the bottom for 2-3 hours	None	Initially negative	From the middle to the extreme lower limit of the littoral zone in situations ranging from full light to deep shade	Bergquist *et al.*, 1970
Halichondria moorei	Creeps on the substratum without rotation for 20-60 hours	None	Negative	Fringes of pools and the water line of rocks in standing pools of the middle littoral zone	Bergquist *et al.*, 1970

under laboratory conditions. In some cases the larvae fuse only when they are crowed together. However, the larvae of *Polymastia* tend to settle on already fixed larvae; and the larvae of *Ophlitaspongia* have a pattern of behavior which leads to the formation of aggregates. The fused postlarvae may develop more rapidly (*Tethya*) and/or show a higher rate of survival (*Ophlitaspongia*) under laboratory conditions than do individual postlarvae. On the other hand, extremely large aggregates of *Lissodendoryx* larvae, containing up to 100 or more larvae, do not metamorphose. *Ophlitaspongia* larvae from morphologically and physiologically distinct populations were observed to fuse together (Fry, personal communication).

The extensive field observations of Burton (1933, 1949) provide circumstantial evidence suggesting that aggregation and eventual fusion of larvae and postlarvae may be of relatively common occurrence under natural conditions. Clusters of small specimens were found scattered over the substratum, and the largest of the specimens in any cluster were up to 20 or more times larger than the smallest. This disparity in size was assumed to result from the fusion of adjacent specimens, but it could have also resulted from some larvae settling earlier than others and achieving more growth. Other possible explanations could also be invoked. Therefore, as Burton has emphasized, the observations are consistent with larval fusion, but they do not demonstrate it. *Hymeniacidon perlevis, Halichondria panicea, Clathrina (Leucosolenia) coriacea, Iophon hyndmani, Dercitus bucklandi,* and *Pachymatisma johnstonia* are among the species of which the larvae apparently aggregate and fuse (see also Fig. 19).

2.4.3.3 Metamorphosis

In *Tetilla (Tethya) serica* there is no larva. The zygotes adhere to the substratum and develop directly into small adult sponges (Watanabe, 1960). Among the other sponges that have been studied, there are basically three different kinds of larvae: parenchymula, amphiblastula, and simple blastula. These larvae differ not only in form but also in the degree of differentiation of their component cells. Consequently development into the adult sponge involves a different set of processes for each kind of larva (see Brien, 1943; Borojevic, 1970, for reviews).

As indicated above, the vast majority of sponges produce a parenchymula larva. In many cases the larvae continue to develop during their free existence, and metamorphosis may actually begin during this period. For example, in *Mycale syrinx* (Wilson, 1935), *Ephydatia fluviatilis* (Brien and Meewis, 1938), *Callyspongia diffusa*

(Sivaramakrishnan, 1951), and *Lissodendoryx similis* (Ali, 1956), the epithelium covering the anterior pole of the larva becomes disorganized during the latter part of the motile phase. In the swimming larva of *Haliclona limbata* spicules and choanocytes appear; the anterior-posterior axis is shortened; and some of the epithelial cells lose their flagellum and migrate towards the interior. Furthermore, other epithelial cells are engulfed in place by amoebocytes which migrate toward the periphery (Meewis, 1939). Not all of the larvae of a species may be released at the same stage of development, and there may be considerable variation in the extent of their differentiation (Brien and Meewis, 1938; Meewis, 1939).

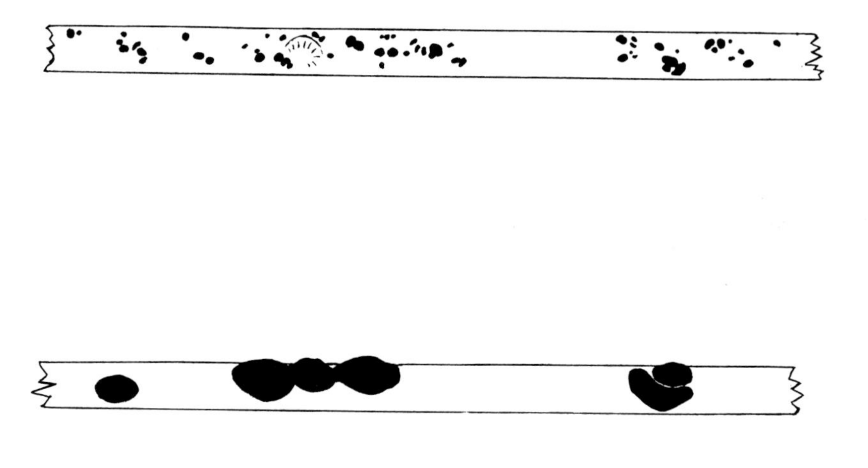

FIG. 19. Blades of eel grass bearing young specimens of *Halichondria bowerbanki* collected in the Mystic Estuary during the second week of July. Note the close proximity and apparent fusion of some of the sponges (Fell, in preparation).

In most species the larvae usually become attached to the substratum by the anterior pole or by a lateral region near the anterior pole (Delage, 1892; Maas, 1893; Brien and Meewis, 1938; Meewis, 1939; Sivaramakrishnan, 1951; Lévi, 1956; Bergquist and Sinclair, 1968). However, the larvae of some sponges appear to become fixed by their side *(Mycale syrinx,* Wilson, 1935; *Lissodendoryx similis,* Ali, 1956) or by the posterior pole *(Mycale fibrexilis,* Wilson, 1894). Following fixation the larvae rapidly spread on the substratum. Amoebocytes (collencytes) at the edge of the site of attachment migrate away from the central mass forming a thin marginal membrane. This membrane eventually consists of a fold of the thin surface epithelium (see below) with a few amoebocytes inside. As the marginal membrane extends radially, the central mass becomes drawn out and flattened (Delage, 1892; Wilson, 1894, 1935; Brien and Meewis, 1938; Ali, 1956). At the same time a network of spongin (collagen) fibers is formed between the basal epithelium of the sponge and the substratum (Lévi, 1964). These events appear to be essential for the normal occurrence of the other metamorphic processes (see Borojević and Lévi, 1965; Borojević, 1966).

The cells of the flagellated epithelium lose their flagellum and migrate into the interior of the larva, while amoebocytes migrate to the periphery where they spread on the surface and form a simple squamous epithelium. This two-way migration has been found to occur during the metamorphosis of all parenchymula larvae, but the fate of the larval flagellated cells appears to be different in different species. In *Ephydatia fluviatilis* (Brien and Meewis, 1938) and *Haliclona limbata* (Meewis, 1939), in which choanocytes are found in the unliberated larva and in the swimming larva, respectively, the flagellated cells have no direct role in the development of the adult sponge. They are engulfed by the archaeocytes and serve only as a nutritive reserve. The choanocytes, which develop precociously, derive from archaeocytes.

In most other species larval flagellated cells develop into choanocytes in the interior of the sponge (see Fig. 20A). In many instances the cells of the larval epithelium are much more numerous than the choanocytes of the young sponge, and large numbers of them are phagocytized by the archoeocytes (Wilson, 1935; Brien, 1943). However, in *Halisarca metschnikovi*, *Halisarca dujardini*, and *Tethya aurantium* there is little, if any, cell loss; and apparently all of the cells of the simple larvae participate directly in the formation of the adult sponges. In most cases numerous flagellated chambers are formed; but in the young specimens of *Halisarca*, a single large chamber is produced (Lévi, 1956).

Especially clear evidence has been provided for the transformation of larval flagellated cells into choanocytes during the metamorphosis of *Mycale contarenii*. First, it was found that a continuity exists between these two types of cells at the ultrastructural level; certain organelles, which are characteristic of choanocytes, appear before all of the characteristic features of the flagellated epithelial cells are lost (Borojević and Lévi, 1965). Second, cultures consisting primarily of larval flagellated cells but also containing some other kinds of cells, rapidly form large numbers of flagellated chambers (Borojević, 1966). However, in cultures of the internal mass of the larva, archaeocytes differentiate into choanocytes. Thus while most, if not all, of the initial choanocytes of the young sponge normally develop from flagellated epithelial cells, the archaeocytes are clearly capable of differentiating into choanocytes under special conditions (Borojević, 1966).

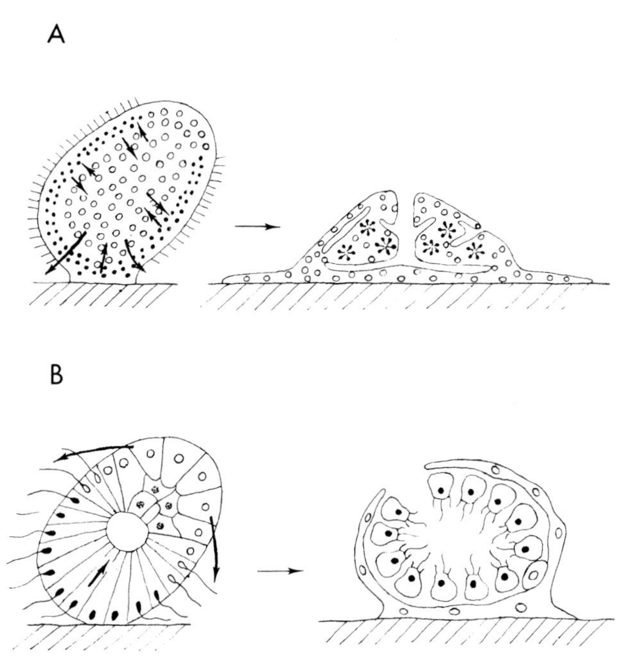

FIG. 20. Metamorphosis of larvae into small sponges: (A) The metamorphosis of a typical parenchymula larva [modified from Maas (1893), Wilson (1935); Lévi (1956); and others] (B) The metamorphosis of an amphiblastula larva of the type of *Leucosolenia* [modified from Minchin (1896)]. Arrows indicate the direction of movement of the cells.

In a large number of species, including *Axinella crista-galli* (Maas, 1893), *Halisarca metschnikovi*, *Halisarca dujardini*, and *Tethya aurantium* (Lévi, 1956), *Haliclona limbata* (Meewis, 1939), *Mycale contarenii* (Borojević, 1966), and *Microciona prolifera* (Simpson, 1968), individual cells are involved in the movements associated with metamorphosis. However, according to Wilson (1935) and Ali (1956) the larvae of *Mycale syrinx* and *Lissodendoryx similis*, respectively, are syncytia in which the movement of nuclei corresponds to the movement of cells in other species. In these cases it seems very probable that the larvae are actually cellular but that the cellular outlines were obscured. In fact some cellular boundaries were observed in these larvae. Finally, according to Sivaramakrishnan (1951), in *Callyspongia diffusa* clusters of larval epithelial cells fuse with single archaeocytes, forming multinucleate masses that move into the interior of the larva. These masses subsequently break up into individual cells, and the former flagellated epithelial cells give rise to the flagellated chambers of the young sponge. Delage (1892) has recorded a similar sequence of events for *Mycale simularis* (*Esperella sordida*), *Reniera densa*, and *Aplysilla sulfurea*, as well as for *Ephydatia* (*Spongilla*) *fluviatilis*. However, it seems almost certain that these authors misinterpreted the phagocytosis of larval flagellated cells (Brien and Meewis, 1938; Brien, 1943; Borojević and Lévi, 1965).

Spaces, which become lined by pinacocytes, appear progressively. These give rise to the incurrent and excurrent canal systems. The canals connect with the flagellated chambers, and pores and an osculum appear. Metamorphosis is then complete.

Oscarella (Halisarca) lobularis and some of the calcareous sponges (the Calcaronea) produce amphiblastula larvae. These larvae attach to the substratum by the anterior pole (Minchin, 1896; Meewis, 1938). The fixed larva of *Oscarella* flattens; and the anterior half invaginates into the posterior half, largely obliterating the central cavity. The cells of the anterior epithelium round up and lose their flagellum, and many of them enter the central cavity where they degenerate. Attachment of the larva is maintained by the marginal region, the inner part of which moves on the substratum towards a central point. In this way a new cavity, the primitive gastral cavity, is enclosed. The cells of its walls progressively align in a columnar epithelium which becomes folded and fragmented and forms the flagellated chambers. Some cells of the superficial layer are also lost, but the changes in this layer are less striking. A depression forms in the upper surface, and cords of cells from the outer layer penetrate under the flagellated chambers. From the invaginated epithelium and cords, the osculum, atrium, and excurrent canals are formed. The incurrent canals also develop from cords of cells which penetrate between the flagellated chambers from the outer epithelium. Finally, other cells from the outer epithelium and the canals move into the interior, forming the mesenchyme (Meewis, 1938). Thus the cells of the anterior half of the larva form the choanoderm, and the cells of the posterior half give rise to all the other elements of the adult sponge.

The metamorphosis of the amphiblastulas of the calcareous sponges is similar to that of *Oscarella*. In the free-swimming larva of

Leucosolenia variabilis, the number of granular cells increases, and the nonflagellated posterior portion becomes much more prominent. Following attachment of the larva to the substratum, the granular cells move down around the flagellated cells, which then form a solid internal mass. The granular cells both form a simple external epithelium which secretes monaxonid spicules and give rise to internal groups of cells which produce triradiate spicules. A central cavity appears in the mass of former flagellated cells, and the latter develop into choanocytes. Later an osculum appears (Minchin, 1896; see Fig. 20B of this chapter).

The larva of *Sycon raphanus* is more advanced in its development than that of *Leucosolenia*. There is substantial overgrowth of the flagellated cells by the granular cells during the free-swimming period (Minchin, 1896). During the metamorphosis many of the flagellated cells degenerate and are phagocytized by the surviving flagellated cells (Duboscq and Tuzet, 1937). Culture experiments indicate that the posterior granular cells alone can form a quasinormal sponge with a flagellated chamber, but that isolated masses of flagellated cells do not survive (Maas, 1906). Therefore the behavior of the posterior granular cells of the amphiblastula is analogous to that of the inner cellular mass of the parenchymula.

Finally, the larva of *Polymastia robusta* and of certain calcareous sponges (the Calcinea) is an undifferentiated blastula; cytodifferentiation apparently begins only after settling of the larva (Borojević, 1967, 1969). During the free existence of the larva of *Ascandra falcata* and of *Clathrina reticulum*, there is first a movement of individual cells from the wall of the blastula into the central cavity. Then the epithelial structure of the wall is completely lost; and the blastula is converted into a solid, unorganized mass of cells. All of the cells lose their flagellum, and the simple cluster of blastomeres settles on the substratum (Borojević, 1969).

When recently released larvae of *Ascandra falcata* are cut into anterior and posterior halves, each half forms a small blastula which later settles and develops into a small but complete sponge. Thus the simple blastula larvae lack the anterior-posterior differentiation of the amphiblastula larvae. There apparently is no migration or sorting out of different cell types during the development of the adult sponge. The cells seem to differentiate according to their position in the fixed embryo (Borojević, 1969).

The time required for metamorphosis varies considerably according to species. For example, it has been reported that in *Haliclona* sp. (Bergquist *et al.*, 1970) and *Microciona prolifera* (Simpson,

1968), metamorphosis is complete within about 24 hours after fixation of the larva; in *Halisarca dujardini* (Lévi, 1956), *Microciona coccinea*, and *Ophlitaspongia seriata* (Bergquist *et al.*, 1970), it is complete within 2 days; in *Mycale syrinx* (Wilson, 1935) and *Halichondria moorei* (Bergquist *et al.*, 1970), it requires 3 days; and in *Lissodendoryx similis* (Ali, 1956), *Halisarca metschnikovi* (Lévi, 1956), and *Leucosolenia variabilis* (Minchin, 1896), it takes about 4, 5, and 6 days, respectively. However, in no case is the temperature given.

Up to this point no mention has been made of gastrulation. Here gastrulation is defined as the rearrangement of the cells of the embryo which establishes the basic organization of the young organism. Accepting this definition, it is reasonable to regard the cellular movements, which occur during the metamorphosis of the larva, as gastrulation (see Meewis, 1938; Brien, 1943; Lévi, 1963). In some sponges, including *Haliclona limbata*, the organization of the young sponge is already largely achieved in the swimming larva. In these cases gastrulation occurs during the development of the larva within the maternal mesenchyme (Meewis, 1939). Finally, in *Sycon raphanus* gastrulation begins during the free-swimming larval stage (Minchin, 1896).

Our knowledge of cytodifferentiation in sponges and of regulatory processes in their development is still very rudimentary (see Borojević, 1970; Lévi, 1970, for general discussions). Techniques which will permit more precise analyses at all levels but particularly at the molecular level are urgently needed.

2.4.3.4 Factors Influencing the Survival of Larvae and Postlarval Stages

Very little is known concerning the factors that influence the survival of larvae and postlarvae. The larvae of many sponges may be subject to predation. According to Bergquist and Sinclair (1968), the larvae of *Microciona coccinea* and of *Ophlitaspongia seriata* are eaten by spionid polychaetes. Sponge larvae may also be carried by currents into situations which are unfavorable for settling and metamorphosis (see Section 2.4.3.2). It has been observed that sponges are eaten by certain molluscs (Burton, 1948; Hartman, 1958), and fishes (Bakus, 1964b) and such predation may be an important factor influencing the survival of young postlarvae (Hartman, 1958).

Few data are available on the longevity of sponges. In Long Island Sound specimens of *Haliclona loosanoffi* die in the autumn, and the sponge survives the winter in the form of gemmules. Hartman (1958) observed that of 81 marked postlarvae, apparently 21 produced gemmules before they died. At Plymouth, England *Sycon coronatum* and *Grantia compressa* appear to be annual sponges (Orton, 1914, 1920). On the other hand, information obtained from the repeated observation of marked specimens of *Halichondria panicea, Hymeniacidon perlevis,* and *Pachymatisma johnstonia* suggests that they may survive for perhaps 2-3 years on the coast of Devonshire, England (Burton, 1949).

Acknowledgments

Appreciation is expressed to the scientists who contributed reprints, bibliographies, and informal communications to assist in the writing of this chapter. The author is also deeply indebted to Dr. Frances Roach, Mrs. Paul Fell, and Miss Ann Huckle for invaluable assistance in some of the original work reported here, to Miss Sibyl Hausman for the drawn illustrations, and to Mrs. John Faigle for some of the Italian translation. Finally, the help of Mrs. Paul Fell in the preparation of the manuscript is gratefully acknowledged.

2.5 References

Ali, M. A. (1956). Development of the monaxonid sponge *Lissodendoryx similis* Thiele. *J. Madras Univ.* **B26,** 553-581.

Annandale, N. (1915). Fauna of the Chilka Lake: Sponges. *Mem. Indian Mus., Calcutta* **5,** 23-54.

Annandale, N., and Kemp, S. (1915). Fauna of the Chilka Lake: Introduction. *Mem. Indian Mus., Calcutta* **5,** 1-20.

Bakus, G. J. (1964a). Morphogenesis of *Tedania gurjanovae* Koltun (Porifera). *Pac. Sci.* **18,** 58-63.

Bakus, G. J. (1964b). The effects of fish-grazing on invertebrate evolution in shallow tropical waters. *Allan Hancock Found. Publ., Occas. Papers* **27,** 1-29.

Bergquist, P. R., and Sinclair, M. E. (1968). The morphology and behaviour of larvae of some intertidal sponges. *N. Z. J. Mar. Freshwater Res.* **2,** 426-437.

Bergquist, P. R., Sinclair, M. E., and Hogg, J. J. (1970). Adaptation to intertidal existence: reproductive cycles and larval behavior in Demospongiae. *Symp. Zool. Soc. London* No. 25, "Biology of the Porifera" (W. G. Fry, ed.), pp. 247-271. Academic Press, New York.

Borojević, R. (1966). Étude experimentale de la différenciation des cellules de l'éponge au cours de son développement. *Develop. Biol.* **14,** 130-153.

Borojević R. (1967). La ponte et le développement de *Polymastia robusta. Cah. Biol. Mar.* **8,** 1-6.

Borojević, R. (1969). Étude du développement et de la différenciation cellulaire d' Éponges Calcaires Calcinéennes (Genre *Clathrina* et *Ascandra*). *Ann. Emb. Morphogen.* **2,** 15-36.

Borojević, R. (1970). Différenciation cellulaire dans l'embryogenèse et la morphogénèse chez les spongiaires. *Symp. Zool. Soc. London* No. 25, "Biology of the Porifera" (W. G. Fry, ed.) pp. 467-490. Academic Press, New York.

Borojević, R., and Lévi, C. (1964). Étude au microscope électronique des cellules de l'éponge: *Ophlitaspongia seriata* (Grant), au cours de la réorganisation après dissociation. *Z. Zellforsch.* **64**, 708-725.

Borojević, R., and Lévi, C. (1965). Morphogénèse expérimentale d'une éponge à partir de cellules de la larve nageante dissociée. *Z. Zellforsch.* **68**, 57-69.

Brien, P. (1932). Contribution à l'étude de la régénération naturelle chez les Spongillidae. *Spongilla lacustris, Ephydatia fluviatilis. Arch. Zool. Exp. Gén.* **74**, 461-506.

Brien, P. (1937). La réorganization de l'éponge après dissociation par filtration et phénomènes d'involution chez *Ephydatia fluviatilis. Arch. Biol.* **48**, 185-268.

Brien, P. (1943). L'embryologie des Éponges. *Bull. Mus. Roy. Hist. Natur. Belg.* **19**, 1-20.

Brien, P. (1967a). Un nouveau mode de statoblastogénèse chez une Éponge d'eau douce africaine: *Potamolepis stendelli* (Jaffé). *Bull. Acad. D. Belg. Cl. Sci.* **53**, 552-572.

Brien, P. (1967b). L'embryogénèse d'une éponge d'eau douce africaine: *Potamolepis stendelli* (Jaffé). Larves des Potamolepides et des Spongillides. Polyphylétisme des Éponges d'eau douce. *Bull. Acad Belg. Cl. Sci.* **53**, 752-777.

Brien, P., and Meewis, H. (1938). Contribution à l'étude de l'embryogénèse des Spongillidae. *Arch. Biol.* **49**, 177-250.

Brøndsted, H. V. (1943). Formbildungsprozesse bei einem sehr primitiven Metazoon, dem Süsswasserschwam, *Spongilla lacustris* (L). *Protoplasma* **37**, 244-257.

Brøndsted, H. V. (1953). The ability to differentiate and the size of regenerated cells after repeated regeneration in *Spongilla lacustris. Quart. J. Microsc. Sci.* **94**, 177-184.

Brøndsted, H. V., and Carlsen, F. E. (1951). A cortical cytoskeleton in expanded epithelial cells of sponge gemmules. *Exp. Cell Res.* **2**, 90-96.

Brown, D. D. (1966). The nucleolus and synthesis of ribosomal RNA during oogenesis and embryogenesis of *Xenopus laevis. Nat. Cancer Inst. Monogr.* **23**, 297-309.

Burton, M. (1928). A comparative study of the characteristics of shallow-water and deep-sea sponges, with notes on their external form and reproduction. *J. Quelsett Microsc. Club.* **16**, 49-70.

Burton, M. (1931). The interpretation of the embryonic and postlarval characters of certain Tetraxonid sponges with observations on the postlarval growth stages in some species. *Proc. Zool. Soc. London,* Parts 1-2, 511-525.

Burton, M. (1933). Observations on the postlarval development in the sponge *Iophon hyndmani* (Bow.) *Ann. Mag. Nat. Hist.* **12**, (10) 196-201.

Burton, M. (1948). The ecology and natural history of *Tethya aurantium* Pallas. *Ann. Mag. Natur. Hist.,* **1**, (12) 122-130.

Burton, M. (1949). Observations on littoral sponges, including the supposed swarming of larvae, movement and coalescence in mature individuals, longevity and death. *Proc. Zool. Soc. London* **118**, 893-915.

Cobb, W. R. (1969). Penetration of calcium carbonate substrates by the boring sponge, *Cliona. Amer. Zool.* **9**, 783-790.

Colussi, A. (1958). Sulla degenerazione di ovociti in *Sycon raphanus* (O. Schmidt) (Calcispongiae). *Ann. Inst. Mus. Zool. Univ. Napoli* **10**, 1-8.

Colwin, A. L., and Colwin, L. H. (1964). Role of the gamete membranes in fertilization. *In* "Cellular Membranes in Development," *22nd Symp. Soc. Study Develop. Growth* (M. Locke, ed.), pp. 233-279. Academic Press, New York.

Connes, R. (1967). Structure et développement des bourgeons chez l'éponge siliceuse

Tethya lyncurium L. Recherches expérimentales et cytologiques. *Arch. Zool. Exp. Gén.* **108**, 157-195.

Curtis, A.S.G. (1970). Problems and some solutions in the study of cellular aggregation. *Symp. Zool. Soc. London* No. 25, "Biology of the Porifera" (W. G. Fry, ed.), pp. 335-352. Academic Press, New York.

Delage, Y. (1892). Embryogénie des Éponges. Développement postlarvaire des Éponges siliceuses et fibreuses marines et d'eau douce. *Arch. Zool. Exp.* Gén. **20**, (2) 345-489.

Dendy, A. (1914). Observations on the gametogenesis of *Grantia compressa. Quart. J. Microsc. Sci.* **60** (2), 313-376.

Devos, C. (1965). Le bourgeonnement externe de l'Éponge *Mycale contarenii* (Martens) (Démosponges). *Bull. Mus. Hist. Natur.* **37**, 548-555.

De Vos, L. (1971). Étude ultrastructurale de la gemmulogenèse chez *Ephydatia fluviatilis* I. Le vitellus - formation -teneur en ARN et glycogène. *J. Microsc.* **10**, 283-304.

Duboscq, O., and Tuzet, O. (1937). L' Ovogenèse, la fécondation, et les premiers stades du développement des éponges calcaires. *Arch. Zool. Exp. Gén.* **79**, 157-316.

Duboscq, O., and Tuzet, O. (1942). Recherches complémentaires sur l'ovogenèse, la fécondation, et les premiers stades du développement des éponges calcaires. *Arch. Zool. Exp. Gén.* **81**, 395-466.

Duboscq, O., and Tuzet, O. (1944). L'ovogenèse, la fécondation et les premiers stades du développement de *Sycon elegans* Bow. *Arch. Zool. Exp. Gén.* **83**, 445-459.

Edmondson, C. H. (1946). Reproduction in *Donatia deformis* (Thiele). *Occas. Pap. Bernice Pauahi Bishop Mus. Honolulu* **18**, 271-282.

Efremova, S. M. (1970). Proliferation activity and synthesis of protein in the cells of fresh-water sponges during development after dissociation. *Symp. Zool. Soc. London* No. 25, "Biology of the Porifera" (W. G. Fry, ed.), pp. 399-413. Academic Press, New York.

Egami, N., and Ishii, S. (1956). Differentiation of sex cells in united heterosexual halves of the sponge, *Tethya serica. Annot. Zool. Japan* **29**, 199-201.

Fauré-Fremiet, E. (1932). Morphogénèse expérimentale (reconstitution) chez *Ficulina ficus* L. *Arch. Anat. Microsc. Morphol. Exp.* **28**, 1-80.

Fell, P. E. (1967). Sponges. In "Methods in Developmental Biology" (F. H. Wilt and N. K. Wessells, eds.), pp. 265-276. Crowell-Collier, New York.

Fell, P. E. (1969). The involvement of nurse cells in oogenesis and embryonic development in the marine sponge, *Haliclona ecbasis. J. Morphol.* **127**, 133-150.

Fell, P. E. (1970). The natural history of *Haliclona ecbasis* de Laubenfels, a siliceous sponge of California. *Pac. Sci.* **24**, 381-386.

Fincher, J. A. (1940). The origin of the germ cells in *Stylotella heliophila* Wilson (Tetraxonida). *J. Morphol.* **67**, 175-197.

Galtsoff, P. S. (1925). Regeneration after dissociation (an experimental study on sponges) II Histogenesis of *Microciona prolifera. J. Exp. Zool.* **42**, 223-251.

Gasic, G. J., and Galanti, N. L. (1966). Proteins and disulfide groups in the aggregation of dissociated cells of sea sponges. *Science* **151**, 203-205.

Gatenby, J. B. (1920). The germ-cells, fertilization and early development of *Grantia (Sycon) compressa. J. Linn. Soc. Zool.* **34**, 261-297.

Gatenby, J. B. (1927). Further notes on the gametogenesis and fertilization of sponges. *Quart. J. Microsc. Sci.* **71**, 173-188.

Gatenby, J. B., and King, S. B. (1929). Note on the nutrient membrane of *Grantia* amphiblastula. *J. Roy. Microsc. Soc.* **49**, 319-320.

Hartman, W. D. (1958). Natural history of the marine sponges of southern New England. *Bull. Peabody Mus.* **12**, 1-155.

Herlant-Meewis, H. (1948). La gemmulation chez *Suberites domuncula* (Olivi) Nardo. *Arch. Anat. Microsc. Morphol. Exp.* **37**, 289-322.

Humphreys, T. (1970). Biochemical analysis of sponge cell aggregation. *Symp. Zool. Soc. London* No. 25, "Biology of the Porifera" W. G. Fry, ed.), pp. 325-334. Academic Press, New York.

Huxley, J. S. (1912). Some phenomena of regeneration in *Sycon;* with a note on the structure of its collar-cell. *Phil. Trans. Roy. Soc. London* **B 202**, 165-190.

Ijima, I. (1901). Studies on the Hexactinellida, Contribution 1 (Euplectelidae). *J. Coll. Sci. Imp. Univ. Tokyo* **15**, 1-299.

Ijima, I. (1903). Studies in Hexactinellida. *J. Coll. Sci. Imp. Univ. Toyko* **18**, 1-124.

John, H. A., Campo, M. S., Mackenzie, A. M., and Kemp, R. B. (1971). Role of different sponge cell types in species specific cell aggregation. *Nature New Biol.* **230**, 126-128.

Jorgensen, O. M. (1917). Reproduction in *Grantia compressa. Rep. Dove Mar. Lab.* **6**, 26-32.

Korotkova, G. P. (1970). Regeneration and somatic embryogenesis in sponges. *Symp. Zool. Soc. London* No. 25, "Biology of the Porifera" (W. G. Fry, ed.) pp. 423-436. Academic Press, New York.

Kumé, M. (1952). Note on the early development of *Tethya serica* Lebwahl, a tetraxonian sponge. *Nat. Sci. Rep. Ochanomizu Univ.* **8**, 97-104.

Leveaux, M. (1939). La formation des gemmules chez les Spongillidae. *Ann. Soc. Roy. Zool. Belg.* **70**, 53-96.

Leveaux, M. (1941). Contribution à l'étude histologique de l'ovogénèse et de la spermatogénèse des Spongillidae. *Ann. Soc. Roy. Zool. Belg.* **72**, 251-269.

Leveaux, M. (1942). Contribution à l'étude histologique de l'ovogénèse et de la spermatogénèse des Spongillidae. *Ann. Soc. Roy. Zool. Belg.* **73**, 33-50.

Lévi, C. (1951a). L'ovipartė chez les spongiaires. *C. R. Acad. Sci. Paris* **233**, 272-274.

Lévi, C. (1951b). Existence d'un stade grégaire transitoire au cours de l'ovogénèse des Spongiaires *Halisarca dujardini* (Johnst.) et *Oscarella lobularis* (O. S.). *C. R. Acad. Sci. Paris* **233**, 826-828.

Lévi, C. (1951c). Remarques sur la faune des Spongiaires de Roscoff. *Arch. Zool. Exp. Gén.* **87**, 10-21.

Lévi, C. (1953). Description de *Plakortis nigra* nov. sp. et remarques sur les Plankinidae (Démosponges). *Bull. Mus. Hist. Nat. Paris* **25 (2)**, 320-328.

Lévi, C. (1956). Étude des *Halisarca* de Roscoff. Embryologie et systématique des Démosponges. *Arch. Zool. Exp. Gén.* **93**, 1-181.

Lévi, C. (1963). Gastrulation and larval phylogeny in sponges. *In* "The Lower Metazoa" (E. C. Dougherty, ed.). pp. 375-382. Univ. of California Press, Berkeley, California.

Lévi, C. (1964). Ultrastructure de la larve parenchymella de Démosponge. I. *Mycale contarenii* (Martens). *Cah. Biol. Mar.* **5**, 97-104.

Lévi, C. (1970). Les cellules des éponges. *Symp. Zool. Soc. London* No. 25, "Biology of the Porifera" (W. G. Fry, ed.), 353-364. Academic Press, New York.

Lévi, C., and Porte, A. (1962). Étude au microscope électronique de l'éponge *Oscarella lobularis* Schmidt et de sa larve amphiblastula. *Cah. Biol. Mar.* **3**, 307-315.

Liaci, L., and Sciscioli M. (1967). Osservasioni sulla maturazione sessuale di un Tetractinellids: *Stelletta grubii* O. S. (Porifera). *Arch. Zool. Italy* **52**, 169-177.

Lutfy, R. G. (1957a). On the origin of the so-called mesoblast cells in the amphiblastula larva of calcareous sponges. *Cellule* **58**, 231-236.

Lutfy, R. G. (1957b). On the placental membrane of calcareous sponges. *Cellule* **58,** 239-246.

Maas, O. (1893). Die Embryonalentwicklung und Metamorphose der Cornacuspongien. *Zool. Jahrb.* **7,** 331-448.

Maas, O. (1896). Erlidigte und strittige Fragen der Schwammentwicklung. *Biol. Zentralbl.* **16,** 231-239.

Maas, O. (1906). Über die Enwirkung karbonatfreier und kalkfreier Salzlösungen auf erwachsene Kalkschwämme und auf Entwicklungsstadien derselben. *Roux' Arch.Entwicklungsmech. Organ.* **22,** 581-599.

MacLennan, A. P. (1970). Polysaccharides from sponges and their possible significance in cellular aggregation. *Symp. Zool. Soc. London* No. 25, "Biology of the Porifera" (W. G. Fry, ed.), pp. 299-324. Academic Press, New York.

Margoliash, E., Schenck, J. R., Hargie, M. P., Burokas, S., Richter, W. R., Barlow, G. H., and Moscona, A. A. (1965). Characterization of specific cell aggregating materials from sponge cells. *Biochem. Biophys. Res. Commun.* **20,** 383-388.

McClay, D. R. (1971). An autoradiographic analysis of the species specificity during sponge cell reaggregation. *Biol. Bull.* **141,** 319-330.

McDougall, K. D. (1943). Sessile marine invertebrates of Beaufort, N. C. *Ecol. Monogr.* **13,** 321-374.

Meewis, H. (1938). Contribution a l'étude de l'embryogénèse des Myxospongidae: *Halisarca lobularis* (Schmidt). *Arch. Biol.* **50,** 3-66.

Meewis, H. (1939). Contribution à l'étude de l'embryogénèse de Chalinidae: *Haliclona limbata. Ann. Soc. Roy. Zool. Belg.* **70,** 201-243.

Meewis, H. (1941). Contribution à l'étude de l'embryogénèse des éponges siliceuses. Développement de l'oeuf chez *Adocia cinerea* (Grant) et *Halichondria coalita* (Bowerbank). *Ann. Soc. Roy. Zool. Belg.* **72,** 126-149.

Miller, R. L. (1966). Chemotaxis during fertilization in the hydroid *Campanularia. J. Exp. Zool.* **162,** 23-44.

Minchin, E. A. (1896). Note on the larva and the postlarval development of *Leucosolenia variabilis* n. sp. with remarks on the development of other Asconidae. *Proc. Roy. Soc. London* **60,** 42-52.

Minchin, E. A. (1897). The position of sponges in the animal kingdom. *Sci. Progr.* **6,** 426-460.

Minchin, E. A. (1900). Porifera. *In* "Treatise on Zoology" (R. Lankester, ed.), pp. 1-178.Black, London.

Moscona, A. A. (1968). Cell aggregation: properties of specific cell-ligands and their role in the formation of multicellular systems. *Develop. Biol.* **18,** 250-277.

Müller, K. (1911). Das Regenerationsvermögen der Sûsswasserschwämme insbesondere Untersuchungen über die bie ihnen vorkommende Regeneration nach Dissociation und Reunition. *Roux' Arch. Entwicklungsmech. Organ.* **32,** 397-446.

Müller, K. (1914). Gemmula Studien an *Ficulina. Wiss. Meer. Untersuch. Abt. Kiel* **16,** 289-313.

Nassonow, N. (1883). Zur Biologie und Anatomie der Clione. *Z. Wiss. Zool.* **39,** 295-308.

Nassonow, N. (1924). Sur l'éponge perforante *Clione stationis* Nason. et le procédé du creusement des galeries dans les valves des huitres. *C. R. Acad. Sci. Russ.* 113-115.

Okada, Y. (1928). On the development of a Hexactinellid sponge *Farrea sollasii, J. Fac. Sci. Imp. Univ. Tokyo* **2,** 1-27.

Orton, J. H. (1914). Preliminary account of a contribution to an evaluation of the sea. *J. Mar. Biol. Ass.* **10,** 312-326.

Orton, J. H. (1920). Sea-temperature, breeding, and distribution in marine animals. *J. Mar. Biol. Ass.* **12,** 339-366.

Pourbaix, N. (1935). Formation histochimique des gemmules d'éponges. *Ann. Soc. Roy. Zool. Belg.* **66**, 33-37.

Pourbaix, N. (1936). Sur le mécanisme d'accumulation des reserves dans les gemmules des Spongillidae. *Mém. Mus. Roy. Hist. Natur. Belg.* 3, 415-419.

Prell, H. (1915). Zur Kenntnis der Gemmulae bei den marinen Schwämmen. *Zool. Anz,* **46,** 97-116.

Rasmont, R. (1956). La gemmulation des Spongillides. IV. Morphologie de la gemmulation chez *Ephydatia fluviatilis* et *Spongilla lacustris. Ann. Soc. Roy. Zool. Belg.* **86,** 349-387.

Rasmont, R. (1961). Une technique de culture des éponges d'eau douce en milieu controlé. *Ann. Soc. Roy. Zool. Belg.* **91,** 147-156.

Rasmont, R. (1962). The physiology of gemmulation in freshwater sponges. *In* "Regeneration" 20th *Symp. Soc. Study Develop. Growth.* (D. Rudnick, ed.), pp. 3-25 Ronald Press, New York.

Rasmont, R. (1963). Le rôle de la taille et de la nutrition dans le déterminisme de la gemmulation chez les Spongillides. *Develop. Biol.* **8**, 243-271.

Rasmont, R. (1968). Chemical aspects of hibernation. *In* "Chemical Zoology II Porifera, Coelenterata, and Platyhelminths" (M. Florkin and B. T. Scheer, eds.), pp. 65-77 Academic Press, New York.

Rasmont, R. (1970). Some new aspects of the physiology of fresh-water sponges. *Symp. Zool. Soc. London* No. 25, "Biology of the Porifera." (W. G. Fry, ed.), pp. 415-422, Academic Press, New York.

Reiswig, H. M. (1970). Porifera: Sudden sperm release by tropical Demospongiae. *Science* **170**, 538-539.

Ruthmann, A. (1965). The fine structure of RNA-storing archaeocytes from gemmules of fresh-water sponges. *Quart. J. Microsc. Sci.* **106**, 99-114.

Sarà, M. (1955). La nutrizione dell' ovcita in Calcispongie Omoceli. *Ann. Ist. Museo. Zool. Univ. Napoli* **7**, 1-30.

Sarà, M. (1961). Ricerche sul gonocorismo ed ermafroditismo nei Porifera. *Boll. Zool.* **28**, 47-60.

Sarà, M., Liaci, L., and Melone, N. (1966). Bispecific cell aggregation in sponges. *Nature (London)* **210**, 1167-1168.

Schmidt, I. (1970). Étude préliminaire de la différenciation des thésocytes d '*Ephydatia fluviatilis* L. extraits mécaniquement de la gemmule. *C. R. Acad. Sci. Paris* **271**, 924-927.

Shore, R. E. (1971). Growth and renewal studies of the choanocyte population in *Hymeniacidon sinapium* (Porifera: Demospongiae) using colcemid and 3-H thymidine. *J. Exp. Zool* **177,** 359-363.

Simpson, T. L. (1963). The biology of the marine sponge *Microciona prolifera* (Ellis and Solander). I. A study of cellular function and differentiation. *J. Exp. Zool.* **154**, 135-151.

Simpson, T. L. (1968). The biology of the marine sponge *Microciona prolifera* (Ellis and Solander). II. Temperature-related, annual changes in functional and reproductive elements with a description of larval metamorphosis. *J. Exp. Mar. Biol. Ecol.* **2**, 252-277.

Siribelli, L. (1962). Differenze nel ciclo sessuale di *Axinella damicornis* (Esper) ed *Axinella verrucosa* 0. Sch. (Demospongiae). *Boll. Zool.* **29**, 319-322.

Sivaramakrishnan, V. R. (1951). Studies on early development and regeneration in some Indian sponges. *Proc. Ind. Acad. Sci.* Sect. B **34**, 273-310.

Storr, J. F. (1964). Ecology of the Gulf of Mexico commercial sponges and its relation to the fishery. *Spec. Sci. Rep. U. S. Fish Wildl. Serv. Fish.* No. 466, 73 p.

Topsent, E. (1887). Contribution a l'étude des Clionides. *Arch. Zool. Exp. Gén.* **5**, 1-166.

Topsent, E. (1888). Notes sur les gemmules de quelques silici-spongidae marines. *C. R. Acad. Sci. Paris* **106**, 1298-1300.

Topsent, E. (1895). Étude sur la faune des Spongiaires du Pas-de-Calais suivie d'une application de la nomenclature actuelle à la monographie de Bowerbank. *Rev. Biol. Nord Fr.* **7**, 6-28.

Topsent, E. (1920). Caractères et affinités des *Thoosa* Hanc. et des *Alectona* Cart. Considérations sur leurs germes à armure. *Bull. Soc. Zool. Fr.* **45**, 88-97.

Topsent, E. (1948). Considérations sur les "plasmodes planctoniques d' Êponges." *Bull. Inst. Oceanogr.* **923**, 1-4.

Trégouboff, G. (1942). Contribution à la connaissance des larves planctoniques d'-Éponges. *Arch. Zool. Exp. Gén.* **82**, 357-399.

Tuzet, O. (1930a). Sur la fécondation de l'éponge siliceuse *Cliona viridis* (Schmidt). *C.R. Acad. Sci. Paris* **191**, 1095-1097.

Tuzet, O. (1930b). Spermatogénèse de *Reniera*. *C. R. Soc. Biol.* **103**, 970-973.

Tuzet, O. (1932). Recherches sur l'histologie des éponges *Reniera elegans* (Bow) et *Reniera simulans* (Johnston). *Arch. Zool. Exp. Gén.* **74**, 169-192.

Tuzet, O. (1947). L'ovogenèse et la fécondation de l'éponge calcaire *Leucosolenia (Clathrina) coriacea* Mont. et de l'Éponge siliceuse *Reniera elegans* Bow. *Arch. Zool. Exp. Gén.* **85**, 127-148.

Tuzet, O. (1948). Les premiers stades du développement de *Leucosolenia botryoides* Ellis et Solander et *Clathrina (Leucosolenia) coriacea* Mont. *Ann. Sci. Natur. Zool.* **10**, 11 103-114.

Tuzet, O. (1964). L'origine de la lignée germinale et la gametogénèse chez les Spongiaires. In "L'Origine de la Lignée Germinale" (E. Wolff, ed.), 79-111 Hermann, Paris.

Tuzet, O., and Connes, R. (1962). Recherches histologiques sur la reconstitution de *Sycon raphanus* O. S. à partir des cellules dissociées. *Vie Milieu* **13**, 703-710.

Tuzet, O., and Pavans de Ceccatty, M. (1958). La spermatogenése, l'ovogenése la fécondation et les premiers stades du développement d' *Hippospongia communis LMK (=H. equina* O. S.) *Bull. Biol. Fr. Belge.* **92**, 331-348.

Tuzet, O. and Paris, J. (1964). La spermatogenése, l'ovogenése, la fécondation et les premiers stades du développement chez *Octavella galangaui. Vie* Milieu **15**, 309-327.

Tuzet, O.; Garrone, R., and Pavans de Ceccatty, M. (1970). Origine choanocytaire de la lignée germinale mâle chez le Démosponge *Aplysilla rosea* Schulze (Dendroceratides). *C.R. Acad. Sci. Paris.* **270**, 955-957.

Vacelet, J. (1964). Étude monographique de l'Éponge calcaire Pharetronide de Méditérranée *Petrobiona massiliana* Vacelet et Lévi. Les pharetronides actuelles et fossiles. *Recl. Trav. Sta. Mar. Endoume* **50**, 1-125.

Van de Vyver, G. (1970). La non confluence intraspécifique chez les spongiares et la notion d'individu. *Ann. Embryol. Morphogen.* **3**, 251-262.

Vaney, C., and Allemand-Martin, A. (1918). Contribution à l'étude de la larve de l' *Hippospongia equina* des côtes de Tunisie. *C.R. Acad. Sci. Paris.* **116**, 82-84.

Warburton, F. E. (1958). The manner in which the sponge *Cliona* bores in calcareous objects. *Can. J. Zool.* **36**, 555-562.

Warburton, F. E. (1961). Inclusion of parental somatic cells in sponge larvae. *Nature (London.)* **191**, 1317.

Warburton, F. E. (1966). The behavior of sponge larvae. *Ecology* **47**, 672-674.

Watanabe, Y. (1957). Development of *Tethya serica* Lebwohl, a tetraxonian sponge I. Observations on external changes. *Natur Sci. Rep. Ochanomizu Univ.* **8**, 97-104.

Watanabe, Y. (1960). Outline of morphological observation on the development of *Tethya serica. Bull. Mar. Biol. Stat. Asamushi Tohoku Univ.* **10**, 145-148.

Wells, H. W., Wells, M. J., and Gray, I. E. (1964). Ecology of sponges in Hatteras Harbor, North Carolina. *Ecology* **45**, 752-767.

Wilson, H. V. (1891). Notes on the development of some sponges. *J. Morphol.* **5**, 511-519.

Wilson, H. V. (1894). Observations on the gemmule and egg development of marine sponges. *J. Morphol.* **9**, 277-406.

Wilson, H. V. (1902). On the asexual origin of the sponge larvae. *Amer. Natur.* **36**, 451-459.

Wilson, H. V. (1907). On some phenomena of coalescence and regeneration in sponges. *J. Exp. Zool.* **5**, 245-258.

Wilson, H. V. (1935). Some critical points in the metamorphosis of the Halichondrine sponge larva. *J. Morphol.* **58**, 285-345.

Wilson, H. V. (1937). Notes on the cultivation and growth of sponges from reduction bodies, dissociated cells, and larvae. In "Culture Methods for Invertebrate Animals" (J. G. Needham, ed.), pp. 137-139. Cornell Univ. Press, Ithaca, New York. Reprinted by Dover, New York. 1959.

Wilson, H. V., and Penney, J. T. (1930). The regeneration of sponges *(Microciona)* from dissociated cells. *J. Exp. Zool.* **56**, 73-147.

Chapter 3

CNIDARIA

Richard D. Campbell

3.1 Introduction

Reproduction of cnidarians provides some of the most significant and interesting avenues for understanding the evolution and functioning of sexual and asexual propagation. This phylum presents some of the simplest animal structures which are diploblastic and definite and clearly on the metazoan plan. Cnidarian gonads are primordial, and embryonic development is prototypic. Yet the frills may also be analyzed, for within the phylum Cnidaria there are animals which have evolved extraordinary complexity. Thus we have an opportunity to examine both the basic reproductive patterns and their elaborations.

An important feature shown by cnidarian reproduction is the high degree of reversibility of morphogenetic processes and substantially less commitment in tissue and cell specialization than that found in development of more advanced phyla.

This chapter is intended to draw together some of the scattered information available on cnidarian reproduction and to provide access to major literature sources. Important previous reviews on all or parts of this subject include: Metschnikoff (1886), Korschelt and Heider (1902), Goette (1907), Kühn (1910, 1913, 1914-1916), Broch (1924a,b), Kükenthal and Krumbach (1925), Dawydoff (1928), Hyman (1940), Uchida and Yamada (1968), and Mergner (1971). For an excellent basic description of Cnidaria see Bouillon (1968).

3.2 Asexual Reproduction

The developmental biology of Cnidaria revolves around asexual reproduction. Most histological and experimental work has concentrated on nonsexual processes. Cnidaria are excellent models for studying animal cyto- and morphogenesis because of their simple forms, high degree of developmental reversibility, and amenability to laboratory investigation. This extensive and important aspect of cnidarian reproduction far exceeds the limitations of this volume. Fortunately large numbers of excellent reviews covering such topics appear each decade. The following paragraphs provide brief statements of asexual reproductive phenomena in Cnidaria with references to reviews elsewhere.

3.2.1 Alternation of Generations

The diversity of types of cnidarian asexual reproduction is most easily organized with reference to the peculiar life cycle phenomenon known as "alternation of generations." Many cnidarian life cycles may be traced through alternating sexually and asexually reproducing forms, and those life cycles which cannot are easily interpreted as derivatives of this basic scheme, shown in Fig. 1. The sexually reproducing form is a swimming, *medusoid* animal. Eggs and spermatozoa produced by medusae* or jellyfish fuse to

*In this review the term medusa will exclusively connote the free-swimming stage of Hydrozoa. The terms jellyfish and scyphomedusa refer only to members of the Scyphozoa. The term medusoid will be used to refer to either or both types of organisms. The term zooid will be used to refer to any medusoid or polypoid unit of cnidarian structure.

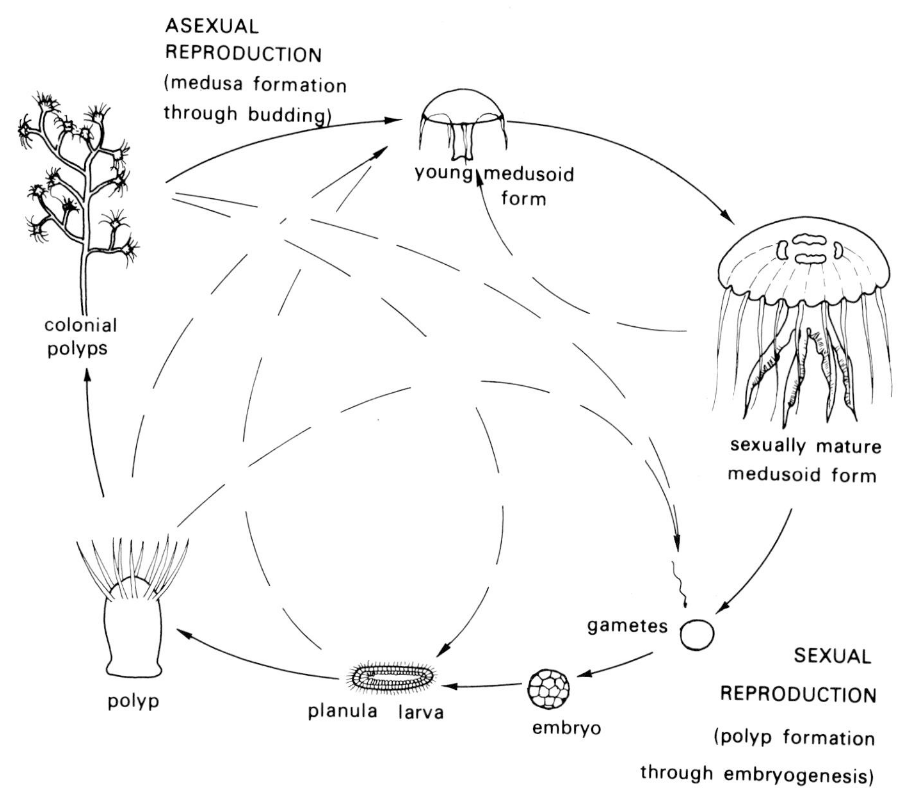

FIG. 1. Basic cnidarian life cycle showing alternation between sexual and asexual "generations." Alternate cycles found in some types are indicated by broken lines; many scyphozoans do not exhibit a polyp stage, for example, and anthozoans have no medusoid form.

form typical zygotes which give rise to planula larvae and then *polyps*. Polyps reproduce medusae asexually by means of buds on their sides. This scheme contrasts strictly from alternation of generations in plants, which involves successions of haploid and diploid forms. Both polyp and medusa "generations" are diploid. Also, the sexually reproducing form of plants (gametophyte) generally arises from a single cell while medusa formation begins with a population of cells and is more properly viewed as a transformed portion of the polyp body wall. Some workers feel that the polyp stage should, therefore, be considered as a larval stage.

Two complications of the basic "alternation of generations" plan shown in Fig. 1 are very abundant. First, asexual reproduction often occurs at stages other than medusoid budding. Most polyp stages reproduce themselves by budding, and some medusae also reproduce by budding medusae. Such homotypic budding often occurs without complete separation of product animals; this may lead to exquisite sessile (e.g., coral) or floating (e.g., siphonophore) colonies. Second, some groups lack either the polyp or medusoid stage from their life cycle. Thus, within the entire class Anthozoa (including the stony corals and sea anemones) no vestige of a medusoid form has been encountered, and a number of representatives of the class Scyphozoa exhibit no polyp "generation." Many hydrozoans also lack either the polyp or medusoid form. Further, individual species or genera tend to accentuate the polyp or medusoid generation, with concomitant simplification of the other.

Our understanding of the cnidarian life cycle and its evolution, interesting in its own right as a chapter in the history of science, is well reviewed and probed by Haeckel (1879), Hyman (1940), Rees (1957), and Hadži (1963).

The reaction of more classical workers to the alternation of generation forms has always been to homologize the forms with one another (see Hyman, 1940) and to find intermediate types (see, e.g., Werner, 1963). More recently there has been interest in studying the developmental mechanisms which separate and maintain the distinctions between polyp and medusoid generations. One of the more advanced investigations of this type is reported by Frey (1968) who finds that dissociated or isolated medusa tissue will regenerate stolon and polyp forms; medusa structures are interpreted as representing a higher developmental state which must be stabilized to prevent reversion.

3.2.2 Budding

Asexual reproduction by budding nearly always represents histological and morphological transformation of existing tissues rather than *de novo* histo- and morphogenesis. Practically all cnidarian tissue consists of two epithelia, ectoderm and endoderm, separated by an acellular (or at least poorly cellularized) layer of mesoglea. These epithelia retain their continuity across most morphological regions and boundaries, so that the formation of new individuals, as in budding, involves mainly changes of tissue folding and readjustments in the numerical levels of cell populations. Thus, it may be viewed as a process of developmental pattern regulation accommodating tissue growth.

Budding usually takes place along stolons in colonial forms and from the polyp side in solitary forms. Some medusae bud directly from the manubrium, bell margin, or tentacular bulbs. Ordinarily, one sees the following histological sequence during budding.

1. One or both epithelia thicken locally to produce a bulge in the tissue.
2. Sometimes ectodermal interstitial cells become more abundant locally.
3. The tissue begins to evaginate at this site.
4. The endoderm acquires longitudinal taeniolae (Hamann, 1882), or ridges, frequently four in number, which establish the future radial symmetry of the zooid.
5. After the outgrowth is papilliform, morphological modeling takes place, including tentacle formation, mouth opening, and zooid shaping.
6. The resulting zooid may separate from its parent or it may remain attached. The polarity of buds is almost always with the future oral end distal.

This sequence of events may show variations from one species to another, mainly in Stages 4 and 5. However, the overall series, and within a species even minute sequences, is so definite that even slight deviations from it indicate abnormality. Thus, abnormal strains of *Hydra viridis* are justly considered to be "nonbudding" since new individuals arise through altered sequences of events (Lenhoff *et al.*, 1969; Moore, 1971). Müller (1964) was able to distinguish between gonozooid and gastrozooid buds at an extremely early time during their formation (early Stage 2). This overall sequence of events relates budding very closely to embryonic development and larval transformation, and sharply distinguishes budding from regenerative (as reviewed by Tardent, 1963) processes.

Budding appears to be mediated principally through epithelial cell behavior, and to some extent probably by gastrocoel fluid pressure and architecture of external supporting structures such as the perisarc. Interstitial cells are now thought to play little if any direct morphogenetic role in this and other morphogenetic processes, although they may be important in patterning. The accumulation of interstitial cells at budding sites is probably largely related to the production of nondividing cell types such as nerves and nematocytes.

3.2.3 Colony Formation

Extensive colonies may arise when budding and growth occur but zooid separation does not. A colony formed in this way (as in most colonial hydroids, siphonophores, corals, and alcyonarians), has a single ectoderm, endoderm, and gastrocoel continuous throughout the entire structure. This situation is, therefore, not analogous to many types of animal colonies where individuals are separate.

3.2.3.1 Colony Growth Patterns

All colonies exhibit patterns in their growth and budding; these patterns may be regular or highly irregular. Colony growth frequently takes place at definite positions within the tissue connecting polyps (such as the stolon network of hydrozoans), and new polyps arise at particular locations. While no thorough consideration of colony patterns exists, Kühn (1909, 1913), Braverman and Schrandt (1966), Hyman (1940), Beklemishev (1969), and Campbell (1972) have summarized some aspects of the problem, and Kükenthal and Krumbach (1925) may be referred to for more extensive but fragmented information.

3.2.3.2 Polymorphism

Polymorphism occurs among different zooids in a colony to such an extent that in some cases, as in reproductive structures, individual zooids approach the status of organs. Cnidarian polymorphism arises by accentuation and suppression of particular morphological portions of the general zooid structure. Thus, nutritive zooids, in widely divergent cnidarian groups, accentuate the mouth and distal (oral) tentacles. Reproductive polyps tend to have these parts poorly developed if at all; they accentuate the proximal portions where budding occurs. The most extreme examples of polymorphic modification include gonophores (see Section 3.3.3) and the buoyant organs of siphonophores (Garstang, 1946; Totton, 1965); here, nearly all semblance to a zooid is lost, with only early developmental similarities indicating their homologies with the usual cnidarian architecture. The subject of polymorphism has also been insufficiently reviewed in recent times; for reference to the extensive older literature and to more recent work, the reader is referred to Hyman (1940), and Berrill (1961).

3.2.4 Asexual Reproduction other than Budding

Several other modes of asexual reproduction occur in the Cnidaria. Although they are generally to be regarded as minor compared to budding, and often simply the outcome of pattern regulation following accidental damage, a few species or groups have evolved a depen-

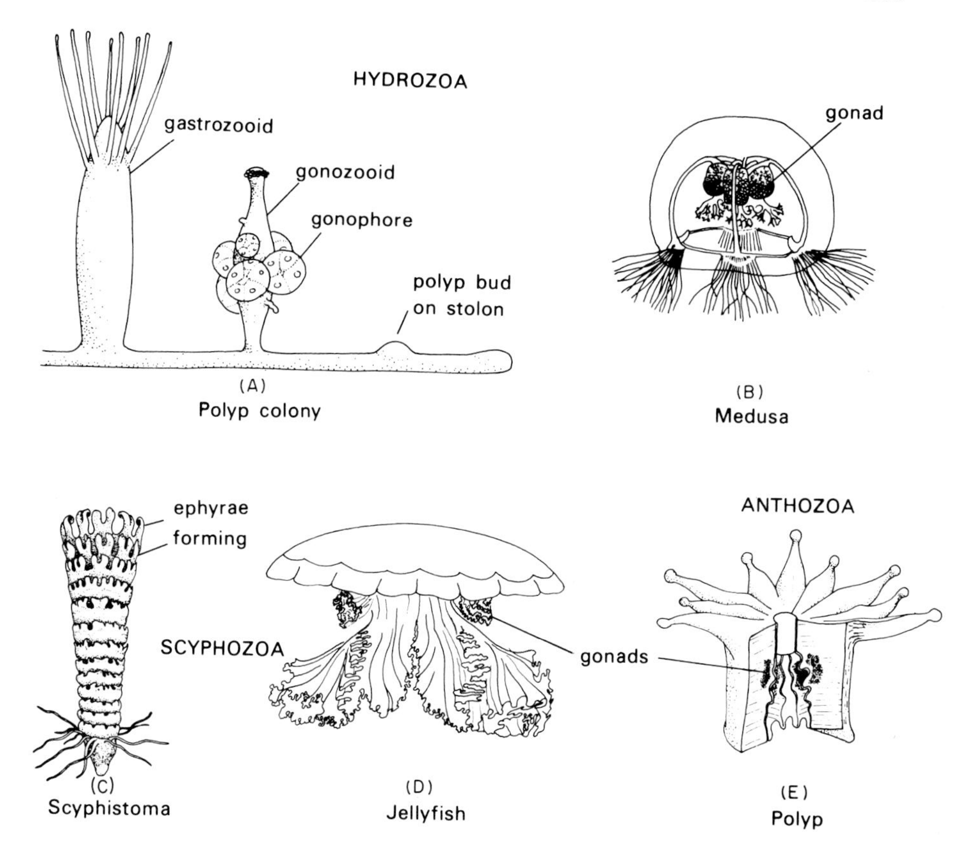

FIG. 2. Some cnidarian reproductive forms. (A) Hydrozoan colony of *Hydractinia echinata*, with gonophores on the sexual zooid, and stolon budding. (B) Hydrozoan medusa *Bougainvillia principis* with gonads (after Russell, 1953). (C) Scyphozoan polyp, the scyphistoma, which is undergoing strobilation to form ephyrae (*Aurelia flavidula*, from Agassiz, 1862). (D) Scyphozoan jellyfish, *Cyanea arctica* with gonads. [From Agassiz (1862).] (E) Anthozoan polyp illustrating position of gonads on septa [modified after Hyman (1940)].

dence upon them. Transverse fission is one of these modes. This seems to be the chief mode of reproduction in the minute tentacleless hydroid *Protohydra* (Muus, 1966) which does not bud. Transverse fission occurs regularly in young anemones of the species *Gonactinia prolifera* (Stephenson, 1935). The most spectacular and important type of transverse fission is strobilation of the scyphozoan scyphistoma larvae (Fig. 2), in which the distal portion of the polyp is flattened and shed as a complete, young jellyfish (ephyra). Scyphistomae may live for years, producing hundreds of ephyrae, (Littleford, 1939; Chapman, 1966).

Longitudinal fission also occurs, and is frequently described because of the bizarre intermediate stages. It is probably rare and of little importance in most cnidarians. However, some corals, such as *Fungia* and brain corals, form polyps or colony subunits by growth and longitudinal subdivision of the oral disc (Hyman, 1940).

Fragmentation of portions of the flattened pedal disc occurs frequently in some sea anemones. In this process, termed pedal laceration, small pieces of the disc margin separate from the surrounding tissue and form a small zooid in a manner resembling regeneration. Sometimes the parent anemone moves, leaving behind it a ring of pedal disc, which then fragments to form a dozen or so individuals (Stephenson, 1928).

3.2.5 Reduction and Dormant Stages

A variety of reduction stages are scattered throughout the phylum Cnidaria. Nearly all cnidarians assume a rounded, vesicular form under adverse conditions, and these can later regenerate [see Berrill, 1953, 1961 (hydrozoan frustules or planuloids); Hyman, 1940]. Examples of more specialized dormant or resistant stages include the following. Resistant podocysts are formed by scyphozoan polyps (Kowalevsky, 1884; Littleford, 1939; Rees, 1957; Chapman, 1966, 1968; Russell, 1970). The pelagic hydrozoan *Margelopsis haeckeli* forms two types of eggs, small summer ones which develop directly into actinula larvae, and large autumnal eggs which are dormant during the winter (Werner, 1956). A variety of other hydrozoans similarly have encystment forms (Rees, 1957), and sexual reproduction in *Hydra* includes obligate encystment in a postcleavage stage.

3.2.6 Asexual Reproduction: Conclusion

Asexual reproduction is of vast significance to cnidarians. It normally takes the form of budding, which resembles transformation of the relatively undifferentiated planula larva into the adult zooid structure. Thus, asexual reproduction may be viewed as the result of continuing embryonic development in new tissue regions. These regions may be systematically propagated in a relatively unpatterned state (e.g., stolons) or may represent a part of a patterned zooid which becomes subdivided. The nature of tissue patterning has always stood out as a central problem in cnidarian development and is often reviewed (Kühn, 1913; Child, 1941; Berrill, 1961; Tardent, 1963; Webster, 1971; Wolpert, 1971). The companion process in budding to patterning is cell and tissue behavior underlying shape changes, and this has recently begun to attract close attention (Campbell, 1968a,c, 1972; Webster, 1971; Beloussov *et al.*, 1972).

3.3 Sexual Reproduction

3.3.1 Sexual Dimorphism

There are no striking examples of sexual dimorphism of somatic body parts among the Cnidaria, but frequently, the sexes may be distinguishable as the gonads ripen. This is commonly attributable to color differences between eggs and spermatozoa, or to size and shape differences between ovaries and testes. For example, in the hydroid *Hydractinia echinata* the yolky eggs are so strongly colored and gonads so abundant that ripe female colonies appear orange while male colonies appear lightly colored. The gonads of the transparent jellyfish *Dactylometra quinquecirrha* confer a bright pink color to males and brown to females (Littleford, 1939). Size and shape differences between ovaries and testes confer significant dimorphic appearances during reproductive seasons in cnidarians such as *Hydra* where the gonads constitute a major bulk of the entire animal, and *Corymorpha palma* where the abundant ovaries appear knobby due to the large ova, while testes appear smooth. Hydrozoans with reduced medusa stages may have dimorphic medusae, in which the female ones are generally more reduced than the males. The complex gonangia of thecate hydroids similarly may be dimorphic. Miller (1970) reports the interesting case of *Gonothyrea loveni* in which the reduced sessile medusa of the female has longer tentacles than that of the male (see Section 3.3.9). Thus, the examples of sexual dimorphism present in the cnidaria are related directly to the developing gametes or the immediately enveloping tissue. Sexual dimorphism seems to have evolved parallel with behavioral complexity; thus, in view of the poorly evolved state of behavior patterns associated with cnidarian reproduction (see Section 3.3.8), it is not surprising that sexual dimorphism is very limited in this phylum.

3.3.2 Sex Determination and Hermaphroditism

Both monoecious and dioecious forms are richly interspersed throughout the various groups of Cnidaria. Hydroids and scyphozoans tend to be dioecious, but beyond this it is difficult to generalize at all. Among the monoecious types, functional hermaphroditism is common. Protandry, a term applied to many cnidarians, frequently refers to local tissue conditions while the whole animal or colony will simultaneously have ripe male and female gametes.

The situation one must deal with in explaining the genetic basis of sex determination is a range of intermediate stages between gonochorism and hermaphroditism. Within a single genus, and sometimes species, representatives may vary from nearly complete separation of the sexes to hermaphroditic or variable sex determination, and there are frequent disputes over whether particular species are monoecious or dioecious. Even in strongly dioecious species one occasionally finds hermaphroditic individuals. No karyotypic sexual dimorphism has been reported (but, see Hargitt, 1920).

The most extensive and informative investigation into cnidarian sex determination is a genetic study on the hydroid *Hydractinia echinata* (Hauenschild, 1954). Clearly, sexual dimorphism is genetically determined in this species. One type of evidence for this was the finding that blastomeres isolated from single embryos always developed into colonies of the same sex (6 embryos were successfully tested; 2 were male and 4 female). Hauenschild investigated the genetic behavior of determinants for "intersexuality"; although *Hydractinia* is strongly dioecious, some male colonies have developing (but never fertile) oocytes in their gonophores. The tendency for such intersexuality is transmitted genetically but not as a single factor, and genetic determinants may be carried without being expressed. Hauenschild also found that the culture methods used affected the expression of this trait. Hauenschild's study supports a conclusion of Bacci (1950) that sex determination is due either to unbalanced polymorphic genetic factors or to environmental factors with varying degrees of stability of expressed characters.

The expressed sex of hydrozoans appears to be determined by the interstitial cells (see Section 3.3.7).

3.3.3 Anatomy of the Reproductive System

In most cnidarians gonads are not separate organs as found in other animals. The germ cells generally are found in interstitial positions of the body tissue which, without the germ cells or before they arrive, exhibit no reproductive specialization. However, the term gonad will be used to refer to areas where gametes are formed. In anthozoans, gametes develop within the gastric mesenteries which are scarcely modified by this event, and in hydrozoans the gametes generally are wedged between ectodermal epithelial cells. In scyphozoans, germ cells develop within the endoderm or within the mesoglea of the gastric septa. Many cnidarians have somewhat elaborated structures in the vicinity of germ cell development, but these clearly represent variations on existing somatic structure rather than the appearance of definite new organs. The body positions of gonads, and some of these structural modifications found in the three classes of cnidarians are described below. The histological structure of the gonads will be discussed in Section 3.3.4. The most systematic examination of gonadal structure remains the early work of Hertwig and Hertwig (1880) (see Broch, 1924a and Hyman, 1940 for further general references).

3.3.3.1 Hydrozoan Gonads.

Many hydrozoans have evolved elaborate polymorphic zooid types, one or several of which may take over the reproductive function and to this extent have specialized gamete-forming structures. The most interesting chapter in the study of cnidarian gonads concerns the evolution of these gamete-bearing polyps. This subject has been carefully considered by many zoologists, perhaps most enlighteningly by Goette (1907), Kühn (1910, 1913) and Kükenthal and Krumbach (1925). Hyman's (1940) treatise provides a brief recent summary.

According to the theory of alternation of generations (see Section 3.2), the medusa form was presumably the original sexually reproducing form, the polyp the asexually reproducing one. Ectodermal gametes develop from interstitial cells of the medusa and accumulate as gonads in the subumbrellar surface, commonly along the manubrium or radial canals. Medusae arise by budding from polyps; in the existing primitive hydroids this budding, as a rule, occurs below the distal tentacle whorl but above its proximal whorl when present (Rees, 1957). But many hydrozoan colonies have sessile gamete-bearing structures, and it is possible to rationalize the homology of these structures by constructing a plausible evolutionary sequence in medusa reduction, stages of which are abundantly illustrated by existing species. Reduced medusae, which nonetheless bear gametes, are called gonophores. Some species may have well-developed medusae or sessile gonophores, depending on the season and environmental conditions (see Berrill, 1953; Yoshida, 1954; Werner, 1963).

The morphogenesis of medusae involves stages illustrated in Fig. 3. Medusa buds are first evident as evaginations of the

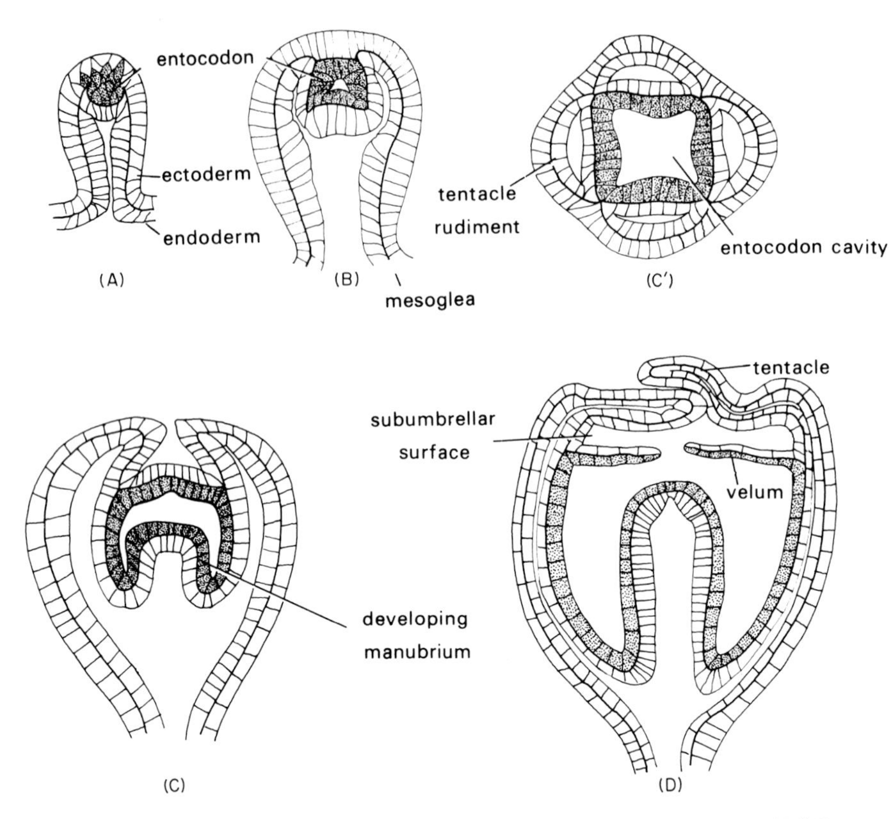

FIG. 3. Development of a medusa through budding seen in four stages. (C') Represents a horizontal cross section through Stage (C). The stippled areas correspond to the entocodon tissue and its derivatives; the presence of this tissue distinguishes medusa and polyp buds. The entocodon forms the entire subumbrellar surface of the medusa which includes the sites of the gonads.

body wall involving both ectoderm and endoderm, with a lumen continuous with that of the polyp gastrovascular cavity. Hamann (1882) made the important observation that the endoderm of such an early bud is organized into four longitudinal ridges, or taeniolae, just as is the endoderm of most young cnidarian zooids. The apical ectoderm thickens and a discrete cell mass separates from it and sinks downward into the endoderm, forming a protrusion into the gastric cavity. This mass of tissue is called the entocodon, and it will give rise to the entire subumbrellar surface of the medusa, including the gonads. Due to the ridged structure of the lateral endoderm, the entocodon is compressed into a square shape as viewed in cross section, and the four clefts of the gastric cavity lateral to this enlarging mass become longitudinal channels. These channels give rise to the radial canals and, by interconnecting laterally, to the marginal canal, of the medusa. The entocodon hollows out, the initial distal surface stretches and eventually ruptures to form the velar ring, and the inner surface of the entocodon and applied endoderm evaginate and perforate to form the manubrium and mouth of the medusa. Medusae are liberated by the pinching off of the bud base, which closes to form the exumbrellar (upper) surface of the medusa. The original apical surface of the medusa bud then remains as a ring, the velum.

Gametogenic cells may often be identified during early medusa development, particularly in species whose medusae are reduced to gonophores. Oogonia may be extremely large at this stage (Fig. 17). These gametogenic cells are usually found in the epidermis, accumulate in the distal portion of the bud, and migrate into or with the entocodon. In this manner they arrive at the site of the gonads, namely the subumbrellar ectoderm. The gonadal tissue may become folded but is otherwise unmodified. As the oocytes become larger,

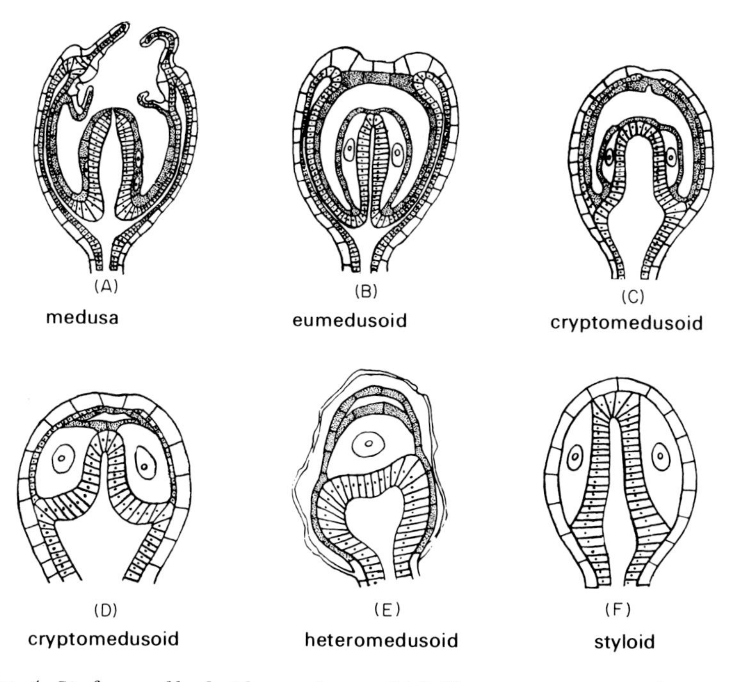

FIG. 4. Six forms of hydroid gonophores which illustrate progressive degrees of evolutionary reduction of the basic medusa structure. The derivatives of the entocodon (stippled) are generally greatly reduced. [Slightly modified from Rees (1957), after Kühn.]

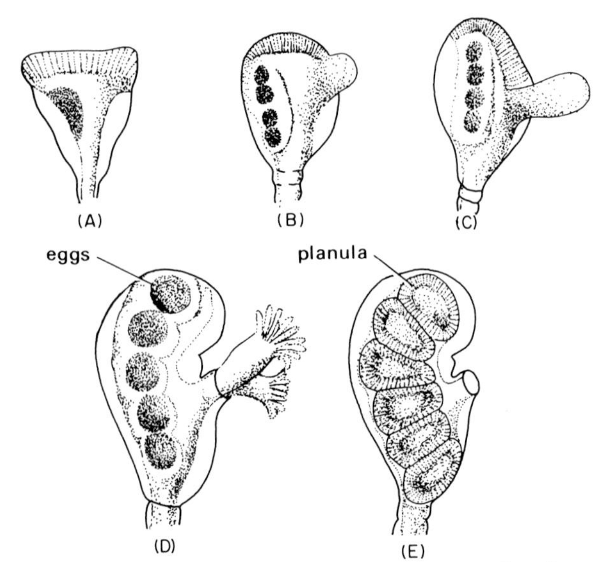

FIG. 5. Female gonangium development in *Halecium sessile*. The primordium is initially symmetrical (A) and similar to that of a nutritive polyp. Oocytes subsequently develop on one surface, while polyps arise from the opposite side and define a pore in the perisarc. Maturation and fertilization of the ova is accompanied by regression of the polyps. [After Kühn (1913).]

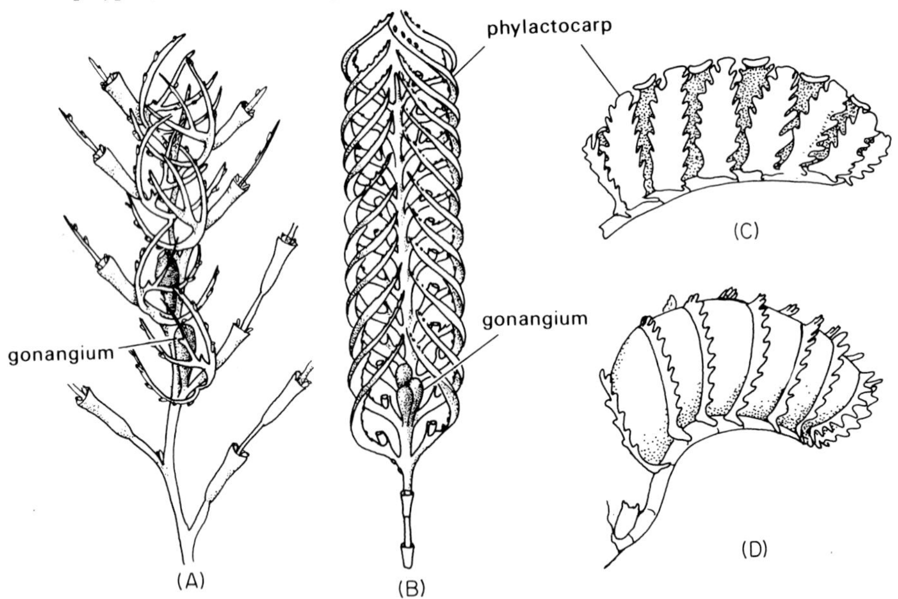

FIG. 6. Four degrees of thecate hydroid colony modification around the gonangium. (A) *Cladocarpus dolichotheca*, (B) *Thecocarpus bispinosus*, (C) and (D) *Aglaophenia filicula*. The gonangia (not visible in C and D) are surrounded with modified branches called hydrocladia or phylactocarps. In the structure (A), termed an open corbula, the phylactocarps form only a loose tangle around the gonangia while in (D) they form a complete capsule. [From Kühn (1913), after Nutting and Allman.]

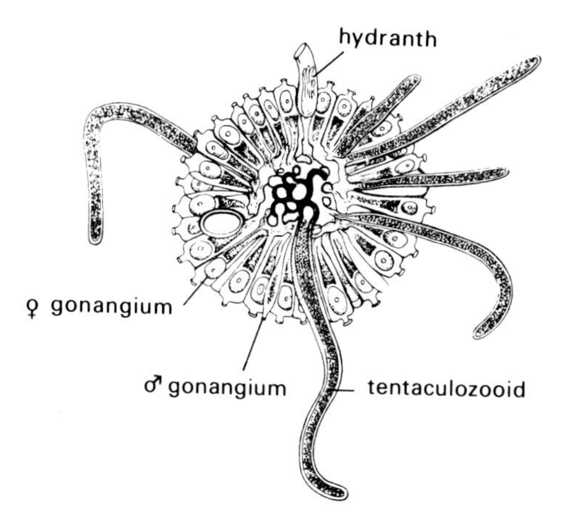

FIG. 7. Coppinia of *Lafoea dumosa* seen in cross section, showing the tight grouping of male and female gonangia, hydranths and tentaculozooids which are presumably protective. [From Kühn (1913), after Nutting.]

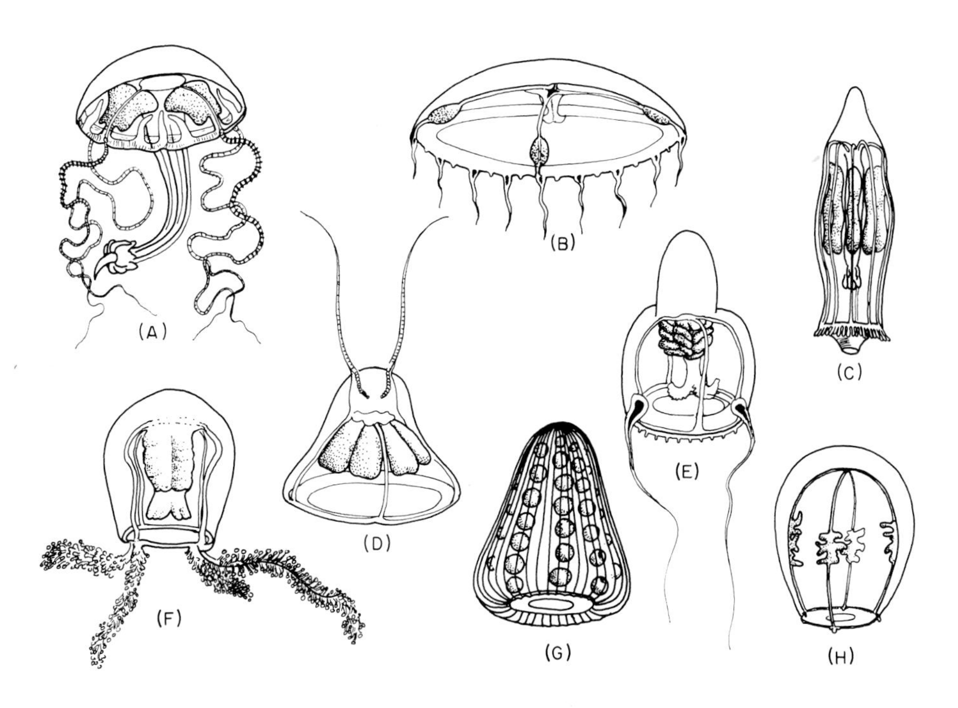

FIG. 8. Hydrozoan medusae showing the varied position and appearance of gonads (strippled) on the subumbrellar surface. (A) *Liriope lutkenii,* (B) *Clytia folleata,* (C) *Aglantha globuligera,* (D) *Solmundella mediterranea,* (E) *Stomatoca rugosa,* (F) *Euchilota paradoxica,* (G) *Eucopella bilabiata,* (H) *Agastra mira.* [(F) After Maas, 1909; others after Mayer, 1910.]

the epithelial cells are greatly distended until they form a thin covering layer with thin processes anchoring on the mesoglea (Fig. 12A,B). In well-developed medusae, gametogenesis does not occur for some time after medusa liberation.

Stages in the evolutionary reduction of medusae to sessile gonophores include (Fig. 4): well-formed medusae which produce gametes very early upon release (e.g., *Obelia*); quite well-formed medusae which, however, remain attached even though swimming pulsations may occur (e.g., *Corymorpha palma*) (Fig. 4A); buds in which radial canals and manubrium are well-developed but no velar opening appears (Fig. 4B); those in which neither radial canals nor manubrium occur but the entocodon (subumbrellar) cavity is present (Fig. 4C); and those in which no entocodon cavity forms (Fig. 4D). In these highly reduced forms it is often unclear from where the entocodon is derived. Some forms have no trace of entocodon at all, the gametes developing between the ectoderm and endoderm (Fig. 4F), or within the endoderm. Hanisch (1970) clarifies the nature of a mesogleal position of gametogenesis in a study of *Eudendrium* spermatogenesis. Spermatogonia in the endoderm approach the mesoglea (which is a well-defined, structured lamella) but do not cross or enter it. Subsequently, the endoderm secretes a second mesoglea between itself and the germ cells. This, if general, would be distinct from the germ cell movement into the mesoglea as occurs in anthozoans (see Fig. 11).

It is speculated occasionally that very reduced hydrozoan gonads which consist simply of accumulations of sex cells in the unmodified polyp body wall (e.g., *Protohydra*, Westblad, 1935; or *Hydra*, Fig. 12A) represent the extreme of this evolutionary sequence.

A concomitant evolutionary sequence involves specialization of polyps, and sometimes groups of polyps, in colonial hydrozoans. In many species, those polyps which bear medusa buds or gonophores lose their nutritive and defensive functions. This often involves the atrophy or poor development of polyp structures which are unrelated to reproduction, such as tentacles and mouth. Reproductive polyps, termed gonozooids or blastostyles, have some attributes themselves of gonads, in that they may be highly modified morphologically and specialized functionally.

A typical athecate hydroid blastostyle has moderately or greatly reduced tentacles and hypostome (Fig. 2A). Thecate blastostyles are frequently much more highly modified, and in addition are surrounded by a gonotheca (perisarc) which may be elaborate (see paragraph on brooding, below). Figure 5 illustrates how the development of such a blastostyle becomes so modified from an early stage that the final structure bears little resemblance to a zooid. Such structures are called gonangia. This tendency of colony modification reaches its highest expression in the thecate hydroids. Figure 6 illustrates four progressive stages (not necessarily an evolutionary sequence) in the modification of the colony portions surrounding gonophores. In the featherlike colonies of thecate hydroids, somatic rays arch over the reproductive elements. In *Lafoea dumosa*, colony structure becomes very dense and specialized around reproductive elements to form a structure called a coppina (Fig. 7). This structure contains both male and female gonangia, as well as nutritional and protective zooids.

Hydromedusae have ectodermal gonads on the subumbrellar surface, frequently with the ectoderm folded or thickened in this area, occasionally to the point of being pendulous. They are generally either on the manubrium wall or along the radial canals (Fig. 8).

3.3.3.2 Scyphozoan Gonads

Jellyfish have simple or elaborate septal structures which function as gonads. The four gastric septa may be infolded by genital pits, depressions arising from the subumbrellar surface and indenting the septa. The gamete forming cells arise and mature within the endoderm or mesoglea surrounding the pits (Fig. 12B). In some cases the bottoms of the genital pits rupture during the breeding season, releasing gametes directly to the exterior. Generally gametes are released into the gastric pouches (gastric cavity) and may be either released through the mouth, before or after fertilization, or brooded somewhere within the complex gastric cavity.

3.3.3.3 Anthozoan gonads

Sea anemone and coral gametes develop directly within the gastric mesenteries (septa). The gonads consist merely of these cells accumulated either at the base of the endodermal epithelium or within the mesoglea. Gametogenesis only occurs in the band between retractor muscles and septal filament. Sulcal septa are sterile. Accumulation and growth of the gametogenic cells causes the local epithelia to rupture, liberating the gametes directly into the gastric cavity. The gametes or developed larvae are then released directly through the mouth, although there are reports of gamete release through pores in the tentacle tips or base (see Uchida and Yamada, 1968).

3.3.3.4 Other reproductive structures

Reproductive structures other than gonads are comprised exclusively of specializations developed by smaller taxonomic groups of cnidarians. Brooding is the major function around which such specializations have arisen. The simplest form of brooding involves retention of eggs and larvae within the gastric cavity; this involves little modification in scyphozoans and anthozoans since the gametes are usually shed directly into the gastric cavity. In many species, brooding consists apparently of no more than harboring free-floating larvae. In *Actinia equina*, brooding is very common but never appears to involve cleavage and blastula stages. Chia and Rostron (1970) hypothesize that early development is external, and that planulae are then taken up by these sea anemones for secondary brooding until after metamorphosis. Thus, males, females, and sexually inactive individuals all contain young larvae. External brooding also occurs following a free-swimming stage in some jellyfish, with planulae sometimes brooding on a parent of a different species (Berrill, 1949).

In *Actinia equina* and some other anthozoans there do not appear to be specialized areas of the gastric cavity for brooding; larvae are found in the main chambers, especially near the animal's base. Other species have special brood pouches. *Epiactis marsupalum*, for example, has indentations on the lower part of its external column which receive brooded larvae from the gastric cavity; here the young anemones grow, one per pouch (Fig. 9B).

Brooding occurs in scattered hydrozoan forms. In *Tubularia* and a number of related forms, early development occurs within the sessile gonophore and the young are freed only at the advanced actinula larval stage. In many thecate hydroids, whose gonophores are often very reduced and are covered with a thick perisarc, brood chambers are formed from distal portions of the gonophore's perisarc. In *Diphasia* (Fig. 9E) this may be larger than the gonadal portion of the gonophore. Other forms have brood chambers constructed of the distal gonophore tissue itself, homologous to a reduced medusa and which may have tentacles (*Gonothyrea*, Fig. 9C). At least some of these hydroids with internal fertilization have evolved chemotactic mechanisms for bringing spermatozoa into the gonophores (see Section 3.3.9). Some cladonemad medusae (e.g., *Eleutheria*, Fig. 9A) have a brood chamber consisting of an extension of the subumbrellar cavity which extends above the gastric cavity. One of the most bizarre cases of brooding among the Hydrozoa is displayed by the solitary *Myriothella*. The lower part of its body column is covered with

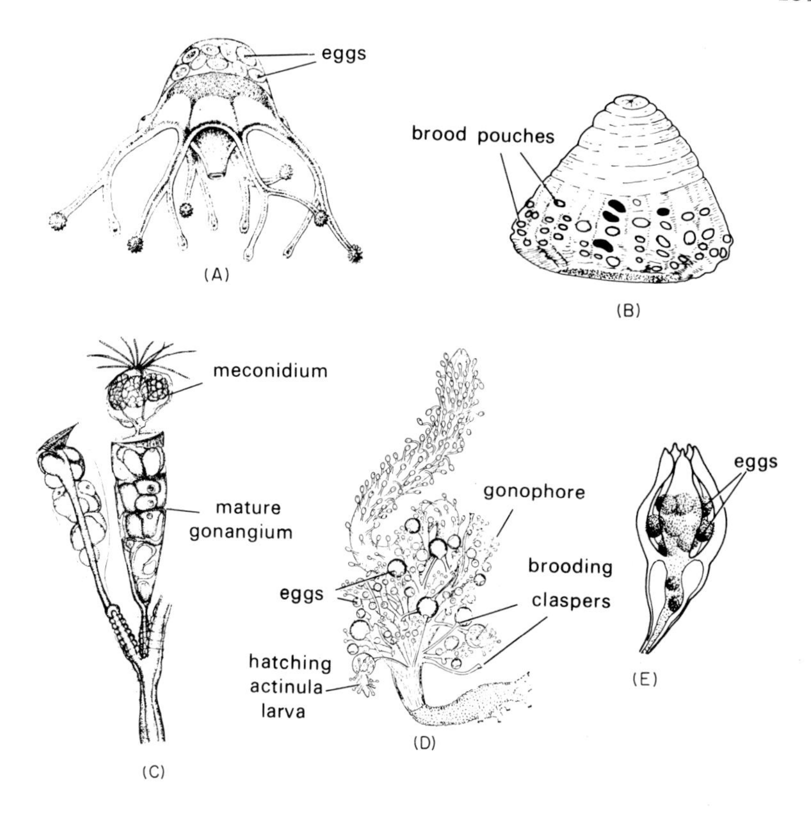

FIG. 9. Brooding in cnidaria. (A) Hydrozoan medusa *Eleutheria dichotoma (Clavatella prolifera)* with eggs in an aboral chamber (after Hincks, 1863). (B) Anthozoan *Epiactis prolifera* in a contracted state showing longitudinal rows of pouches along the lower surface of the column, each of which contains one young anemone. When the anemones are extended the tentacles of the brooded individuals protrude. [From Carlgren (1893).] (C) Hydroid *Gonothyrea loveni* with two female gonangia shown. The one on the left is immature; the one on the right bears a meconidium, a medusa modified as a brood chamber, containing embryos. [From Miller (1970).] (D) Hydroid *Myriothela phrygia* whose proximal tentacles are modified claspers which take eggs from the gonophores and brood them until they hatch into actinula larvae. [After Allman (1871-1872).] (E) Hydroid *Diphasia* brood chamber at distal end of gonophore, made of leaves of hydrotheca. [From "The Invertebrates," Vol. I, by L. Hyman. Copyright 1940 by The McGraw-Hill Book Company, New York. Used with permission of McGraw-Hill Book Company.

clublike claspers, the tips of which grab the eggs from the gonophores and brood them throughout development until the actinula larval stage (Fig. 9D).

3.3.4 Origin of Germ Cells

Germ cells and their accessory cells arise from interstitial cells. In the scyphozoans and anthozoans they are endodermal in origin. In the Hydrozoa they appear to be chiefly ectodermal in origin, but in at least a few cases they are thought to arise in the endoderm (Kühn, 1913), and in *Clava multicornis*, Goette (1907) asserts that spermatogonia are derived ectodermally and oogonia endodermally. The site of origin of these cells has been hotly contested in the literature, largely because they seem to arise from the same interstitial cells as those which form nerves, nematocytes and other specialized cells. (This itself is a most complex issue; Weismann, 1883, who pioneered much work on cnidarian reproduction championed the idea that the germ cell line is always distinct from the cells forming other, somatic cells. The current opposite view, by no means proven, has been well stated by Brien, 1953.)

These germinal interstitial cells seem to be of wide distribution within the body, and it is generally impossible to assess the origin of a cell type which cannot be distinguished from other interstitial cells, at least by classic histological techniques employed.

Gametogenic cells are first recognizable as gonial cells by their slightly larger size, more lightly staining nucleus, prominent nucleolus and position within the animal. At this time they still have all the ultrastructural appearances of typical undifferentiated interstitial cells: abundant free ribosomes, only scattered mitochondria and rough endoplasmic reticulum, and no Golgi apparatus.

In medusa or gonophore budding it is generally believed that special interstitial cells, sometimes segregated for some time in the colony, migrate into the bud and that these subsequently give rise to gametes. A typical pathway involves cell migration from the base to the tip of the gonophore, migration into the entocodon, and from there into future manubrial or gonadal tissue (Fig. 10). Actually this would *a priori* seem to require no active cell migration since interstitial cells in the vicinity of a young bud would be carried up to the tip by tissue growth and displacements as the bud enlarges (Fig. 10A-C). Then at the tip, the cells could be simply part of the entocodon, because this is generally ectodermal in origin (Fig. 10D). Once in the entocodon the interstitial cells are already in the gonadal precursor tissue, since the manubrium ectoderm arises from entocodon. Most investigators have not considered these tissue movements, and in their descriptions, it is very difficult to separate active from passive cell movement.

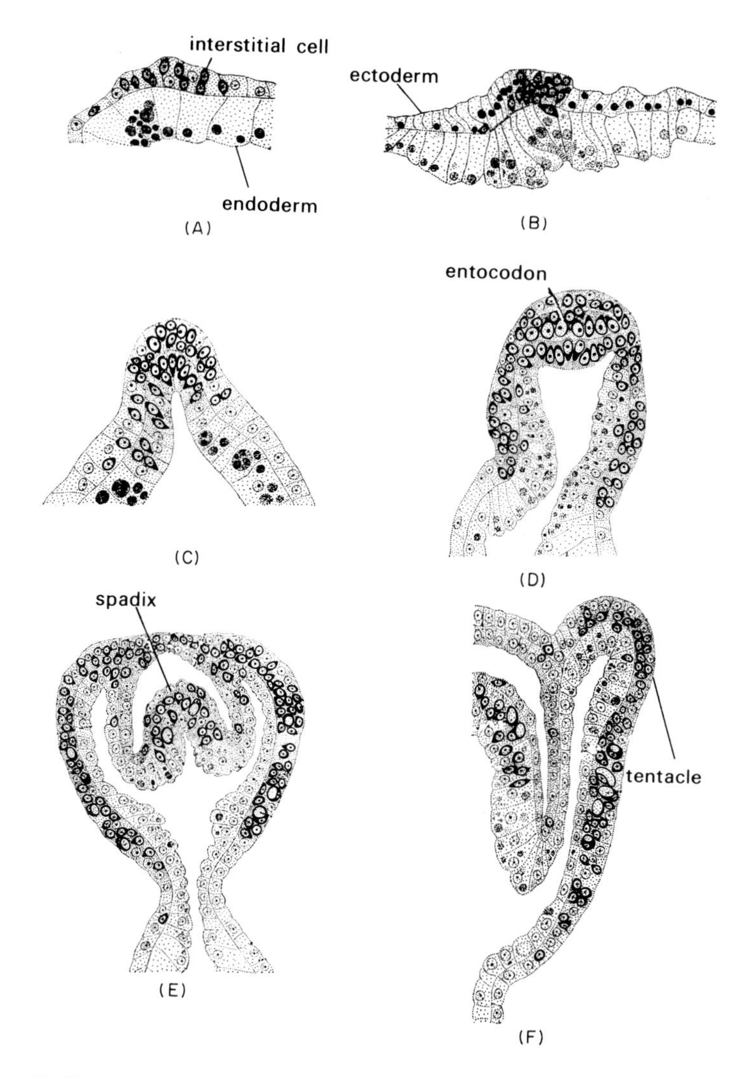

FIG. 10. Positions of interstitial cells during budding in the hydroid *Eleutheria dichotoma*. Interstitial cells (black) are initially present at the site of budding (A,B) and a few early migrate into the endoderm (B). Many are contained in the entocodon and spadix (C,D) and are in position for the future gonads on the manubrium. Some of these interstitial cells form nematocytes (E,F, black cells with open white spaces) and others give rise to gametes. [From Weiler-Stolt (1960).]

There are, however, two commonly described patterns of interstitial cell movements, presumed to be intermediates in primordial

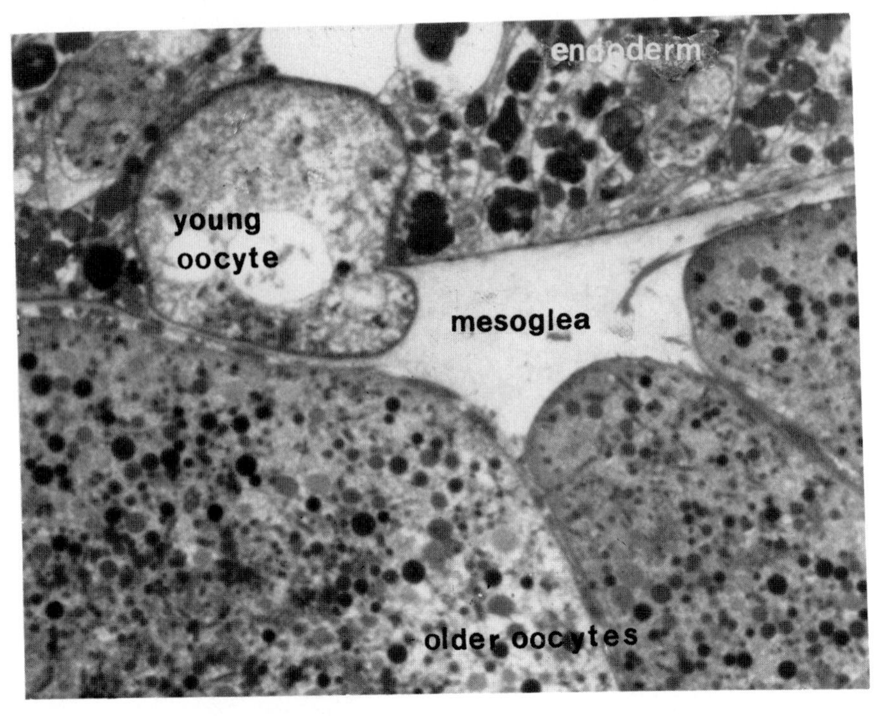

FIG. 11. Young oocyte (center) of the sea anemone *Peachia quinquecapitata* migrating into the septal mesoglea (Me) among more mature oocytes. (1925X) [From Spaulding (1972).]

germ cell migration, which definitely require active cell migration. One of these involves movement of cells across the mesoglea. This is frequently cited in histological investigations (see Weiler-Stolt, 1960). The other involves indirect migration pathways to the gonads, the one most frequently reported being interstitial cell movement into the endoderm at the base of the gonophore or medusa bud, distal migration within the endoderm, and then transfer into the entocodon. Again, from histological evidence it is very difficult to reconstruct such processes with any certainty. One of the most analytic attempts to do so was made by March (1915) in *Corymorpha palma*. She found that the number of interstitial cells within the gonophore bud remains practically unchanged, about 20 per bud, throughout the early development of the gonophore. During this time, however, the center of distribution of these cells changes markedly, from proximal ectoderm, to proximal, then distal endoderm, and finally to the entocodon. Thus it seems that interstitial cells which are going to give rise to gametes have the capacity to migrate actively, and that they show considerable variation in the pathways of migration from the polyp body wall to the gonads. For entry into the voluminous literature of the histological origin of sex cells, the reader is referred to Goette (1907), Kühn (1910, 1913), and Weiler-Stolt (1960).

The embryonic origin of germ cells in scyphozoans and anthozoans has not been determined. Anthozoan germ cells are recognizable as large interstitial cells in mesentery endoderm prior to gametogenesis, and from there they move into the mesoglea (Fig. 11).

The limited accessory cells in cnidarian gonads consist entirely of cells from the gametogenic lineage and the epithelial cells comprising the gonads.

3.3.5 Gametogenesis

Cnidarian gametogenesis is similar to the basic pattern in most animals, the major unusual feature being the lack of specialized accessory cells in many cases, and the process of oocyte fusion during oogenesis in some species. Almost certainly the complexities of these processes, as with gonadal structure, are severely underestimated at the present time. Rather little electron microscopic analysis is available, and this is particularly needed in the area of intimate cell relations during gametogenesis.

3.3.5.1 Spermatogenesis

Spermatogenesis may occur synchronously within one gonad or one gonad region, or it may occur in continuing sequence with a typical spatial progression of cells during maturation. The anthozoan *Actinia equina* provides an example of continuous spermatozoan formation with spermatocytes at the exterior of the cords of developing sex cells and mature spermatozoa towards the interior. The lumen in this case is eccentric, and sperm release involves periodic rupturing of the tubule wall and covering endoderm (Fig. 12C). In hydrozoans, polarization tends to be baso-apical within the testis ectoderm, that is with spermatogonia near the mesolamella and spermatids and spermatozoa at the epithelial surface where they are released. Despite this orderly progression, many athecate hydroids have short-lived gonads and the process of gametogenesis should therefore be considered as continuous only until the large number of gonial cells are used up. These gonophores are replaced by others which usually form systematically above them and are moved proximally by column tissue displacements (e.g., in the genera *Hydra* and *Hydractinia*). In the very reduced heteromedusoid and styloid

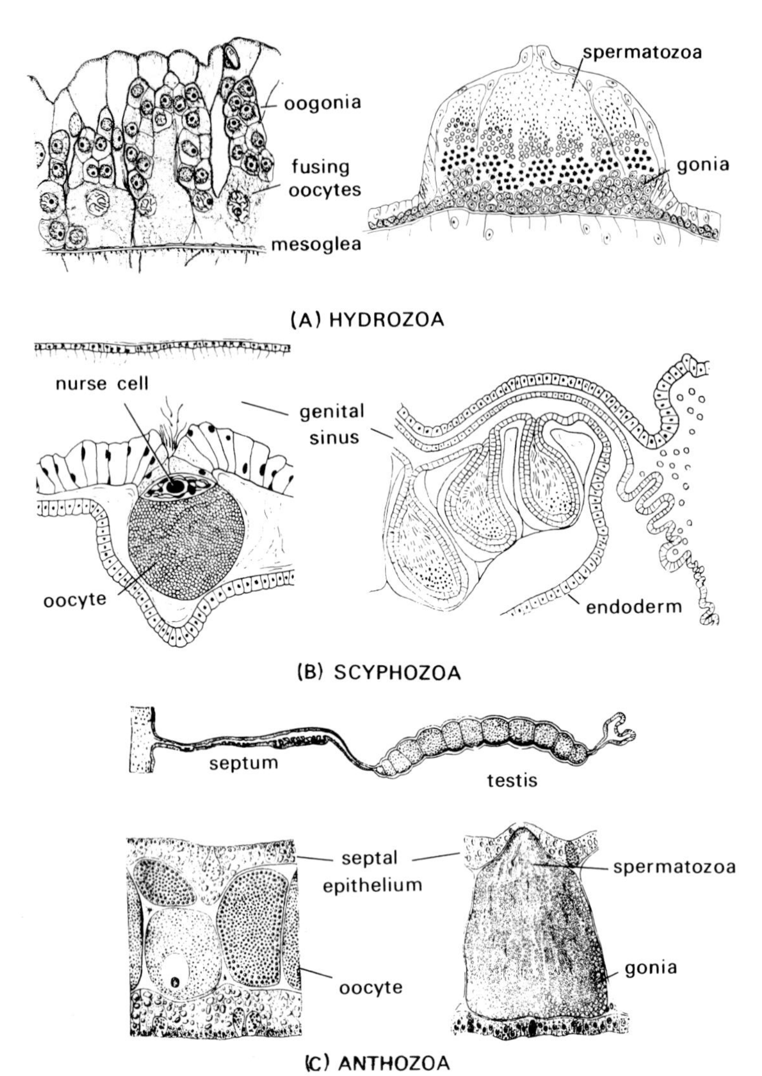

FIG. 12. Histological structure of cnidarian gonads. On left are ovaries, on right testes. (A) Hydrozoa, represented by *Hydra attenuata* [from Brien and Reniers-Decoen (1951)]; Gonads consist of gametogenic cells clustered between ectodermal epithelial cells.Gametes are shed through the epithelial surface. (B) Scyphozoa, represented by *Cyanea capillata* ovary and *Aurelia aurita* testis [from Widersten (1965)]; gonial cells in the mesoglea are lifted in folds of the endoderm into which may project the genital sinus. Gametes are released into the sinus. (C) Anthozoa, represented by *Sagartia parasitica* septum (top) and testis, and *Cerianthus membranaceus* ovary [from Hertwig and Hertwig (1879)]. The gonads consist of gametogenic cells within the mesoglea of the outer portion of the septa (but not extending into the septal filament), and are thus lined on both sides by endoderm. Eggs and spermatozoa are released into the gastric cavity.

gonophores of thecate hydrozoans where gametes develop in the endoderm or mesoglea and are released into the gastrovascular lumen, polarization occurs progressively from the mesolamellar surface to the luminal surface of the endoderm (e.g., Lunger, 1971).

Cnidarian spermatogenesis, as observed in a great many light microscopic studies (see Hegner, 1914; Franzén, 1956), is typical of animal spermatogenesis in general. Spermatogonia are cells of modest size with very large nuclei. Chromatin is conspicuously condensed along the envelope surface giving the nucleus a clear appearance. The spermatogonium has a prominent nucleolus. Meiotic divisions occur rapidly in the usual manner. The resulting nuclei are small and dense. Subsequent cytoplasmic maturation includes posterior migration of a pair of centrioles from which a flagellum arises, and accumulation of several mitochondria in the pericentriolar region. The cytoplasm is greatly reduced in extent during maturation.

Electron microscopic analysis of cnidarian spermatogenesis is slowly appearing (Hanisch, 1970; Roosen-Runge and Szollosi, 1965; Lunger, 1971; Stagni and Lucchi, 1970; Weissman *et al.*, 1969; Summers, 1970). At this level of observation the process also appears fairly typical (Fig. 13). Meiosis is rapid. Synaptinemal complexes occur (Stagni and Lucchi, 1970). The spermatid has well-developed endoplasmic reticulum draped over the nucleus. The nucleus first constricts in width and finally in length to a blunt conical shape. Weissman *et al.* (1969) trace the early polarization of nucleus and cytoplasm. A small number of mitochondria become spherical and attach very intimately to the posterior margin of the nuclear envelope, surrounding the centrioles. The most surprising finding is the absence of Golgi activity in acrosome formation. The Golgi elements are very active, but there is no clear evidence of acrosome formation (Stagni and Lucchi, 1970; Lunger, 1971; Hanisch, 1970; Summers, 1970). Lunger does report small vesicles at the spermatid tip, perhaps derived from the Golgi bodies, but which bear no particular resemblance to acrosomal bodies. *Pennaria* spermatozoa (Summers, 1970) have a rosette of 30-40 Golgi-derived vesicles which surround an anterior protrusion on the nucleus. These are not released as the spermatozoan enters the egg jellycoat; these vesicles may represent a primitive form of what evolved into the acrosome.

Another feature of spermatogenesis which has emerged from the electron microscopical studies is that the centriolar satellite bodies and rootlets are rather elaborate. A detailed account of the appearances and attachments of centriolar satellites during spermatogenesis has been provided by Szollosi (1964). Spermatids develop striated rootlets by which the distal centriole (which serves as the flagellar

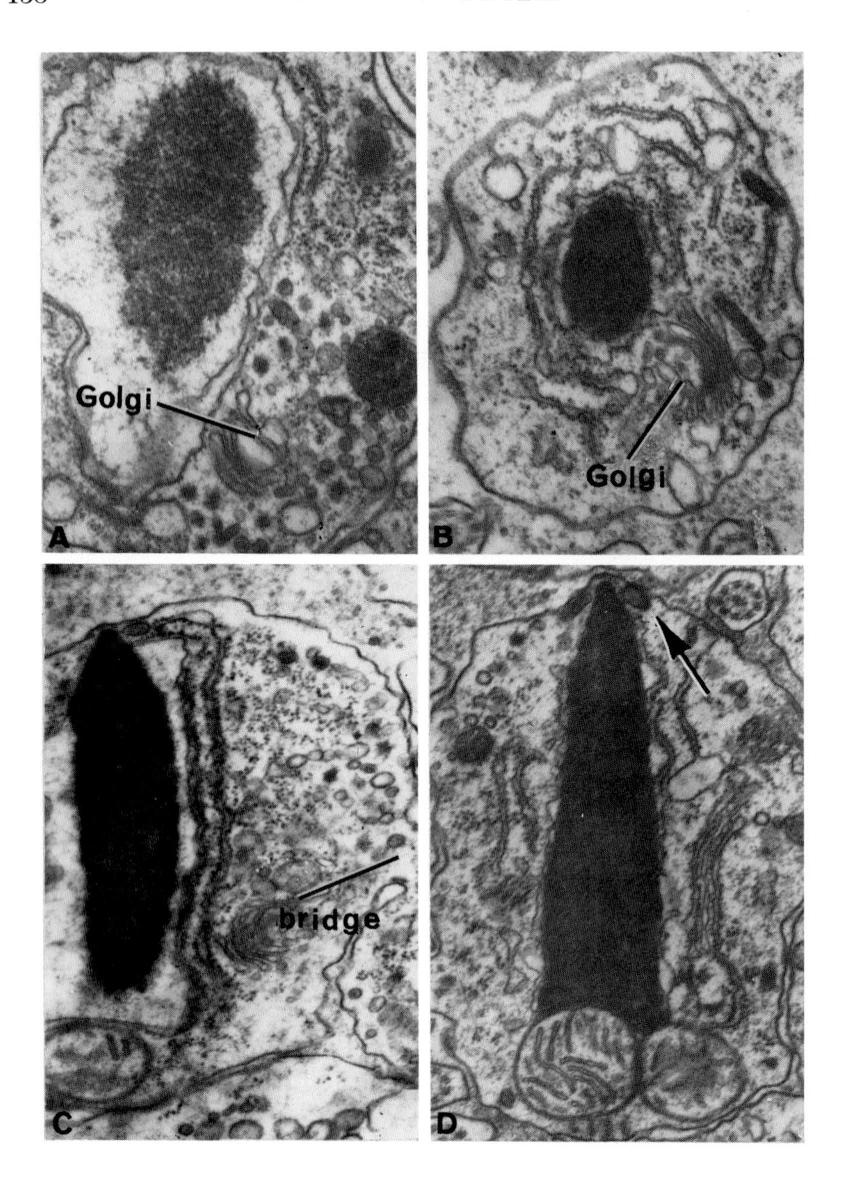

FIG. 13. Spermatogenesis in the hydroid *Campanularia flexuosa*. There is extensive elaboration of the Golgi apparatus during the period of nuclear compaction and elongation. The nearly mature spermatid (D, arrow) has apical vesicles which may represent primitive acrosomal bodies. Note the presence of intercellular bridges which are present between clusters of developing spermatocytes and spermatids (C, "bridge"). [From Lunger (1971).]

basal body) is attached to the most distal region of the midpiece plasma membrane (Fig. 14). Each of the nine centriolar triplets has a rootlet which, midway towards the plasma membrane, ramifies into a series of smaller rootlets. These are thought perhaps to aid in providing mechanical attachment between the flagellum and midpiece, and have also been described in other hydroid spermatozoa (Summers, 1970; Lunger, 1971) and in scyphozoan spermatozoa (Afzelius and Franzén, 1971). Kessel (1968) also reported elaborate centriolar rootlets in oocytes, and Hanisch (1970) has given a detailed description of other centriolar rootlet elaborations and attachments in spermatids and spermatozoa of *Eudendrium;* he also provides a very detailed account of the ultrastructure of spermatogenesis.

Hydrozoan spermatocytes tend to be clustered into groups or "nests." This is due partly to their being wedged between epithelial cells. In the hydroids *Campanularia* (Lunger, 1971) and *Hydra* (Burnett *et al.*, 1966; Schincariol *et al.*, 1967; Stagni and Lucchi, 1970) spermatocytes and spermatids within a nest are joined by intercellular bridges and therefore represent a syncytium. This may partly explain synchrony in the development of clustered cells. Hanisch (1970) states that bridges are absent in *Eudendrium.* It is not

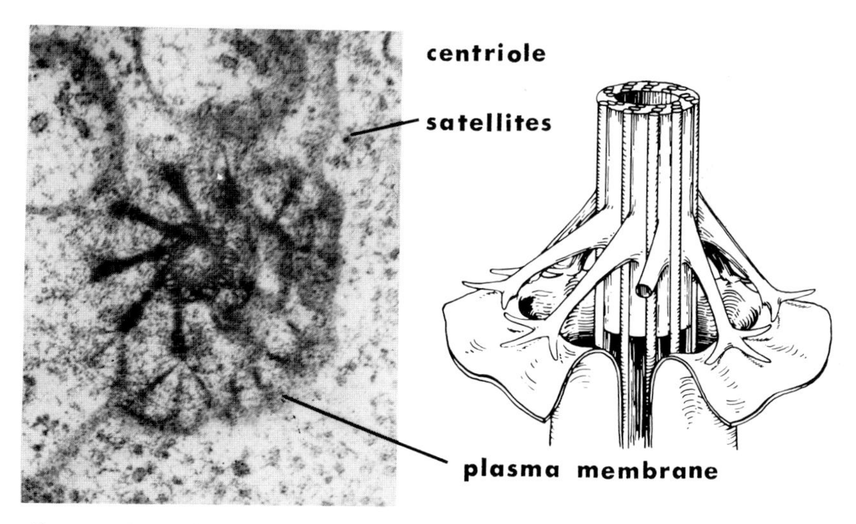

FIG. 14. Centriole and satellite fibers in hydrozoan spermatozoa. Left: section through the distal end of a centriole of *Campanularia flexuosa* illustrating the nine rootlets arising between the triplet microtubules. These rootlets branch once or twice before inserting onto the membrane. (45,000X) [From Lunger (1971).] Right: Reconstruction of the centriole and satellite structures from *Philalidium*, illustrating their relations to the plasma membrane. [From Szollosi (1964).]

known how spermatocytes are thus joined; however, intercellular bridges may be general in many animal phyla (Fawcett, 1961).

Nothing is known about the role played by enveloping epithelial cells as accessory cells. Electron microscopists comment on the intimate relations between the epithelial cells and the spermatids, and the structural cells presumably aid at least in removal of excess spermatid cytoplasm. One of the most interesting questions in cnidarian spermatogenesis is to what extent the evolution of accessory cells is foreshadowed here.

Mature cnidarian spermatozoa are small and conservative in structure (Fig. 15) (see Franzén, 1966, for review; also Afzelius and Franzén, 1971). Little cytoplasm is included other than a small number (2-5) of round mitochondria and the centrioles which make up the midbody. Small vesicles rather than a well-developed acrosome are present in the apical cap. All spermatozoa actively swim by means of a long (generally 30-90 μm) flagellum which has the usual ultrastructural architecture.

3.3.5.2 Oogenesis

Oogenesis is simplest in the Anthozoa, where the oocytes move into the septal mesoglea at an early stage and appear to mature quite

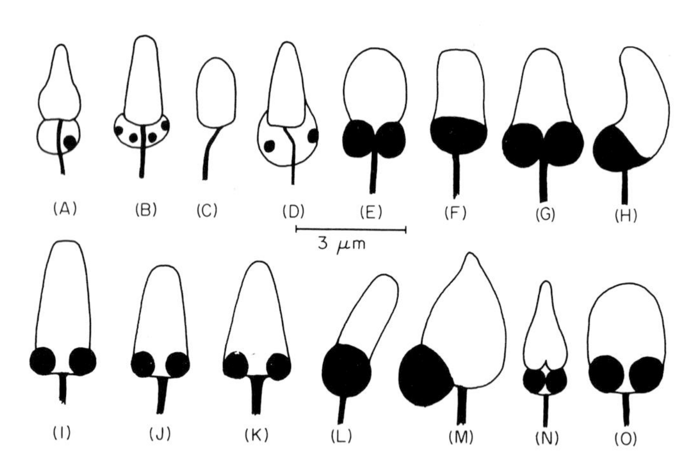

FIG. 15. Cnidarian spermatozoa. Only heads are depicted; black indicates mitochondria. (A) *Caryophyllia smithii*, (B) *Bolocera tuediae*, (C)*Funiculina quadrangularis*, (D) *Sagartia troglodytes* (A-D from Franzén, 1966), (E, F) *Polyorchis karafutoensis* (Nagao, 1970), (G) *Phialidium* sp. (Roosen Runge and Szollosi, 1965), (H) *Tubularia radiata* (Nagao, 1965), (I) *Hydra fusca* (Schincariol *et al.*, 1967), (J) *Hydra attenuata* (Stagni and Lucchi, 1970), (K) *Pennaria tiarella* (Summers, 1970), (L) *Muricea fruticosa* (Grigg, 1970), (M) *Tealia crassicornis* (Retzius, 1904), (N) *Limnocnida tanganyica* (Bouillon, 1957), (O) *Corymorpha palma* (original).

independently (Fig. 11, 12C). As the oocyte grows it becomes filled with components normally found in developing eggs: abundant yolk granules, ribosomes, annulate lamellae, and mitochondria.

In hydrozoans, oogenesis is associated with more complicated intercellular interactions than is spermatogenesis, perhaps representing the beginning exploratory evolution of follicle or nurse cells. Most hydrozoan gonads begin with more oocytes than will eventually ripen, and frequently only a single ovum develops in each gonad. The other oocytes are expended in the process of feeding the successful oocyte. This process is highly variable in different species (see Kühn, 1913). In simpler cases there is little or no apparent growth of any except the definitive oocytes; the others degenerate, apparently at the expense of the growing cells. In other species there may be extensive development of all oocytes but eventually most of them degenerate. In a number of hydrozoans (e.g., *Millepora,* Mangan, 1909; *Hydra,* Brien and Reniers-Decoen, 1950; and *Tubularia,* Nagao, 1965) the oocytes actually fuse successively to form a single symplastic large oocyte. In these cases there is one nucleus which survives and enlarges while the others become highly pycnotic and apparently inactive. The adjunct nuclear masses, called "pseudocells," do remain, however, and have received considerable attention because of their conspicuousness in histological preparations. They remain throughout embryogenesis, and their fate and possible functions remain obscure.

In some hydrozoan species, oocytes become nourished at the expense of ectodermal or endodermal epithelial cells, again apparently by cell fusion (e.g., in *Eudendrium,* Congdon, 1906). A striking demonstration of epithelial cells nourishing oocytes is provided by the transfer of symbiotic intracellular algae, zoochlorellae and zooxanthellae. These symbionts normally live in the cnidarian epithelial cells and are entirely absent from interstitial cells, from which oocytes arise. In *Hydra* (Brien and Reniers-Decoen, 1950; Valkanov, 1967) the algae are transferred during growth of the oocytes; this involves release of algae by the endodermal epithelial cells, movement of algae across the mesoglea and layer of ectodermal epithelial cell muscular processes, and entrance into the fusing oocytic syncytium. In *Millepora* the oocytes are between the ectoderm and endoderm; zooxanthellae and other food materials are transferred from the endoderm after the mesoglea loses its distinctiveness. However, light microscope studies fail to provide necessary detail on cell boundaries, and electron microscope observations on this process so far have been rudimentary. In the thecate hydroid *Aglaophenia helleri* (Müller-Calé and Krüger, 1913) algal transfer occurs before oocyte growth, but after germ cell arrival in the gonophores.

Widersten (1965) provides a recent analysis of scyphozoan oogenesis, including a description of cells spcialized enough to be termed nurse cells. These are initially endodermal epithelial cells adjacent to the oogonia. When a germ cell migrates into the mesoglea, a few epithelial cells retain a tight and intimate contact with it and are pulled slightly below the cell layer, yet remain epithelial cells (Fig. 12B). These cells are described as nourishing the growing oocyte.

Electron microscope analyses of cnidarian oogenesis are just beginning to appear (Kessel, 1968; see footnote of Glätzer, 1970; Spaulding, 1971). Kessels's detailed study of vitellogenesis in trachyline medusa oogenesis indicates great elaboration of Golgi bodies and endoplasmic reticulum in association with yolk synthesis which is apparently entirely endogenous. However, oocytes were found to produce highly infolded channels from the cell surface which associated with the yolk forming machinery, and a well-developed vitelline membrane was found in older oocytes; these observations may reflect an incipient evolutionary stage of oocyte nourishment by, and interaction with, somatic cells.

Mature cnidarian eggs may be covered with microvilli (Szollosi, 1970) or spines (Gemmill, 1920; Spaulding, 1972), both of which contain cytoplasmic microfilaments which Szollosi has related to cleavage mechanics. A micropyle may be present in eggs which have elaborate coats. Most eggs, however, are smooth and appear naked. A number of investigators describe eggs as being covered by a "delicate membrane" (e.g., McMurrich, 1890; Nagao, 1965). However, these species seem to have eggs coated with a transparent "jelly layer," 20-50 μm thick (e.g., Ballard, 1942). The egg of the leptomedusa *Bougainvillia multitentaculata* is invested with thousands of nematocytes, forming a solid layer over it (Fig. 16) (Szollosi, 1969). These presumably arise from the maternal tissue, and remain in place throughout cleavage. However, by the time of planula formation they have sunken down between adjacent ectodermal cells, in the histological position normal for nematocytes.

3.3.6 Gametogenic Cycles

Cnidarian gametogenesis frequently exhibits circadian and seasonal periodicities. Apparent tidal or lunar reproductive cycles were found in corals by Marshall and Stephenson (1933).Variations within the day are closely coupled with spawning and are discussed in Section 3.3.7. Seasonal reproductive cycles are commonly exhibited by almost all cnidarians so far studied, only a few breeding continuously or randomly throughout the year.

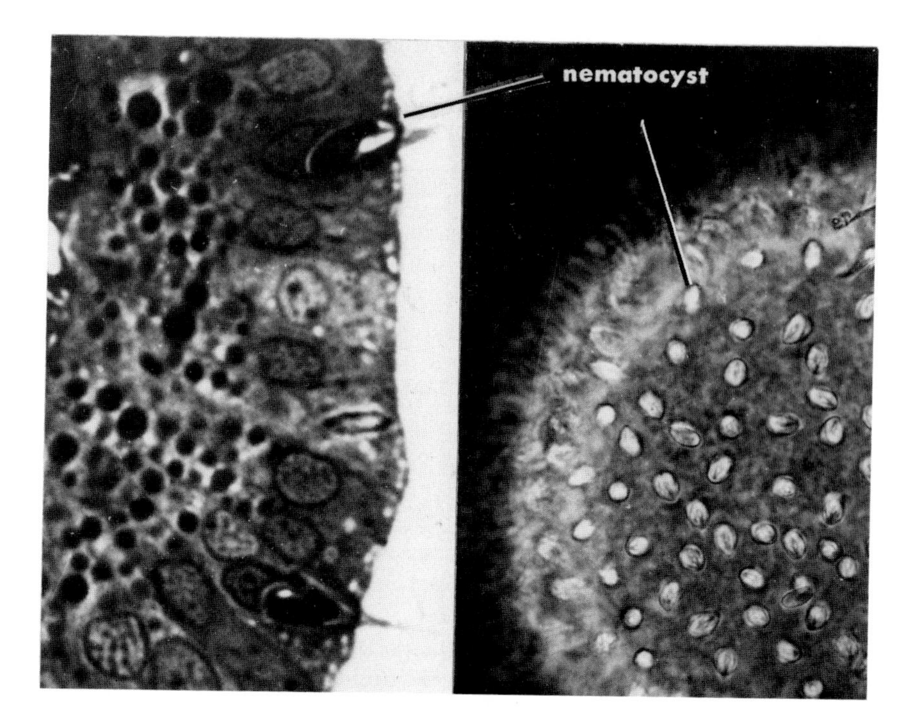

FIG. 16. "Armed egg" of the Hydromedusa, *Bougainvillia multitentaculata.* The egg (photograph at right, 650X) has numerous nematocytes containing nematocysts on its surface. During cleavage and blastulation the nematocytes intercalate between the embryo cells in positions typical for nematocytes. Photograph at left is section through a planula showing nematocytes which have been contributed maternally (2050X). [From Szollosi (1969).]

Table I (p. 172) summarizes breeding seasons of a few species. In the short-lived hydromedusan forms, breeding periods are nearly the same as seasons of animal abundance. In the long-lived polyp forms and some scyphomedusae, seasonality must be determined by visual inspection of the gonads (in the case of opaque anthozoans with internal gonads, this generally requires a dissection or histological preparation). Some of the most extentive data on gonadal cycles in cnidarians comes from studies by Marshall and Stephenson (1933) on corals, Ford (1964) on sea anemones, and Grigg (1970) on gorgonians.

3.3.7 Factors Influencing Gametogenesis

Gametogenetic control has been studied experimentally in the Class Hydrozoa; in other classes one is limited, in probing factors which influence gametogenesis, to drawing correlations between seasonal environmental variations and reproductive periods.

The factors conducive to the formation of gonozooids in several colonial hydroids have been found to include nutrition, stagnation, colony age, colony density, and temperature. Most experiments in this area involve growing colonies, from small polyp or stolon explants, under a variety of laboratory conditions. One finds generally that cultures which are poorly fed will develop without gonozooids or gonophores. This has been interpreted as a result of competition for nutrients among the several growing regions of a colony (see Crowell, 1957, whose experiments have provided models for more recent investigations). The development of sex polyps seems to involve processes which compete poorly with processes such as stolon tip growth. On the other hand, it is a common observation that once a colony enters into a sexual phase reproductive functions may continue at the expense of the rest of the colony, which may become nutritionally exhausted. Even under conditions of adequate feeding, the onset of reproduction may be accompanied by an apparent decrease in colony growth rate (Braverman, 1963; Roosen-Runge, 1970).

Another common factor found to favor gonozooid development is colony age, although it is not clear which attribute of age is most important. Mackie (1966) found that *Tubularia* hydranths form gonophores early in development and gametogenesis begins when the hydranth is about 2 weeks old. Gonophore development is thus linked to age (see also Berrill, 1952b, Nagao, 1965). Gonophores also arise very early during *Tubularia* regeneration (Campbell and Campbell, 1968).

Initial polyps of polymorphic colonies are always nutritive. Hauenschild (1954) concluded that sexuality in *Hydractinia* arose only after the colony completely covered available substratum, and when no substratum was provided, the free-floating colonies underwent precocious sexuality. In the sister genus, *Podocoryne*, Braverman (1963) found that sexuality arose in colony regions according to local age, even though the substratum was not filled. The thickness of the stolon mat, which increased with age, appeared to be an important factor. Explants containing a matted stolon network always develop reproductive polyps; however, they arise only from the mat, never from the linear stolons growing out from the explant. In many hydroids the stolon system never becomes densely anastomosing; gonozooids are formed only after several weeks of growth, and then only in the more central portions of the colony (e.g., Roosen-Runge, 1970).

Another plausible determinant is local colony density, and some work indicates that stagnation (as would perhaps be promoted by high colony density) can quite directly induce sexuality. This work began with Loomis' reproducible induction of the sexual state in hydra by high pCO_2 levels in a defined culture solution (see Loomis, 1959). Similar studies were reported with other hydrozoans (*Hydractinia*, Müller, 1969a; *Podocoryne*, Braverman, 1962). However, the role of CO_2 in inducing sexuality has remained highly controversial, and colony crowding may favor the development of reproductive structures even when the water is circulated and exchanged (West and Renshaw, 1970).

Each species seems to have an optimal temperature for development of reproductive structures (Nishihira, 1968c; Uchida and Yamada, 1968). Berrill (1952a, 1953) has described the way in which temperature differentially affects the somatic and reproductive development in hydroids. Temperature has at least two dissociable effects on sexual reproduction in *Sertularella miurensis* (Nishihira, 1968c). Temperatures above about 10°C are required for planula formation (and, therefore, presumably gametogenesis), which results in extensive planula release during August. Colonies cannot produce gonophores when the temperature is above about 18°C. Thus, after the August reproduction there is a month or two when the colonies are quiescent. As the water temperature declines in the autumn the newly formed colonies begin to form gonophores, and there is a release of planulae in November. After that the water is too cold to support gametogenesis, although colonies have gonophores throughout the winter. Thus high temperature stimulation of gametogenesis and inhibition of gonophore formation results in two reproductive seasons, August and November.

The subject of a hormonal type of control of cnidarian sexuality has interested a number of workers because if such a system could be demonstrated, it should be present in an extremely simple form. Charniaux-Cotton (1965) has joined a few others in considering such a situation plausible. One type of interesting experimentation leading to this conclusion involves induction of sexuality by tissue grafts taken from sexual animals. Brien (1964) found that a hydra could be induced to enter into a sexual state by a small graft from a sexually active animal. Müller (1964) also showed this in the colonial *Hydractinia;* in this case a piece of gonozooid was grafted into a gastrozooid and the host became a gonozooid. Inductive potency was highest in the distal polyp region. Müller subsequently (1969a) reported extraction of a chemical substance from gonozooids which could transform gastrozooids. Colcemid, α-ketoglutarate and isocitrate also transformed gastrozooids. Brändle (1970) has carried out a similar set of investigations on the related genus, *Podocoryne carnea*; by means of extracts or grafts of gonozooid hypostome she was able to transform a gastrozooid into a gonozooid which produced medusae.

An interesting sidelight of the grafting studies is that if the tissues used are from opposite sexes, the male character always dominates (Pirard, 1961; Müller, 1964; Brien, 1965; Tardent, 1966). This can fit into a model involving hormones (Brien, 1964), but Müller (1964) explains this sex reversal in *Hydractinia* on the basis of physical competition of germ cells for tissue space, with the smaller male cells able to expel the female counterparts into the polyp gastrovascular cavity (Fig. 17).

Tardent (1966) and Müller (1967) found that x-irradiation of the donor male tissue prevented its reversing the sex of a female host hydra; this is taken as strong evidence that sexual dimorphism is determined autonomously by interstitial cells, which are x-ray sensitive.

It appears that there are several distinct stages in the development of sexuality which are under quite different types of control. First is the initiation of reproductive polyps; this appears to be affected by general metabolic and growth factors. Induction of gametogenesis is probably separately controlled, and late stages in gametogenesis (see Section 3.3.9) are often controlled separately. The uncertain picture resulting from experimental work is due partly to a lack of discrimination between these various stages of sexuality. Schmid and Tardent (1969) have found, for example, that gametogenesis in the medusa bud of *Podocoryne carnea* continues even if the gonads are

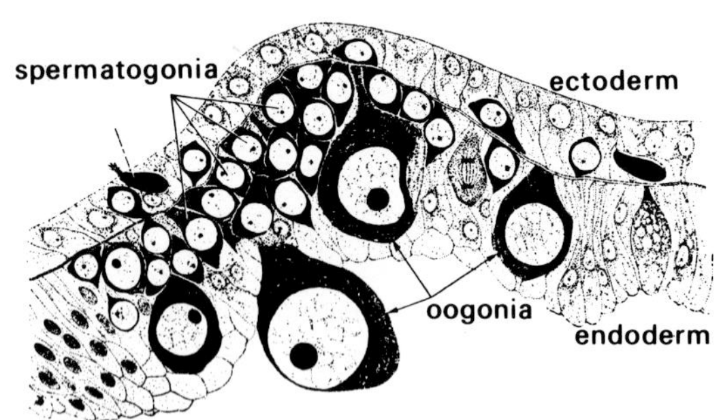

FIG. 17. Competition between oogonia and spermatogonia in an intersexual gonozooid of *Hydractinia echinata* produced by tissue grafting. In such intersexual polyps the male phenotype always eventually dominates. In this figure the spermatogonia appear to be forcing the oogonia out of the endoderm and into the gastric cavity where they die. [From Müller (1964).]

dissociated into small fragments which reorganize to form polyps.

Certain direct controls over particular gametic maturation events have also been found. These generally appear to function in entraining gametogenesis to environmental variables to synchronize spawning. Exposure to light stimulates germinal vesicle breakdown in appropriately developed oocytes in *Hydractinia echinata* (Ballard, 1942), and darkness initiates the same process in *Spirocodon saltatrix* (Yoshida, 1954). In *Hydractinia epiconcha* (Yoshida, 1954), a dark period as short as 10 minutes stimulates germinal vesicle breakdown.

A number of scyphozoan and anthozoan species are protandric, suggesting that age has a direct influence on gametogenesis. In sea anemones where different septa harbor relatively independent gonads, protandric sequences are seen in individual septa. The entire polyp may thus simultaneously have male, hermaphroditic, and female gonads on successively older septa (Chia and Rostron, 1970).

3.3.8 Reproductive Behavior

Various forms of behavior have been evolved to aid in reproductive success; most of these are involved quite directly with bringing gametes together, and are more or less restricted to small taxonomic groups of cnidarians. No behavioral patterns are sufficiently elaborate and direct to be termed courtship.

One of the simpler and most wide-spread behavioral traits aiding cnidarian reproduction is aggregation of individuals prior to reproduction. In medusae and jellyfish, aggregation frequently leads to enormous, dense swarms or blooms. These are manifested both in dense swimming aggregates and in spectacular masses stranded together on beaches. The mechanisms by which they aggregate are not known. However, the slow rates at which they swim and the low organization of the nervous system make it unlikely that swarming results directly from animals detecting and swimming toward other individuals of a species. It is much more probable that swarming is affected by physical factors, such as winds and water currents, and that rudimentary behavior tends to keep individuals of one species in a particular hydrographic stratum. For fuller accounts of swarming, the reader is referred to discussions and references cited in Russell (1953, 1970).

Most polyp stages are quite sessile, and polyp aggregation, which frequently occurs, arises both by selective settling and by asexual reproduction. Some sea anemones, however, have developed behavior patterns associated with spawning. These sometimes involve the female moving short distances up to the male, stimulated by male spawning (Uchida and Yamada, 1968). Some of the alcyonarians (e.g., *Renilla*) are particularly mobile and actively move in response to hydrographic conditions (J. Morin, personal communication) and may remain aggregated.

3.3.9 Spawning

3.3.9.1 MECHANISMS

Spawning mechanisms are very simple in the Cnidaria, in reflection of the elementary organization of the gonads and individuals. In all cnidarians, the gonodal tissue ruptures, largely due to distension by the gametes, and the sex cells are thereby released directly into the seawater (Hydrozoa) or into the gastrovascular system (Scyphozoa and Anthozoa). In the latter two classes, gamete shedding is generally effected by simple emptying of the gastrovascular cavity through the mouth by contraction of body musculature. In some cases gametes may be shed through other pores (tentacle tips, aboral pores, genital

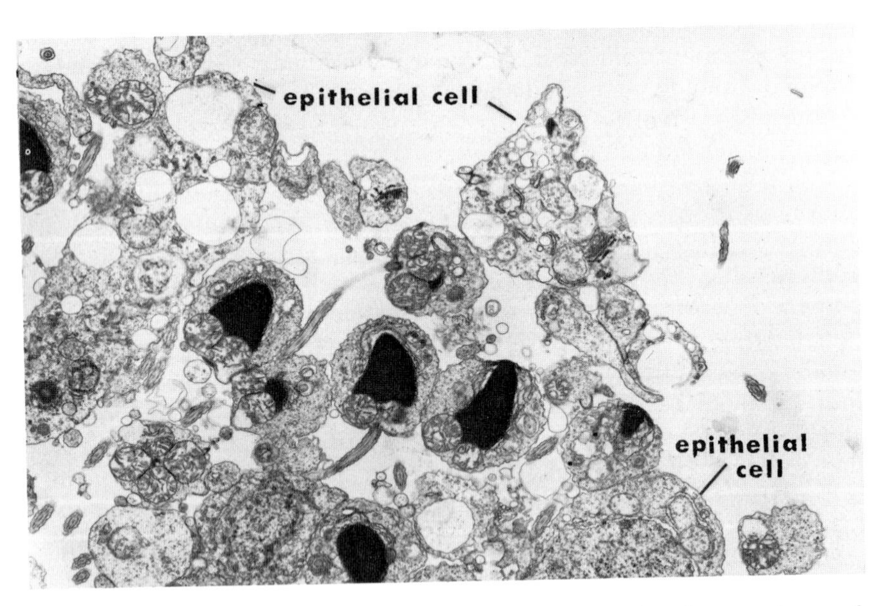

FIG. 18. Spermatocyte release in the medusa *Phialidium*. This electron micrograph shows the wall of a testis during rupture of the contacts between adjacent epithelial cells with spermatocytes emerging into the seawater. (4600X) [From Roosen-Runge and Szollosi (1965).]

pores) and in these cases the mechanisms of expulsion are not known. Ciliary action also has been implicated in gamete shedding by some species of Anthozoa, where eggs come out the mouth along the siphonoglyph. Roosen-Runge and Szollosi (1965) made a detailed study of spermatozoan shedding in the hydromedusa *Phialidium*. The testis ruptures by breaking the adhesive contacts between adjacent supporting epithelial cells, first at a highly localized but apparently random region (Fig. 18). Rupture then spreads across the testis within a few minutes, to include almost the whole testis surface. Within 4 minutes after all spermatozoa have escaped, the surface has healed again.

A detailed description of spawning in *Hydractinia epiconcha* was reported by Yoshida (1954). In this species the gonads are within a medusoid gonophore which is sessile but may detach from the blastostyle before or after spawning. Testis rupture releases spermatozoa, but in contrast to *Phialidium* studied by Roosen-Runge and Szollosi, the gonads degenerate after a single discharge. Egg release involves a complex sequence of events including initial freeing of ova into the subumbrellar cavity, presumably by an enzymatic mechanism, perforation of the gonophore velar plate, and expulsion of the eggs by pulsations of the bell.

3.3.9.2 Synchronization and Coordination

Spawning coordination is probably the most important behavioral trait exploited by all classes of Cnidaria to ensure fertilization. In some cases female spawning is reported to be in direct response to male spawning (e.g., in *Sagartia*, Nyholm, 1943; *Halcampa*, Uchida and Yamada, 1968). Synchrony is more generally induced by similar individual response to physical environmental factors. Numerous species respond to variations in light intensity, spawning either shortly after darkening, after a long period of darkening, or after a dark period followed by light. Roosen-Runge and Szollosi (1965) found that periodic (twice daily) gamete release in the medusa *Phialidium* appears to be controlled by an endogenous circadian rhythm which in nature becomes entrained to morning and evening times. Even isolated testes retained this periodicity and sensitivity to light for several days. These authors and Ballard (1942) postulate that a spawning inducer accumulates during periods in the light. Yoshida (1954) has made a detailed study of the relation between the onset of darkness and spawning in *Hydractinia epiconcha*. Many processes are triggered by darkness, including oocyte maturation, gonad rupture, gonophore detachment and pulsation, gonophore swelling, and separation of the egg jellycoat. Isolated gonophores responded in the same manner as attached ones. This work more or less followed the lines of, and confirmed, Ballard's (1942) classic investigation of *Hydractinia echinata*. Light stimulus applied to the isolated gonophore caused the germinal vesicle to move to an eccentric position, completion of meiosis, and shedding 55 minutes later. Once started, the process proceeds independently of light.

Several interesting elaborations are known which aid sperm movement to the egg. In some hydroids, both thecate and athecate, chemotactic direction of sperm migration to the female gonad has been discovered by Miller (1966a,b) (Fig. 19). This is particularly exciting since it represents the first well-documented case of spermatozoan chemotaxis in animals. Chemically isolatable attractants are released by the female gonophores and have been shown to directly affect the spermatozoan swimming pattern. The substances isolated have molecular weights under 1000, are heat stable, and show some but not perfect species specificity (Miller, 1966c).

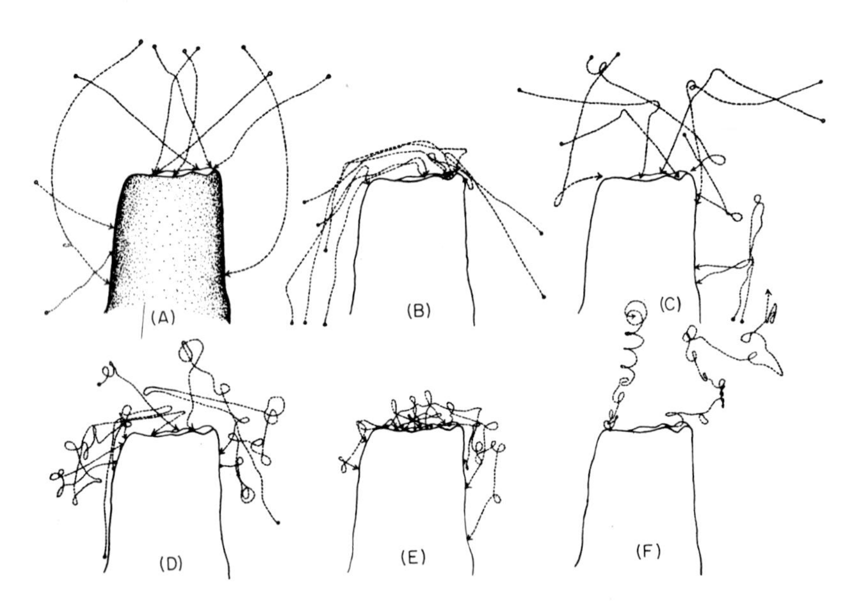

FIG. 19. Plotted trails of *Campanularia flexuosa* sperm attracted to the distal end of a mature female gonangium of *C. flexuosa*. The open circles represent the start of the trail. Six categories of sperm behavior are shown in A-F, with the gonangium in the center of each figure. Some of these patterns show strong chemotaxis (B-E), (A) shows sperm which were originally directed towards gonangium, and (F) illustrates trails of two sperm which swam away from gonangium. Right and left directions are reversed in this illustration. [From Miller (1966a).]

Another interesting adaptation related to sperm movement, also found by Miller (1970), is sperm guidance along tentacles of the meconideum (reduced medusa) (Fig. 9C), which harbors ripe eggs and later broods the embryos (see Section 3.3.3) in *Gonothyrea loveni.* Spermatozoa which encounter these tentacles stick to them and subsequently move along the tentacle surface in a 2-4 μm layer of extracellular material on the tentacle surface. Again there is some, but not complete, species specificity. Details of sperm movement are provided by Miller (1970).

3.3.9.3 Breeding Period

There is great variation in the length and season of breeding in the Cnidaria. Table I lists breeding period data for representative types. Further information on the environmental control of the breeding time may be found in Section 3.3.7.

3.4 Development

3.4.1 Embryonic Development

Embryonic development within the Cnidaria reflects the overall primitiveness, with particular lines of evolutionary specialization, which characterizes cnidarian structure and gametogenesis. Some aspects of representative species are summarized in Table II (pp. 174-175).

Mergner (1971) has provided an excellent account of cnidarian embryogenesis and experimental studies related to this subject.

3.4.1.1 Fertilization

Eggs are fertilized after oocyte meiosis is completed, that is, after polar body formation. In some cases polar bodies are seen at the egg surface; often they are never found even upon careful searching. In some cases a fertilization membrane is formed. Metschnikoff (1886) and Roosen-Runge (1962) provide some of the most detailed studies of fertilization. The zygotes tend to be generalized in structure and behavior. The nucleus is almost always situated near the animal pole, which affects cleavage as described below. Fertilization is generally external but frequently internal (in the gastrovascular cavities of Anthozoa and Scyphozoa, and in the gonophore lumen of Hydrozoa), with the young zygote expelled immediately or retained for brooding.

TABLE I
Reproductive Seasons

Species	Reproductive season	Geographical region considered	Sex	Place of development	References
Class Hydrozoa					
Order Hydroida					
Rathkea octopunctata	June–July	Coast of Greenland	Dioecious	External	Berrill, 1952a
Tubularia radiata	September–November	Japan	Dioecious	Gonophores	Nagao, 1965
Syncoryne mirabilis	March–May	Atlantic	Dioecious	External	Berrill, 1953
Sertularella miurensis	August, November	Japan	Dioecious	Gonotheca	Nishihira, 1968c
Polyorchis karafutoensis	June–July	Japan	Dioecious	External	Nagao, 1970
Bougainvillia superciliaris	June–August	Atlantic	Dioecious	Manubrial folds	Costello *et al.*, 1957
Order Trachylina					
Tesserogastria musculosa	All year	Norway	Dioecious	External	Hesthagen, 1971
Class Scyphozoa					
Dactylometra quinquecirrha	Summer	Chesapeake Bay	Dioecious	Internal	Littleford, 1939
Aurelia aurita	Summer	Atlantic	Dioecious	Internal	Costello *et al.*, 1957
Order Rhizostomeae					
Cassiopea andromeda	April–August	Red Sea	Dioecious	External	Gohar and Eisawy, 1961
Class Anthozoa					
Subclass Alcyonaria					
Muricea fruticosa	Winter	California	Dioecious	External	Grigg, 1970
Muricea californica	Winter	California	Dioecious	External	Grigg, 1970
Heteroxenia fuscens	Spring–Fall	Red Sea	Monoecious	Internal	Gohar and Roushdy, 1961
Subclass Zoantharia					
Sagartia troglodytes	August	Atlantic	Dioecious	External	Nyholm, 1943
Metridium dianthus	Summer	Atlantic	Dioecious	External	Costello *et al.*, 1957
Actinia equina	All year, especially March, December	Great Britain	Dioecious	Internal	Chia and Rostron, 1970
Anthopleura elegantissima	September–October	California	Dioecious	External	Ford, 1964
Flavia doreyensis	Mainly December	Great Barrier Reef	Monoecious	–	Marshall and Stephenson, 1933
Porites haddoni	January–July	Great Barrier Reef	–	Internal	Marshall and Stephenson, 1933
Pocillopora bulbosa	monthly, all year	Great Barrier Reef	–	Internal	Marshall and Stephenson, 1933

3.4.1.2 Hydrozoan embryogenesis

Hydrozoan embryonic development is of a very primitive type. Cleavage tends to be total and equal except where eggs are very yolky (e.g., *Aglaura*). Where the animal pole is recognizable and where blastomere rearrangement is not extensive, the first two cleavage planes are mutually perpendicular and include the polar axis; the third cleavage is equatorial. Thus it is usually of the radial type (Fig. 20A), and in cases it may lead to particularly regular tiers of blastomeres even at the 64-cell stage. Generally, however, cleavage tends to be much less regular. The marked irregularity of hydrozoan cleavage patterns relative to those of other animals appears partly due to less well-defined adhesion between blastomeres. Thus, blastomeres may often not adhere tightly in a mass, and will frequently roll into new positions between cleavage divisions (Fig. 20B). Less commonly, cleavage is unequal. An example of "anarchic" cleavage in the medusa *Oceania armata* (Fig. 20C) is given by Metschnikoff (1886).

One peculiarity of cleavage found in many cnidarian embryos is asymmetric, or unilateral, furrowing. In this situation (Fig. 21), the cleavage furrow begins at one side of the blastomere (generally at the animal pole) and moves progressively across the cell. As the furrow cuts deeper, the divided portions of the blastomere readhere as though being "zipped together" (Dan, 1960). This arises due to the eccentric position of the nuclei, which lie near the animal pole (Rappaport, 1963). Cleavage is thus completed at the animal pole minutes or hours before the furrow reaches the vegetal pole, and often after the next cleavage begins. This gives rise to peculiar horseshoe-shaped cleavage stages (Fig. 21) (Dan and Dan, 1947; Rappaport, 1963; Nagao, 1965; Uchida and Yamada, 1968). Unilateral furrowing also leads to important cytoplasmic and blastomere movements, particularly rotation, which greatly modify the basic radial cleavage pattern (Uchida and Yamada, 1968).

Hydrozoan segmentation stages, at least up to the 16-cell stage, have been found to be completely regulative in all cases examined (Zoja, 1895; Hargitt, 1904a, Maas, 1905; Hauenschild, 1954; Rappaport, 1969), Unfortunately, there has been little experimental work testing for mosaicism in the highly evolved siphonophore eggs; these have very asymmetrical development and heterogeneous cytoplasm characteristic of determinate eggs; Carré (1969) found some degree of mosaicism at an advanced stage of development but not during cleavage.

There are two common blastula types: (1) the *stereoblastula*,

DEVELOPMENT

Species	Egg		Cleavage	Blastula	Gastrulation		Reference
	Size	Characters			Mode	Product	
Class Hydrozoa							
Order Hydroida							
Tubularia radiata	500-600 μm	Very yolky	Irregular	Nearly solid, irregular	Multipolar delamination	Actinula larva	Nagao, 1965
Hydractinia echinata	160-170 μm	Yolky: green or red	Radial–irregular	Nearly solid, irregular	Multipolar delamination and ingression	Planula	Teissier and Teissier, 1928
Polyorchis karafutoensis	140-160 μm	Moderately yolky	Radial	Coeloblastula	Multipolar proliferation	Planula	Nagao, 1970
Spirocodon saltatrix	60-70 μm	Transparent	Radial with blastomere rotation	Coeloblastula	Unipolar ingression	Planula	Uchida and Yamada, 1968
Order Trachylina							
Gonionema depressum	70 μm	Brown, opaque	Radial with blastomere rotation	Coeloblastula	Multipolar delamination	Planula	Uchida and Yamada, 1968
Tesserogastria musculosa	140 x 240 μm	Oval, opaque, yellow-brown	Total, equal	Coeloblastula		Planula	Hesthagen, 1971
Order Siphonophora							
Lensia conoidea	500 μm	Transparent	Radial	Coeloblastula	Secondary delamination	Planula	Carré, 1967

Class Scyphozoa							
Order Stauromedusae							
Thaumatoscyphus distinctus	50 μm	Pale yellow	Radial – irregular	Stereoblastula	Unipolar ingression	Planula (nonciliated)	Uchida and Yamada, 1968
Haliclystus octoradaiatus	30 μm	Slightly yolky	Radial	Stereoblastula	Unipolar ingression	Planula (nonciliated)	Wietrzykowski, 1912
Order Coronatae							
Nausithoë punctata	230 μm	Yolky	Radial	Stereoblastula	Invagination	Planula	Komai, 1935
Order Semaeostomeae							
Aurelia aurita	150-230 μm	Yolky	Radial	Coeloblastula	Invagination or ingression	Planula	Berrill, 1949
Dactylometra quinquecirrha	70-190 μm	Yellow, yolky	Radial	Stereoblastula		Planula	Littleford, 1939
Order Rhizostomeae							
Mastigias papua	100 μm		Radial	Coeloblastula containing some yolk	Invagination	Planula	Uchida, 1926
Cassiopea andromeda	140-170μm	Cemented into egg mass				Planula	Gohar and Eisawy, 1961
Class Anthozoa							
Subclass Alcyonaria							
Heteroxenia fuscescens	300 μm	Yolky, centrolecithal	Irregular, partly syncytial				Gohar and Roushdy, 1961
Subclass Zoantharia							
Metridium marginatum	120-160 μm	Pink opaque	Radial with increasing irregularity	Coeloblastula sometimes containing yolky debris	Multipolar delamination	Planula with stomodaeum	McMurrich, 1890
Peachia quinquecapitata	120 μm	Yolky		Coeloblastula	Invagination	Planula	Spaulding, 1972
Sagartia troglodytes	80-90 μm		Radial (regular)	Coeloblastula (much yolk inside)	Invagination	Planula	Nyholm, 1943

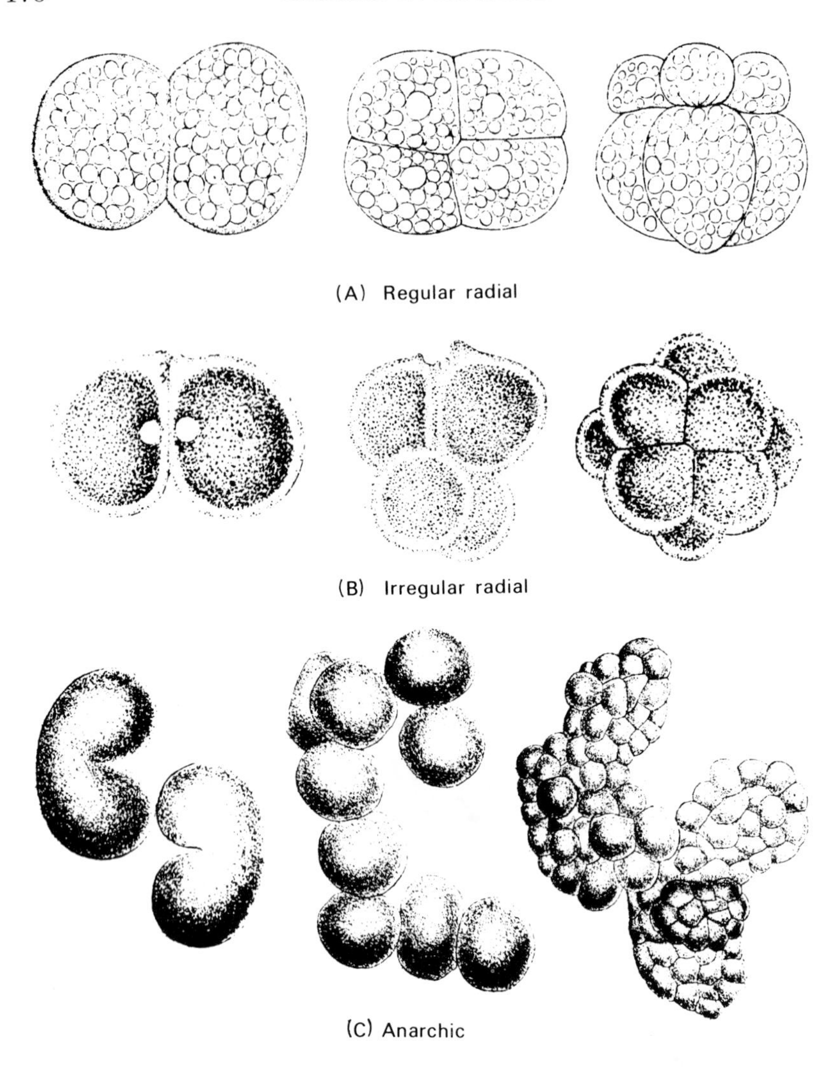

FIG. 20. Three types of embryonic cleavage in Hydrozoa, as seen beginning from the two-cell stages. (A) Regular radial pattern, exhibited by *Aglaura hemistoma.* (B) Irregular radial cleavage in *Rathkea fasciculata.* (C) Anarchic cleavage, as found in *Oceania armata.* [All after Metschnikoff (1886).]

which is solidly packed with cells, containing at most a few, small, irregular interstices resembling a blastocoele, and (2) the *coeloblastula,* in which the blastomeres form a single layered epithelium enclosing a large blastocoele. Some hydrozoan eggs which are telolecithal cleave by a superficial mechanism and result in a *periblastula* (e. g., *Eudendrium,* Hargitt, 1904b).

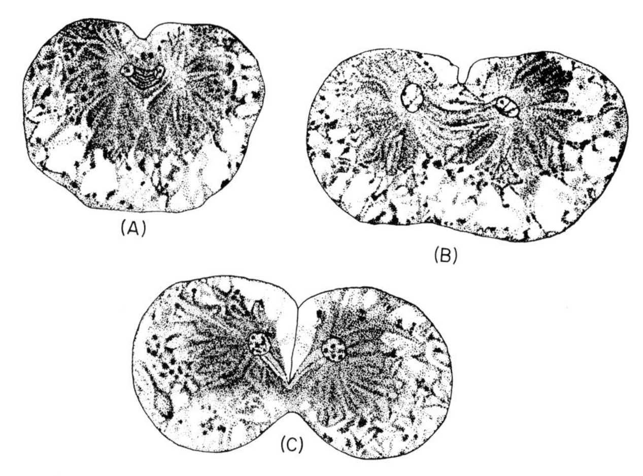

FIG. 21. Unilateral, asymmetric cleavage common in cnidarian development, as exhibited by *Spirocodon saltatrix*. The furrow begins at the edge of the blastomeres (at the animal pole during the first cleavage), and progressively and sometimes slowly cuts across the blastomere. One result of this is that the whole blastomeres or some cytoplasm are rotated, contributing to irregularity in the subsequent cleavage pattern. [From Dan and Dan (1947).]

Gastrulation in the Class Hydrozoa is diverse, apparently primitive, and never involves simple invagination. Generally, gastrulation is multipolar. The following represent the chief patterns by which embryonic tissue layers become segregated (Table II; Fig. 22):

Unipolar ingression (Fig. 22A). In a small number of forms, endoderm segregation occurs clearly by inward migration of individual cells at the vegetal pole of a coeloblastula. This process appears to involve a decrease in individual cell adhesion for surrounding cells, rounding, and finally severing the cell's attachment at the surface of the blastula.

Multipolar ingression (Fig. 22B). Widespread ingression of individual cells occurs in a few hydrozoans, such as *Solmundella mediterranea* (Dawydoff, 1928) and *Bougainvillia* (Gerd, 1892). This may begin at a very early cleavage stage, as in *Solumundella.*

Multipolar delamination (Fig. 22C). Endoderm commonly arises by radial cleavage of blastomeres, resulting in one daughter cell remaining at the embryo's surface (ectoderm) and one daughter cell within the interior (endoderm). In a number of hydrozoans this mechanism of gastrulation occurs exclusively; the finest example available is from the trachymedusa family Geryonidae (Metschnikoff, 1886). This means of gastrulation presents the opportunity for systematic segregation of cortical and interior ooplasms into ectoderm

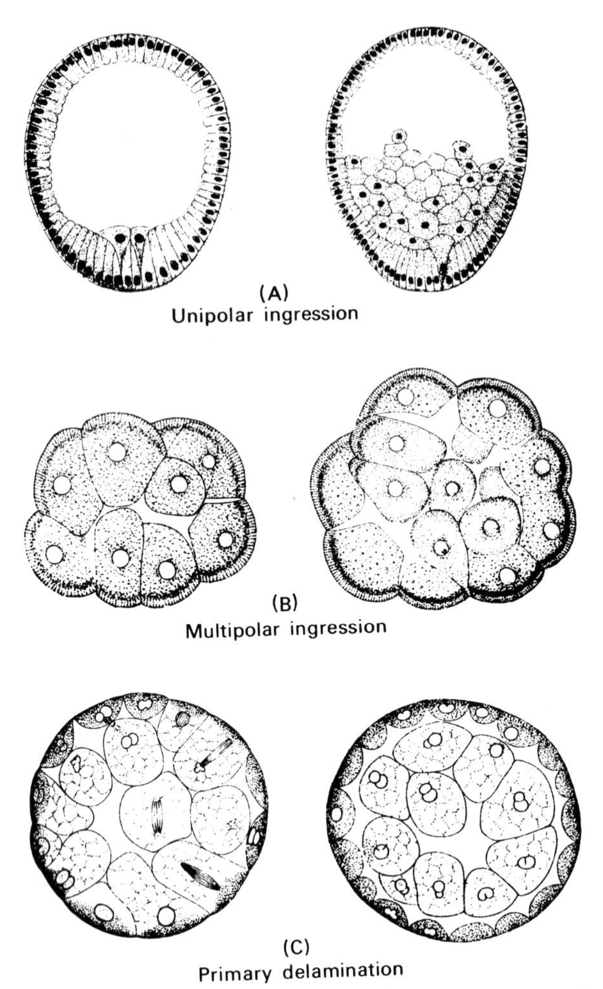

FIG. 22. Major modes of cnidarian gastrulation. (A) *Octorchis gegenbauri.* (B) *Aeginopsis mediterranea.* (C) *Geryonia proboscidalis.* [(A-C) From Metschnikoff (1886).] (D) *Hydra grisea* [from Brauer (1891)]. (E) *Clava squamata* [from Harm, (1902)].

and endoderm, respectively. Unfortunately, the 8-cell stage blastomere separation experiments of Zoja (1895) involved embryos before endodermal delamination began.

Mixed pattern (Fig. 22D). The most frequent type of hydrozoan gastrulation represents a combination of multipolar ingression and delamination. This tends to appear irregular, and in different species one of the two modes may predominate. Tannreuther's (1908) description of hydra development serves as a good representation of this mode of gastrulation.

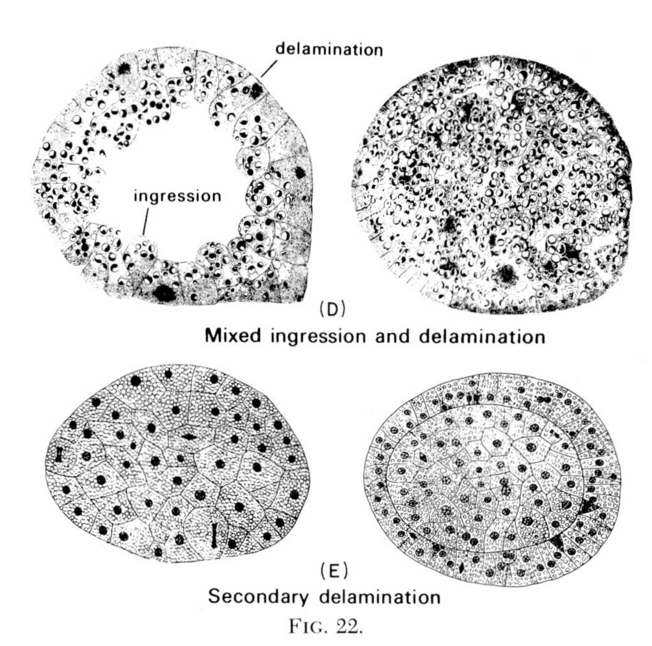

FIG. 22.

Secondary delamination (Fig. 22E). In many cases where cleavage results in a stereoblastula, tissue segregation occurs late in development by a more or less simple histological rearrangement leading to the outer cell forming a coherent epithelium, distinct from the inner cell mass. Simultaneously a mesolamella appears at this histological boundary. This process is sometimes called secondary delamination since the cells already have their approximate final positions. This is well-illustrated by the development of *Clava squamata* (Harm, 1902). This method of gastrulation is also present in the well-studied genus *Tubularia* (Nagao, 1965), and is the normal mode of gastrulation within the Order Siphonophora (Totton, 1965). Here the peripheral ooplasm is peculiarly granular and distinct from the yolky central yolk-laden cytoplasm. During cleavage the interior cytoplasmic mass remains syncytial or cleaves into a few large blastomeres. The granular cytoplasm remains in the ectoderm, and to some extent in the endodermal layer of small cells which becomes early defined in the antereo-ventral side. At this stage the larva halves are unable to regulate complete development when separated (Carré, 1969), suggestive of local cytoplasmic determination of tissue fate.

A modification of this method of gastrulation is frequently designated syncytial delamination; here the central cells are syncytial at the time of the secondary delamination, so that the ectoderm which forms rests on an uncleaved cytoplasmic core. This occurs in animals with very telolecithal eggs and superficial cleavage. The syncytial central mass may represent secondary cell fusion after an initial normal cleavage (e.g., *Turritopsis nutricula,* Brooks and Rittenhouse, 1907).

In secondary delamination the formation and development of the endoderm is a complex and controversial matter (see Dawydoff, 1928; Woltereck, 1905a).

The origin of particular cell types during embryogenesis has been only briefly described in a few cases. Interstitial cells are thought to arise throughout large areas of the embryo. In *Tubularia* their origin has been ascribed to asymmetric divisions of large endodermal cells, with the small interstitial cells thus produced migrating into the ectodermal layer before mesoglea is synthesized (Nagao, 1965). Weiler-Stolt (1960) also describes the origin of interstitial cells as endodermal, with subsequent migration into the ectoderm. Cnidocyte formation begins after the origin of interstitial cell production (Weiler-Stolt, 1960; Bodo and Bouillon, 1968) and may occur first in the endoderm before interstitial cells migrate to the ectoderm (Fig. 23). Korn (1966) has reviewed the origin of the nervous system.

Important sources for hydrozoan embryogenesis include Metschnikoff, 1886; Kühn, 1910; Dawydoff, 1928; and Bodo and Bouillon, 1968.

3.4.1.3 Scyphozoan Embryogenesis

Early embryogenesis in scyphozoans curiously presents a more regular but highly disputed picture than that of hydrozoans. The fertilized egg forms a thin fertilization membrane and undergoes an

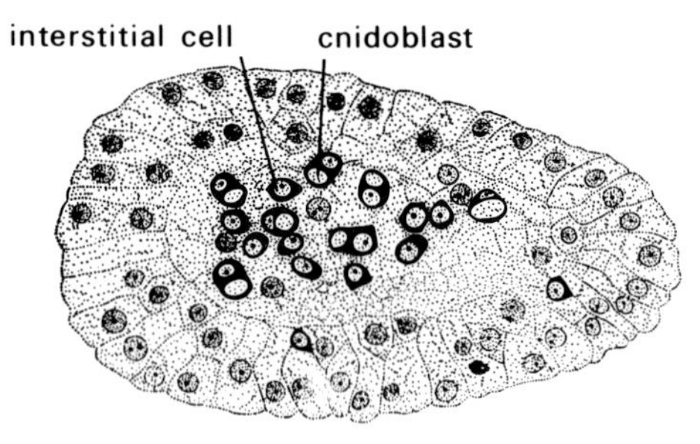

Fig. 23. Origin of interstitial cells and nematoblasts in the embryonic endoderm of the hydrozoan *Eleutheria dichotoma.* Interstitial cells (black) subsequently migrate into the ectoderm as shown by cell at right side of the embryo. (700X) [From Weiler-Stolt (1960).]

undistinguished cleavage which resembles a radial type with some blastomere rearrangement. The first two divisions are always total and generally equal. Thereafter cleavage becomes progressively less regular and synchronized.

The late stages of blastulation and gastrulation have provoked much disagreement among different investigators, often studying the same species. Some forms regularly form strict stereoblastulae while others have well-defined blastocoeles. Gastrulation presents a spectrum from orderly invagination to polar ingression. Multipolar delamination is also said to occur (Hyde, 1894). Berrill (1949) has produced the most orderly, although highly generalized, discussion of the different gastrulation forms. According to Berrill, jellyfish with small eggs (e.g., *Haliclystus*) cleave to form stereoblastulae and exhibit ingressive gastrulation. In some cases this has components of epiboly (e.g., Uchida and Yamada, 1968). Scyphozoa with large eggs (e.g., *Pelagia*) exhibit typical polar invagination. *Aurelia* has been the most extensively studied, and the discrepancies between various authors' descriptions are due (according to Berrill, 1949) to the great variation in egg size shown by this species. Small eggs form endoderm irregularly while large ones by invagination. Intermediate sized eggs show considerably more complex intermediate forms of gastrulation.

Gastrulation can lead either to a planula-like larva, a polyp-like scyphistoma stage, or directly into a young medusoid form termed an ephyra. Another possible ontogenetic form is a polyembryonic tissue mass. This last form has been described in *Haliclystus* (Wietrzykowski, 1912) as an overgrown solid gastrula, off of which bud planulae. The planula larvae of Scyphozoa resemble those of some trachymedusae in being solid and vermiform with only a single column of endodermal cells. Planulae settle and metamorphose into scyphistomae.

3.4.1.4 Anthozoan Embryogenesis

Anthozoan early development is rather uniform as far as is known, the major variation being the amount of yolk in the eggs which in turn renders cleavage less total and regular, and gastrulation less obviously a form of invagination. The subclass Zoantharia (including the common sea anemones) often have eggs exhibiting typical radial cleavage followed by invagination. Cleavage occurs by means of a progressing furrow, as described for hydrozoan development.

In *Renilla* (Wilson, 1883) early nuclear division may occur without egg cleavage; then at the 8, 16, or 32 cell stage the cytoplasm suddenly segments fully. This is apparently related to the yolky nature of these eggs. There is also a tendency for the large, heavily yolk-laden anthozoan eggs to shed yolky deposits into the blastocoele which blocks to a certain extent the normal course of invagination (McMurrich, 1890; Dawydoff, 1928). In both subclasses of Anthozoa, eggs are often so yolky that cleavage may be partly superficial and gastrulation of the resulting stereoblastulae occurs by delamination (McMurrich, 1890; Gohar and Roushdy, 1961) (Fig. 24).

Planulae result directly from elongation and hollowing of the embryo after gastrulation. At this time, or in some forms after settling, an oral invagination gives rise to the pharynx which is thus lined with ectoderm. Two other events occurring during the development of a planula larva are (1) formation of mesenchyme, which arises by a loss of compaction of the basal portion of the ectoderm, (2) formation of septa, which involves the longitudinal folding of endoderm into ridges containing an extension of mesoglea.

Anthozoan embryogenesis is best treated by Gemmill (1920).

3.4.2 Larvae

3.4.2.1 Larval Types

Most Cnidaria exhibit larval forms during development and these are generally planktonic. The most developmentally generalized and abundant type is the planula (Fig. 25A); an elongate ciliated larva with ectoderm and endoderm, and frequently a gastrovascular cavity. Relatively long-lived (i. e., up to weeks) planula types contain all the

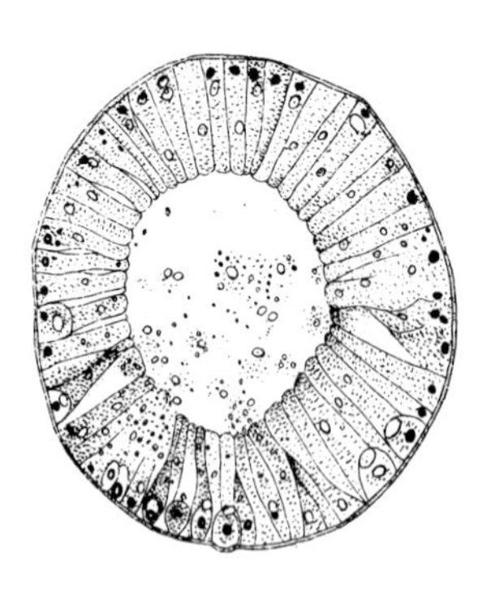

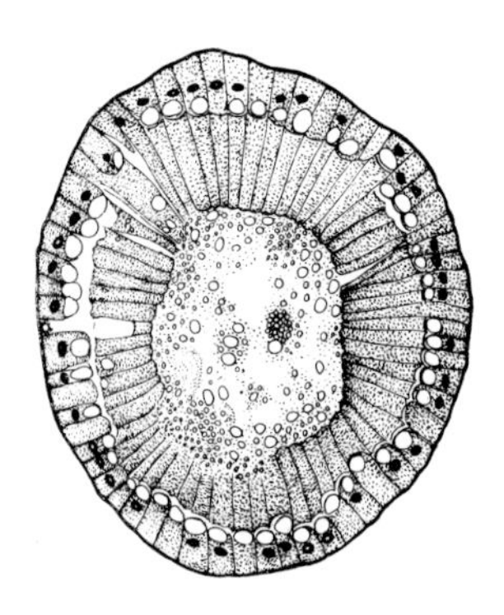

Fig. 24. Two stages in Anthozoan gastrulation by delamination in very yolky eggs. Much debris is also shed into the blastocoele. [From McMurrich (1890).]

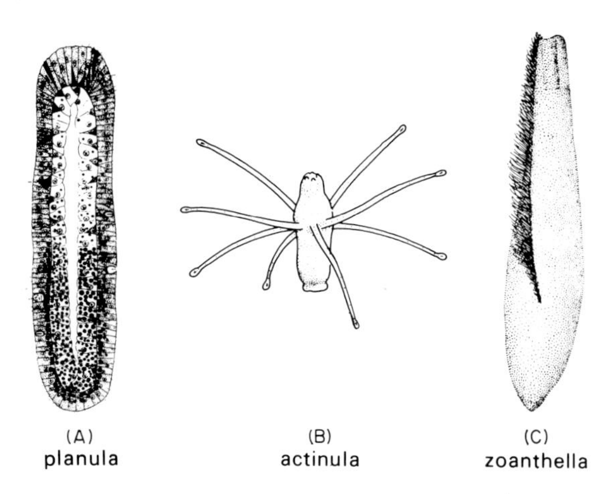

FIG. 25. Three cnidarian larvae. The planula Larva (A) is the most common larval form and can lead a long, independent existence in some cases (from Wulfert, 1902). The actinula larva (B) is found in some groups of hydrozoans, including *Myriothella* (Fig. 9) and is further developed than is the planula. The zoanthella larva (C) occurs during the life cycle of some Anthozoa; It possesses six complete septa and also represents a postplanula stage [after Heath (1906)].

basic cnidarian cell types: epithelial, contractile, nervous and sensory, gland, nematocyst, and digestive. Some planulae are apparently capable of feeding through a posterior (oral) pore. A few planulae are not free-swimming (e. g., *Haliclystus*, Thiel, 1966). Widersten (1968) has extensively reviewed planula structure.

The three classes of cnidarians have additional distinctive larval forms. Within the Hydrozoa, the planula larva is the prevalent one except in the Siphonophora and Trachylina. In these orders, the larval forms are temporary intermediate stages of metamorphosis from the planula, and will thus be described in Section 3. 4. 2. 2. One larval form deserving mention is the actinula larva of certain gymnoblastic hydroida (e. g., *Tubularia*). This is an 8-tentacled ciliated pear-shaped larva (Fig. 25B) which may float freely for some time before settling, and which greatly resembles some capitate hydroids. The actinula larva is thought by some to represent an ancestral form and has played an important role in discussions of cnidarian phylogeny (see Rees, 1957). When the actinula larva settles, it anchors by its aboral end and its tentacles become the polyp's proximal whorl.

The major larval form of scyphozoans is the scyphistoma, which is analogous to the polyp generation of the Hydrozoa. This larval type is discussed in Section 3.2.

In the Anthozoa, the two prevalent larval types, which develop from planulae, are really intermediate stages in the metamorphosis of the planula. They may be very long lived, however, and have played considerable part in discussions of anthozoan evolution. These two larval stages are called the Edwardsian and Halcampoides forms. Their structure is essentially that of mature anemones and they are defined by the number of septa which have developed. In the Edwardsian larva there are eight septa (four couples); in the Halcampoides larva there are twelve septa arranged in six complete pairs. In some genera (e. g., *Peachia* and *Edwardsia*) these larval stages are prolonged and spent as parasites on medusae. The Cerinula larvae of varied Cerianthids (Van Beneden, 1898) are strongly bilateral forms of Edwardsia larvae with four tentacles. The Zoanthella larvae (Fig. 25C) of the Zoantheria (Dawydoff, 1928) represent Halcampoides stages, but lack tentacles and have a single strongly developed longitudinal band of cilia for swimming.

3.4.2.2 Metamorphosis

Among hydrozoans and scyphozoans with polyp stages, planula metamorphosis is stimulated by settling, and is typically rapid. Other forms of these two orders usually undergo metamorphosis directly into the medusoid form. Anthozoan planulae which are brooded often metamorphose in the brooding chambers before release, while ovoviviparous species have planulae which settle before metamorphosis.

In the following analysis of diverse patterns of cnidarian metamorphosis, I intend to show that broad similarities may be found which offer the most comprehensive generalization about the unity of reproduction and development of the phylum. After this brief overview, particular cases of settling and metamorphosis will be reviewed.

When planulae undergo metamorphosis, two histological changes are almost universally represented. These are: (1) the oral pole undergoes invagination, vesiculation, or some homologous activity; and (2) the distal endoderm becomes ridged longitudinally. The apical invagination gives rise to the pharynx in anthozoans and the subumbrellar surface in hydrozoan medusae. It gives rise to certain complex oral structures in aberrant hydrozoan polyps, and to floats and swimming bells in Siphonophores. It is generally absent in scyphozoans; however, this has always been a controversial point (Dawydoff, 1928). The endodermal ridging pattern establishes the most obvious features of radial symmetry and organization, including placement of oral tentacles, positions of septa and the pattern of mouth folds and lappets. It seems to establish the radial organization of the gastric cavities and canals in medusoid forms.

These two histological activities also occur during asexual reproduction (budding, regeneration, etc.) and allow the most general comparative interpretations of cnidarian form and development to be drawn.

The transformation of planulae into the five major zooid types is described very briefly below. For fuller synopses of, and literature references to, embryological analyses of these events the following works should be consulted: Metschnikoff, 1886; Kühn, 1910, 1913; Kükenthal and Krumbach, 1925; Dawydoff, 1928; Berrill, 1949; Carré, 1967; Bodo and Bouillon, 1968.

a. Hydrozoan polyp (Fig. 26A). This is one of the simplest transformations of the planula larva, and greatly resembles budding (Section 3.2.2). Generally the planula settles by its anterior end and the oral tip of the polyp originates from the posterior end. Very early in this process the apical endoderm acquires the taeniolate organization, so that the gastrovascular cavity offers a stellate appearance in cross section. Mouth and tentacle formation follow; the mouth forms where the ectoderm becomes extremely thin and finally pierced, and the tentacles represent evaginations around the hypostome. The tentacles may appear simultaneously or in definite or indefinite order, depending on the species. The radial positions of the first tentacles are always coincident with the intertaeniolate radii. Subsequently, other structures arise which are typical for the species, such as gonophores, proximal tentacles, buds, or stolons. For more detailed accounts of this and variant modes of planula-polyp transformation, see Hamann (1882), Kühn (1913) and Dawydoff (1928).

Scyphozoan polyp (scyphistoma) (Fig. 26B). Early scyphistoma larval development from the planula greatly resembles hydrozoan polyp formation; the planula attaches to the substratum by its anterior end, the endodermal ridges arise in the now distal portions, and mouth and radial tentacles (i.e., those alternating with endodermal ridges) appear. At this point the scyphistoma becomes more complex. The taeniolae become so tall that the endodermal cell bases no longer reach the vicinity of the ectoderm; the endoderm is actually folded, with a branch of mesoglea running into it, and these walls are called septa. This process begins at the oral pole and progresses proximally. The mesoglea also becomes considerably thicker. Then a cluster of cells from the hypostomal ectoderm grow downward as a column into the mesoglea of each septum. These cells give rise to muscles. Later an invagination from the ectoderm follows into each

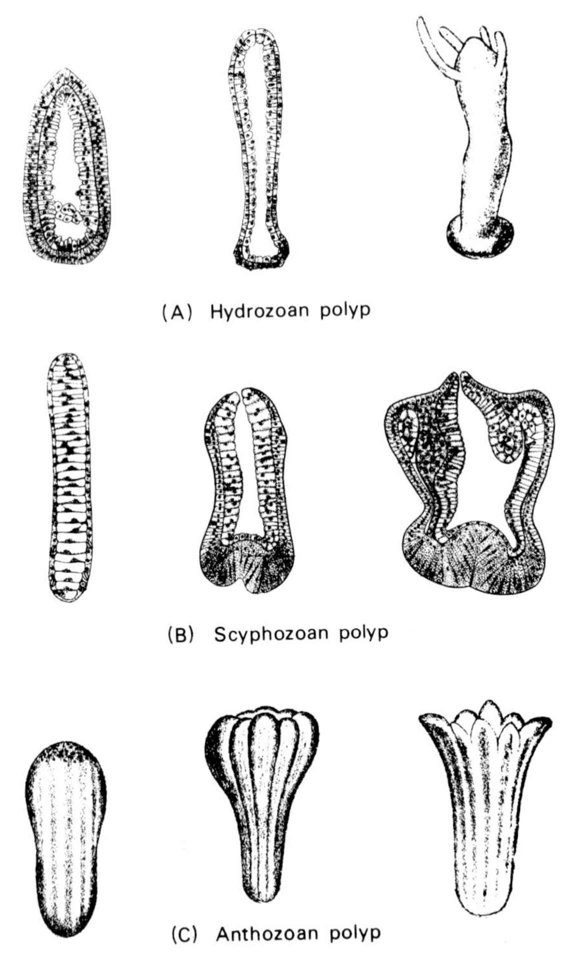

FIG. 26. Cnidarian metamorphosis patterns (see text). (A) *Cordylophora lacustris* (from Morgenstern, 1901). (B) *Haliclystus octoradiatus* [from Wietrzykowski (1912)]. (C) *Lebrunia coralligens* [from Duerden (1899)]. (D) *Liriope mucronata* [from Metschnikoff (1886)]. (E) *Epibulia aurantiaca* [from Pflugfelder (1962), after Metschnikoff)]. (F) *Pelagia noctiluca* [from Metschnikoff (1886)].

muscle cord, giving rise to the genital pits (funnels). Thus scyphozoan polyp development exhibits most features of hydrozoan polyp formation, but introduces additional rather major complications. The above description is based on *Lucernaria;* more details of this and of other scyphozoan polyp development can be found in Hein (1900) and Dawydoff (1928).

Anthozoan polyp (Fig. 26C). Anthozoan polyp formation from a planula is similar to early stages in scyphistoma development. An

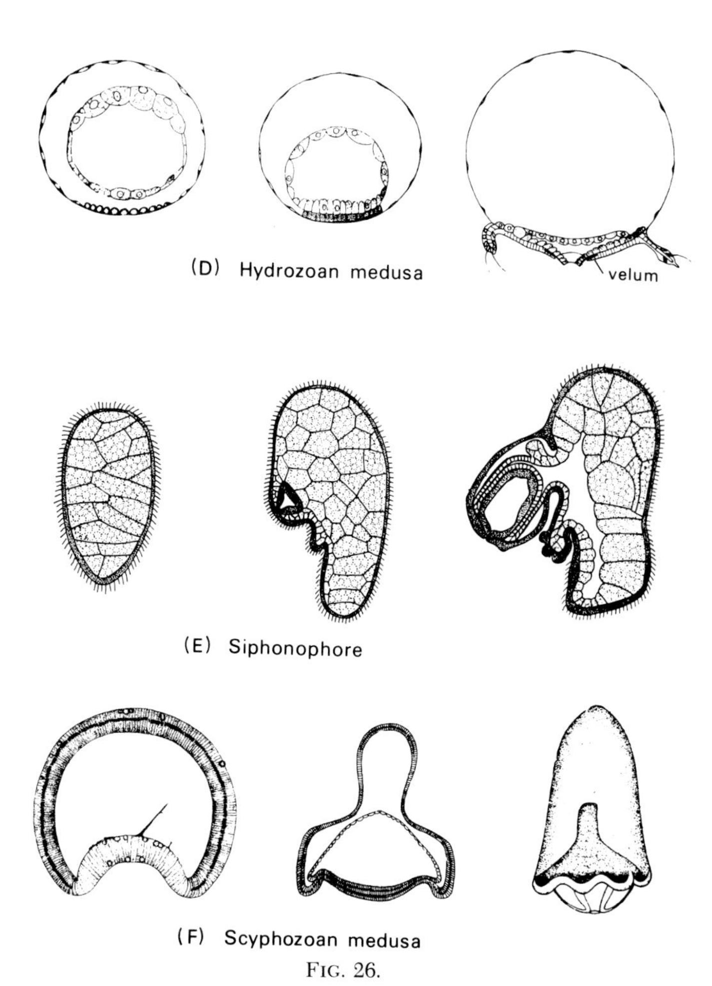

FIG. 26.

important novel feature of anthozoans, however, is that concurrent with taeniola initiation, the apical ectoderm, and perhaps endoderm, invaginates. This gives rise to the mouth and ectoderm-lined pharynx. This is in contrast to a piercing mouth formation in the other two classes. Tentacles arise normally. Taeniolae arise and transform into mesenteries, as in scyphozoans; in this case they extend from lateral wall to the pharynx, and only lower along the column do they have an inner margin free. The formation of taeniolae has not been well studied in anthozoans; there is some question as to the original number, but two septa arise prominently early and extend far proximally. These are the two directives, which establish the plane of bilateral symmetry in anthozoans.

Anthozoan planula transformation thus exhibits features of great similarity to that in the classes Scyphozoa and Hydrozoa, with the prominent innovation of the pharynx. Further information concerning this development may be found in McMurrich (1890), Gemmill (1920), and Dawydoff (1928).

Hydrozoan medusa (Fig. 26D,E). Some representatives of the Trachylina and Siphonophora have direct development of a medusoid form from a planula. There is great variety in these aberrant developments; two types will be mentioned here. The "planula" of the trachyline family Geryonidae (Fig. 26D) consists of spherical shells of ectoderm and endoderm, separated by intervening mesoglea. By some unknown mechanism the two tissue shells come into contact at a side which will become the oral surface. The layers form a thickened plate here, from which arise a ring of tentacles and mouth. A circular ridge forms around the mouth which gradually becomes a circular flap, the velum. It is not known when the taeniolate organization of the manubrium appears. This and other examples of direct medusa development are most elegantly analyzed by Metschnikoff (1886) and Woltereck (1905b).

In the complex colonial Siphonophores, the first structure to be formed is the float organ and this should be regarded as a homologue of a medusa. Float development (Fig. 26E) resembles the way a typical hydromedusa arises from a budding polyp; the ectoderm (in this case on the planula side) invaginates similarly to entocodon formation and the endoderm is restricted to form four radial canals. The homologue of the velum is the outer covering of ectoderm. In other Siphonophores the initial development may be quite modified but is invariably centered around apical ectodermal invagination, usually with radial compartmentalization of surrounding endoderm.

For details on Siphonophora development the reader should consult Dawydoff (1928), Garstang (1946), Totton (1965), and Carré (1967, 1969).

Scyphomedusa (Fig. 26F). A few jellyfish arise by direct transformation of planulae, without intervening scyphistoma stages. In the development of *Pelagia noctiluca* (most carefully studied species) the blastopore is retained as the mouth and the surrounding ectoderm and endoderm form a thickened plate which will become the subumbrellar surface of the jellyfish. A marginal ridge then forms surrounding the plate, along which eight lobes develop to give rise to the lappets. This development thus resembles that of some trachyline medusae, but is even simpler and, perhaps, more primitive. For further information on scyphomedusan formation see Metschnikoff (1886) and Berrill (1949).

3.4.2.3 Planula settling

Planula larvae probably usually settle within a few days of being formed. However, if they are not presented with conditions favorable for settling, they will not undergo metamorphosis, and eventually (from days to months, depending on the species) they will die. The fact that planula metamorphosis begins shortly (sometimes a few minutes) after settling indicates that the two activities are closely associated. It appears that planula settling is a behavioral event which occurs when a particular set of conditions are met, and that some part of the settling process triggers metamorphosis to begin. Settling involves adhering to a substratum and this is aided by sticky gland cell secretions (Bodo and Bouillon, 1968) and sometimes by nematocysts; since the nervous system may be closely related to nematocytes (Schneider, 1890; Slautterback, 1967), it is possible that a nervous impulse mediates a triggering of metamorphosis when nematocyst discharge accompanies settling.

Many species of planulae exhibit considerable discrimination in selecting specific settling sites. The planula of the hydrozoan *Proboscidactyla flavicirrata* settles only on the rim of a sabellid worm tube, where it develops into a colony (Campbell, 1968b). The settling process is initiated by the planula getting swept into the ciliary currents of a sabellid worm's tentacles. Upon contact with the tentacles, the planula discharges nematocysts which anchor it to the tentacles. The planula then becomes sticky, presumably due to gland cell secretory activity, and is rubbed off the tentacles onto the rim of the tube, by periodic retractions of the worm. Metamorphosis ensues shortly afterwards.

An interesting series of investigations into planula settling concern *Hydractinia echinata*, a hydroid which grows almost exclusively on snail shells inhabited by hermit crabs. Schijfsma (1935) carried out some early important experiments dispelling the idea that there was a chemattractant which directed the planula in its settling. Cazaux (1958) concluded that *Hydractinia echinata* planula settling is determined by physical conditions, namely the presence of a hard substratum across which water is briskly flowing. Thus Cazaux was able to get a high incidence of settling and metamorphosis on clean shells or glass surfaces by establishing a water flow across them (Fig. 27), while settling did not occur under similar conditions when the water was still. Müller (1969b) has reexamined settling in this species and concurs with Korn (1966) that settling does not generally occur under conditions of very clean water and substrata. Müller was able to isolate four species of bacteria which, in pure culture, could stimulate

FIG. 27. Stimulation of settling of *Hydractinia* planulae. The planulae normally settle only on hermit crab shells inhabited by hermit crabs. Cazaux was able to induce settling on empty shells (small shells in figure) by having them towed (left) or carried (right) by hermit crabs. These experiments support Cazaux' conclusion that movement of the shell is more important than occupancy by a crab. [From Cazaux (1961).]

Hydractinia planulae to settle and metamorphose. A heat-labile causative agent could be extracted from the bacteria which possessed the inducing activity. Different bacterial strains had different activity, and the medium on which they were grown, bacterial density, and planula age all affected the planula response. Müller describes the behavior of the planula, gliding along substrata in response to both light and oxygen conditions, and settling where bacterial density or activity was sufficient.

Settling thecate and athecate hydroid planulae *Sertularella miurensis* and *Coryne uchidai*, respectively, show marked preferences for algae of certain species (Nishihira, 1965, 1967a,b, 1968a,b,c). Both the physical form and the chemical composition of the frond surface seem to be important in planula discrimination. Spaulding (1972) reports that planulae of the parasitic anthozoan, *Peachia quinquecapitata*, enter their medusa hosts' gastrovascular cavities, perhaps by sticking to the oral lobes and being swept in by the medusa manubrial cilia. After metamorphosing within their host, the larvae burrow to the outside or come out the mouth, attach to the subumbrellar surface by means of nematocysts, and eat the medusa gonads. Planulae would not grow or metamorphose free of the host. Chia and Rostron (1970) find that *Actinia* larvae will not undergo metamorphosis if taken out of the gastric cavity of adults which are brooding them; within the cavity they undergo metamorphosis without settling.

Many investigators have examined the factors stimulating scyphistoma larvae to undergo strobilation, thus forming the sexual generation of the jellyfish. Food, temperature and light intensity all may have marked effects on strobilation frequency, but the interrelations between these and other environmental factors are poorly understood and obviously complex. Work on the stimulation of strobilation has been very well reviewed by Russell (1970, pp. 16-22). *Mastigias papua* scyphistomae strobilate only following infection by intracellular symbiotic zooxanthellae (Sugiura, 1964). Iodine can also stimulate strobilation (Paspalev, 1938; Spangenberg, 1967), and Spangenberg (1971) reports that thyroxine is synthesized and stimulates strobilation in *Aurelia* scyphistomae, hinting that a primitive hormonal system may be involved.

Acknowledgments

Preparation of this review was supported in part by NIH Research Development Award K04-GM42595 and NSF Research Grant GB 29284. Dr. Lary V. Davis greatly helped in reviewing the literature for this chapter.

3.5 References

Afzelius, B. A., and Å. Franzén (1971). The spermatozoon of the jellyfish *Nausithoë*, *J. Ultrastruct. Res.* **37**, 186-199.

Agassiz, L. (1862). "Contributions to the Natural History of the United States", Vol. III, 301 pp. Little, Brown, Boston, Massachusetts.

Allman, G. J. (1871-2). "A Monograph of the Gymnoblastic or Tubularian Hydroids", Parts I, II, 450 pp. Ray Society, London.

Bacci, G. (1950). Alcuni problemi dell'ermafroditismo negli Invertebrati. *Boll. Zool. (Suppl.)* **17**, 193-212.

Ballard, W. W. (1942). The mechanism for synchronous spawning in *Hydractinia* and *Pennaria*. *Biol. Bull.* **82**, 329-339.

Beklemishev, W. N. (1969). "Principles of Comparative Anatomy of Invertebrates"(J. M. MacLennan and Z. Kabata, translators), Vol. 1, Promorphology; Vol. 2, Organology. 490, 529 pp. Oliver and Boyd, Edinburgh.

Beloussov, L. V., Badenko, L. A., Katchurin, A. L., and Kurilo (Filetcheva). L. R. (1972). Cell movements in morphogenesis of hydroid polyps. *J. Embryol. Exp. Morphol.* **27**. 317-337.

Berrill, N. J. (1949). Developmental analysis of scyphomedusae. *Biol. Rev.* **24**, 393-410.

Berrill, N. J. (1952a). Growth and form in gymnoblastic hydroids. II. Sexual and asexual reproduction in *Rathkea*. III. Hydranth and gonophore development in *Pennaria* and *Acaulis*. IV. Relative growth in *Eudendrium*. *J. Morphol.* **90**, 1-32.

Berrill, N. J. (1952b). Growth and form in gymnoblastic hydroids. V. Growth cycle in *Tubularia*. *J. Morphol.* **90**, 583-601.

Berrill, N. J. (1953). Growth and form in gymnoblastic hydroids. VII. Growth and reproduction in *Syncoryne* and *Coryne*. *J. Morphol.* **92**, 273-302.

Berrill, N. J. (1961). "Growth, Development and Pattern," 555 pp. Freeman, San Francisco, California.

Bodo, F., and J. Bouillon (1968). Étude histologique du développement embryonnaire de quelques hydroméduses de Roscoff: *Phialidium hemisphaericum* (L.), *Obelia sp.* Péron et Lesueur, *Sarsia eximia* (Allman), *Podocoryne carnea* (Sars), *Gonionemus vertens* Agassiz. *Cah. Biol. Mar.* **9,** 69-104.

Bouillon, J. (1957). Étude Monographique du Genre *Limnocnida* (Limnomeduse). *Ann. Soc. Roy. Zool. Belgique* **87,** 254-500.

Bouillon, J. (1968). Introduction to coelenterates. *In* "Chemical Zoology" (M. Florkin and B. T. Scheer, eds.), Vol. 2, pp. 81-147. Academic Press, New York.

Brändle, E. (1970). Bedeutung der kolonialen Komponenten für die Bildung und Differenzierung der Medusen von *Podocoryne carnea* M. Sars. Ph.D. Thesis, Univ. of Zürich, 286 pp.

Brauer, A. (1891). Ueber die Entwicklung von Hydra. *Z. Wiss. Zool.* **70,** 169-216.

Braverman, M. H. (1962). Studies in hydroid differentiation. I. *Podocoryne carnea* culture methods and carbon dioxide induced sexuality. *Exp. Cell Res.* **27,** 301-306.

Braverman, M. H. (1963). Studies on hydroid differentiation. II. Colony growth and the initiation of sexuality. *J. Embryol. Exp. Morphol.* **11,** 239-253.

Braverman, M. H., and Schrandt, R. G. (1966). Colony development of a polymorphic hydroid as a problem in pattern formation. *Symp. Zool. Soc. London* **16,** 169-198.

Brien, P. (1953). La pérennité somatique. *Biol. Rev.* **28,** 308-349.

Brien, P. (1964). Contribution à l'étude de la biologie sexuelle chez les hydres d'eau douce. Induction gamétique et sexuelle par méthode des greffes en parabiose. *Bull. Biol. Fr. Belg.* **97,** 214-283.

Brien, P. (1965). Considération à propos de la reproduction sexuée des invertébrés. *Année Biol. Ser.* **4,** 353-365.

Brien, P. and M. Reniers-Decoen (1950). Etude d'*Hydra viridis* (Linnaeus) (La blastogénèse, la spermatogénèse, l'ovogénèse). *Ann. Soc. Roy. Zool. Belg.* **81,** 33-110.

Brien, P., and M. Reniers-Decoen (1951). La Gametogénèse et l'Intersexualité chez *Hydra attenuata* (Pall). *Ann. Soc. Roy. Zool. Bel.* **82,** 285-327.

Broch, H. (1924a). Hydroida. *In* "Handbuch der Zoologie, Band 1." (W. Kükenthal and T. Krumbach, eds), pp. 422-458. De Gruyter, Berlin.

Broch, H. (1924b). Trachylina. *In* "Handbuch der Zoologie, Band 1" (W. Kükenthal and T. Krumbach, eds.), pp. 459-484. De Gruyter, Berlin.

Brooks, W. K., and Rittenhouse, S, (1907). On *Turritopsis nutricula* (McCrady). *Proc. Boston Soc. Natur. Hist.* **33,** 429-460.

Burnett, A. L., Davis, L. E., and Ruffing, F. E., (1966). A histological and ultrastructural study of germinal differentiation of interstitial cells arising from gland cells in *Hydra viridis*. *J. Morphol.* **120**, 1-8.

Campbell, R. D. (1968a). Cell behavior and morphogenesis in hydroids. *In Vitro* **3,** 22-32.

Campbell, R. D. (1968b). Host specificity, settling, and metamorphosis of the two-tentacled hydroid *Proboscidactyla flavicirrata*. *Pac. Sci.* **22,** 336-339.

Campbell, R. D. (1968c). Holdfast movement in the hydroid *Corymorpha palma:* Mechanism of elongation. *Biol. Bull.* **134,** 26-34.

Campbell, R. D. (1972). Development. *In* "Biology of the Cnidaria" (H. M. Lenhoff and L. Muscatine, eds.), Academic Press, New York (in press).

Campbell, R. D., and Campbell, F. (1968). Tubularia regeneration: Radial organization of tentacles, gonophores, and endoderm. *Biol. Bull.* **134**, 245-251.

Carlgren, O. (1893). Über das Vorkommen von Brutraumen bei Actinien. *Öfvers Kungl. Vetens. Akad. Förhand.* **50**, 231-238.

Carré, D. (1967). Etude du développement larvaire de deux Siphonophores: *Lensia conoidea* (Calycophore) et *Forskalia edwardsi* (Physonecte). *Cah. Biol. Mar.* **8**, 233-251.

Carré, D. (1969). Etude du développement larvaire de *Sphaeronectes gracilis* (Claus, 1873) et de *Sphaeronectes irregularis* (Claus, 1873), Siphonophores Calycophores. *Cah. Biol. Mar.* **10**, 31-34.

Cazaux, C. (1958). Facteurs de la morphogénèse chez un hydraire polymorphe, *Hydractinia echinata* Flem. *C. R. Acad. Sci. Paris* **247**, 2195-2197.

Cazaux, C. (1961). Signification et origine de l'association entre Hydractinie et Pagure, Rôle des tropismes larvaires dans le developpement de l'Hydraire. *Bull. Sta. Biol. Arcachon* **13**, 1-5.

Chapman, D. M. (1966). Evolution of the Scyphistoma. *Symp. Zool. Soc. London* **16**, 51-75.

Chapman, D. M. (1968). Structure, histochemistry and formation of the podocyst and cuticle of *Aurelia aurita*. *J. Mar. Biol. Ass. U. K.* **48**, 187-208.

Charniaux-Cotton, H. (1965). Hormonal control of sex differentiation in invertebrates. *In* "Organogenesis" (R. L. DeHaan and H. Ursprung, eds.), pp. 701-740. Holt, New York.

Chia, F.-S., and Rostron, M. A. (1970). Some aspects of the reproductive biology of *Actinia equina* (Cnidaria:Anthozoa). *J. Mar. Biol. Ass. U. K.* **50**, 253-264.

Child, C. M. (1941). "Patterns and Problems of Development," 811 pp. Univ. of Chicago Press, Chicago, Illinois.

Congdon, E. D. (1906). Notes on the morphology and development of two species of *Eudendrium*. *Biol. Bull.* **11**, 27-46.

Costello, D. P., Davidson, M. E., Eggers, A., Fox, M. H., and Henley, C. (1957). Methods for Obtaining and Handling Marine Eggs and Embryos. 247 pp. Mar. Biolog. Lab., Woods Hole, Massachusetts.

Crowell, S. (1957). Differential responses of growth zones to nutritive level, age, and temperature in the colonial hydroid *Campanularia*. *J. Exp. Zool.* **136**, 63-90.

Dan, K. (1960). Cyto-embryology of echinoderms and amphibia. *Int. Rev. Cytol.* **9**, 321-367.

Dan, K., and Dan, J. C. (1947). Behavior of the cell surface during cleavage. VIII. On the cleavage of medusan eggs. *Biol. Bull.* **93**, 163-188.

Dawydoff, C. (1928). "Traité d' Embryologie comparée des Invertébrés." 930 pp. Masson, Paris.

Duerden, J. E. (1899). The *Edwardsia*- stage of the Actinian *Lebrunia*, and the Formation of Gastro-coelomic Cavity. *J. Linnean Soc. Zool.* **27**, 269-316.

Fawcett, D. W. (1961). Intercellular bridges. *Exp. Cell Res. Suppl.* **8**, 174-187.

Ford, C. E. (1964). Reproduction in the aggregating sea anemone, *Anthopleura elegantissima*. *Pac. Sci.* **18**, 138-145.

Franzén. Å. (1956). On spermiogenesis, morphology of the spermatozoan and biology of fertlization among invertebrates. *Zool. Bidr. Uppsala* **31**, 355-482.

Franzén, Å (1966). Remarks on spermiogenesis and morphology of the spermatozoon among the lower Metazoa. *Ark. Zool.***19**, 335-342.

Frey, J. (1968). Die Entwicklungsleistungen der Medusen-knospen und Medusen von *Podocoryne carnea* M. Sars nach Isolation und Dissoziation. *Arch. Entwicklungsmech. Organ.* **160**, 428-464.

Garstang, W. (1946). The morphology and relations of the Siphonophora. *Quart. J. Microsc. Sci.* **87**, 103-193.
Gemmill, J. F. (1920). The development of the sea anemones *Metridium dianthus* (Ellis) and *Adamsia palliata* (Bohad). *Phil. Trans. Roy. Soc. London B.* **209**, 352-375.
Gerd, W. (1892). Zur Frage über die Keimblätterbildung bei den Hydromedusen. *Zool. Anz.* **15**, 312-316.
Glätzer, K. H. (1970). Das Membranbesatz der Mikrovilli einer Planula von *Corydendrium parasiticum* (L.) (Hydrozoa, Athecata) *Cytobiologie* **2**, 408-412.
Goette, A. (1907). Vergleichende Entwicklungsgeschichte der Geschlects-individuen der Hydropolypen. *Z. Wiss. Zool.* **87**, 1-335.
Gohar, H.A.F., and Eisawy, A.M. (1961). The development of *Cassiopea andromeda* (Scyphomedusae). *Publ. Mar. Biol. Sta. Al-Ghardaqa (Red Sea)* **11**, 147-190.
Gohar, H. A. F., and Roushdy, H. M., (1961). On the embryology of the Xeniidae (Alcyonaria) (with notes on the extrusion of the larvae). *Publ. Mar. Biol. Sta. Al-Ghardaqa (Red Sea)* **11**, 45-70.
Grigg, R. W. (1970). Ecology and Population Dynamics of the Gorgonians, *Muricea californica* and *Muricea fruticosa*. Coelenterata: Anthozoa, Ph. D. Thesis, Univ. of California, San Diego, 278 pp.
Hadži, J. (1963). "The Evolution of the Metzaoa," 499 pp. Pergamon, Oxford.
Haeckel, E. (1879). Das System der Medusen. Erster theil einer monographie der Medusen (Craspedotae). *Denkschr. Med. Naturwiss. Ges. Jena* **1**, 1-360.
Hamann, O. (1882). Der Organismus der Hydroidpolypen. *Jen. Z. Naturwiss.* **15** (N. F. vol. 8), 473-544.
Hanisch, J. (1970). Die Blastostyle und Spermienentwicklung von *Eudendrium racemosum* Cavolini. *Zool. Jahrb. Anat.* **87**, 1-62.
Hargitt, C.W. (1904a). The early development of *Pennaria tiarella* McCr. *Arch. Entwicklungsmech. Organ.* **18**, 453-488.
Hargitt, C. W. (1904b). The early development of *Eudendrium*. *Zool. Jahrb. Abt. Anat.* **20**, 257-276.
Hargitt, G. T. (1920). Germ cells of coelenterates. III. *Aglantha digitalis*. *J. Morphol.* **20**, 593-608.
Harm, K. (1902). Die Entwicklungsgeschichte von *Clava squamata*. *Z. Wiss. Zool.* **73**, 1-55.
Hauenschild, C. (1954). Gentische und Entwicklungsphysiologische Untersuchungen ueber Intersexualitaet und Gewebevertraeglichkeit bei *Hydractinia echinata* Flemm. (Hydroz. Bougainvill.). *Arch. Entwicklungsmech. Organ.* **147**, 1-41.
Heath, H. (1906). A new species of Semper's larva from Galapagos Islands. *Zool. Anz.* **30**, 171-175.
Hegner, R. W. (1914) "The Germ Cell Cycle in Animals," 346 pp. Macmillan, New York.
Hein, W. (1900). Untersuchungen über die Entwicklung von *Aurelia aurita*. *Z. Wiss. Zool.* **67**, 401-438.
Hertwig, O., and Hertwig, R. (1879). "Die Actinien, Anatomisch und Histologisch mit besonderer Berücksichtigung des Nervenmuskelsystems," 224 pp. Fischer, Jena.
Hertwig, O. and R. Hertwig (1880). Der Organismus der Medusen und Seine Stellung zur Keimblaettertheorie. *Denkschr. Med.-Naturwiss. Ges. Jena*, **2(1)**, 1-70.
Hesthagen, I. H. (1971). On the biology of the bottom-dwelling trachymedusa *Tesserogastria musculosa* Beyer. *Norw. J. Zool.* 19, 1-19.
Hincks, T. (1863). "A History of the British Hydroid Zoophytes," Vol. II, 338 pp. Van Voorst, London.

Hyde, I. H. (1894). Entwicklungsgeschichte einiger Scyphomedusen. *Z. Wiss. Zool.* **58,** 531-565.

Hyman, L. H. (1940). "The Invertebrates," Vol. I, Protozoa through Ctenophora, 726 pp. McGraw-Hill, New York.

Kessel, R. G. (1968). Electron microscope studies on developing oocytes of a coelenterate medusa with special reference to vitellogenesis. *J. Morphol.* **126,** 211-247.

Komai, T. (1935). On *Stephenoscyphus* and *Nausithoë*. *Mem. Coll. Sci. Kyoto Univ. B* **10,** 289-339.

Korschelt, E. and K. Heider (1902). Lehrbuch der vergleichenden Entwicklungsgeschichte der wirbellosen Thiere. Lief. I-III. G. Fischer, Jena, 1509 pp.

Korn, H. (1966). Zur ontogenetischen Differenzierung der Coelenteratengewebe (Polyp-Stadium) unter besonderer Berücksichtigung des Nervensystems. *Z. Morphol. Oekol. Tiere* **57,** 1-118.

Kowalevsky, A. (1884). Zur Entwicklungsgeschichte der *Lucernaria*. *Zool. Anz.* **7,** 712-717.

Kühn, A. (1909). Sprosswachstum und Polypenknospung bei den Thecaphoren. Studien zur Ontogenese und Phylogenese der Hydroiden. *Zool. Jahrb. Abt. Anat.* **28,** 387-476.

Kühn, A. (1910). Die Entwicklung der Geschlechtsindividuen der Hydromedusen. Studien zur Ontogenese und Phylogenese der Hydroiden. II. *Zool. Jahrb. Abt. Anat.* **30,** 43-174.

Kühn, A. (1913). Entwicklungsgeschichte und Verwandtschaftsbeziehungen der Hydrozoen. I. Teil: Die Hydroiden. *Ergeb. Fortschr. Zool.* **4,** 1-284.

Kühn, A. (1914-16). Coelenterata. *In* "Klassen und Ordnungen des Tier-Reichs" (H. G. Bronn, ed.), Vol. 2 (Abt. 2, Lief. 22-36), pp. 371-538. Winterische Verlag, Leipzig.

Kükenthal, W., and Krumbach, T., (eds.) (1925). "Handbuch der Zoologie, Band 1," pp. 419-901. De Gruyter, Berlin.

Lenhoff, H. M., Rutherford, C., and Heath, H. D. (1969). Anomalies of growth and form in Hydra. Polarity, gradients, and a neoplasia analog. *Nat. Cancer Inst. Monogr.* **31,** 709-737.

Littleford, R. A. (1939). The life cycle of *Dactylometra quinquecirrha*, L. Agassiz in the Chesapeake Bay. *Biol. Bull.* **77,** 368-381.

Loomis, W. F. (1959). Feedback control of growth and differentiation by carbon dioxide tension and related metabolic variables. *In* "Cell, Organism and Milieu" (D. Rudnick, ed.), pp. 253-293. Ronald Press, New York.

Lunger, P. D. (1971). Early stages of spermatozoan development in the colonial hydroid *Campanularia flexuosa*. *Z. Zellforsch.* **116,** 37-51.

Maas, O. (1905). Experimentelle Beiträge zur Entwicklungsgeschichte der Medusen. *Z. Wiss. Zool.* **82,** 601-610.

Maas, O. (1909). Japanische Medusen. München Abh. Akad. Wiss. Suppl. **1,** 1-52.

Mackie, G. O. (1966). Growth of the hydroid *Tubularia* in culture. *Symp. Zool. Soc. London* **16,** 397-410.

Mangan, J. (1909). The entry of Zooxanthellae into the ovum of *Millepora*, and some particulars concerning the medusae. *Quart. J. Microsc. Sci.* **53,** 697-709.

March, L. M. (1915). A study of germ cells of *Corymorpha palma*. *Univ. Kansas Sci. Bull.* **9,** 247-258.

Marshall, S. M., and Stephenson, T. A. (1933). The breeding of reef animals. Part 1. The corals. *Great Barrier Reef Expidit. 1928-29, Sci. Rep.* **3(8),** 219-245.

Mayer, A. G. (1910). "Medusae of the World", Vols. I-III, 735 pp. Washington, D. C.: Carnegie Institution.

McMurrich, J. P. (1890). Contributions on the morphology of the Actinozoa. II. On the development of the Hexactiniae. *J. Morphol.* **4,** 303-330.

Mergner, H. (1971). Cnidaria. *In* "Experimental embryology of marine and freshwater invertebrates" (G. Reverberi, ed.), pp. 1-84. North-Holland Publ., Amsterdam.

Metschnikoff, E. (1886). Embryologische Studien an Medusen. "Ein Beitrag zur Genealogie der Primitiv-Organe," 159 pp, plus atlas. Alfred Hölder, Vienna.

Miller, R. L. (1966a). Chemotaxis during fertilization in the hydroid *Campanularia. J. Exp. Zool.* **162,** 23-44.

Miller, R. L. (1966b). Chemotaxis during fertilization in the hydroids *Tubularia* and *Gonothyrea. Amer. Zool.* **6,** 509 (Abstract).

Miller, R. L. (1966c). Gel filtration of the chemotactants of the hydroids *Campanularia, Tubularia* and *Gonothyrea. Amer. Zool.* **6,** 611 (Abstract).

Miller, R. L. (1970). Sperm migration prior to fertilization in the hydroid *Gónothyrea loveni. J. Exp. Zool.* **175,** 493-503.

Moore, L. B. (1971). Non-budding hydra strains: characterization and bud induction. Ph.D. Thesis, Univ. of California, Irvine, 90 pp.

Morgenstern, P. (1901). Untersuchungen über die Entwicklung von *Cordylophora lacustris* Allman. *Z. Wiss. Zool.* **70,** 567-591.

Müller, W. A. (1964). Experimentelle Untersuchungen über Stockentwicklung, Polypendifferenzierung und Sexualchimären bei *Hydractinia echinata. Arch. Entwicklungsmech. Organ.* **155,** 181-268.

Müller, W. A. (1967). Differenzierungspotenzen und Geschlechtsstabilität der Zellen von *Hydractinia echinata. Arch. Entwicklungsmech. Organ.* **159,** 412-432.

Müller, W. A. (1969a). Determination der Geschlechtspolypen von *Hydractinia echinata.* Eine Biologische und Chemische Analyse. *Arch. Entwicklungsmech. Organ.* **164,** 37-47.

Müller, W. A. (1969b). Auslösung der Metamorphose durch Bakterien bei den Larven von *Hydractinia echinata. Zool. Jahrb. Anat.* **86,** 84-95.

Müller-Calé, K., and Krüger, E. (1913). Symbiotische Algen bei *Aglaophenia helleri* and *Sertularella polyzonias. Mitth. Zool. Stat. Napoli,* **21,** 83-112.

Muus, K. (1966). Notes on the biology of *Protohydra leuckarti* Greef (Hydroidea, Protohydridae). *Ophelia* **3,** 141-150.

Nagao, Z. (1965). Studies on the development of *Tubularia radiata* and *Tubularia venusta* (Hydrozoa). *Publ. Akkeshi Mar. Biol. Stat.* **15,** 9-35.

Nagao, Z. (1970). The Metamorphosis of the Anthomedusa, *Polyorchis karafutoensis* Kishinouye. *Publ. Seto Mar. Biol. Lab.* **18,** 25-31.

Nishihira, M. (1965). The association between hydrozoa and their attachment substrata with special reference to algal substrata. *Bull Mar. Biol. Stat. Asamushi* **12,** 75-92.

Nishihira, M. (1967a). Observations of the algal selection by larvae of *Sertularella miurensis* in nature. *Bull. Mar. Biol. Stat. Asamushi,* **13,** 35-48.

Nishihira, M. (1967b). Dispersal of the larvae of a hydroid, *Sertularella miurensis. Bull. Mar. Biol. Stat. Asamushi,* **13,** 49-56.

Nishihira, M. (1968a). Experiments on the algal selection by the larvae of *Coryne uchidai* Stechow (Hydrozoa). *Bull. Mar. Biol. Stat. Asamushi* **13,** 83-89.

Nishihira, M. (1968b). Brief experiments on the effect of algal extracts in promoting the settlement of the larvae of *Coryne uchidai* Stechow (Hydrozoa). *Bull. Mar. Biol. Stat. Asamushi* **13,** 91-101.

Nishihira, M. (1968c). Dynamics of natural populations of epiphytic hydrozoa with special reference to *Sertularella miurensis* Stechow. *Bull. Mar. Biol. Stat. Asamushi* **13,** 103-124.

Nyholm, K-G. (1943). Zur Entwicklungsbiologie der Ceriantharien und Aktinien. *Zool. Bidr. Uppsala* **22,** 87-248.

Paspalev, G. V. (1938). Uber die Entwicklung von *Rhizostoma pulmo* Agass. *Arb. Biol. Meeresst. Varna* **7**, 1-17.
Pflugfelder, O. (1962). "Lehrbuch der Entwicklungsgeschichte und Entwicklungsphysiologie der Tiere," 347 pp. Fischer, Jena.
Pirard, E. (1961). Induction sexuelle et intersexualité chez une hydre gonochorique *(Hydra fusca)* par méthode des greffes. *C. R. Acad. Sci. Paris* **253**, 1997-1999.
Rappaport, R. (1963). Unilateral furrowing in *Hydractinia* and *Echinorachnius. Anat. Record* **145**, 273-274 (Abstract).
Rappaport, R. (1969). Division of isolated furrows and furrow fragments in invertebrate eggs. *Exp. Cell Res.* **56**, 87-91.
Rees, W. J. (1957). Evolutionary trends in the classification of capitate hydroids and medusae. *Bull. Brit. Mus. Natur. Hist. Zool.* **4**, 453-534.
Retzius, G. (1904). Zur Kenntnis die Spermien der Evertebraten, I. *Retzius' Biologisches Untersuchungen N. F.* **11**, 1-32.
Roosen-Runge, E. C. (1962). On the biology of sexual reproduction of hydromedusae, genus *Phialidium* Leuckhart. *Pac. Sci.* **16**, 15-24.
Roosen-Runge, E. C. (1970). Life cycle of the hydromedusa *Phialidium gregarium* (A. Agassiz, 1862) in the laboratory. *Biol. Bull.* **139**, 203-221.
Roosen-Runge, E. C. and D. Szollosi (1965). On biology and structure of the testis of *Phialidium leuckhart* (Leptomedusae). *Z. Zellforsch.* **68**, 597-610.
Russell, F. S. (1953). "The Medusae of the British Isles. Anthomedusae, Leptomedusae, Limnomedusae, Trachymedusae and Narcomedusae," 530 pp. Cambridge Univ. Press, London and New York.
Russell, F. S. (1970). "The Medusae of the British Isles. II. Pelagic Scyphozoa with a Supplement to the First Volume on Hydromedusae." University Press, Cambridge, 284 pp.
Schijfsma, K. (1935). Observations on *Hydractinia echinata* (Flem) and *Eupagurus bernhardus* (L.). *Arch. Neerland. Zool.* **1**, 261-314.
Schincariol, A. L., Habowsky, J. E. J., and Winner, G. (1967). Cytology and ultrastructure of differentiating interstitial cells in spermatogenesis in *Hydra fusca. Can. J. Zool.* **45**, 590-593.
Schmid, V., and Tardent, P. (1969). Zur gametogenese von *Podocoryne carnea* M. Sars. *Rev. Suisse Zool.* **76**, 1071-1078.
Schneider, K. C. (1890). Histologie von *Hydra fusca* mit besonderer Berücksichtigung des Nervensystems der Hydroidpolypen. *Arch. Mikorosk. Anat.* **35**, 321-379.
Slautterback, D. B. (1967). The cnidoblast-musculoepithelial cell complex in tentacles of hydra. *Z. Zellforsch.* **79**, 296-318.
Spangenberg, D. B. (1967). Iodine induction of metamorphosis in *Aurelia. J. Exp. Zool.* **165**, 441-450.
Spangenberg, D. B. (1971). Thyroxine induced metamorphosis in *Aurelia. J. Exp. Zool.* **178**, 183-194.
Spaulding, J. G. (1971). Preliminary observations on the structure of the oocyte, ovum and blastula of the anemone *Peachia quinquecapitata.* Manuscript in preparation.
Spaulding, J. G. (1972). The life cycle of *Peachia quinquecapitata,* an anemone parasitic on medusae during its larval development. *Biol. Bull.* **143**, 440-453.
Stagni, A. and Lucchi, M. L. (1970). Ultrastructural observations on the spermatogenesis in *Hydra attenuata. In* "Comparative Spermatology" (Baccio Baccetti, ed.), pp. 357-362, Academic Press, New York.
Stephenson, T. A. (1928). "The British Sea Anemones, Vol. I" 148 pp. Ray Society. London.
Stephenson, T. A. (1935). "The British Sea Anemones. Vol. II" Ray Society. London.

Sugiura, Y. (1964). On the life-history of rhizostome medusae II. Indispensability of zooxanthellae for strobilation in *Mastigias papua*. *Embryologia* **8,** 223-233.

Summers, R. G. (1970). The fine structure of the spermatozoon of *Pennaria tiarella* (Coelenterata). *J. Morphol.* **131,** 117-129.

Szollosi, D. (1964). Structure and function of centrioles and their satellites in the jellyfish *Phialidium gregarium*. *J. Cell Biol.* **21,** 465-479.

Szollosi, D. (1969). Unique envelope of a jellyfish ovum: the armed egg. *Science* **163,** 586-587.

Szollosi, D. (1970). Cortical cytoplasmic filaments of cleaving eggs: a structural element corresponding to the contractile ring. *J. Cell. Biol.* **44,** 192-209.

Tannreuther, G. W. (1908). The development of hydra. *Biol. Bull.* **14,** 261-280.

Tardent, P. (1963). Regeneration in the Hydrozoa. *Biol. Rev.* **38,** 293-333.

Tardent, P. (1966). Experimente zur Frage der Geschlechtsbestimmung bei *Hydra attenuata* (Pall.). *Rev. Suisse Zool.* **73,** 357-381.

Teissier, L., and Teissier, G. (1928). Les principales étapes du développement d'-*Hydractinia echinata* (Flem.). *Bull. Soc. Zool. Fr. Paris* **52,** 537-547.

Thiel, H. (1966). The evolution of the scyphozoa. *Symp. Zool. Soc. London* **16,** 77-117.

Totton, A. K. (1965). "A Synopsis of the Siphonophora." 230 pp. Brit. Mus. (Natur. Hist.), London.

Uchida, T. (1926). The anatomy and development of the rhizostome medusa, *Mastigias papua* L. Agassiz, with observations on the phylogeny of Rhizostomae. *J. Fac. Sci. Imp. Univ. Tokyo Zool.* **1,** 45-95.

Uchida, T., and Yamada, M. (1968). Cnidaria. *In* "Invertebrate Embryology" (M. Kumé and K. Dan, eds.; J. C. Dan, transl.), pp. 86-116. Publ. for the U. S. Nat. Library of Med., Washington, D. C., by Nolit Publ. House, Belgrade.

Valkanov, A. (1967). Beiträge zur Histologie des grünen Süsswasserpolypen *(Chlorohydra viridissima)*. *Zool. Anz.* **178,** 137-150.

Van Beneden, E. (1898). Die Anthozoen der Plankton - Expedition. *Ergebn. Plankton Expedit.* **2,** (k,3), 222 pp.

Webster, G. (1971). Morphogenesis and pattern formation in hydroids. *Biol. Rev.* **46,** 1-46.

Weiler-Stolt, B. (1960). Ueber die Bedeutung der Interstitiellen Zellen für die Entwicklung und Fortpflanzung Mariner Hydroiden. *Arch. Entwicklungsmech. Organ.* **152,** 398-454.

Weismann, A. (1883). "Die Entstehung der Sexualzellen bei den Hydromedusen. Zugleich ein Beitrag zur Kenntnis des Baues und der Lebenserscheinungen dieser Gruppe," 295 pp. Fischer, Jena.

Weissman, A., Lentz, T. L., and Barrnett, R. J. (1969). Fine structural observations on nuclear maturation during spermiogenesis in *Hydra littoralis*. *J. Morphol.* **128,** 229-240.

Werner, B. (1956). Der Zytologische Nachweis der Parthogenetischen Entwicklung bei der Anthomeduse *Margelopsis haeckeli* Habtlaub. *Naturwissenschaften* **43,** 541-542.

Werner, B. (1963). Effect of some environmental factors on differentiation and determination in marine Hydrozoa, with a note on their evolutionary significance. *Ann. N. Y. Acad. Sci.* **105,** 461-488.

West, D. L., and Renshaw, R. W. (1970). The life cycle of *Clytia attenuata* (Calyptoblastea: Campanulariidae). *Mar. Biol.* **7,** 332-339.

Westblad, E. (1935). Neue Beobachtungen ueber *Protohydra*. *Zool. Anz. Leipzig* **111,** 152-158.

Widersten, B. (1965). Genital organs and fertilization in some scyphozoa. *Zool. Bidrag. Uppsala* **37**, 45-58.

Widersten, B. (1968). On the morphology and development in some cnidarian larvae. *Zool. Bidrag. Uppsala* **37**, 139-182.

Wietrzykowski, W. (1912). Recherches sur le développement des Lucernaires. *Arch. Zool. Exp. Gén. Ser. 5.* **10**, 1-95.

Wilson, E. B. (1883). The development of *Renilla. Phil. Trans. Roy. Soc. London* **174**, 723-815.

Wolpert, L. (1971). Positional information and the spatial pattern of cellular differentiation. *Symp. Soc. Exp. Biol.* **25**, 391-415.

Woltereck, R. (1905a). Beitraege zur ontogenie und Ableitung des Siphonophorenstocks mit Anhang zur Entwickelungsphysiologie der Agalmiden. *Z. Wiss. Zool.* **82**, 611-637.

Woltereck, R. (1905b). Bemerkungen zur Entwickelung der Narcomedusen und Siphonophoren. *Verh. Deutsch. Zool. Ges. Vers.* **15**, 106-122.

Wulfert, J. (1902). Die Embryonal Entwicklung von Gonothyrea. *Z. Wiss. Zool.* **71**, 296-327.

Yoshida, M. (1954). Spawning habit of *Hydractinia epiconcha*, a hydroid. *J. Fac. Sci. Tokyo Univ. Zool.* **7**, 67-78.

Zoja, R. (1895). Sullo sviluppo dei blastomeri isolati dalle nova di alcune meduse (e di altri organismi). *Arch. Entwicklungsmech. Organ.* **1**, 578-595.

Chapter 4

CTENOPHORA

Helen Dunlap Pianka

4.1 Introduction

Five orders are recognized among the Ctenophora: Cydippida, Cestida, Lobata, Beroida, and Platyctenea.* The phylum is characterized by biradial symmetry, and members possess eight radially arranged rows of ciliated combs, a biradially structured gastrovascular system, and, except for the beroids, two tentacles (Fig. 1). All are marine, free-living, and planktonic or at least pelagic, but the platyctenids *Coeloplana, Tjalfiella, Vallicula,* and *Savangia* lose their larval combs and as adults assume a creeping or sessile existence. One platyctenid, *Gastrodes,* is apparently parasitic.

Much of this chapter is based on the author's study of oogenesis and spawning in several species of pelagic ctenophores collected at

*The division of the phylum into classes Nuda (Beroida) and Tentaculata (all other orders), based on the absence or presence of tentacles (Hyman, 1940), is of doubtful validity (Komai, 1963).

the Friday Harbor Laboratories, San Juan Island, Washington (Dunlap, 1966). Emphasis in this study was on one species of the lobate genus *Bolinopsis* (possibly *B. microptera*, A. Agassiz, 1865), but accompanying observations were made on the cydippids *Pleurobrachia bachei* (A. Agassiz, 1865), another species of *Pleurobrachia* (possibly *P. pigmentata*, A. Agassiz, 1865), *Dryodora glandiformis* (L. Agassiz, 1860), and one species of *Beroë* (possibly *B. cyathina*, A. Agassiz, 1865). Unless otherwise noted, mention of these species in the chapter refers to this research.

4.2 Asexual Reproduction

All adult ctenophores thus far studied are able to regenerate lost parts, and many can regenerate completely from small fragments (Mortensen, 1912, 1913; Krempf, 1921; Zirpolo, 1930; Okada, 1931; Tanaka, 1931a,b; Coonfield, 1936, 1937; Dawydoff, 1938a; Tokin, 1961; Freeman, 1966, 1967). However, this ability is known to be used in asexual reproduction only in the platyctenid genera *Coeloplana* (Krempf, 1921; Tanaka, 1931a,b), *Planoctena* (Dawydoff, 1938a), and *Vallicula* (Freeman, 1967). The process is similar in all species, and has been compared with pedal laceration in anemones (Hyman, 1940). Small pieces break off from the edge of the adult body, usually as the animal is creeping about, and often at night (Tanaka, 1931b), and each piece regenerates completely to form a new individual.

Sexual and asexual reproduction take place simultaneously in *Coeloplana gonoctena* from the South China Sea (Krempf, 1921), *C. bocki* from Japan (Tanaka, 1932), and *Vallicula multiformis* from Bermuda (G. Freeman, personal communication), but occur at different times in the Japanese species *Coeloplana mitsukurii* and *C. echinicola* (Tanaka, 1931a,b). In the South China Sea, Krempf (1921) noted a year-round occurrence of laceration in *C. gonoctena*, although Dawydoff (1938a) collected lacerating *Planoctena* only in July and August. Tanaka (1931a,b) reports that, in Japan, asexual reproduction of *C. echinicola* and *C. mitsukurii* occurs only from June through mid-August, when the water temperature ranges from 22° to 29°C. Later, when water temperatures exceed 32°C, both species disappear from the littoral zone. They reappear with cooler autumn waters but now reproduce sexually rather than asexually. Spontaneous laceration is inhibited in the laboratory at temperatures above

30°C, and regeneration of experimentally excised pieces of tissue takes longer than at lower temperatures (Tanaka, 1931a).

Nutrients probably affect the natural occurrence of asexual reproduction. Freeman (1967) observed that fed specimens of *Vallicula* lacerate about four times more often than those not fed.

4.3 Sexual Reproduction

4.3.1 Hermaphroditism

All ctenophores are hermaphrodites, most simultaneously, but protandry occurs among platyctenids. Protandry is best documented for *Coeloplana gonoctena*, an ectocommensal on *Alcyonium* (Krempf, 1921; Dawydoff, 1938b). None of Krempf's specimens had both mature ovaries and mature testes. Those with testes were frequently quite small, abundant, and could be found at any time of year, while those with ovaries were always large, rare, and were collected only near the end of summer. Among other species of *Coeloplana*, animals frequently have mature testes but immature ovaries (Komai, 1922; Dawydoff, 1938b). Mature ovaries have never been observed in species of *Ctenoplana*, *Planoctena*, and *Savangia*, although mature testes are repeatedly described (Willey, 1896; Dawydoff, 1933, 1936, 1950; Komai, 1934). Although these observations suggest protandry, there could well be an earlier maturation of testes, with eventual simultaneous hermaphroditism. Moreover, some platyctenids are probably not protandrous. Apparently *Coeloplana bocki*, one of the better known species in the order, is usually a simultaneous hermaphrodite (Komai, 1922).

In several species of *Coeloplana*, there are possible seminal receptacles, which would allow storage of sperm until ovaries mature (see Section 4.3.2.5).

In the platyctenid *Gastrodes*, candidates for oocytes have been described but none for developing or mature sperm (Section 4.3.4.2.2). However, the genus is poorly known (Section 4.4.2.2.2), and there is no reason as yet to suspect protogyny or a separation of the sexes.

4.3.2 Anatomy of the Reproductive System

Ctenophores are at the tissue grade of construction (Hyman, 1940). Thus, strictly speaking, their reproductive system consists of gametogenic tissues associated with various specialized epithelia. However, in their degree of specialization, organization, and compaction, gonadal tissues in some cases may merit consideration as organs.

4.3.2.1 Location and Anatomy of the Gonads

Gametogenic tissues occur in the eight meridional canals underlying the comb rows, or in their homologues or derivatives. Typically each canal contains tissues of both sexes, with ovaries being perradial and bordering the major (tentacular and sagittal) planes of the animal, while testes are interradial and face the minor planes (Fig. 1). This arrangement holds throughout the phylum, except that in cestids, some platyctenids, and temporarily in certain larvae and juveniles (Section 4.3.6.5), entire meridional canals (either the four subtentacular or the four subsagittal) are sterile.

4.3.2.1.1 *Cydippida.* Gonads of cydippids, comprising the most primitive order of ctenophores, are simple generally uninterrupted tracts of tissue extending along a major part of each of the eight meridional canals (Fig. 1) (A. Agassiz, 1874; Hertwig, 1880; Chun, 1880, 1898; Komai, 1918; Krumbach, 1927). Gonads occur in only the four subtentacular canals of *Euchlora rubra* (Chun, 1880; Komai and Tokioka, 1942), *Charistephane fugiens* and *Tinerfe cyanea* (Chun, 1880, 1898). However, these could be juveniles, perhaps of other species (Mayer, 1912; Komai, personal communication; see Section 4.3.6.5), especially since *Euchlora filigera* (Chun, 1880) and *Tinerfe lactea* (Mayer, 1912) have gonads in all eight canals.

4.3.2.1.2 *Cestida.* Sexual tissues in cestids are limited to the four subsagittal canals bordering the aboral rim of the animal's flattened vermiform body. In *Cestum*, gonads form continuous tracts of tissue along the full length of these canals, while in the smaller *Velamen* (*"Vexillum"*) they are interrupted, resulting in a number of genital swellings alternating with sterile sections (Fig. 2) (Fol, 1869; Chun, 1880; Bigelow, 1912).

4.3.2.1.3 *Lobata.* Some lobates become sexual as larvae (Section 4.3.6.5), and the larval gonads resemble those of adult cydippids, although they occur only in subsagittal canals (Fig. 9D). Gonads of adult lobates are found in both subsagittal and subtentacular meridional canals, and are of two general types. In *Leucothea* (=*Eucharis*) and perhaps in certain *Mnemiopsis* species they are limited to blind pockets lateral to the canals at the level of each comb plate (Fig. 3A) (Chun, 1880; Mayer, 1912; Komai, 1918; L. Baker, personal communication). In *Deiopea*, *Bolinopsis* (= *Bolina*), and *Mnemiopsis leidyi*, gonads occur mainly or only in the stretches of the canals between

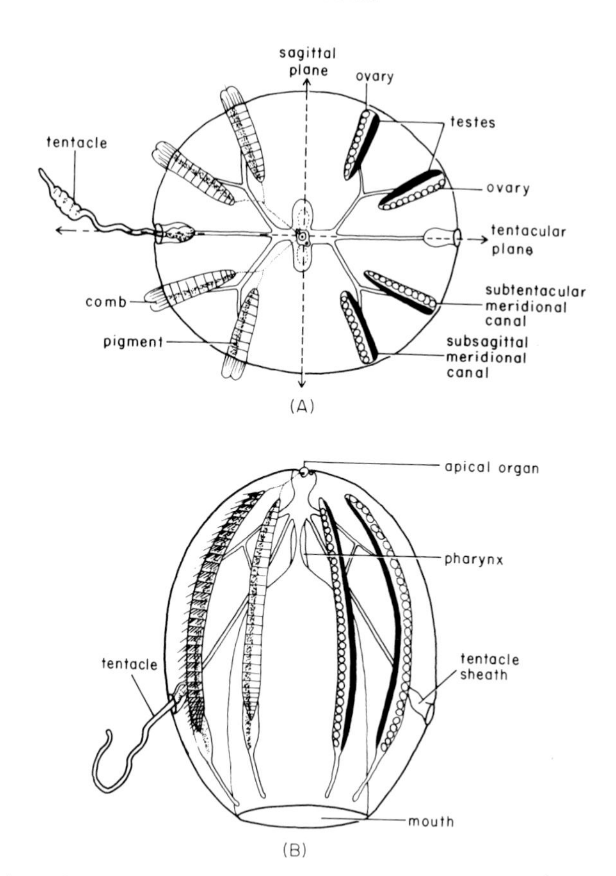

FIG. 1. *Dryodora glandiformis*, a primitive cydippid ctenophore, showing the basic biradial symmetry and plan of construction of all ctenophores, and the gross structure of the gonads among members of the order Cydippida. Pigment occurs only over the testes in this species. Gonopores are located along the gonads at the level of the tentacle sheath openings (see Fig. 19B-C). Testes shown as solid color; ovaries depicted as circles. (A) View from aboral pole. (B) View from end of sagittal axis. (Drawn from life. Tentacles and combs not shown on the right; gonads omitted on left.)

combs, and the lateral comb pockets are sterile (*Deiopea, Mnemiopsis leidyi*), poorly developed (*Bolinopsis microptera*) (Fig. 3B), or absent (*B. vitrea* = "*hydatina*") (Chun, 1892; Krumbach, 1927; Freeman and Reynolds, 1973). In *Bolinopsis* at least, gonadal tissues appear to expand with growth of the animal and tend to fuse beneath the combs, thus secondarily assuming the simple cydippid condition (Section 4.3.3) (Chun, 1892).

4.3.2.1.4 *Beroida*. Among beroids, gonadal tissues either form simple tracts in the meridional canals proper, or they extend to varying degrees into the lateral branches characteristic of beroid

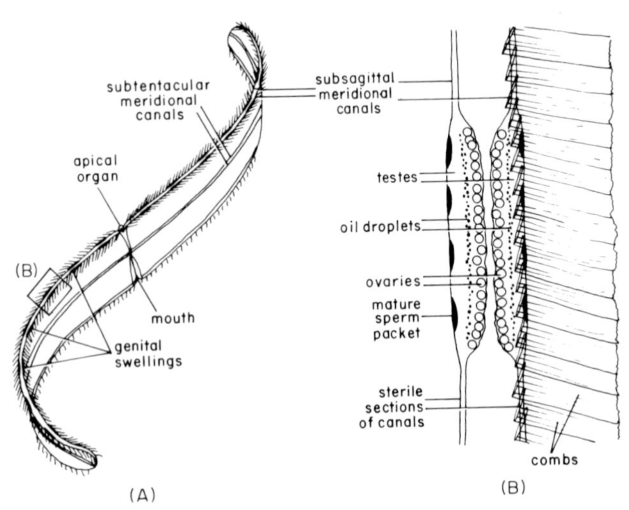

FIG. 2. The cestid *Velamen parallelum*. (A) Lateral view of whole animal, showing gonads as periodic genital swellings along the four subsagittal canals. The two pairs of subsagittal canals extend in opposite directions from the apical organ. (B) Enlargement of the genital swellings in two adjacent canals. Combs not shown over the canal to the left. [Redrawn from Chun (1880).]

meridional canals, or they develop in separate sexual diverticulae external to these branches (L. Agassiz, 1860; Hertwig, 1880; Chun, 1880; von Ledenfeld, 1885; Bigelow, 1912; Komai, 1918; Krumbach, 1927). In some, testes alone are housed in such special diverticulae, while ovaries occupy the proximal parts of the regular lateral branches (Fig. 3C) (L. Agassiz, 1860; Komai, 1918). Sexual tissues have also been observed in branches of the marginal canal bordering the mouth (Komai, 1918). The literature often contains varying descriptions of the gonads of apparently the same species, and it is likely that at least in some beroids there is a continuing expansion of the gonadal tissues with growth of an individual. Thus, gametogenic tissues may first occur only in the meridional canals proper, but later proliferate into the lateral branches, or sexual diverticulae may form in older specimens. However, such ontogenetic changes in gonadal structure have not been documented, and may not be of universal occurrence among beroids. The existing state of confusion in the taxonomy of *Beroë* (see Mortensen, 1912; Mayer, 1912; Bigelow, 1912; Nelson, 1925; Greve, 1970), adds to difficulties involved in attempts at assessing trends within the order.

4.3.2.1.5 *Platyctenea*. The variability of platyctenid gonads is in accord with a probable polyphyletic origin of this order (Komai, 1963). One species, *Tjalfiella tristoma*, retains the simple cydippid

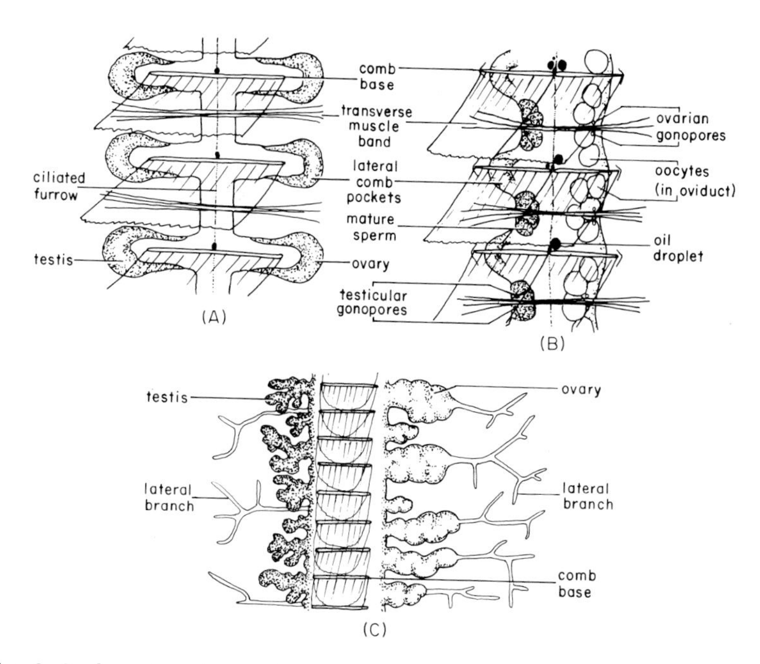

FIG. 3. Surface view of part of a comb row and its underlying meridional canal, containing the gonads, in two lobates (A and B) and a beroid (C). Testes to the left, ovaries to the right. (A) *Leucothea ("Eucharis") multicornis*. Gonads lie in blind pockets lateral to each comb. [Redrawn from Chun (1880).] (B) *Bolinopsis microptera*. Gonads occur along the length of the canal, and lateral comb pockets develop on the interradial side (left), and contain immature testicular tissue. (Drawn from life.) (C) *Beroë* sp. (*"Idyia roseola"*) Branches lateral to the meridional canals are sterile on the interradial (left) side, but contain ovarian tissues in their proximal parts on the perradial side. Testes lie in diverticulae external to these lateral branches. Ovarian and testicular tissues also occur in the meridional canal itself. Pigment is more highly concentrated over testes than ovaries. [Redrawn from L. Agassiz (1860).]

condition, with gonads forming uninterrupted masses of tissue in each of the eight reduced meridional canals (Fig. 4A) (Mortensen, 1912). An entirely different arrangement, resembling that of some lobates and beroids, occurs in *Lyrocteis imperatoris*, in which both ovaries and testes develop as numerous (up to 600) complex diverticulae along the meridional vessels, which themselves are sterile (Fig. 4D and E) (Komai, 1942). The reproductive system of *Vallicula multiformis* has not been adequately described, but both ovarian and testicular tissues are localized in distinct pockets among the anastomosing branches of the peripheral gastrovascular system, with one individual possessing up to 15 ovaries and 30 testes (G. Freeman, personal communication).

Throughout the rest of the platyctenids, ovaries have been ob-

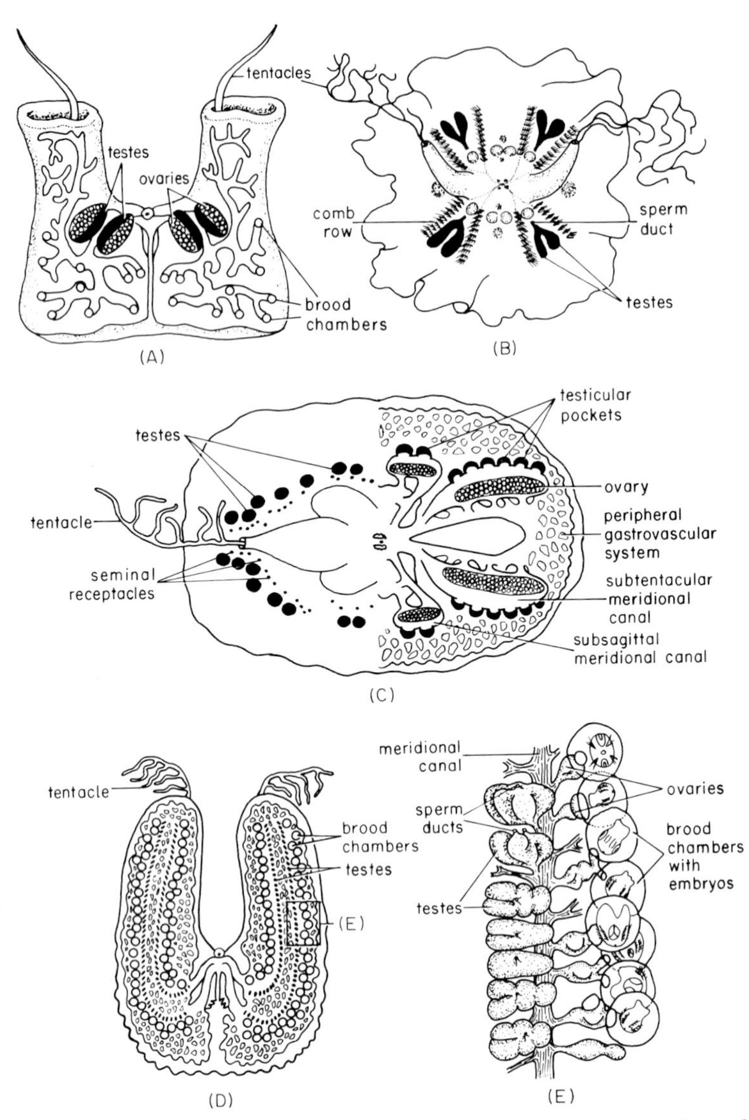

FIG. 4. Gonadal structure among platyctenids. (A) *Tjalfiella tristoma*. Lateral view of animal showing four of the eight "genital organs," each containing an ovary (circles) and a testis (solid). Brood chambers shown as circles at the blind ends of the peripheral gastrovascular branches. [Redrawn from Komai (1942).] (B) *Planoctena yuri*. Aboral view, showing the four pairs of testes, each joined by a common sperm duct, lying along the interradii between comb rows. Ovaries have not been observed in this species. [Redrawn from Dawydoff (1936).] (C) *Coeloplana bocki*. Aboral view. Left half drawn to show general appearance of testes and seminal receptacles in surface view of whole animal. Right side shown with epidermis and tentacle removed, and

served only in several *Coeloplana* species, as simple continuous tracts of tissue (Fig. 4C) (Komai, 1922; Tanaka, 1932). Testes have been described for most species and are always more complicated. Thus, in many *Coeloplana* species they form from 15 to 30 pockets along the meridional trunks (Fig. 4C) (Krempf, 1921; Komai, 1922; Dawydoff, 1938b). These pockets are more numerous along the subtentacular canals, and may be absent from the shorter subsagittal canals (Komai, 1922). Among species of *Ctenoplana* testes occur only along the subtentacular canals, in the form of *Coeloplana*-like pockets in *Ctenoplana maculomarginata* (Komai, 1934; see Fig. 218B in Hyman, 1940) and as four large compact testes in *Ctenoplana korotneffi* (Willey, 1896). Similar compact testes, joined in pairs aborally, are found along the subsagittal canals of *Savangia* (Dawydoff, 1950), and along all eight canals of *Planoctena yuri* (Fig. 4B) (Dawydoff, 1933, 1936).

Gonads have not been observed in *Gastrodes*, although possible oocytes in the ventral epidermis have been described (Section 4.3.4.2.2).

4.3.2.2 Histology of the Gonadal Tissues

A meridional canal is typically bordered on the inner side by a low ciliated epithelium, and on the outer (or, in platyctenids, dorsal) side by a high, sparsely ciliated epithelium composed of complex digestive cells (Fig. 5 and 6). The sexual tissues are closely associated with the latter, occurring generally between it and the gelatinous body parenchyme. Gonadal pockets and diverticulae (Section 4.3.2.1) have not been adequately examined, but they usually contain a cavity communicating with the meridional canal and are lined entirely by apparent digestive cells. The digestive epithelium associated with the ovary is often more complex than that covering the testis.

Basal extensions of the digestive cells, the interstitial processes, ramify throughout the ovarian tissues and to a lesser extent into the testis. These are doubtlessly functional in transport and distribution of nutrients to developing gametes and their associated cells, especially the ovarian nurse cells (Section 4.3.4.2.1) and the oviducal gland

ovaries and testicular pockets diagrammatically represented. [Modified from Komai (1922).] (D and E) *Lyrocteis imperatoris*. (D) Lateral view showing numerous testicular diverticulae and brood chambers along meridional canals (ovaries not drawn.). (E) Enlargement of part of one meridional canal, with testicular and ovarian diverticulae, and brood chambers containing developing embryos, shown as dissected away from surrounding tissues. [Redrawn from Komai (1942).]

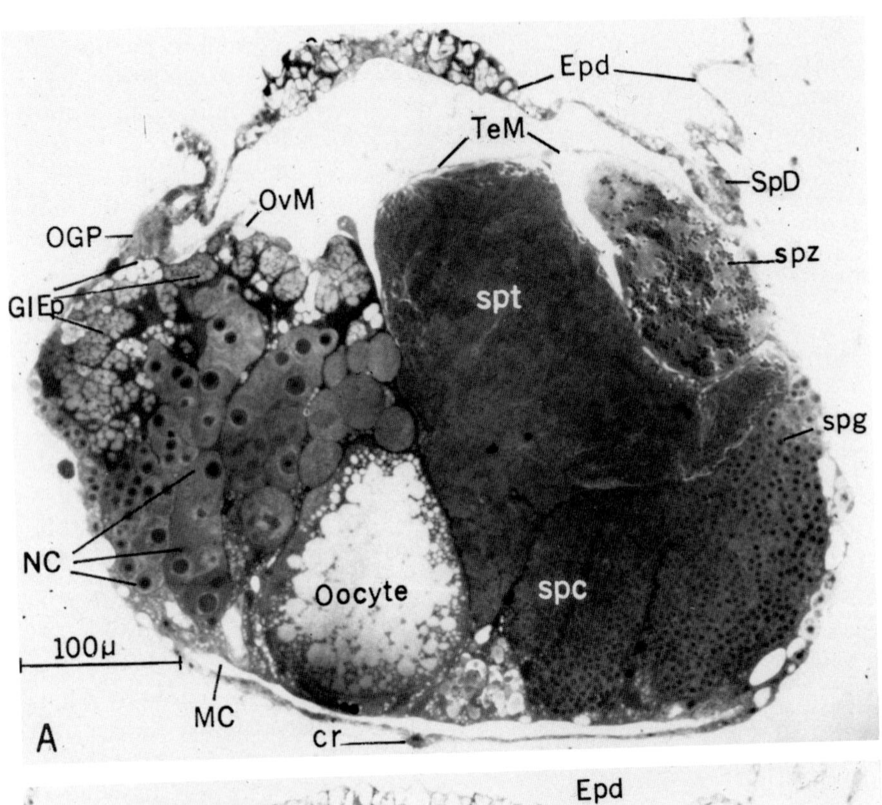
Epd
TeM
SpD
OvM
OGP
spz
GlEp
spt
spg
NC
Oocyte
spc
100μ
MC
A
cr

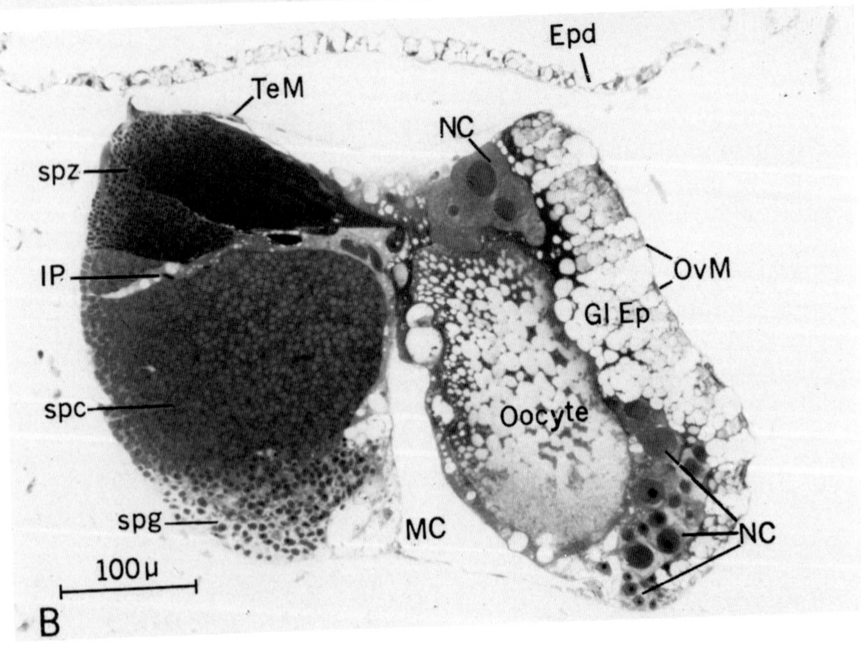
Epd
TeM
NC
spz
IP
OvM
Gl Ep
spc
Oocyte
spg
MC
NC
100μ
B

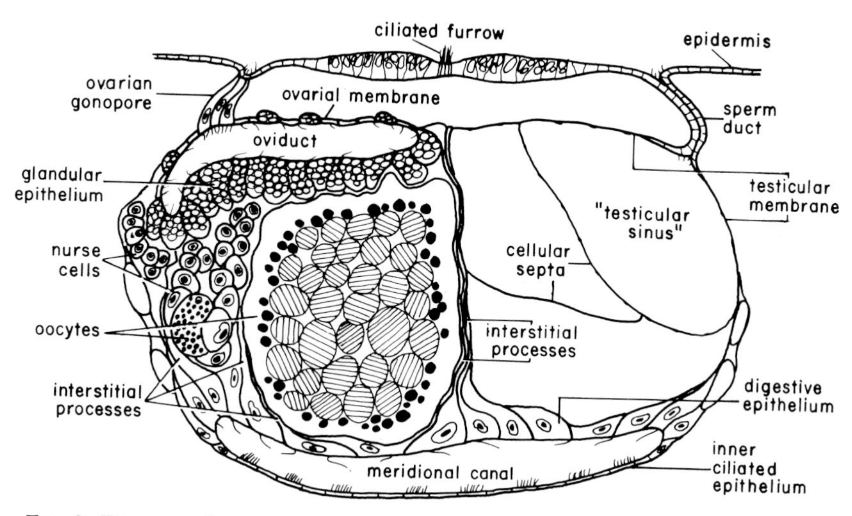

FIG. 6. Diagram of a representative cross section of a meridional canal and gonadal tissues, taken at the level of the gonopores, in *Bolinopsis microptera*. (See Fig. 5A.) Ovary on the left separated from testis on right by the long central interstitial processes. Developing and mature sperm not shown. Mature sperm would occupy the "testicular sinus."

cells (Section 4.3.2.3.2). Similar cell processes separate the gonadal tissues from the body parenchyme, and in some species evidently form septa between packets of maturing spermatocytes and spermatids (Section 4.3.4.1) (Fig. 5B).

Separation of ovarian and testicular tissues is accomplished merely by such thin interstitial processes in many species *(Bolinopsis, Pleurobrachia bachei)* (Fig. 5A and 6). In others *(Dryodora, P. pigmentata),* the ovary and testis are separated by the canal lumen as well (Fig. 5B) (Section 4.3.4). In still others there are varying degrees of isolation of ovaries and testes as described above (Section 4.3.2.1.).

The oviduct epithelia (4.3.2.3.2) lie external (*Bolinopsis, P. bachei*) or lateral *(Dryodora, P. pigmentata)* to the ovarian tissues (Fig. 5 and 6). Covering the testis in a comparable position is a thin epithelium, the "testicular membrane" (4.3.2.3.1). Pigment, commonly observed in living animals over the testes, as in *Dryodora* (Fig. 1), occurs apparently within large cells closely associated with the testicular

◀FIG. 5. Photomicrographs of cross sections of a meridional canal and gonadal tissues of the lobate *Bolinopsis microptera* (A) and the cydippid *Dryodora glandiformis* (B). Material fixed with Osmium tetroxide, hence yolk in the oocytes is not preserved. OGP, ovarian gonopore cells; OvM, ovarial membrane; GlEp, glandular epithelium of the oviduct; NC, nurse cells; SpD, sperm duct; TeM, testicular membrane; spg, spermatogonia; spc, spermatocytes; spt, spermatids; spz, spermatozoa; IP, interstitial processes of digestive cells; MC, meridional canal lumen; cr, cell rosette (see Hyman, 1940); Epd, epidermis. (Compare Fig. 5A with Fig. 6.)

membrane or with interstitial processes.

4.3.2.3 Gonoducts and Gonopores

Evidence for gonopores and gonoducts comes from observations of spawning (Section 4.3.8.1), and from histological studies, and is available for representatives of all orders except the Cestida, which have not received adequate attention.

4.3.2.3.1 *Sperm Ducts.* Sperm ducts have been observed in the platyctenids *Coeloplana* (Krempf, 1921; Komai, 1922; Dawydoff, 1938b), *Ctenoplana* and *Planoctena* (Willey, 1896; Dawydoff, 1933, 1936), *Lyrocteis* (Komai, 1942), and *Savangia* (Dawydoff, 1950), and in the lobate *Bolinopsis* (Dunlap, 1966) (Fig. 4B and E, 5A, 6). Where histological observations have been made (*Coeloplana, Ctenoplana, Lyrocteis, Bolinopsis*) each duct is revealed as a narrow ciliated tube spanning the parenchyme from the testis to the epidermis, opening through the latter by a pore (Fig. 6) (see also Fig. 220A in Hyman, 1940).

Actual spermiation through these ducts has been observed in living animals only in *Bolinopsis* (Section 4.3.8.1). Komai (1922, 1942) observed evident sperm release through the ducts in histological sections of *Coeloplana* and *Lyrocteis*. However, their structure and location leave no doubt that these are functional sperm ducts in all cases. In the cydippids *Dryodora* and *Pleurobrachia*, ducts have not been obtained in sections, but their existence is implicated by spawning observations (Section 4.3.8.1).

The number of ducts varies with the species. There are up to 600 in an individual *Bolinopsis* (with two between each pair of combs; Fig. 3B) or *Lyrocteis* (with one or two ducts for each testicular diverticulum; Fig. 4E) (Komai, 1942). At the other extreme are *Ctenoplana, Planoctena,* and *Savangia*, in which each large testis or pair of testes is served by one or two ducts, hence an individual has two to eight ducts (Fig. 4B)(Willey, 1896; Komai, 1934; Dawydoff, 1936, 1950). Similarly, one duct, or at the most a few, occurs for each of the eight testicular tracts of *Dryodora* (Section 4.3.8.1) (Fig. 19). Intermediate numbers of ducts occur in *Coeloplana* and certain *Ctenoplana* species, with one for each of the 10-50 testicular pockets (Komai, 1922, 1934).

The entire process of sperm release may involve more than the sperm ducts alone. The duct epithelium is continuous with the testicular membrane (Fig. 6) (Willey, 1896; Komai, 1922, 1942), and mature sperm occupy a space, the "genital sinus" of earlier literature, bounded on the outside by the testicular membrane and separated from the immature part of the testis by a cellular septum (Section 4.3.4.1). Sperm could be released into this "testicular sinus" before being discharged through the ducts, thus accounting for visible changes which occur just prior to spawning (Section 4.3.8.1). Although it requires further investigation, the testicular sinus and the sperm ducts would then be analogous with, respectively, the oviduct and ovarian gonopores described next.

4.3.2.3.2 *Oviducts and Ovarian Gonopores.* These have been both discovered and described in the species of *Bolinopsis*, *Dryodora*, *Beroë*, and *Pleurobrachia* at Friday Harbor (Dunlap, 1966). Each oviduct is a flattened tube external or lateral to the ovarian tract, paralleling the length of the meridional canal. It is bordered on its inner side, closely applied to the ovary, by a complex epithelium composed of two types of sparsely ciliated gland cells (Fig. 5 and 6). One type of gland cell is more numerous and contains a mucopolysaccharide secretion (Fig. 16). The other is rarer and contains a vaguely granular secretion. On its outer side, bordering the parenchyme, the oviduct is lined with a low ciliated epithelium, interspersed with small mucous cells (Fig. 6). This outer epithelium is equivalent to the "ovarial membrane" of earlier literature (e.g., Komai, 1922).

In *Dryodora*, each oviduct is a long continuous tube provided with one gonopore (Fig. 19B-C) (Section 4.3.8.1). In both species of *Pleurobrachia* and in *Bolinopsis*, the glandular epithelium is reduced beneath each comb, and here the ducts are functionally discontinuous during spawning (Fig. 19D-E). In these latter species each duct opens by many gonopores, with one pore between each pair of combs in the *Pleurobrachia* species (Fig. 19E), and two in *Bolinopsis* (Fig. 3B).

A closed gonopore is represented in sections by a strand of about six to eight generally cylindrical cells forming a highly ciliated part of the ovarial membrane. These cells span the parenchyme and connect with a slight ciliated invagination of the epidermis (Fig. 7A, also Fig. 5A and 6). Electron micrographs of these cell strands in *Bolinopsis* reveal numerous cytoplasmic filaments, similar to those in nearby smooth muscle cells.

During spawning, oocytes become ameboid and ovulate through the glandular epithelium into the oviduct, where secretions of the gland cells adhere to the vitelline membrane and form the egg's jelly coat (Fig. 7B) (Section 4.4.1.2). Oocytes are then released through the gonopore which forms in the center of the cell strand and its connected epidermal invagination (see Section 4.3.8.1).

Although oviducts have not previously been recognized as such, illustrations of the ovaries of many ctenophores (Hertwig, 1880;

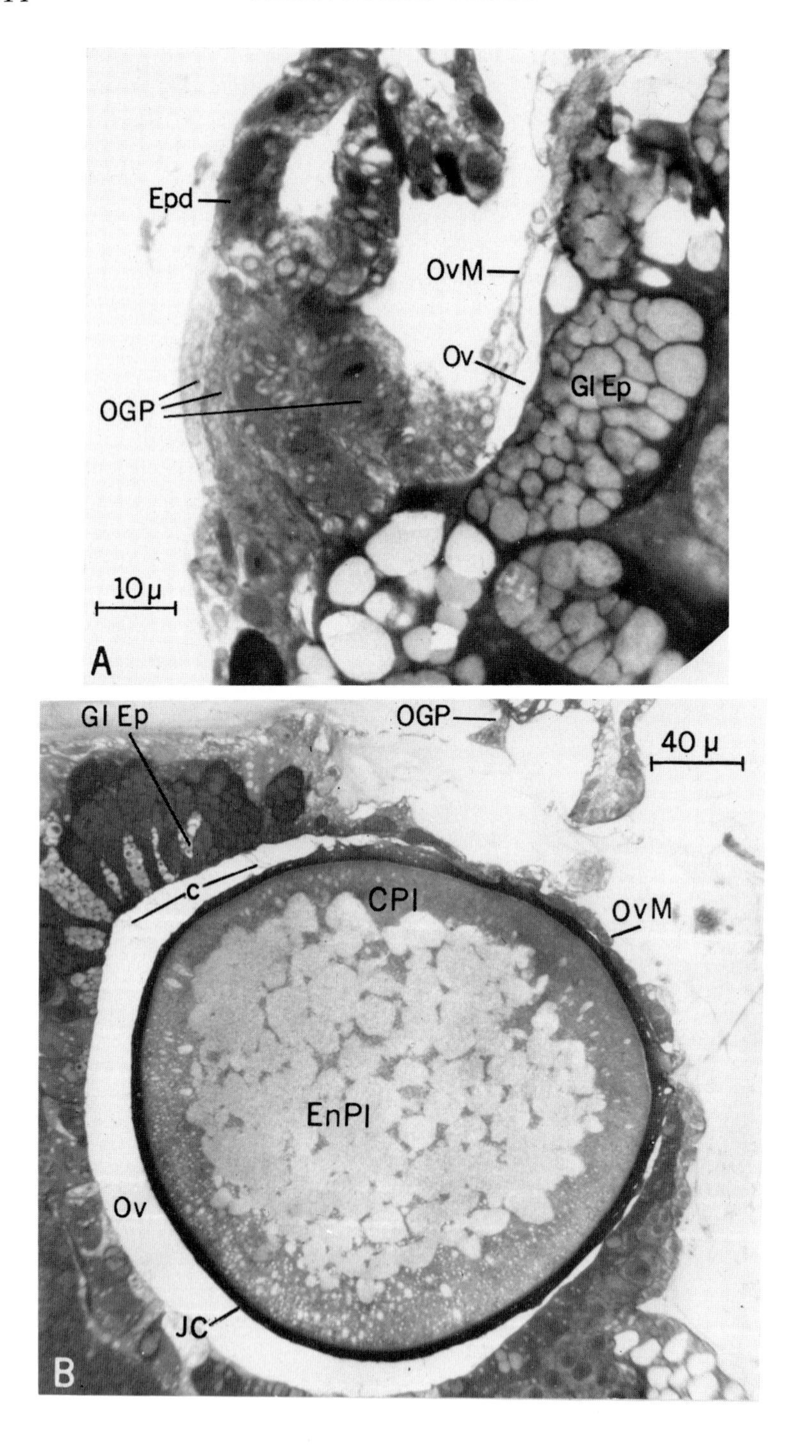
Epd
OvM
Ov
Gl Ep
OGP
10 μ
A
Gl Ep
OGP
40 μ
c
CPl
OvM
EnPl
Ov
JC
B

Chun, 1892, 1898; Garbe, 1901; Komai, 1922) indicate that they are probably of universal occurrence throughout the phylum. In previous studies, the gland cells and ovarial membrane have been illustrated, although the lumen, which is narrow and slitlike except during spawning, has been usually ignored (see Fig. 8). Chun (1892) predicted that the gland cells produce the egg's jelly coat, although he also believed that these cells remain attached to the oocytes as the latter ovulate into the meridional canal lumen (but see Section 4.3.8.1).

That ovarian gonopores occur in many species is evident from observations of spawning (Section 4.3.8.1). In sections, the cluster of gonopore cells appears comparable to an entire sperm duct (Fig. 5A and 6); and the "Verbindungsstränge", or cell strands, connecting both ovaries and testes with the overlying epidermis, as described for many ctenophores (Hertwig, 1880; Chun, 1892, 1898; Komai, 1922) (Fig. 8), doubtlessly refer to both. The problematical "Genitalsäckchen" in *Callianira* (see Section 4.3.2.5 and 4.3.3) might well be comparable to the ciliated epidermal invaginations connecting with the ovarian gonopore cells of *Bolinopsis*.

Problems remain in the case of *Coeloplana bocki*, in which Komai (1922) reported egg release through the mouth, and possibly also in *C. gonoctena*, in which Krempf (1921) suggested that eggs are shed through the ventral epidermis. Krempf does not provide an adequate description of the ovary of *C. gonoctena*, but the gonopores could open through the ventral epidermis. In *C. bocki*, there is a distinct oviduct with the component epithelia, and cell strands (Fig. 8), possibly ovarian gonopore cells, which in certain cases connect with the epidermis. Disregarding the mechanism of final egg release (see Section 4.3.8.1), oocytes doubtlessly first ovulate into the oviduct where they acquire their gelatinous coating (Section 4.4.1.8). In *C. bocki*, the cell strands might therefore represent "vestigial" ovarian gonopores.

4.3.2.4 Brood Chambers

Internal brood chambers have been observed in the platyctenids

◀Fig. 7. (A) Enlargement of a section through the ovarian gonopore cells of *Bolinopsis microptera* (see Fig. 5A and 6). The ciliated epidermal invagination (Epd) connects with one end of the gonopore cells (OGP), while the inner ends of these cells border the oviduct lumen (Ov). The oviduct is lined with an inner glandular epithelium (G1Ep) and an outer ovarial membrane (OvM). Note the two types of gland cells. (B) Mature *Bolinopsis* oocyte within the oviduct, covered with a jelly coat (JC). The two main zones of the ooplasm, the endoplasm (EnPl) and cortical plasm (CPl), are evident. Yolk in the former is not preserved. Ov, lumen of oviduct; c, cilia in oviduct; GlEp, glandular epithelium of oviduct; OGP, part of ovarian gonopore cells near their connection with the epidermis.

Tjalfiella tristoma (Mortensen, 1912) and *Lyrocteis imperatoris* (Komai, 1942) (see Section 4.4.1.8). Those in *Tjalfiella* are arranged irregularly beneath the lateral epidermis, sometimes in a double layer, and each is a thin-walled sac abutting against a blind end of the ramifying gastrovascular system (Fig. 4A). Thus the brood chambers are separated from the gonads by much of the canal system. In *Lyrocteis*, on the other hand, the chambers occur in rows paralleling the series of ovaries and each is evidently an expansion of the distal part of an ovarian diverticulum (Fig. 4D-E).

Rankin (1956) observed embryos of *Vallicula multiformis* developing within four thick-walled "spherical chambers" at the junction of the tentacular and transverse canals. However, it is not known whether these chambers function primarily in brooding.

At least in *Lyrocteis* the brood chambers may well be homologues of oviducts in other ctenophores (Section 4.3.2.3.2). Ovoviviparity could be accomplished by having sperm enter through the ovarian gonopores, and fertilization occur within the oviducts, which would then be functional uteri. Brooded embryos of *Coeloplana mitsukurii* might also be located in oviducts, although Tanaka (1932) reports that embryos develop in the "gastrovascular spaces" (Section 4.4.1.8). In any case, in both *Lyrocteis* and *C. mitsukurii* embryos upon hatching are reported to "rupture" through the body epidermis (Tanaka, 1932; Komai, 1942), which could involve ovarian gonopores.

4.3.2.5 Seminal Receptacles

Possible seminal receptacles have been described in several *Coeloplana* species (Komai, 1920, 1922; Krempf, 1921; Tanaka, 1932). These are located above the ovaries, or at least above the perradial sides of the meridional canals, and in living animals are visible as a series of opaque white or pigmented spots on the dorsal side of the animal, paralleling the series of testes (Fig. 4C). In sections, each is flask-shaped, with an ampullar swollen blind inner end lying near the ovarial membrane, and a narrow ciliated duct leading to the outside through a pore in the epidermis (Fig. 8).

These have been interpreted as seminal receptacles by Komai (1920, 1922), who observed them in the Japanese species *Coeloplana bocki*, *C. willeyi*, and *C. mitsukurii*. In these species the ampullae are either empty, as occurs in most individuals with immature gonads, or contain masses of mature sperm, as usually occurs in specimens with mature ovaries and testes (Fig. 8), and occasionally in those with mature testes but immature ovaries.

Both the origin and fate of the stored sperm are unknown. Komai

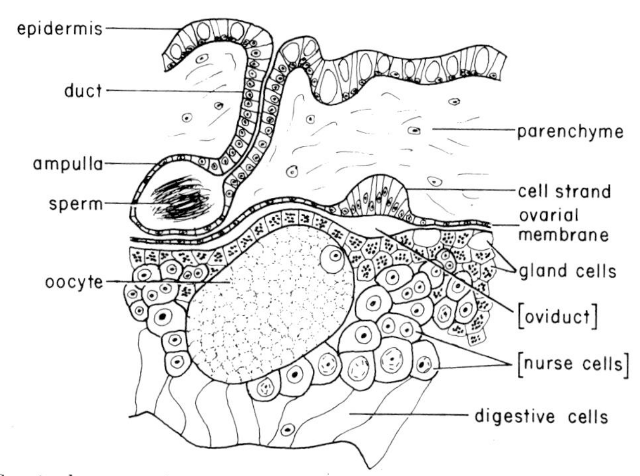

FIG. 8. Seminal receptacle of *Coeloplana bocki*, and its associated ovary, redrawn and modified from Komai (1922). The ampulla of the receptacle is filled with mature sperm. The nurse cells were interpreted by Komai as young oocytes. The "cell strand" could represent a proximal part of a cluster of ovarian gonopore cells. The oviduct and its epithelia, which Komai did not recognize, are exaggerated in this redrawing.

(1922) proposes that they originate from another individual, and penetrate the scant tissue between the receptacle and the ovary, to fertilize the oocytes of the same individual. It seems as likely that the sperm originate from the testes of the same animal, although the mechanism of collection is unknown. Transfer of sperm from one individual to another in concentrations observed in the receptacles would be difficult in the absence of a mating act, unless they are transferred in bundles, for which there is no evidence. [Sperm do develop in bundles in many ctenophores, and in *Ctenoplana korotneffi* mature sperm form organized masses (see Section 4.3.4.1). However, the only possible evidence of seminal receptacles in any *Ctenoplana* species is Komai's (1934) observation of empty receptaclelike structures in *C. maculomarginata*.] Sperm from the receptacle could then penetrate the ovarial membrane and fertilize the oocytes within the oviduct, of which Komai was unaware (Fig. 8) (Section 4.3.2.3.2). On the other hand, sperm could be released from the receptacle through its duct, perhaps much as they are released from the testes (Section 4.3.8.1), to fertilize oocytes of the same or another animal externally. In any case, if the structures in *Coeloplana* are seminal receptacles, their occurrence is probably related to protandry, or an earlier maturation of the testes (Section 4.3.1) and are necessary for an individual to store sperm until ovaries mature (Ghiselin, 1969) (see Section 4.4.1.1).

Krempf (1921), in the Indochinese species *C. gonoctena*, interpreted apparently identical structures as isolated small testes ("follicules testiculaires") occurring in addition to the main large testes ("testicules composés"). This is based on what he interpreted to be stages in the development of each as an endodermal outpocketing of the meridional canal epithelium which, complete with solid duct, migrates through the parenchyme to connect secondarily with the epidermis. He maintained that sperm arise from the epithelium of the ampulla during migration, and become fully mature before connection is made with the outside through a hollowing out of the duct. Although Krempf's observations and interpretation are difficult to accept, and he did not present convincing evidence of spermatogenic activity within the ampullae, further research is required before his findings are rejected. Komai did not observe stages in early development of the receptacles, although he did note changes concomitant with increasing maturity of the gonads, with the cells lining the ampullae becoming vacuolate and losing their initial ciliation.

Certain ciliated epidermal invaginations ("Genitalsäckchen") overlying a limited part of the ovary of the cydippid *Callianira bialata* (Hertwig, 1880; Chun, 1898), and the partially ciliated slitlike invaginations above the gonads of *Tjalfiella* (Mortensen, 1912) have both been compared with *Coeloplana's* putative seminal receptacles (Komai, 1922; Hyman, 1940). However, there is no evidence that sperm are ever contained within these structures, and they could be involved in gamete release (Section 4.3.2.3), or they might perform a sensory function, as realized by Hertwig (1880), and supported by Chun (1898) and Mortensen (1912).

4.3.3 Origin of the Germ Cells and Gonads

Recognizable germ cells in ctenophores are always closely associated with the endoderm. Although they could have migrated there during early embryonic development, from either the mesoderm (Schneider, 1904) or the ectoderm, there is no convincing evidence of either. There is, however, considerable support for an endodermal origin of the gametes and of all gonadal tissues from the studies of Chun (1880, 1892, 1898), Garbe (1901), Mortensen (1912), Krempf (1921), and Komai (1922, 1942). Hertwig's (1880) once controversial interpretation of the "Genitalsäckchen" of *Callianira* and the "Verbindungsstränge" of many cydippids (see Section 4.3.2.3 and 4.3.2.5) as evidence of an ectodermal origin of the gonads may be considered amply refuted (Chun, 1892, 1898; Mortensen, 1912; Komai, 1922). The situation in the platyctenid *Gastrodes*, which may differ from that in all other ctenophores, is treated in Section 4.3.4.2.2.

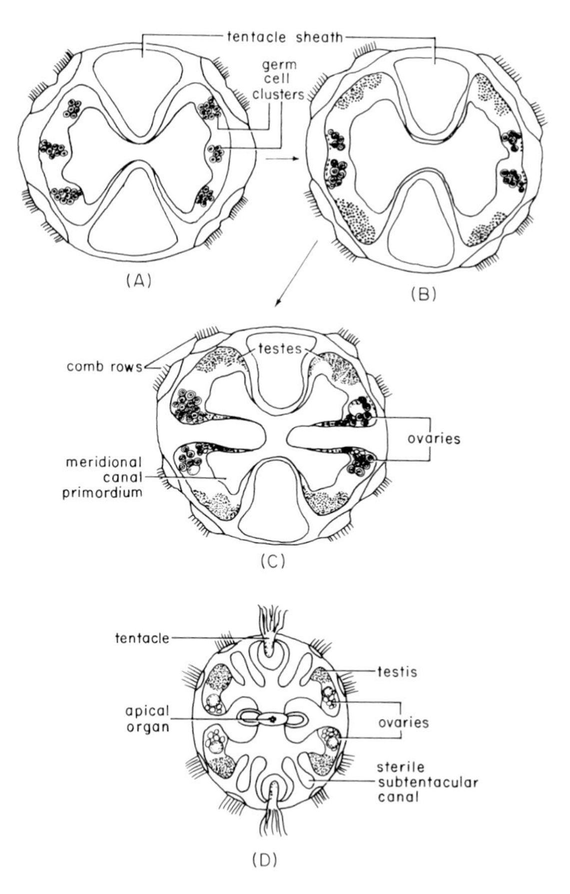

FIG. 9. Diagrammatic representations of the development and structure of larval gonads in two *Bolinopsis* species, shown as in equatorial cross sections of cydippid larvae. (A), (B), and (C) Stages in development of larval gonads of a *Bolinopsis* larva (*"Pleurobrachia rhodopsis"*). [Redrawn from Garbe (1901).] Subtentacular canals were not described in this form. (D) Sexual larva of *Bolinopsis vitrea* (*"Bolina hydatina"*) showing gonads in four subsagittal canals. See text for explanation. [Redrawn from Chun (1892).]

Germ cells are first recognized in the cydippid larva as clusters of cells associated with the outer endodermal lining of the meridional canal primordia. The development of larval gonads (Section 4.3.6.5) from these cells has been described briefly by Chun (1892) and in greater detail by Garbe (1901). Garbe observed three initial germ cell clusters in the oral extremes of each of the two meridional canal primordia, which proliferate aborally into three germ bands (Fig. 9A). The middle band of each triplet splits (Fig. 9B) and is separated by the division of each canal primordium (Fig. 9C). The resulting germ bands facing the sagittal plane form ovaries while those facing the tentacular plane form testes. Chun (1892) observed in another species that germ cells also occur in the subtentacular canal primordia (which Garbe did not describe in his species), but remain quiescent until after metamorphosis (Fig. 9D) (see Section 4.3.6.5). He did not observe the three initial germ cell clusters described by Garbe, but noted two clusters in each of the eight meridional canal primordia, in both larvae and juveniles (Chun, 1880, 1892).

During metamorphosis of *Bolinopsis vitrea* the larval gonads are reduced to isolated clusters of germ cells, lying in the meridional canals between adjacent combs, which later expand and fuse beneath the combs to form the uninterrupted condition of the adult gonads (Section 4.3.2.1.3).

Development of adult ctenophore gonads from germ cell clusters has not been adequately studied. However, Chun (1880, 1892), Garbe (1901) and Komai (1922) present some evidence that the ovarial and testicular membranes, the oviducal gland cells, and the sperm ducts and ovarian gonopore cells are all descended from the germ cells, although they were uncertain of the significance of most of these structures and hence did not give them the attention they deserve. For example, Komai (1922) observed stages in formation of sperm ducts and the ovarian cell strands of *Coeloplana bocki*, and describes each as growing out from their respective gonads and secondarily connecting with the epidermis. Finally, there is no doubt that the true nurse cells of the ovary (Section 4.3.4.2.1) are of direct germ cell origin.

4.3.4 Cytodifferentiation of the Gametes

Several distinct stages of gametogenesis are included in a representative cross section of the gonads of most ctenophores. The relative positions of these stages depends on the structure of the particular gonad (Section 4.3.2.1). In general, gonial stages are located near the junction of the inner ciliated and outer digestive epithelia of the meridional canals. On the ovarian side this corresponds also to the junction of the inner and outer oviducal epithelia (Fig. 5 and 6). Mortensen (1912) described a "germinal zone" in *Tjalfiella* which appears to be a continuation of the inner epithelium of the meridional canal.

In some species (e.g., *Pleurobrachia bachei, Bolinopsis*), maturing oocytes and spermatocytes approach each other towards the midline of the gonadal band in section (Fig. 5A, 6). In *Dryodora* and *P.*

pigmentata the ovary and testis are tilted, so that gonial stages occur towards the center of the animal, and maturing gametes approach the epidermis (Fig. 5B). In species with gonadal diverticulae and pockets *(Coeloplana, Lyrocteis, Ctenoplana)* (see Section 4.3.2.1), gonia are found proximally within the diverticulae and mature gametes distally, near the gonoducts, where present, or at least near the epidermis (Willey, 1896; Komai, 1922, 1942).

The state of gonadal development is generally the same for an entire animal (Section 4.3.6.4); however, this has not been adequately investigated. Komai (1922) observed that the more peripheral testicular pockets of *Coeloplana bocki* (comparable to the more oral gonads in other ctenophores) are often immature, while those towards the animal's center (aboral pole in other ctenophores) contain ripe spermatozoa. This probably reflects a continuing peripheral addition of testicular pockets with growth.

4.3.4.1 Spermatogenesis

Although testes of many ctenophores, encompassing all five orders, have been cursorily examined (Fol, 1869; Chun, 1880, 1892, 1898; Hertwig, 1880; Willey, 1896; Mortensen, 1912; Krempf, 1921; Komai, 1922, 1942), details of spermatogenesis remain largely unknown.

Spermatogonia occur in a localized part of the testis (see above) and give rise to batches of developing spermatocytes (Fig. 5 and 6). These occur in large packets in which all component cells are in the same stage of development. As the cells mature, presumably near the onset of spermiogenesis, each packet becomes separated from the less mature packets by a cellular septum, formed by interstitial processes of the digestive epithelium in some species (*P. bachei, Dryodora*, Fig. 5B), but by separate small cells in others (*Bolinopsis microptera*, Fig. 5A and 6; probably also in *Coeloplana bocki*, Komai, 1922; *Lyrocteis*, Komai, 1942; and *Ctenoplana*, Willey, 1896). These septa are continuous with the testicular membrane surrounding the outer part of the testis (see Section 4.3.2.2 and 4.3.2.3.1).

Komai (1922, 1942) described certain large nonsexual cells among developing sperm in several *Coeloplana* species, and proposed that they are testicular nurse cells. They have not been observed in other ctenophores, and it is not clear whether they are distinct from the cells of the septa just described.

Electron micrographs reveal narrow open intercellular bridges connecting developing spermatocytes of *Bolinopsis* (Fig. 10), similar to those observed in other animal testes (Fawcett, 1961; Nagano, 1961). Up to four interconnected spermatocytes have been obtained

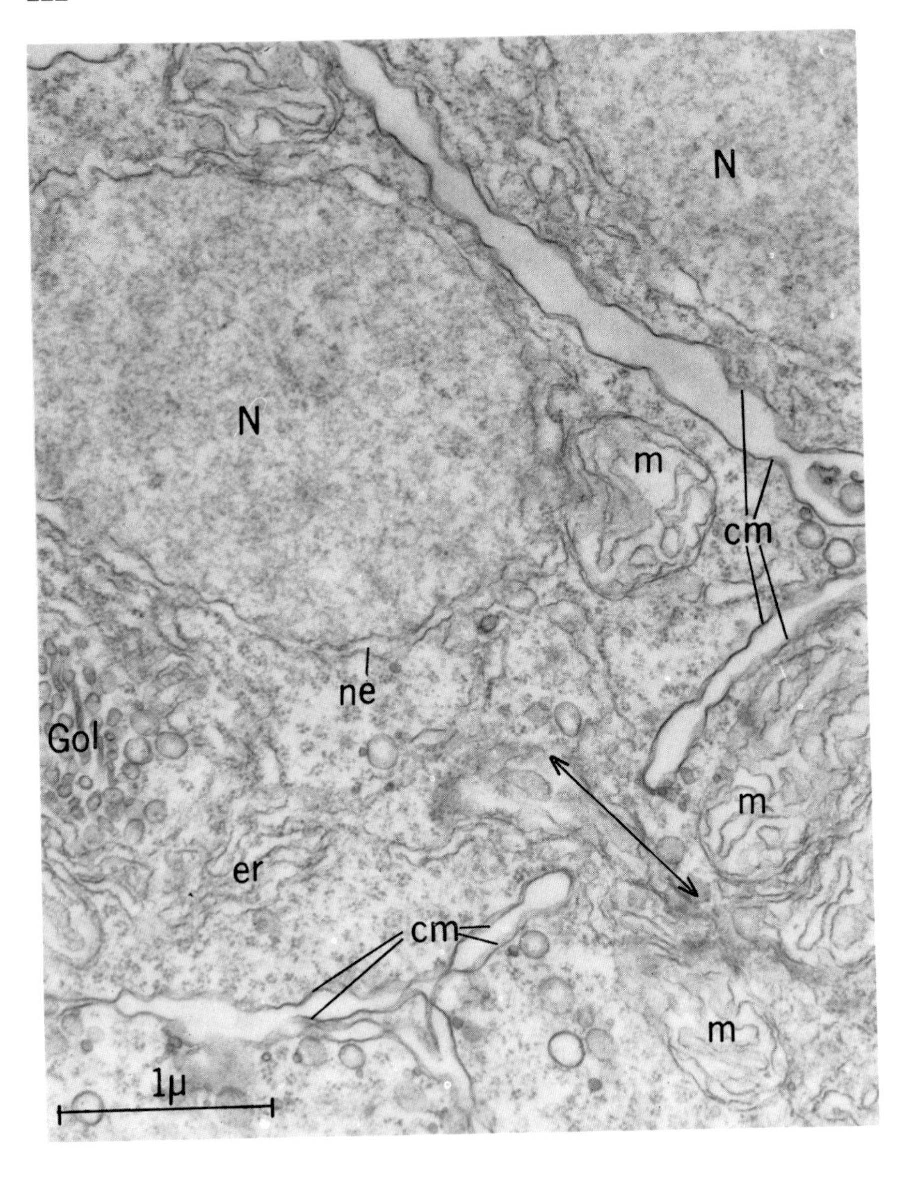

FIG. 10. Electron micrograph of an intercellular bridge connecting two developing spermatocytes of *Bolinopsis microptera*. Arrow passes through the cytoplasmic connection. N, spermatocyte nucleus; ne, nuclear envelope; Gol, Golgi body; er, endoplasmic reticulum; m, mitochondria; cm, cell membranes.

in a single ultrathin section, and therefore a much larger number might be involved in a syncytium.

Mature sperm often form large wedge-shaped bundles, in which component gametes parallel one another (Chun, 1880, 1892, 1898). In *Dryodora* the tails of the sperm forming a wedge appear embedded in the gastrodermal cells (Fig. 5B). Willey (1896) described large pyriform bundles of ripe sperm in *Ctenoplana korotneffi*, with tails pointing both towards the center of the bundle and radiating outwards. These bundles lie at the inner opening of the sperm duct.

Sperm heads are pyriform or spheroid in pelagic ctenophores (*Beroë ovata*, Chun, 1880; *Bolinopsis*, *Pleurobrachia*, and *Dryodora*). They are distinctly filiform in the sessile *Coeloplana* (Komai, 1922; Krempf, 1921) and the alternately pelagic and creeping *Ctenoplana* (Willey, 1896).

4.3.4.2 Oogenesis

Oogenesis has been adequately studied only in *Bolinopsis microptera* (Dunlap, 1966). Relatively cursory studies have revealed that in all major respects the process is similar in *Pleurobrachia bachei*, *P. pigmentata*, *Dryodora*, and *Beroë* at Friday Harbor. That oogenesis is similar throughout the phylum is also evident from examination of published illustrations of the ovaries of other species: *Beroë ovata* (Chun, 1892), *Callianira bialata* (Chun, 1898), and *Coeloplana bocki* (Komai, 1922) (Fig. 8), and from Krempf's (1921) brief description of oogenesis in *C. gonoctena*. In treatments of still other species, the ovaries have been distinctly immature or else the illustrations and descriptions lack detail (e.g. *Velamen*, Fol, 1869; *Hormiphora*, Hertwig, 1880; *Euchlora* and *Lampetia*, Chun, 1880; *Tjalfiella*, Mortensen, 1912; *Lyrocteis*, Komai, 1942).

Nurse cells have been identified in many ctenophores and probably occur universally throughout the phylum, although they have previously been interpreted as young oocytes. Chun (1898) recognized that "oocytes" of *Callianira bialata* develop in packets containing one large and numerous small "oocytes." In a brief description of the ovaries of *Coeloplana gonoctena* and other unspecified pelagic ctenophores, Krempf (1921) mentioned that "modified oocytes" are "sacrificed" in development of the definitive oocytes. Krempf proposed that the former are "phagocytized" by "their more vigorous neighbors." As described next, ctenophore oocyte-nurse cell systems are highly organized, and the processes involved are far more complex and controlled than the term phagocytosis suggests.

A similarity in oocyte development throughout the phylum might be expected in view of the nearly identical structure of all ctenophore eggs (Section 4.4.1.3). At least in *Bolinopsis*, nurse cell activity

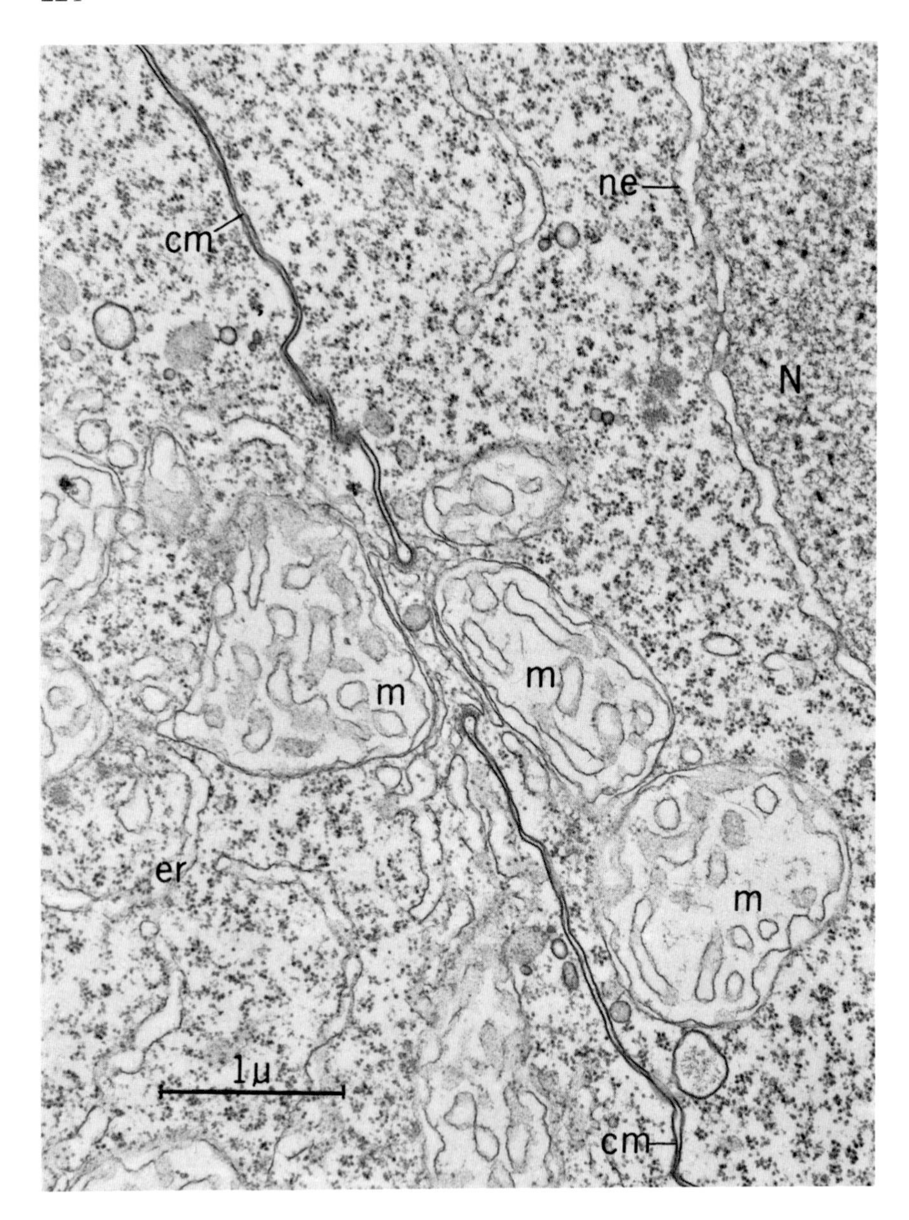

FIG. 11. Electron micrograph of an intercellular bridge, between two young nurse cells of *Bolinopsis,* occluded on either side by mitochondria (m). Some cytoplasmic elements, notably endoplasmic reticulum lamellae, are often continuous through such a bridge opening, but there is no evidence of a flow through the bridge. N, nurse cell nucleus; ne, nuclear envelope; cm, cell membranes of connected nurse cells; er, lamellae of endoplasmic reticulum. (Osmium tetroxide fixation.)

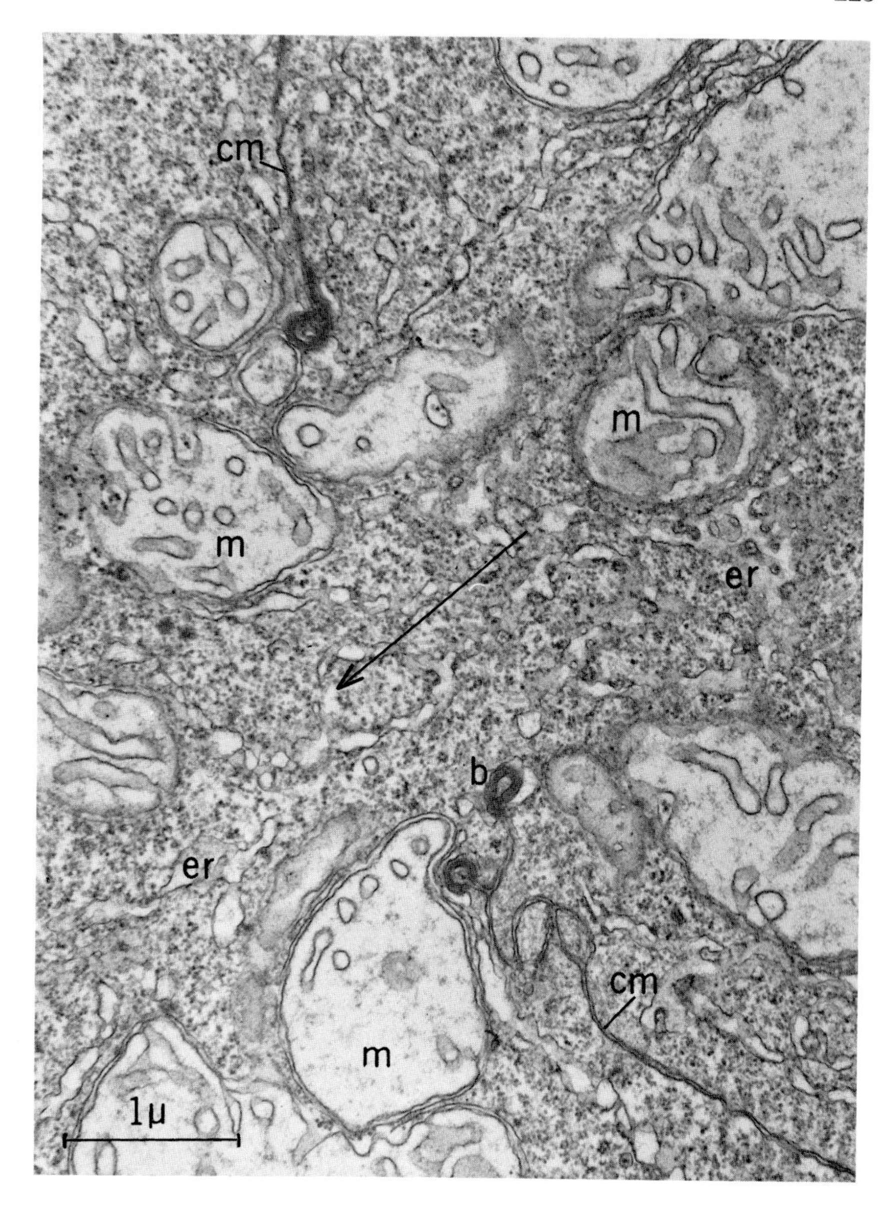

FIG. 12. Electron micrograph of an open intercellular bridge between two mature nurse cells of *Bolinopsis*, showing the flow of cytoplasm through the bridge in direction of the arrow. The border (b) of a bridge often becomes divided as the bridge opens, as in this section. m, mitochondria; er, endoplasmic reticulum lamellae; cm, cell membranes. (Glutaraldehyde fixation.)

is intimately involved in formation of the segregated ooplasm unique to and characteristic of eggs of all ctenophores. One possible significant exception occurs in *Gastrodes* (Section 4.3.4.2.2).

4.3.4.2.1 *Oogenesis in Bolinopsis microptera.* Throughout its development, each *Bolinopsis* oocyte is associated with approximately 100 nurse cells, divided into three unequal clusters. Each such oocyte-nurse cell complex is a syncytium, in which component cells are connected by cytoplasmic bridges, basically similar to those between spermatocytes (Section 4.3.4.1) and among ovarian nurse cells and oocytes of some insects (Meyer, 1961; Koch *et al.*, 1967) and annelids (unpublished observations on *Tomopteris* oocytes and nurse cells). These bridges doubtlessly represent arrested division furrows, and hence are evidence that a complex originates from a single oogonium through multiple mitotic divisions with incomplete cytokineses. There are three bridges per cell, except for the most distal nurse cells within each cluster, which have one or two bridges (Fig. 13). This pattern and the approximate number of nurse cells in each of the three clusters suggests a sequence of eight synchronous divisions giving rise to an oocyte-nurse cell complex, starting with an oogonium, with each cell dividing twice after being produced, and then ceasing to divide. The resulting syncytium would contain one central oocyte (one of the two oldest cells) and three nurse cell clusters composed of, respectively, 54, 33, and 20 cells.

Consistent changes in the intercellular bridges correlate with major stages in development of the oocyte-nurse cell complex. Bridges connecting immature and developing nurse cells, either with other nurse cells or with the oocyte, are narrow (0.5-0.8 μm diameter) and occluded on either side by large cytoplasmic organelles, usually mitochondria (Fig. 11). This occlusion functionally isolates nurse cells during their development, and an occluded bridge often connects cells in distinctly different stages of development. Once a nurse cell attains maximum size, the bridge connecting it with the oocyte (either directly or through intervening nurse cells) widens up to 5 μm in diameter, and its cytoplasm passes out toward and eventually into the oocyte (Fig. 12; also Fig. 13, 14, 15, and 16).

The development of an oocyte-nurse cell complex may be divided into four stages: previtellogenesis, early and late vitellogenesis, and corticogenesis (Fig. 13). Initially the most distal nurse cells may be still dividing, and all cells of an early previtellogenic complex are similar, with the oocyte distinguished primarily by its apically situated nucleus containing a basal nucleolus (Fig. 13A).

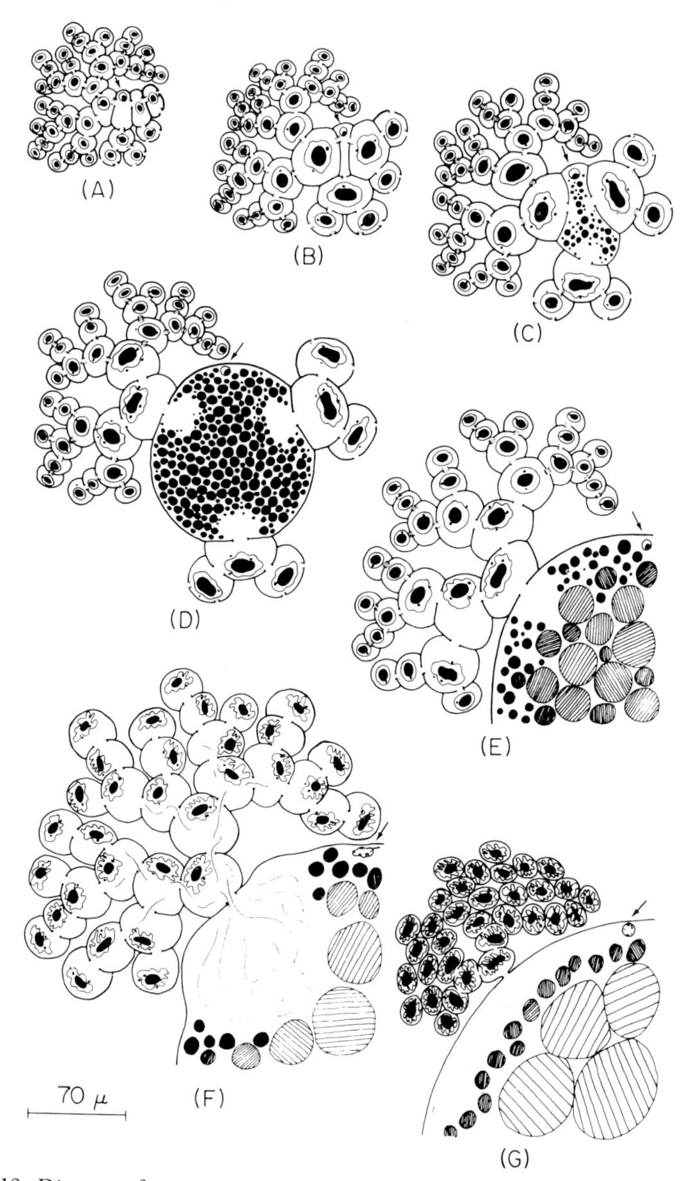

FIG. 13. Diagram of major stages of oogenesis in *Bolinopsis:* previtellogenesis (A-B), early vitellogenesis (C-D), late vitellogenesis (E), and corticogenesis (F-G). Small arrows point toward the oocyte nucleus. In (A) through (D) one entire nurse cell cluster is shown with the oocyte, while each of the other two clusters is represented by its three most proximal cells. Only one complete cluster is depicted in (F) and (G). See text.

Nurse cell development begins prior to any oocyte development. Nurse cells develop sequentially, beginning with those adjacent to the oocyte and continuing distally within each cluster. Briefly, nurse cell growth involves an increase in cell diameter from about 9 to 30 μm, of nuclear diameter from 4 to 16 μm, and nucleolar diameter from 2 to 7 μm. During this growth the nucleolus forms numerous buds, there is an increase in both the number and concentration of nuclear envelope pores, and the nuclear envelope becomes folded. The cytoplasm becomes filled with organelles, including extensively developed rough endoplasmic reticulum, numerous free ribosomes, mitochondria, and discrete small Golgi bodies. There is a distinct perinuclear zone devoid of large organelles (Fig. 14). The endoplasmic reticulum forms interconnected lamellae generally concentric about the nucleus, and often continuous with small stacks of annulate lamellae and with sworls of endoplasmic reticulum reminiscent of so-called "yolk nuclei" of many animal oocytes (Raven, 1961). Once cytoplasm begins to flow out of a nurse cell, its nuclear envelope becomes even more highly folded about the nucleolus, and the nucleus is displaced to one side of the cell (Fig. 13, 14, 15A).

All nurse cells undergo the same general structural changes, although the more distal cells do not become quite as large as the proximal cells, and their cytoplasm is never as densely packed with organelles. Also, the complex sworls of endoplasmic reticulum are not as large or compact in the distal cells, and mitochondria are generally small and spheroid, in contrast to the larger vermiform mitochondria of proximal cells. Feulgen preparations indicate that at least the seven most proximal nurse cells in each cluster (a total of 21) become highly polyploid. Chromatin in all nurse cells is diffuse and reticulate.

Nuclei of oocytes with nurse cells are characteristically smaller and less active than nuclei of solitary oocytes (Raven, 1961; Davidson, 1968). However, *Bolinopsis* affords the first reported example of an oocyte nucleus that does not grow at all. In sharp contrast with nurse cell nuclei, the oocyte nucleus is small (4-5 μm in diameter) and inconspicuous throughout oogenesis (Fig. 13, 14, 15A). It remains near the apical pole, within a micron or two of the oolemma, and its chromosomes are in the polarized synaptic stage of the first meiotic prophase. There is a single small nucleolus (Fig. 13A-E and 14), and evidence of slight nucleolar budding. Electron micrographs reveal that the nuclear envelope is largely devoid of pores, except for a few which are consistently associated with a small stack of cytoplasmic annulate lamellae.

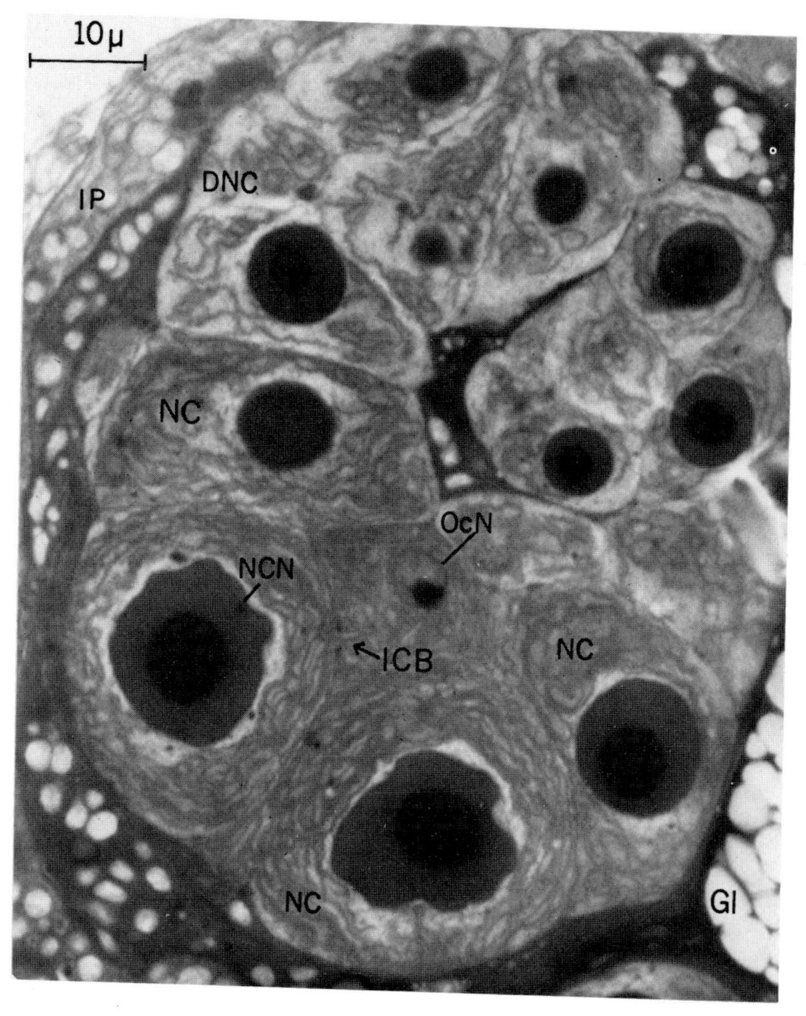

FIG. 14. Photomicrograph of a section of an oocyte-nurse cell complex at a stage between those portrayed in Fig. 13B and C, at the onset of inflow of cytoplasm from the most proximal nurse cells through an open intercellular bridge (ICB). Note the difference between the mature nurse cell nuclei (NCN) and the oocyte nucleus (OcN). A pale perinuclear zone surrounds the mature nurse cell nuclei. The distal nurse cells (DNC) are relatively immature. IP, interstitial process; Gl, Gland cells (of oviduct). (Osmium tetroxide fixation, hence initial yolk bodies in oocyte are not preserved.)

In fact, all the major synthetic functions of oogenesis in *Bolinopsis* apparently are performed by the nurse cells, and not by the oocyte. This is evident from the striking difference between the oocyte nucleus and the nurse cell nuclei, and also from the fact that oocyte growth is largely and perhaps entirely dependent on the acquisition of nurse cell cytoplasm. An oocyte does not begin to enlarge significantly until its three most proximal nurse cells are fully differentiated, the three oocyte bridges have widened, and nurse cell cytoplasm begins to flow into the oocyte (Fig. 13C, 14).

Thus, previtellogenesis involves growth of the most proximal nurse cells, but not of the oocyte (Fig. 13A, B). The onset of vitellogenesis is simultaneous with initial inflow of nurse cell cytoplasm (Fig. 13C; also Fig. 14, in which yolk is not preserved). Early vitellogenesis encompasses oocyte growth to a sphere of about 100 μm in diameter, filled with small yolk bodies (Fig. 13D, 15A), during which only the nine most proximal nurse cells (three per cluster) are contributing cytoplasm.

Yolk first appears as tiny preyolk bodies, resolved only in electron micrographs. Rarely, these bodies have been observed in the nurse cell cytoplasm before it is passed into the oocyte, but usually they are seen first in the transferred cytoplasm within the oocyte. Larger yolk bodies, visible in light micrographs, are observed only in the oocyte (Fig. 13C). All yolk bodies are bound by a single unit membrane. The composition of the yolk is poorly known. It cannot be preserved by most fixation methods (Fig. 5, 7B, 15), but is adequately preserved with isotonic glutaraldehyde (Fig. 16, 17). All yolk bodies formed during early vitellogenesis are highly PAS-positive. The yolk is therefore probably a hydrated protein-polysaccharide complex. There is no evidence of lipids. Preyolk bodies are frequently associated with Golgi complexes. A tentative hypothesis of yolk formation is that its protein component is synthesized by the rough endoplasmic reticulum in the nurse cells, and is packaged into preyolk bodies by the

FIG. 15. (A) Photomicrograph of a section of an oocyte-nurse cell complex at a stage corresponding approximately to that in Fig. 13D, at the peak of vitellogenesis. Yolk bodies are represented in this section by pale spots in the oocyte, since the yolk is not preserved (osmium tetroxide fixation). The oocyte nucleus (OcN) is barely visible as a dark sphere in the oocyte periphery. Parts of the three nurse cell clusters are included in this section, and an intercellular bridge (ICB) is seen opening into the oocyte. Note the small patch of nurse cell cytoplasm within the oocyte near the entrance of the bridge, and contrast with Fig. 15B. Note the large nuclei and nucleoli of the proximal nurse cells (PNC). The distal nurse cells (DNC) are still immature. IP, interstitial process; spc, spermatocytes in adjacent testis. (B) Photomicrograph of an intercellular bridge (arrow) opening into an oocyte in a stage comparable to that in Fig. 13F, including a bulging patch of cortical plasm (CPl). The irregular lines in the cortical plasm and nurse cell (NC) represent lamellae of endoplasmic reticulum, and their orientation reflects the lines of flow of the cytoplasm. The nurse cell nucleus is not included in this section. Golgi bodies (G) are evident as large dark spots, and mitochondria (m) as tiny unstained bodies. The subcortical plasm (SbCPl) is represented by denser cytoplasm; yolk here and in the endoplasm (EnPl) is not preserved. (Osmium tetroxide fixation.)

Golgi, which could contribute the polysaccharide component. Electron micrographs suggest that many preyolk bodies fuse to form the

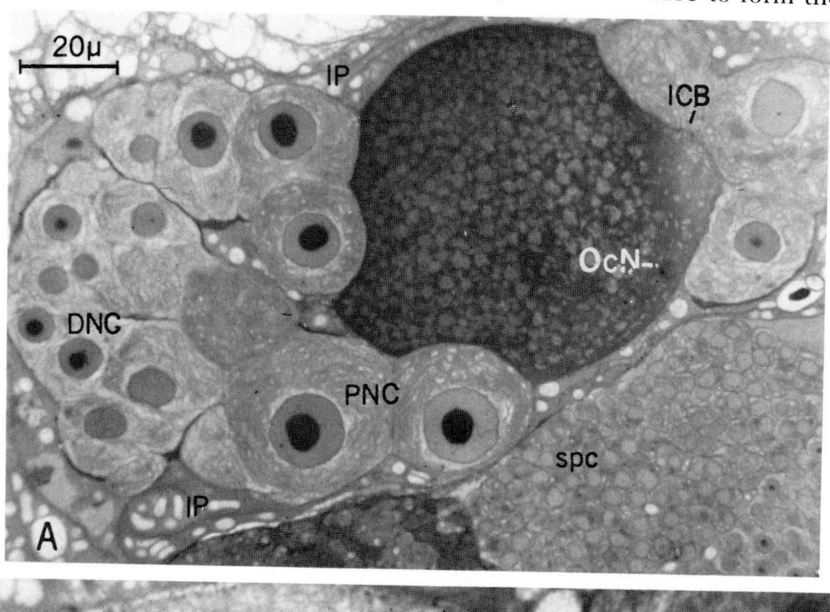

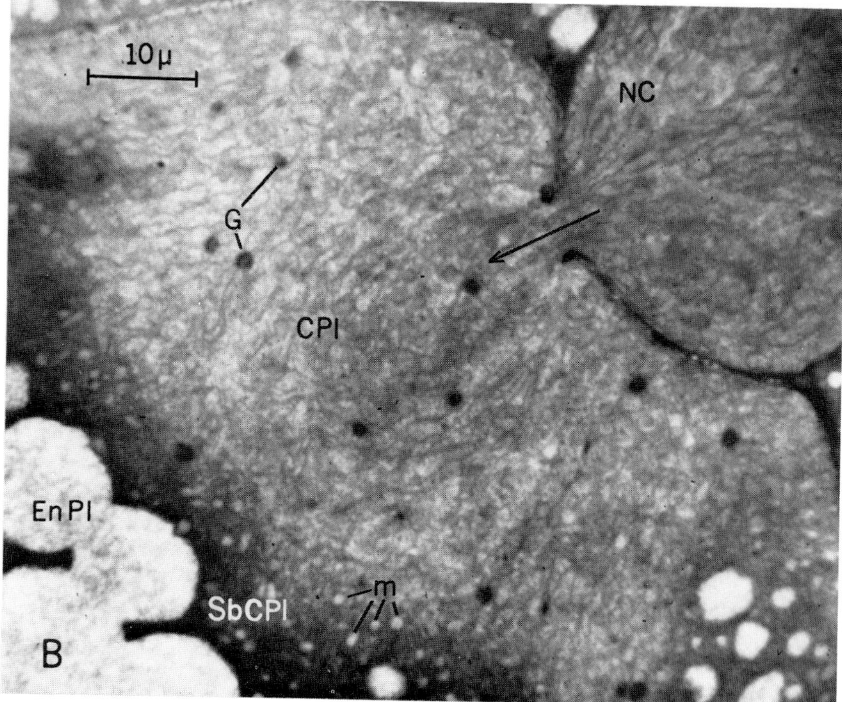

larger yolk spheres characteristic of the oocyte at the end of early vitellogenesis.

Later vitellogenesis involves further growth of the oocyte to about 150 μm in diameter (Fig. 13E). The central yolk spheres enlarge, their contents becoming progressively pale and evidently diluted, resulting perhaps from hydration. Peripheral yolk bodies (presumably those most recently formed) remain relatively dense and small. These retain their strong reaction to the PAS procedure, while the central yolk bodies become progressively PAS-negative (Fig. 16). There is further addition of nurse cell cytoplasm from twelve more cells (four in each cluster), and yolk body formation evidently continues in the oocyte periphery. The oolemma now becomes thrown into numerous irregular microvilli, which could function in uptake of nutrients or fluid, or could also be involved in formation of the vitelline membrane, which appears at this stage as a vaguely fibrillar material among the microvilli (Fig. 17). Whether the membrane is produced by the oocyte, or by the interstitial processes surrounding the oocyte, is not known.

During corticogenesis, at which time the oocyte attains its final diameter of 200–210 μm, the approximately 80 remaining distal nurse cells, which have only begun to differentiate until now, mature rapidly and apparently simultaneously. The rest of the bridges open, and cytoplasm passes from all nurse cells through the branching system of bridges into the oocyte (Fig. 13F). This cytoplasm first collects in three patches which push aside the peripheral yolk bodies and form bulges at the oocyte surface (Fig. 15B and 16). It later spreads out evenly over the peripheral yolk, forming the definitive cortical plasm (Fig. 13G). There is no further vitellogenesis, although yolk bodies may continue to swell. The peripheral yolk bodies remain relatively small, dense, and PAS-positive (Fig. 16). Nurse cells now shrink sequentially, beginning distally and continuing proximally within each cluster, and then separate from each other and from the oocyte by completion of the long-arrested cytokineses. Their remnants, now nearly devoid of cytoplasm, consist of shrunken cell membranes, nuclear envelopes highly folded about still large nucleoli, and degenerating chromatin (Fig. 13G). These disintegrate among the ovarian interstitial processes.

▶Fig. 16. Photomicrograph of an oocyte during corticogenesis, comparable to the stage in Fig. 13G, fixed with glutaraldehyde and stained by the PAS procedure to demonstrate carbohydrates. The endoplasm (EnPl) is unstained, while the peripheral yolk bodies in the subcortical plasm (SbCPl) are strongly PAS-positive and thus have a high carbohydrate content. The mucous secretion of the oviduct's glandular epithelium (GlEp) also stains strongly. Note the absence of a reaction in the nurse cells (NC) and in the cortical plasm (CPl). DgC, digestive cells lining the meridional canal.

The segregated ooplasm of a full-grown *Bolinopsis* oocyte is composed of three distinct zones: (1) a central mass of large, pale yolk spheres with scant intervening cytoplasm, (2) a narrow intermediate

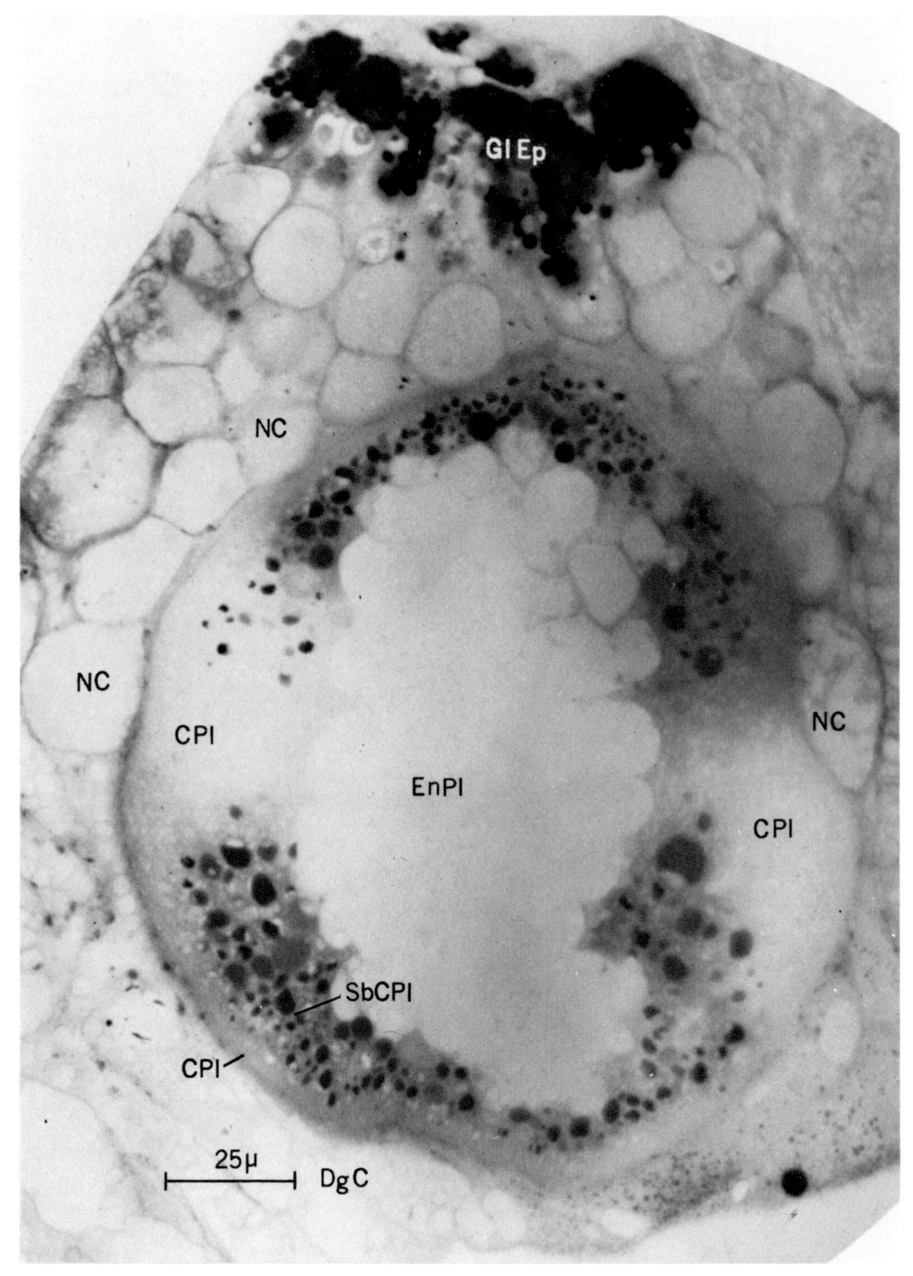

zone surrounding the first and composed of dense, small, PAS-positive peripheral yolk bodies dispersed in relatively abundant cytoplasm, and (3) an outermost layer of yolk-free cytoplasm composed predominantly of lamellae of endoplasmic reticulum oriented parallel to the oolemma (Fig. 17, also 13G). The first zone constitutes the "endoplasm" of embryological studies, and the last forms at least part of the cortical plasm, or "cortex" or "ectoplasm" (Section 4.4.1.3). The intermediate zone has not been described previously, since it is not readily visible in living eggs, and is clearly distinguished only after fixation with glutaraldehyde. (This zone is also present in mature oocytes of *Dryodora* and *Pleurobrachia*, and hence probably occurs in all ctenophore eggs.) In previous studies it has apparently been included in the cortical plasm. Thus, Reverberi (1957) determined in eggs of *Leucothea* ("*Eucharis*") and *Beroë* that mitochondria are most concentrated in the cortical plasm. At least in *Bolinopsis* mitochondria are apparently equally concentrated in the intermediate zone, among the peripheral yolk bodies (Fig. 17). Because of its association with the outer cortical plasm during embryonic development, this new zone will be called the "subcortical plasm."

In *Bolinopsis* the endoplasm is formed early in oogenesis from the nine or so most proximal nurse cells, the subcortical plasm later from these and the twelve more distal nurse cells, while the cortical plasm proper is derived primarily from the 80 or so most distal nurse cells. These last have remained relatively immature until late in oogenesis, which could be of significance to early embryonic development. For example, relatively short-lived molecules could be synthesized by these nurse cells, transported to the cortical plasm, and utilized in embryonic differentiation before disintegrating.

The very existence of nurse cells, especially in such large numbers, is in itself evidence that oogenesis in ctenophores is rapid (see Davidson, 1968). However, the time required for oogenesis, and for

▶FIG. 17. Electron micrograph of the edge of a mature *Bolinopsis* oocyte in the oviduct, comparable to that in Fig. 7B except that this oocyte has been preserved with glutaraldehyde, and hence the difference between the yolk in the endoplasm (EnPl) and in the subcortical plasm (SbCPl) is apparent. The cortical plasm (CPl) contains mainly lamellae of endoplasmic reticulum oriented more or less parallel to the oocyte cell membrane (oolemma). Mitochondria (m) occur among these lamellae and also among the dense peripheral yolk bodies of the subcortical plasm. Only one yolk body of the endoplasm (EnPl) is included in this micrograph. The vitelline membrane (VM) is visible as a layer of fibrillar material on the outer surface of the oolemma. Two gland cells (Gl), one of each type composing the inner oviducal epithelium (see Section 4.3.2.3.2), are in the upper part of the micrograph. Several cilia (c) are also included. (Compare with Fig. 13G.)

the various stages, is unknown. Greve (1970) presents evidence suggesting that oocyte development in *Beroë gracilis* and *Pleurobrachia pileus* from the North Sea requires only about 2 days (see Section 4.3.6.3). In *Bolinopsis*, corticogenesis is evidently more rapid than

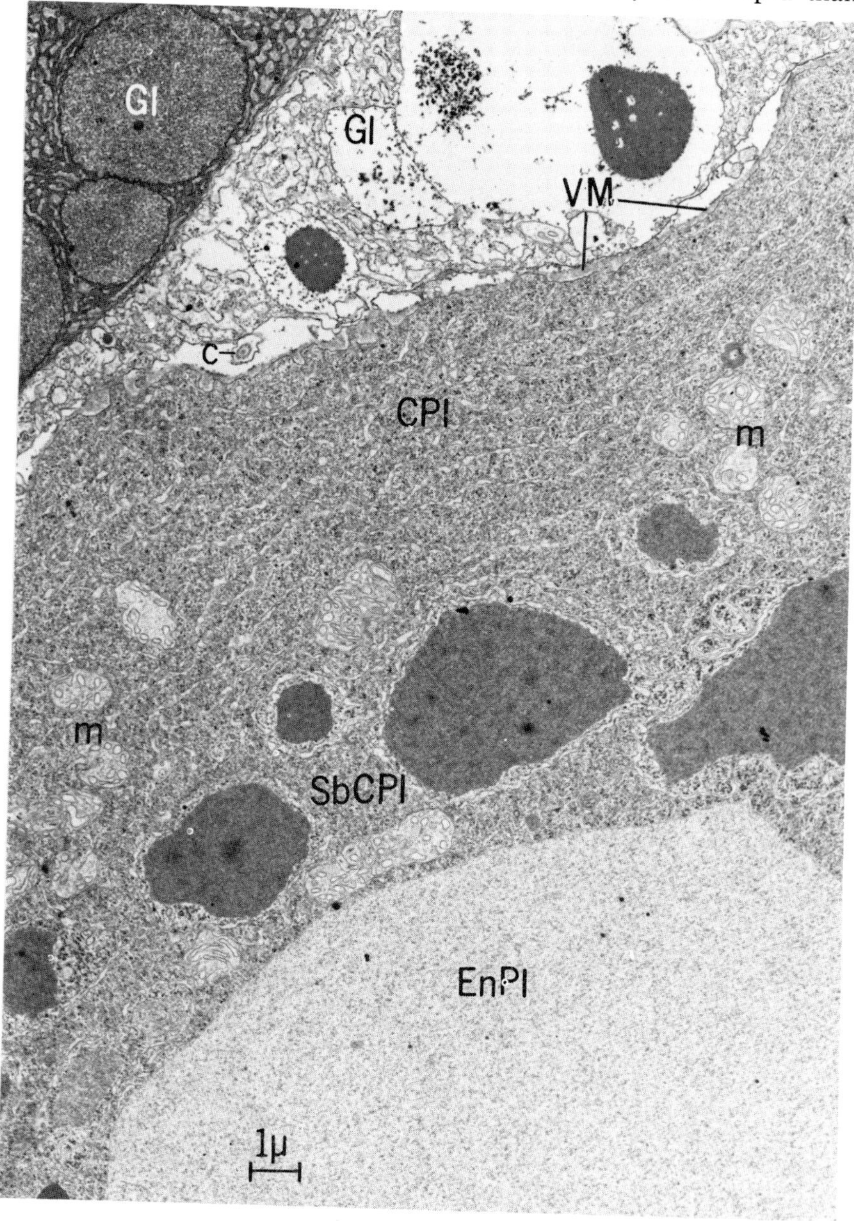

vitellogenesis, since stages in the development of the distal nurse cells are rarely encountered in sections, and the inflowing cortical plasm forms patches bulging at the oocyte surface. This contrasts with vitellogenesis, in which all intermediate stages of differentiation of the proximal nurse cells are commonly encountered, and the inflowing cytoplasm forms small, nonbulging patches (compare Fig. 13D and E, and 15A with Fig. 13F, 15B, and 16). However, differences in the size of the patches could also reflect different rates of assimilation of the nurse cell cytoplasm by the oocyte.

Although meiosis occurs some time after spawning in *Beroë ovata* (Yatsu, 1912), in *Bolinopsis* the first polar body is formed during ovulation into the oviduct (Section 4.3.2.3.2). It appears at the oocyte's apical end, where the nucleus has been located throughout oogenesis, the end which first breaks through the oviduct's glandular epithelium. The second polar body is usually present when the oocyte is released through the ovarian gonopore, hence it could be given off either in the oviduct or during extrusion through the gonopore. In most ctenophores the first polar body divides and all three polar bodies then remain within the vitelline membrane, attached to the oocyte by a thin cytoplasmic strand. In *Mnemiopsis*, however, the polar bodies are not attached to the egg, but are associated with the vitelline membrane (Freeman and Reynolds, 1973).

4.3.4.2.2 *Oogenesis in Gastrodes.* The only ctenophore in which oogenesis appears to differ significantly from that in *Bolinopsis* is the peculiar platyctenid *Gastrodes*, of which there are two described species, *G. parasiticum* and *G. komaii*, both occurring for part of their life cycle and both possibly parasitic in the tunicate *Salpa* (Korotneff, 1888, 1891; Komai, 1922; Dawydoff, 1937). Gonads have not been observed in *Gastrodes*, and neither have candidates for spermatocytes or spermatozoa. Korotneff and Komai interpreted as oocytes certain large cells occurring among the endodermal cells of the planula larva (Section 4.4.2.1.2) (Fig. 21), and later among the ectodermal epidermal cells of the ventral everted pharynx (Fig. 18). Komai found these "egg-cells" in all specimens, with large ones occurring only in the larger individuals. The smaller egg-cells are found near the animal's center, the larger ones often form a circle paralleling the body's circumference. As they enlarge, their nuclei and nucleoli also enlarge, and the cells become pyriform and occasionally ameboid and bulge into the underlying parenchyme, remaining attached to the epidermis by a stalk. Apparently there is some differentiation of the cytoplasm into central and superficial layers, but there is no yolk, and certainly the largest in no way resemble

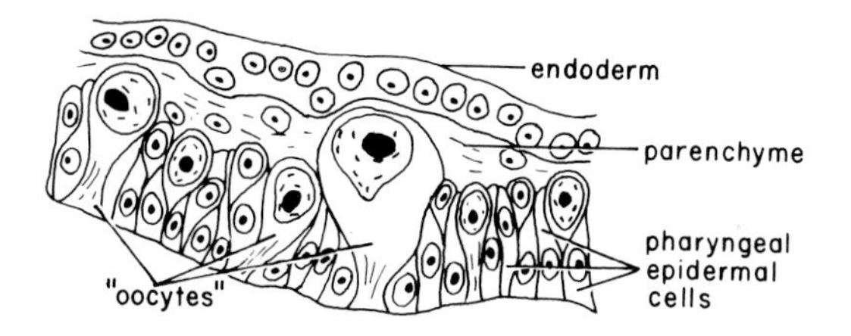

FIG. 18. *Gastrodes parasiticum:* part of a section of the ventral pharyngeal epidermis, showing possible oocytes in several stages of development. [Redrawn from Komai (1922).]

mature oocytes of other ctenophores. Komai (1922) noted the similarity between these cells and oocytes of Actinae. He also observed apparent stages in their release from the ventral epidermis directly to the outside.

Since no one has observed the ovum or cleavage stages of *Gastrodes*, the identity of these "egg-cells"is not certain. The earliest known developmental stage is the planuloid larva (Section 4.4.2.1.2). In view of the parasitic nature of *Gastrodes*, and the occurrence of the planula larva, both egg structure and oogenesis might be expected to differ from that in other ctenophores. Komai states that "The presence of egg cells in the ectoderm seems to be a secondary adaptation to its parasitic life. The primordial germ cells initially found in the endoderm of the planula larva migrate into the pharyngeal ectoderm where they can get nourishment directly from the host" (1963, p. 186). Since sperm have not been observed, it has been proposed that *Gastrodes* is parthenogenetic (Korotneff, 1888, 1891; Komai, 1922). It is also possible that these large cells are not oocytes, and that the sexual stage of *Gastrodes* is yet to be found (Rankin, 1956) (Section 4.4.2.2.2).

4.3.5 Gametogenic cycles within populations

Little is known about reproductive cycles in ctenophores, and most is related to breeding periods (Section 4.3.8.3). Most ctenophores appear to be annual organisms, at least in coastal waters. L. Agassiz (1860), after 12 consecutive years of observations, reported that pelagic ctenophores from the New England coast are invariably small and sexually immature in early summer, enlarge and develop gonads by late July and August, and spawn in late August and September. He proposed that adults die in late autumn, and the young spend the winter in deeper waters. Mortensen (1912) questioned Agassiz's interpretation, noting that, in the North Sea, *Beroë* are often large in

January. Esterly (1914) also presented evidence suggesting annuality in *Pleurobrachia bachei* from the southern California coast. He found large individuals in winter and spring, but never in June or July, implying that in contrast with east coast species the adults survive the winter and succumb after spawning in the late spring.

At Friday Harbor, spring breeding populations of the most common species, *Bolinopsis microptera* and *Pleurobrachia bachei*, consist of small to medium-sized individuals, but as summer progresses animals three to four times larger are encountered. These are not always reproductive, and in fact the largest specimens collected in late summer usually have poorly developed gonads, suggesting further spawning (Section 4.3.8.3). Any one individual probably spawns off and on throughout the summer and early autumn, after the major spring spawning, continuing to grow in the meantime. The size of reproductive specimens collected in late spring and summer varies considerably, with *Bolinopsis* ranging from 10 to 100 mm (polar axis), *P. bachei* from 5 to 20mm, and *Dryodora* from 5 to 15 mm (see Section 4.3.6.5). At least the largest individuals may succumb before spring, and the spring breeding population would then consist of survivors of the previous season's spawnings. All of this suggests a life span of about one year, but it could be less. Also, size alone may not always be a good criterion of age, since Greve (1970) reports that growth rates among individuals of a population are highly variable, and that at least *Beroë cucumis* shrink if starved. Freeman (1967) also documented a decrease in size of starved *Vallicula*.

Krempf (1921) provides evidence for annuality in the protandrous *Coeloplana gonoctena* (Section 4.3.1). In early summer all individuals are small and have mature testes, while near the end of summer they are large and some develop ovaries. Most adults apparently die after laying eggs.

4.3.6 Factors Influencing Gametogenesis

4.3.6.1 Temperature

There is no evidence that a change in temperature controls the seasonal development of adult ctenophore gonads, although in those species which spawn in late summer or autumn (Section 4.3.8.3), there may be a correlation between warmer waters and spawning. At Friday Harbor, ctenophores are reproductive in early spring, well before the temperature of the surface water has changed appreciably (Section 4.3.8.3).

Chun (1892) suggested that temperature determines the development of larval gonads (Section 4.3.6.5), since sexually mature larvae were collected as soon as surface waters warmed in early summer, and could be found as long as it remained warm. He also speculated that paedogenesis does not occur in *Cestum veneris* because this species descends to colder waters in summer, and larvae therefore never contact warmer waters. However, *Bolinopsis* larvae have been found to be sexual in cold waters (Garbe, 1901; Krumbach, 1927), hence Chun's correlation may be incidental.

4.3.6.2 LIGHT

Seasonality in the onset of gametogenesis, evident from a distinct seasonality of spawning in certain species (Section 4.3.8.3), might be initially a reaction to light, specifically to change in day length. Ctenophores possess probable photoreceptors, located in the floor of the apical organ (Chun, 1880; Horridge, 1964) (see Section 4.3.8.2). G. Freeman (personal communication) has tentative evidence in *Vallicula* of control over gametogenesis by the apical organ (Section 4.3.6.4). Nutrients would of course also be necessary for extensive oocyte development, but the main food supply of most ctenophores (zooplankton, Hyman, 1940) is also regulated indirectly by light (through phytoplankton).

4.3.6.3 NUTRIENTS

Greve (1970) presents incomplete but intriguing evidence that nutrients stimulate gametogenesis in *Pleurobrachia pileus* and *Beroë gracilis* from the North Sea. Under generally favorable laboratory conditions, adults of these species produce fertilized eggs about two days after initiation of excessive feeding.

J. Hirota (personal communication) has preliminary evidence that inadequate nourishment may stimulate gametogenesis in juvenile *Pleurobrachia bachei* (or perhaps adequate nourishment inhibits gametogenesis). Thus, if an individual is growing rapidly, gonads usually do not develop until it has reached adult size (see Section 4.3.6.5).

If nutrients do affect gametogenesis, their influence could be direct, since the interstitial processes of the digestive cells are in direct contact with developing gametes (Section 4.3.2.2). However, the effect could also be indirect, perhaps through an endocrine intermediate.

4.3.6.4 Endocrines

There is some evidence of endocrine regulation of reproduction in ctenophores. First, the generally synchronous development of all gonads in an individual (Section 4.3.4) is indirect evidence of endogenous control of gametogenesis. Secondly, G. Freeman's experiments (personal communication) suggest that the apical organ affects gametogenesis in the platyctenid *Vallicula*. Removal of this organ results in a more extensive development of both ovaries and testes than in controls. Under the conditions of the experiment, animals without apical organs were sexual for a greater number of days than controls. Freeman proposes that the apical organ normally inhibits gametogenesis, possibly through a neurosecretory intermediate. In pelagic ctenophores nerves do connect the apical organ with the full length of each comb row (Heider, 1927; Horridge and Mackay, 1964). However, S. Tamm (personal communication) has seen no evidence of neurosecretory cells in the floor of the apical organ of *Beroë*.

4.3.6.5 Age (Paedogenesis)

Paedogenesis, or reproduction by larvae or juveniles, has been documented in several ctenophores, and may be of wide occurrence throughout the phylum. Chun (1880, 1892) first observed the phenomenon in the lobates *Leucothea multicornis* and *Bolinopsis vitrea* from the Mediterranean. Larvae of these species develop ovaries and testes, similar to adult gonads, although occurring only in subsagittal canals (Fig. 9D). Following larval spawning, the gonads regress to clusters of germ cells, which after "metamorphosis" (Section 4.4.2.2.1) again multiply to form adult gonads (Section 4.3.3). Eggs produced by larvae are one-half the diameter of those spawned by adults, although the nuclei are the same size and ooplasmic structure is similar (Section 4.4.1.3).

Chun (1880, 1892) employed the term "dissogeny" to describe the phenomenon of two periods of sexuality, one in the larva and one in the adult, separated by a period surrounding metamorphosis during which the gonads regress.

There are several other reports of larval sexuality, in two *Bolinopsis* species, one from Trieste, the other from the North Sea (Garbe, 1901; see Mortensen, 1912) (Section 4.3.3). Also, Chun's "*Charistephane fugiens*" (Section 4.3.2.1.1) is almost certainly a sexual larva or juvenile of a lobate or cestid.

Paedogenesis in the form of *juvenile* reproduction has also been observed in several ctenophores (Chun, 1880, 1898). At Friday Harbor, I have noted that *Bolinopsis* are often reproductive when only 10 mm long, (polar axis including lobes), or one-tenth maximal adult size. Similarly *P. bachei* from Friday Harbor are sexual at about 5 mm (polar axis), one-fifth maximum adult size, and *Dryodora* at 5 mm, or one-third full polar diameter.

This tendency towards juvenile reproduction has also been observed by J. Hirota at Scripps, who has raised two generations of *P. bachei* in captivity, and has followed individuals through at least 100 days, measuring growth and egg production at 2 day intervals. His preliminary data (personal communication) show that although most animals first spawn eggs at approximately 60 days of age and an average of 8 mm diameter (tentacular axis), a few individuals also spawn at about 40 days of age and an average of 2 mm diameter. This early reproductive period lasts for only a week or 10 days, and results in production of no more than 150 eggs, of which fewer than 50% hatch. Reproduction at 8 mm, however, spans growth up to about 14 mm, and lasts up to 44 days (with no indications of impending senility), during which 12,000-17,600 eggs are produced, and over 80% hatch. Hirota has tentative evidence that slower growth is correlated with early reproduction, and larger rapidly growing individuals bypass early spawning (Section 4.3.6.3). Larvae resulting from early reproduction differ morphologically from those from late reproduction, although Hirota has not yet studied the difference in detail. The status of the gonads between early and late reproduction is not known.

Greve (1970) also followed ctenophores from eggs through adulthood in the laboratory, but did not observe any period of early reproduction. He reports that *P. pileus* from the North Sea [tentatively synonymized with *P. bachei* by Moser (1909) and Bigelow (1912)] usually becomes sexual at a diameter of 5.5-11 mm (polar axis). *Beroë gracilis* first reproduces at a length of 5.5-15 mm, after which there is a decrease in growth rate. At 25-30 mm, *B. gracilis* ceases to grow, but continues to reproduce if adequately fed.

In summary, although ctenophores may usually become sexual at a more or less specific age or size, there is a tendency for individuals to reproduce earlier, in larval and juvenile stages. This is in spite of the fact that offspring of such precocious reproduction may be fewer, smaller, or less viable. All of this suggests that mortality among pelagic ctenophores must be high, and that any one individual's chances of reproducing at a larger size, and then leaving a greater number of more viable offspring, are slim. Both the ultimate ecological reasons and the controlling mechanisms involved in early gametogenesis require investigation. Factors which may have led to paedogenesis in ctenophores include an unpredictable food supply, as suggested by the work of Hirota and Greve (see Section 4.3.6.3), the innate fragility of most pelagic ctenophores (despite their regenerative abilities; see Section 4.2), and extensive predation, especially by the cannibalistic *Beroë* (Chun, 1892; Greve, 1970). The latter two factors were proposed by Chun (1892) as being the ultimate reasons for dissogeny. It may be significant that paedogenesis has yet to be reported in any *Beroë* species. It is also interesting that half-larvae of *Leucothea* and *Bolinopsis*, obtained experimentally from half-embryos (Section 4.4.1.4), develop gonads in their two subsagittal canals, suggesting that development of larval gonads is intrinsic in these species (however, see Section 4.3.6.1).

4.3.6.6 Parasites

Several species of trematodes and nematodes are parasitic in the parenchyme and pharynx of both larval and adult ctenophores (Chun, 1880; Mortensen, 1912; Komai, 1918), with no reported effect on the animals. Also, *Euchlora* (Chun, 1880) and *Beroë* (Mayer, 1912) harbor unicellular algae in their meridional canals, often among developing gametes, but effects on the animal and specifically on gametogenesis have not been studied.

4.3.7 Reproductive Behavior

Swarms of pelagic ctenophores have been frequently observed wherever a particular species is common, and occasionally a rare species may suddenly appear in a swarm (L. Agassiz, 1860; A. Agassiz, 1865, 1874; Chun, 1880, 1892, 1898; A. Agassiz and Mayer, 1899; Mayer, 1912; Easterly, 1914; Bigelow, 1915; Komai, 1918; Nelson, 1925). Whether swarming is correlated with reproduction has not been adequately documented, although A. Agassiz (1894) reported that the water around swarming individuals is at times filled with eggs. Swarming has also been correlated with simultaneous swarms of prey organisms *(Calanus)* (Chun, 1898). Also, swarming frequently, though not exclusively, occurs in autumn (Chun, 1880; Nelson, 1925) (see Section 4.3.8.3).

Although swarming prior to spawning would increase the probability of cross fertilization, the significance of outbreeding in ctenophores is debatable (see Section 4.4.1.1).

4.3.8 Spawning

4.3.8.1 Mechanisms of Spawning

In most and perhaps all ctenophores, except certain ovoviviparous species (Section 4.4.1.8), gametes are released through the epidermis above the gonads, probably always through ducts and pores (Section 4.3.2.3). Mayer (1912) reported of the cydippid *Tinerfe lactea* that "When mature the eggs are cast out through the side walls of the eight meridional canals (p. 17)." This rather ambiguous report has gone largely unnoticed, and Totton (1954) is usually credited (Hyman, 1959) with the first observation of egg release through the epidermis, as observed in *Beroë ovata*, a *Pleurobrachia* species, and the cydippid *Lampetia pancerina.* At Friday Harbor I have repeatedly observed release of both oocytes and sperm through consistently located pores above the gonads in *Bolinopsis microptera, Dryodora glandiformis,* and *Pleurobrachia bachei,* and oocyte release alone in *P. pigmentata* (Dunlap, 1966). Similar observations of gamete release have been made for *Mnemiopsis leidyi* at Woods Hole (Freeman and Reynolds, 1973) and a *Beroë* species near San Diego (J. Hirota, personal communication).

Most early reports state that gametes are released first into the gastrovascular canal system, are there fertilized, and the zygotes are discharged through the mouth (Chun, 1880, 1892, 1898; Krumbach, 1927; Hyman, 1940). This might be understood in view of my observations of *Pleurobrachia bachei,* in which both oocytes and sperm frequently break out into the meridional canals, especially if the specimen is damaged or roughly handled. This invariably happens during preparation of the tissues for fixation. If such an animal is maintained, oocytes are often fertilized and the zygotes develop as they are swept about in the canal system. However, there is a high percent of abnormality, and the embryos either disintegrate or are discharged through the aboral excretory pores (Fig. 1A), but never through the mouth. This usually happens to a small degree when an animal is spawning normally in the laboratory; whether it also happens in nature is unknown. It could be a result of the fragility of meridional canal tissues. More likely there are connections between each meridional canal and the oviduct and "testicular sinus" (Section 4.3.2.3), as suggested by Chun (1892) for *Bolinopsis vitrea,* through which gametes or zygotes pass.

Reports of "egg-laying" through the mouth in the platyctenid *Coe-*

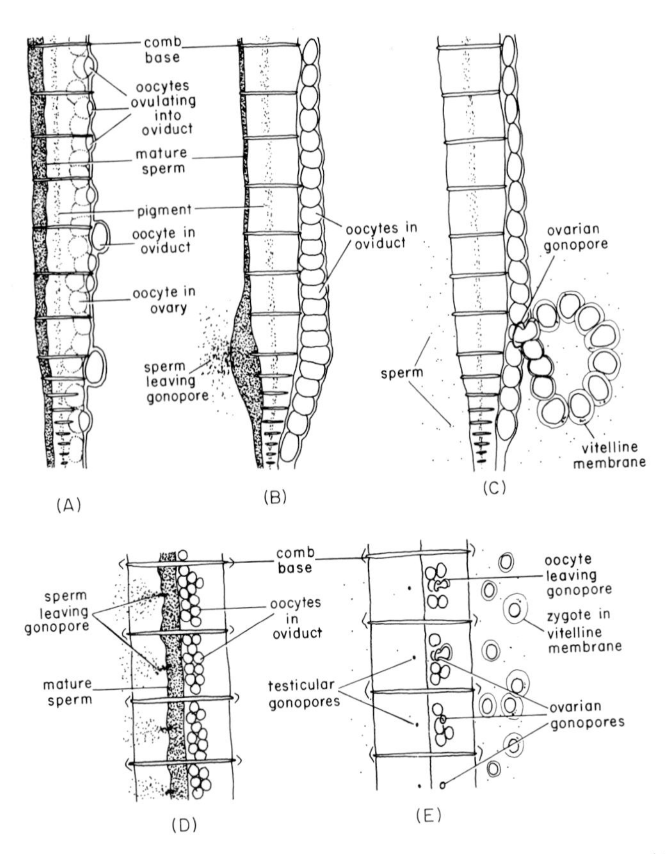

FIG. 19. Spawning in ctenophores, diagrammed in successive stages. (A–C) *Dryodora glandiformis*. (D–E) *Pleurobrachia bachei*. In (A) oocytes are ovulating into the oviduct, which lies lateral to the meridional canal in this species. The oviduct in *P. bachei* lies external to the canal, hence ovulation is not easily diagrammed and is not shown. (B) and (D) are comparable stages, as are (C) and (E). See text for details. (Drawn from life, surface view. Combs not included.)

loplana bocki (Komai, 1922) and through the ventral epidermis of *C. gonoctena* (Krempf, 1921) might well be valid, especially since both species brood their embryos beneath the ventral creeping sole (Section 4.4.1.8). Connections between the oviducts and canal system could be widespread in platyctenids (Section 4.3.2.3.2), since embryos are often brooded in the canal system (Section 4.4.1.8). In any case, further consideration of spawning in these platyctenids, and of the possibility of spawning through the mouth, must await future research.

Gamete release as observed in living *Dryodora* and *P. bachei* is diagrammed in Fig. 19. Certain preliminary changes occur in gonads of all species indicating that spawning is imminent. Oocytes become clearly outlined, correlated with their ovulation into the oviduct and being coated with a refractile layer of mucus (Section 4.3.2.3.2 and 4.4.1.2) (Fig. 19A and D, also 3B). Mature sperm concomitantly become apparent as opaque white bands *(Pleurobrachia* and *Dryodora)* (Fig. 19A and D), or similarly white packets *(Bolinopsis)* (Fig. 3B). Whether the change in sperm reflects a process similar to ovulation is unknown (see Section 4.3.2.3). The increased opacity could reflect a change in water content of the sperm.

In *Dryodora*, following the above changes, the rows of both oocytes and sperm are collected and compressed along each canal towards the level of the tentacle sheath openings and of the gonopore openings (Fig. 19B, also 1B). Oocytes at the ends of each oviduct become elongate, while those near the gonopores become flattened and discoid.

Sperm release precedes oocyte release. Sperm are discharged through tiny pores over the opaque sperm masses, at one point between each pair of combs in *Pleurobrachia* (Fig. 19D), at two points between each pair of combs in *Bolinopsis* (Fig. 3B), and from one general area, apparently including several pores between adjacent combs, in *Dryodora* (Fig. 19B) (see also Section 4.3.2.3).

Several spurts of sperm release, each involving a regular sequence of events, are required to empty an animal of all mature sperm. First, combs over the entire animal simultaneously cease beating. Then, in *Bolinopsis*, the transverse muscle bands between each pair of combs (Fig. 3B) all contract briefly, causing a temporary invagination of each entire comb row. At a comparable time in *Dryodora*, a single muscular girdle constricts the animal at the level of the gonopores. No similar muscular contraction takes place in *Pleurobrachia*. Following relaxation of these muscles, discharged sperm form white clouds above the testicular pores (Fig. 19B and D), and are then immediately dispersed by a sudden resumption of rapid comb beating. This sequence is repeated several times over a period of about five minutes until all sperm are emitted.

Oocyte release immediately follows completion of sperm release. In positions above the ovaries exactly comparable in each species to the positions of the testicular pores, the ovarian gonopores open, each appearing as a distinct circular aperture, 50–100 μm in diameter (Fig. 3B and 19E). In *Dryodora* one gonopore forms for each oviduct (Section 4.3.2.3.2) and is about 200 μm in diameter (Fig. 19C). Oocytes become constricted and evidently ameboid as they leave the pores. Usually a single oocyte is about half-way out before resumption of comb activity completes its release. As in sperm release, oocyte release occurs in bursts, each involving periodic cessation of comb beating, contraction of the body musculature, extrusion of gametes, and resumption of comb activity dispersing the gametes. Each such burst usually results in extrusion of one, or at the most two oocytes per pore. All ovarian gonopores remain open throughout oocyte release, and close several minutes following its completion. Release of all oocytes requires from 5 to 10 minutes in *Bolinopsis* and *Pleurobrachia*, and up to half an hour in *Dryodora*, under laboratory conditions.

Oocytes in *Dryodora* leave the pores one or two at a time, but remain attached through confluence of the jelly coats, resulting in eight strings of 20–40 oocytes each. Strings of eggs have also been reported for *Bolinopsis infundibulum* (A. Agassiz, 1894) and a *Pleurobrachia* species (Kowalewsky, 1866), suggesting that oviducal structure (Section 4.3.2.3.2) and the method of spawning are similar to those of *Dryodora*.

In freshly collected undamaged specimens all spawning events occur simultaneously over the entire animal, while damaged individuals show little coordination, and some ripe gametes are often retained.

The muscular contractions accompanying both oocyte and sperm release may be involved in opening gonopores, or forcing gametes out, or perhaps both. Similar contractions reportedly occur in many ctenophores in response to simple tactile stimuli (Hyman, 1940). The transverse muscle bands in *Bolinopsis* are closely associated with both sperm ducts and the oviducal ovarian gonopore cells (Section 4.3.2.3). No comparable muscles occur in *Pleurobrachia*, doubtlessly correlated with its relatively stiff gelatinous body. However, L. Agassiz (1850) described cell strands, in surface view of *P. pileus*, radiating out from points which must represent gonopores. These strands could be muscles although Agassiz interpreted them as nerves. However, oocytes also have some motility of their own, utilized during ovulation (Section 4.3.2.3.2). The cytoplasmic filaments observed in the ovarian gonopore cells of *Bolinopsis* (Section 4.3.2.3.2) may be contractile and function in opening the ovarian gonopores. Finally, especially in *Dryodora*, there is a suggestion that the oviduct, and perhaps also the tubular "testicular sinus," contracts, which could force the gametes out.

At Friday Harbor, a large *Bolinopsis* sheds up to 2000 oocytes at a time, a *Dryodora* from 250 to 300, and a single *P. bachei* up to 1000 oocytes in a single spawning. J. Hirota (personal communication) reports that *P. bachei* maintained in the laboratory (see Section 4.3.6.5) spawn from 50 to 1700 eggs every 2 days.

4.3.8.2 Synchronization and Coordination of Spawning

The time of spawning in an individual ctenophore can usually be controlled by manipulating its environment with respect to light or dark. At Friday Harbor, freshly collected specimens placed in the dark for an undetermined minimal amount of time, usually overnight, will spawn as described (Section 4.3.8.1) after being returned to the light. Nearly all *Bolinopsis* thus treated initiate sperm release in about 1½ hours and go through subsequent events of spawning nearly synchronously. (The time lag is slightly longer for *Dryodora*, and has not been determined for *Pleurobrachia*.) At Woods Hole the situation appears reversed, since *Mnemiopsis* will spawn in the laboratory in response to initiation of dark treatment, releasing fertilized eggs about eight hours after being placed in the dark (Freeman and Reynolds, 1973).

Such predictable reactions to changes in light suggest that light controls the time of spawning in nature, thus synchronizing spawning among individuals. This would increase chances of fertilization and especially of cross fertilization (however, see Sections 4.4.1.1 and 4.3.7). It also suggests that in nature *Bolinopsis* and *Dryodora* at Friday Harbor spawn shortly after dawn, while *Mnemiopsis* at Woods Hole spawns shortly after midnight, with the trigger in the latter case being sunset (Freeman and Reynolds, 1973).

Horridge (1964) described the ultrastructure of possible photoreceptors in *Pleurobrachia pileus*, first observed by Chun (1880), occurring in the floor of the apical organ. These could well be involved in the spawning response to light. Also, pigment, which could be light sensitive, is often localized over the gonads, in some only over testes (e.g., *Dryodora*, Fig. 1 and 19; also in *Mnemiopsis*, Freeman and Reynolds, 1973), in others mainly over the testes (*Beroë*, Fig. 3C).

4.3.8.3 Breeding Periods

Ctenophores reportedly breed year round in the Mediterranean (Naples) (Kowalewsky, 1866; Chun, 1880), although there is a peak of spawning during spring and summer. Only large specimens of *Cestum veneris* are reproductive in winter at Naples, while even the smallest individuals of other species spawn during colder months (Chun, 1880). Larval lobates evidently become sexual only during warmer months (Section 4.3.6.1 and 4.3.6.5). In the Adriatic (Trieste), spawning is reportedly limited to summer (Chun, 1880). Farther north, in the southern North Sea, at least some pelagic ctenophores reproduce in the spring (Künne, 1939).

Spawning along the Atlantic coast of North America occurs primarily in late spring or early summer from Tortugas, Florida (Mayer, 1912) north to southern Massachusetts (A. Agassiz, 1874). At Woods Hole, Freeman and Reynolds (1973) report that *Mnemiopsis* is reproductive from mid-July through August, and that the peak of reproduction appears to occur during the first week of August. From Cape Cod and north along the Maine coast, spawning is limited to late summer and early autumn (A. Agassiz, 1874; L. Agassiz, 1860) (see also Section 4.3.5).

Data presented by Esterly (1914) suggest that *Pleurobrachia bachei* from the coast of southern California has a major spawning in late spring or early summer. Hirota (personal communication) obtained eggs in the plankton as early as March from the same general area. In the coastal waters of Washington State, at Friday Harbor, spawning occurs mainly from mid-April through mid-June in all species, and in *Dryodora* is limited to these months. *Bolinopsis* and *P. bachei* continue to be reproductive, although less predictably, throughout the summer and early autumn. Thus, nearly all specimens of these two species collected during spring have ripe gonads, while less than 50% of those collected in August are reproductive. However, adequate collections have not been made from September through April, and there could be another major spawning, perhaps in autumn, or spawning might continue throughout the winter. As is true of many localities, difficulties involved in obtaining specimens from surface waters during rough winter weather, and from deeper waters at any time, may lead to the erroneous assumption that spawning does not occur at times, especially in winter.

Finally, off the Japanese coast, *Beroë cucumis* spawns as early as April (Komai, 1918), while the platyctenid *Coeloplana bocki* reproduces sexually in late summer and autumn (Komai, 1922). *C. gonoctena* also reproduces sexually in late summer and early autumn, in the South China Sea (Krempf, 1921) (see Section 4.2).

4.4 Development

4.4.1 Embryonic Development

Because of the unique segregated ooplasm, the equally unique biradial pattern of cleavage, and the highly determinate development of the embryo, there has been considerable interest in ctenophore embryology. Most of the early literature is adequately reviewed and summarized by Korschelt and Heider (1895) and MacBride (1914) and this and later literature are treated by Dawydoff (1928), Korschelt (1936), and Hyman (1940). Although a few more recent studies (Reverberi, 1957, 1966, 1971; Reverberi and Ortolani, 1963, 1965; Farfaglio, 1963; LaSpina, 1963; Ortolani, 1964; Freeman and Reynolds, 1969, 1972, 1973; Freeman *et al.*, 1971) have covered and partially clarified certain aspects of polarity, cell lineage, and mesoderm formation, and have shed new light on the results of earlier workers in experimental analysis of determinate development, most facets of ctenophore embryology remain to be satisfactorily described and analyzed.

Cleavage, gastrulation, and early larval development are remarkably similar throughout the phylum, with differences becoming increasingly apparent during postgastrular differentiation, and being most striking in postlarval transformations.

4.4.1.1 Fertilization

In the species at Friday Harbor, fertilization apparently occurs when an oocyte is extruded through the ovarian gonopore into the seawater (Section 4.3.8.1). Cytological details of fertilization are not available. Yatsu (1912) reported that more than one sperm may enter the egg. A small projection, probably representing a fertilization cone, has been observed on the egg surface near the site of polar body emission in both *Coeloplana bocki* (Komai, 1922) and *Bolinopsis microptera*.

Pelagic ctenophores are self-fertile. Nearly 100% normal development has been repeatedly obtained from self-fertilized eggs of a single *Bolinopsis microptera*, *Pleurobrachia bachei*, and *Dryodora glandiformis* isolated from the time of collection (Dunlap, 1966), and the same has been observed for a *Beroë* at Florida (G. Hendrix, personal communication). A more rigorous demonstration of self-fertilization has been obtained by J. Hirota (personal communication), who isolated juvenile *P. bachei* in filtered seawater and obtained a high percent of normal offspring from such single individuals.

Certainly the sequence of events of spawning (Section 4.3.8.1) would tend to favor self-fertilization: an individual first sheds sperm, briefly disperses them, and releases oocytes into its own sperm "suspension." Self-fertilization may be of widespread occurrence in nature among pelagic ctenophores. When in swarms, cross-fertilization might also occur, perhaps synchronized by reaction to light (Section 4.3.7 and 4.3.8.2).

In platyctenids, how and when fertilization takes place is unknown, and is complicated in many by the occurrence of seminal receptacles (Section 4.3.2.5), brooding and ovoviviparity (Section 4.3.2.4 and 4.4.1.8), sessility, and frequent gregariousness (Dawydoff, 1938b; Komai, 1922, 1942). It remains possible that outbreeding may be significant among platyctenids.

4.4.1.2 Vitelline Membrane and Jelly Coat

A thin and tough vitelline membrane envelops all ctenophore eggs and embryos. It has been previously but erroneously maintained, based on the assumption that oocytes ovulate into the meridional canal lumen, that both this membrane and the jelly coat covering it are derived from transformed gastrodermal cells and ovarian gland cells (Chun, 1892; Mortensen, 1912; Komai, 1942). I have determined in the Friday Harbor species that the vitelline membrane is distinctly noncellular and is formed at the oocyte surface during the latter half of oogenesis (Section 4.3.4.2.1) and that the jelly coat is secreted by the oviducal gland cells (Section 4.3.2.3.2 and 4.3.8.1).

In the species at Friday Harbor, the vitelline membrane begins to rise from the oocyte surface immediately upon contact with seawater during normal spawning (Section 4.3.8.1) (Fig. 19). It is not known whether sperm entry is necessary for lifting the membrane, nor whether an oocyte can be fertilized after the membrane has separated from the oocyte. The function of the membrane is probably mainly protective. However, it might also serve in prevention of polyspermy, since numerous sperm are often trapped in it, especially when sperm concentration is high.

The vitelline membrane is closely applied to the surface of the brooded eggs and embryos of *Coeloplana bocki* (Komai, 1922) and *Tjalfiella tristoma* (Mortensen, 1912). A space of varying depth separates the membrane from the egg surface in all other ctenophores, both oviparous and ovoviviparous, and is filled with a watery substance, probably seawater taken up during lifting of the membrane.

A thin coat of jellylike mucopolysaccharide is applied to the vitelline membrane in all known cases (except possibly the internally brooded embryos of *Tjalfiella* and *Lyrocteis;* Section 4.4.1.8). Yatsu noted that this jelly coat is thicker in fertilized than unfertilized eggs (see Reverberi and Ortolani, 1965). Confluence of the jelly coats causes eggs to form strings in certain species (Section 4.3.8.1), and dissolution of the mucus approximately a day following spawning allows the eggs to separate.

4.4.1.3 The Zygote

The ctenophore zygote is spherical and ranges in diameter from 120 μm (*P. bachei*) to over a millimeter (*Beroë ovata*), with most being in the order of 200-500 μm (excluding the vitelline membrane) (A. Agassiz, 1874; Chun, 1880, 1892; Yatsu, 1912; Komai, 1918, 1922). Nuclear diameters vary, from 10 or 12 μm in the 210 μm egg of *Bolinopsis microptera*, to 50 μm in the 250-300 μm egg of *B. vitrea* and *Leucothea multicornis* (Chun, 1880, 1892). Of considerable interest is Chun's (1892) report that while zygotes formed by sexual larvae of the latter two species are only 130-140 μm in diameter, or about half that of adult eggs, the nuclei are the same size as those of adult eggs, or 40-50 μm (see Section 4.3.6.5).

The ctenophore zygote is basically similar to the mature ovarian oocyte (Section 4.3.4.2.1), and is characterized by a distinctly segregated ooplasm. In the living state, only two zones are visible: an inner "endoplasm" composed of translucent yolk spheres, and an outer thin cortical plasm ("ectoplasm") which is devoid of yolk, appears granular, and is often vaguely yellow or pale green, but is not bioluminescent (Freeman and Reynolds, 1969, 1973). The two plasms have a number of distinct cytochemical properties; notably, mitochondrial enzymes are concentrated in the cortical plasm (Reverberi, 1957). Mitochondria are also scattered among the central yolk spheres (unpublished observations on *Bolinopsis* and *Pleurobrachia* development) (Section 4.3.4.2.1). Reverberi observed certain "anomalous" eggs of *Beroë* and "*Eucharis*" (= *Leucothea*), which did not develop, and which had spots of mitochondria-rich cytoplasm on their surface rather than the usual even cortical layer of mitochondria. These spots could well represent patches of cortical plasm which failed to spread out over the oocyte surface during the last stage of oogenesis (Section 4.3.4.2.1).

The zygote nucleus has not been adequately described. It lies in the cortical plasm near the site of polar body attachment (Section 4.3.4.2.1), and is often slightly flattened parallel to the cell membrane. Usually a single nucleolus is present. The polar bodies are situated at the vegetal (future oral) pole of the embryo (Reverberi and Ortolani, 1963).

4.4.1.4 Cleavage

Cleavage is unilateral, holoblastic, unequal, and (after the second division) biradial, and highly determinate. Early divisions are usually quite rapid, each requiring from 5 to 10 minutes. The first two cleavages begin at the animal pole, and are meridional and equal. The first transects the zygote along the future sagittal plane, and the second along the future tentacular plane.

Yatsu (1911, 1912) demonstrated experimentally that the ctenophore zygote is initially regulative, with any nucleated half cleaving and developing as a whole embryo, but becomes determined at the onset of the first cleavage. After the first cleavage has begun, any experimentally produced half of the embryo (if cut in the oral-aboral axis), or either of the first two blastomeres behaves in isolation as a half embryo and eventually forms only four comb rows. Each of the first four blastomeres divides as a quarter embryo, later developing only two comb rows (Chun, 1880; Driesch and Morgan, 1895; Fischel, 1897, 1898, 1903; Ziegler, 1898; Yatsu, 1911, 1912; Freeman and Reynolds, 1973).

The third cleavage begins near the animal pole and divides each of the four blastomeres along a plane at 45° and hence oblique to the first division plane. It is also unequal, resulting in an eight-cell stage composed of two rows, of four blastomeres each, paralleling the future tentacular plane (Fig. 20). Those cells at the ends of the tentacular axis, which are smaller and slightly aboral, are the "end" cells, while the larger more oral blastomeres are the "middle" cells. The difference in size between end and middle cells varies among species.

As first suspected by Yatsu (1912) and recently confirmed by Farfaglio (1963), Reverberi and Ortolani (1963), and Freeman and Reynolds (1969, 1973), only the four end macromeres of the eight-cell stage, in isolation, are able to form comb plate rows and do so in the intact embryo. These results disagree with those of earlier workers that each of the octants of the embryo gives rise to one comb row.* Further, Freeman and Reynolds (1971, 1973) have shown that in *Mnemiopsis* the middle macromeres give rise to the light-producing cells (photocytes) of the larva, while the end cells are not involved in photocyte production and in isolation develop into larvae which do not bioluminesce (see Sections 4.4.1.6.5 and 4.4.1.5). Experiments involving centrifugation of eggs to remove all or most of the yolk have shown that both middle and end macromeres cleaving from egg fragments lacking yolk and containing only cortical plasm are able to form comb plate cilia, but not photocytes. If a middle macromere

*Reverberi and Ortolani (1965) suggest that the earlier workers mistook the large balancer cilia of the developing apical organ, which are formed by the middle blastomeres (see Section 4.4.1.6.3), to be cilia of developing comb plates.

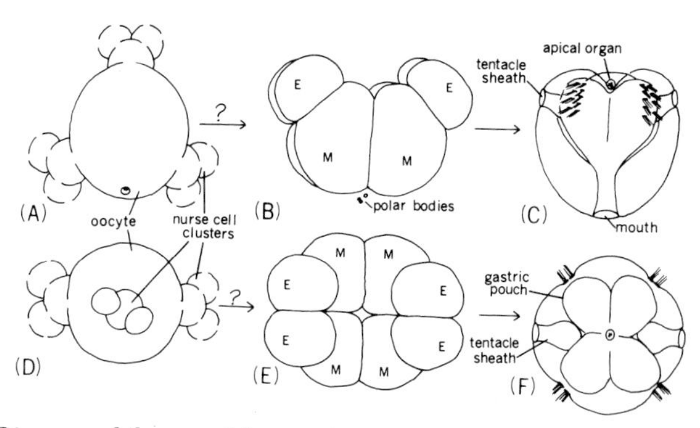

FIG. 20. Diagram of the possible correlation between the biradial symmetry of the oocyte-nurse cell complex (A) and (D) and that of the eight-cell stage of the embryo (B) and (E) and of the cydippid larva (about 1 day old) (C) and (F) of *Bolinopsis microptera*. (A) through (C) viewed from the sagittal axis; (D) through (E) viewed from the aboral pole. The polarity relationships between the embryo and larva (B) to (C) and (E) to (F) have been established (Reverberi and Ortolani, 1963), and the identity of the oocyte's basal-apical axis with the embryo's animal-vegetal axis is also known (Section 4.3.4.2.1). However, whether the loci of the two lateral nurse cell clusters bear any relationships to the ends of either the tentacular (as shown) or sagittal axis has not been determined, and thus (A) to (B) and (D) to (E) are partly hypothetical. The "end" and "middle" blastomeres of the embryo are labelled, respectively, "E" and "M."

contains some yolk it does not develop comb plate cilia, but can produce photocytes (Freeman and Reynolds, 1971, 1973). In any case, the abilities to form comb plate cilia and photocytes, although experimentally alterable, are normally segregated at the eight-cell stage into the end and middle macromeres, respectively.

By the eight-cell stage the defining characteristics of biradial symmetry of the embryo have been established: the oral-aboral axis, the tentacular plane, and the sagittal plane. This symmetry is reflected not only in the configuration of blastomeres but also in their respective developmental potentials, as just described. We may recall that during oogenesis in *Bolinopsis* the three nurse cell clusters join the oocyte at three points: two lateral and one more or less basal to the apical nucleus (Fig. 13; Section 4.3.4.2.1). This results in a biradial symmetry of the oocyte-nurse cell complex, which could well correspond with that of the embryo and adult (Fig. 20), although a causative correlation with developmental potentials of the blastomeres is not being suggested. In fact, Freeman and Reynolds (1972) demonstrated that segregation of developmental potentials with respect to comb plate versus photocyte formation does not begin to occur until the second cleavage.

The fourth cleavage is perpendicular to the first three, hence is horizontal, and is also distinctly unequal, with each of the eight blastomeres giving off a micromere towards the aboral pole. Until now the endoplasm and cortical plasm have been generally equally distributed among blastomeres, each being a small edition of the zygote. However, with the fourth cleavage there is a partial separation of the plasms, and the eight micromeres contain predominantly cortical plasm, while the eight macromeres contain mainly endoplasm. The fate of the "subcortical plasm" (Section 4.3.4.2.1) has not been adequately studied, but preliminary observations on *Bolinopsis* development suggest that much of it comes to lie at the 16-cell stage in the micromeres. Farfaglio (1963) demonstrated that ability to form combs is limited at the 16-cell stage to the four end micromeres, and later to their descendents, again in disagreement with the conclusions of earlier workers that cortical plasm bestows comb-forming potential on a blastomere. Each of the four end micromeres is responsible for the formation of two comb rows.

Through the 16-cell stage divisions are synchronous in all quadrants, but during the rest of cleavage, divisions of the middle cells and their descendents lag behind those of the end cells (Komai, 1922; Farfaglio, 1963). The macromeres give off two more generations of micromeres, and then become quiescent until late in gastrulation. The micromeres continue to divide, and eventually form a thick annular ring of small cells (the "aboral micromeres"), aboral to the two rows of macromeres.

4.4.1.5 GASTRULATION

Gastrulation occurs first by epiboly, or an oral migration of the aboral micromeres over the surface of the macromeres, and ends with invagination, or emboly, in which the macromeres turn inward and carry some micromeres to the inside of the embryo. During epiboly those micromeres destined to form comb rows divide and migrate more rapidly than the rest, forming four thickened ribs of smaller cells over the embryo's surface.

Near the end of epiboly, the macromeres divide once equally and then once unequally, with the divisions of the end cells now lagging behind those of the middle cells. The latter division results in formation of 24 "gastrular micromeres," 16 arising from the middle cells, and eight from the end cells (Farfaglio, 1963). These gastrular micromeres evidently contain much of the cortical plasm remaining in the macromeres. They appear on the oral side and are brought to the inside of the embryo by subsequent invagination of the macromeres.

As described by Metchnikoff (1885) for the cydippid *Callianira*, these gastrular micromeres, after being carried inward to the aboral side of the gastric cavity, proliferate to form a cross. The cells of the long arms, corresponding to the tentacular axis of the embryo, eventually form the tentacular and body musculature, while those of the shorter sagittal axis bud off and develop mesenchymal parenchyme cells. Reverberi and Ortolani (1963) and Ortolani (1964) followed the fate of these cells by marking them with chalk granules. They present evidence supporting Metchnikoff's description and suggest further that the former cells arise from end gastrular micromeres while the latter arise from middle gastrular micromeres. However, Freeman and Reynolds (1973) have demonstrated in some detail that in *Mnemiopsis* at least most of these oral micromeres are progenitors of the photocytes (Section 4.4.1.4 and 4.4.1.6.5). Hence the evidence that the gastrular micromeres give rise to mesodermal musculature and collenchyme, and therefore that the mesoderm in ctenophores is of endodermal origin, is placed in doubt, although it remains possible that some of these micromeres may form mesenchymal elements. In any case, although ectoderm is clearly derived from the aboral micromeres and endoderm from the macromeres, the problem of the origin of mesoderm in ctenophores (see Hyman, 1940) is still with us. There is evidence that at least some mesoderm may arise from ectodermal cells budded off at the oral region (Hyman, 1940).

4.4.1.6 Organogenesis

The main structures of the cydippid larva are the combs (motility), tentacles (prey-catching), an apical sense organ (orientation), a stomodeal pharynx (feeding, settling), and a simple gastrovascular system (digestion).

4.4.1.6.1 *Comb Rows.* Each comb is formed by fusion of numerous long cilia arising from several rows of columnar ("polster") cells (Afzelius, 1961; Horridge and Mackay, 1964). Combs first appear near the end of epibolic gastrulation, in four short irregular rows, each row containing from four to seven combs. The rows gradually separate orally and become more regular and parallel, and new combs are added at the oral ends of the rows, usually only after hatching but in some (e.g., *Coeloplana bocki,* Komai, 1922) prior to hatching. The cilia of the combs are first short, and beat irregularly and individually, and later elongate and beat synchronously so that the embryo can swim about within the membrane (see Section 4.4.1.2).

In the cydippid larvae of the platyctenids *Tjalfiella* (Mortensen,

1912), *Coeloplana gonoctena* (Krempf, 1921), *C. bocki* (Komai, 1922), and *Lyrocteis* (Komai, 1942), all of which lose their combs during metamorphosis (Section 4.4.2.2), the cilia of the combs are of exaggerated length, possibly related to the short larval stage. In *Lyrocteis* and *Tjalfiella* each pair of comb rows is recessed into an epidermal pocket, reminiscent of the temporary invagination of entire comb rows in many adult ctenophores (Section 4.3.8.1) (Dawydoff, 1933; Hyman, 1940).

4.4.1.6.2 *Tentacles.* Tentacle primordia appear following gastrulation as two oval ectodermal thickenings, several cell layers deep, at the ends of the tentacular axis. Each primordium invaginates, and the sides of the invagination form the tentacle sheath while the tentacle stem sprouts from the blind tip of the invagination. The tentacles are primarily ectodermal, and Farfaglio (1963) has presented evidence for derivation of ectodermal primordia from the second generation of end aboral micromeres (Section 4.4.1.4). Tentacles are invested with mesodermal musculature (Section 4.4.1.5). In nearly all ctenophores they are covered with colloblast cells, which proliferate from ectodermal cells at the base of the tentacle stem. In the peculiar *Euchlora rubra*, nematocysts are present in place of colloblasts. Whether *Euchlora* constructs these nematocysts, or whether they originate from prey medusae, is still uncertain (see Hyman, 1959; Komai, 1963). Studies of embryonic and larval development of this species should clarify the problem and are greatly needed.

Beroids never develop tentacles as larvae or as adults, but as embryos apparently possess both the ectodermal and mesodermal tentacle rudiments (Metchnikoff, 1885), suggesting that the presence of tentacles is the primitive condition in the phylum.

4.4.1.6.3 *Apical Organ.* The apical organ also develops from an ectodermal thickening, composed of a single layer of columnar cells, forming a slight depression at the aboral pole. S. Tamm is currently studying the ultrastructure and development of this organ in *Beroë*. Tamm (personal communication) has observed that cilia forming the dome ("veil") appear first, at the rim of the ectodermal thickening. Following completion of the dome, whole cells, each containing a statolith granule, are released from the floor of the apical thickening, adhere together, and float freely in the dome cavity. Four groups of long balancer cilia grow into the cavity and eventually contact the statolith cells, which then form folds about the balancer cilia and thus adhere to them.

Ortolani (1964) demonstrated that the first generation of middle micromeres give rise to the balancer cilia, certain cilia of the apical organ floor, and those of the polar fields (Reverberi and Ortolani,

1965). Apparently each middle micromere is responsible for the formation of one balancer. The statolith itself and the transparent dome have a different but undetermined origin (Reverberi and Ortolani, 1965; Reverberi, 1971).

4.4.1.6.4 *Stomodeum.* The stomodeum is derived from descendents of the third generation of micromeres from the end macromeres (Reverberi and Ortolani, 1963), at least in pelagic ctenophores, and is formed by ectoderm drawn inward during invagination of the macromeres (Section 4.4.1.5). In platyctenids, however, there is a second invagination after the blastopore closes (Komai, 1922).

The stomodeum gives rise to the pharynx and esophagus of the larva and adult. In platyctenids the pharynx reaches a high degree of complexity, forming numerous folds in the larva and during metamorphosis opening out to become the ventral creeping sole of the adult (Section 4.4.2.2.1) (Mortensen, 1912; Komai, 1922, 1942).

4.4.1.6.5 *Gastrovascular system.* From the end of the esophagus, the gastrovascular system is entirely of endodermal origin. After gastrulation, it develops first as an irregular cavity in the center of the macromeres, expanding into four pouches, which may be narrow (*Coeloplana, Beroë*) or spacious (*Lyrocteis,* most pelagic ctenophores). Each pouch occupies a quadrant of the body and lies beneath a pair of comb rows (Fig. 20). At the time of hatching, there are varying degrees of constriction of each pouch into two, each resulting pocket being a primordium of one meridional canal (see Fig. 9). These are joined at the center of the animal by the stomach which communicates orally with the esophagus and aborally with the outside through two tiny excretory pores. The tentacular canals are represented by two blind pouches, extending orally from the stomach.

Bioluminescence, an ability displayed by all adult ctenophores, first appears in the larva when the comb plate cilia begin to develop (Section 4.4.1.6.1) (Freeman and Reynolds, 1969, 1973). Freeman and Reynolds (1973) have shown that the loci of luminescence are the four developing gastric pouches, and that the photocytes responsible for the luminescence (Section 4.4.1.4) are in intimate association with the endodermal lining of the pouches.

4.4.1.7 Hatching

Larvae hatch by rupturing the vitelline membrane. At least in *Coeloplana,* larvae apparently leave the membrane oral end first Krempf, 1921; Komai, 1922). Komai describes a prehatching behavior of the larva of *C. bocki,* suggestive of attempts to break the membrane with the mouth. Mayer (1912) illustrates a similar event in *Mnemiopsis leidyi.*

The time of hatching in pelagic ctenophores is usually about 1-2 days after fertilization (Mayer, 1912; Greve, 1970; J. Hirota, personal communication).

4.4.1.8 Brooding

Only platyctenids have been reported to brood their embryos. Ovoviviparity, or retention of the embryos within the adult organism, occurs in *Tjalfiella* (Mortensen, 1912), *Coeloplana mitsukurii* (Tanaka, 1932), *Lyrocteis* (Komai, 1942), and *Vallicula* (Rankin, 1956; G. Freeman, personal communication). Brood chambers of *Tjalfiella* and *Lyrocteis* are described in Section 4.3.2.4.

In *Coeloplana bocki* (Komai, 1920, 1922) and *C. gonoctena* (Krempf, 1921) embryos are brooded under the creeping sole of the adult, within a mass of gelatinous material which is sticky in *C. bocki* and relatively solid in *C. gonoctena.* Embryos of *C. mitsukurii* are retained within the "gastrovascular spaces" of the adult (Tanaka, 1932) (see Section 4.3.2.4). In *Vallicula,* G. Freeman (personal communication) observed embryos developing both in the gastrovascular canals and outside the animal, while in the same species Rankin (1956) noted embryos being brooded both beneath the adult, attached to the ventral epidermis, and within the four "spherical chambers" (Section 4.3.2.4).

In both *Coeloplana bocki* (Komai 1922) and *C. mitsukurii* (Tanaka, 1932) embryos brooded by a single individual are in approximately the same stage of development, hence the eggs were probably fertilized at the same time. Several stages are brooded simultaneously within a single *Lyrocteis* (Komai, 1942). In *Tjalfiella,* Mortensen (1912) reported that all stages may be simultaneously brooded within the same adult.

The number of embryos brooded by a platyctenid individual is generally fewer than the number spawned in one spawning by a single pelagic ctenophore, which varies from 250 to 2000 eggs (Section 4.3.8.1). A single *Tjalfiella* broods up to 35 embryos at a time (Mortensen, 1912), while a *Coeloplana bocki* broods from 10 to 200 but usually no more than 50 (Komai, 1922). One *Lyrocteis* may contain up to 400 developing embryos (Komai, 1942).

There is no evidence of a difference in initial size of eggs which are to be brooded and those which develop free from the parent organism (see Section 4.4.1.3). Embryos of both *Tjalfiella* and *Lyrocteis* enlarge prior to hatching, possibly a result of ovoviviparity, although there is no evidence that they receive nourishment from the parent.

Escape of brooded embryos has been observed in living animals in

Coeloplana mitsukurii (Tanaka, 1932) and *Lyrocteis* (Komai, 1942). In both species the larvae leave through tiny holes in the epidermis (see Section 4.3.2.4).Whether hatching from the vitelline membrane coincides with or precedes this escape is not known.

4.4.1.9 Factors Influencing Embryonic Development

Nothing is known about factors affecting ctenophore embryonic development, except that I have noted that warming above about 15-18°C tends to hasten cleavage and early development but also leads to a high percent of abnormality, in the species at Friday Harbor. Freeman and Reynolds (1973) report that cleavage in *Mnemiopsis* is inhibited by temperatures much above 24° C, but that postgastrular development is not affected by such high temperatures.

4.4.2 Larvae

4.4.2.1 Larval Types

4.4.2.1.1 *Cydippid.* All ctenophores except *Gastrodes* have a cydippid larva (Fig. 20), although the use of the term "larva" to describe this stage is, strictly speaking, valid only for the order Platyctenea, since true metamorphosis occurs only in this group (Section 4.4.2.2). Greve (1970) contends that the Lobata cydippid stage is also rightfully a larva, since the larval tentacles are lost, and the adult tentacles develop anew.

The cydippid larva of the platyctenids is their only free-swimming dispersal stage and is short-lived (Section 4.4.2.2). Neither larvae nor any embryonic stage of *Ctenoplana* and *Planoctena* has been observed. Specific differences, foreshadowing adult modifications, occur among cydippid larvae of each of the five orders, and usually allow identification to order (Hyman, 1940).

4.4.2.1.2 *"Planula."* The youngest stage of the platyctenid *Gastrodes* thus far observed bears a remarkable resemblance to the planula larva of the Cnidaria (Fig. 21) (Komai, 1922, 1963). It has a solid endodermal core, containing possible primordial germ cells (Section 4.3.4.2.2), and is covered with a uniformly ciliated epidermis. Komai (1922) observed one such larva in the process of boring through the test of the tunicate *Salpa* (Section 4.4.2.2.2). Because the planuloid form could represent structural modifications related to the penetration of the "host," perhaps phylogenetic significance should not be placed on its occurrence in a ctenophore, at least until more is known of earlier embryonic stages, which have never been observed (see also Section 4.3.4.2.2).

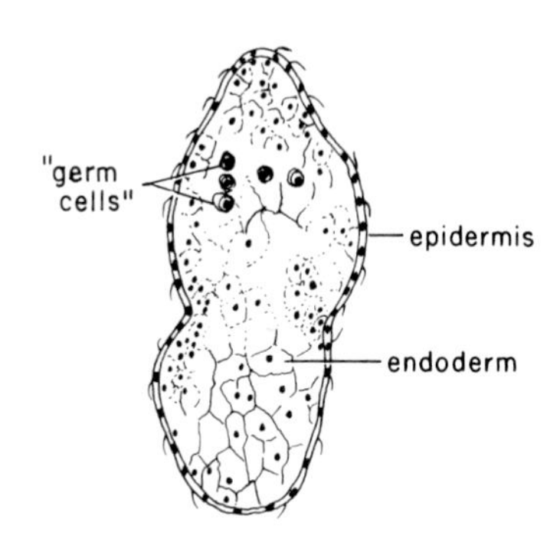

FIG. 21. Longitudinal section through a "planula" larva of *Gastrodes parasiticum*, showing the ciliated epidermis, and the "primordial germ cells" among the cells of the solid endoderm. [Redrawn from Komai (1922).]

4.4.2.2 METAMORPHOSIS AND SETTLING

4.4.2.2.1 *Cydippid.* Varying degrees of transformation of cydippid larvae into the adult form occur among ctenophores. The least change occurs in the Cydippida and Beroida, with considerably more taking place in the Lobata and Cestida (Hyman, 1940). However, only in the Platyctenea is the change from larva to juvenile rapid and spectacular, and deserving of the term metamorphosis (see Korschelt and Heider, 1895; Korschelt, 1936; Hyman, 1940).

The process of metamorphosis in a platyctenid has been described only in *Coeloplana bocki* (Komai, 1922). After hatching, the larva swims about for a while, in jerky movements reminiscent of copepods, evidently because of the relatively long comb cilia (Section 4.4.1.6). Possibly such movement allows a more effective dispersal, or is related to predator escape. The larva then alternates between swimming and settling briefly on its expanded everted pharynx, and finally gives up swimming altogether, concomitant with degeneration and eventual loss of its combs. Then specific characters of the adult begin to develop (Mortensen, 1912; Komai, 1922, 1942).

4.4.2.2.2 *"Planula."* The planuloid larva of *Gastrodes parasiticum*, having entered the mantle of the host *Salpa*, becomes cup shaped, develops eight comb rows, an apical organ, rudimentary tentacles, and a gastrovascular system consisting first of four pouches and later of eight subequal pouches reminiscent of those of a freshly settled juvenile *Coeloplana bocki* (Komai, 1922, 1963; see Fig. 221 in Hyman, 1940). Komai (1922) considers this last stage, the oldest to occur in *Salpa*, as the sexual stage, and interprets as oocytes certain large cells in the ventral epidermis (Fig. 18; Section 4.3.4.2.2). The only reported observation of any later stage is that of Dawydoff (1937), who described changes taking place in a single *Gastrodes komaii* liberated from a *Salpa*. During the course of 3 days this specimen flattened out like *Coeloplana*, lost its combs, and assumed a creeping existence, after which it died. *Gastrodes* has been considered a parasite within *Salpa*, but Dawydoff's observations indicate that there is more to be known about its life cycle. Rankin (1956) suggests that "the 'parasitic' stages are making use of an ingenious method of transport to new grounds and that the truly adult stage will yet prove to be sessile" (p. 69).

4.4.2.3 Factors Influencing Growth, Metamorphosis, and Settling

Very little is known about the biology of ctenophore larvae. That cydippid larvae begin feeding soon after hatching is evident from observations of food particles in the digestive cells of recently hatched larvae (Chun, 1880; Komai, 1922). J. Hirota (personal communication) reports that larvae of *P. bachei*, if adequately fed, grow to four or five times their initial diameter by the time they are morphologically juveniles, which requires about 18 days in the laboratory.

Greve (1970) reports that there is a high degree of mortality (99%) in *Pleurobrachia pileus* after hatching in the laboratory, but factors involved here are not known.

The behavior of the metamorphosing larva of *Coeloplana* (Section 4.4.2.2) suggests that it is "testing" the substrate, but there are no published observations relevant to this problem. The degree of "host" specificity observed in many sessile ectocommensal platyctenids (Mortensen, 1912; Krempf, 1921; Komai, 1922, 1942; Dawydoff, 1938b) also points to some degree of substrate preference in the larva, since the adults have little or no motility.

Komai's (1922) observation that the length of the free-swimming larval stage of *Coeloplana bocki* varies from 4 hours to more than a day indicates that the time of settling and metamorphosis is not intrinsic, but may be affected by external factors.

Acknowledgments

This chapter would not have been completed without the patience, help, and encouragement of my husband, Eric. He critically read several drafts of the manuscript, and any clarity is doubtlessly a result of his efforts. The research at Friday Harbor was supervised by Dr. Richard A. Cloney, whose advice and excellent training are gratefully acknowledged, and was supported by a National Science Foundation predoctoral fellowship.

4.5 References

Afzelius, B. A. (1961). The fine structure of the cilia from ctenophore swimming plates. *J. Biophys. Biochem. Cytol.* **9**, 383-394.

Agassiz, A. (1865). North American Acalephae. Order Ctenophorae Eschscholtz. *Ill. Cat. Mus. Comp. Zool. Harvard*, no. 2, 7-40.

Agassiz, A. (1874). Embryology of the Ctenophorae. *Mem. Amer. Acad. Arts Sci.* **10,** 357-398.

Agassiz, A., and Mayer, A. G. (1899). Acalephs from the Fiji Islands. *Bull. Mus. Comp. Zool. Harvard* **32,** 176-177, pl. 15-16.

Agassiz, L. (1850). Contributions to the natural history of the Acalephae of North America. II. On the beroid medusae of the shores of Massachusetts. *Mem. Amer. Acad. Arts Sci.* **4,** 313-374.

Agassiz, L. (1860). "Contributions to the Natural History of the United States", Vol. 3, 301 pp. Little, Brown, Boston.

Bigelow, H. B. (1912). The ctenophores of the "Albatross" expedition. *Bull. Mus. Comp. Zool. Harvard* **54,** 369-404.

Bigelow, H. B. (1915). Exploration of the coast water between Nova Scotia and Chesapeake Bay, July and August, 1913, by the U. S. Fisheries schooner "Grampus." Oceanography and Plankton. *Bull. Mus. Comp. Zool. Harvard* **59,** 149-359.

Chun, C. (1880). Die Ctenophoren des Golfes von Neapel. *Fauna Flora Golfes Neapel* **1,** 1-311.

Chun, C. (1892). Die Dissogonie, eine neue Form der geschlechtlichen Zeugung. *Festsch. Zum siehenzigsten Geburtstage Rudorf Leuckarts* 77-108, pl. 9-13.

Chun, C. (1898). Die Ctenophoren der Plankton-expedition. *Ergeb. Plankton Exped.* 2 1-33, pl. 1-3.

Coonfield, B. (1936). Regeneration in *Mnemiopsis leidyi* Agassiz. *Biol. Bull.* **71,** 421-428.

Coonfield, B. (1937). The regeneration of plate rows in *Mnemiopsis leidyi* Agassiz. *Proc. Nat. Acad. Sci. U. S.* **23,** 152-158.

Davidson, E. (1968). "Gene Activity in Early Development," 375 pp. Academic Press, New York.

Dawydoff, C. (1928). "Traité d'Embryologie Comparée des Invertébrés," 930 pp. Masson, Paris.

Dawydoff, C. (1933). Quelques observations sur la morphologie externe et la biologie des *Ctenoplana. Arch. Zool. Exp. Gén.* **75,** 103-128.

Dawydoff, C. (1936). Les Ctenoplanidae des eaux de l'Indochine Française. Etude systématique. *Bull. Biol. Fr. Belg.* **70,** 456-486, pl. 15.

Dawydoff, C. (1937). Les Gastrodes des eaux indochinoises et quelques observations sur leur cycle évolutif. *C. R. Acad. Sci. Paris* **204,** 1088-1090.

Dawydoff, C. (1938a). Multiplication asexuée, par lacération, chez les Ctenoplana. *C. R. Acad. Sci. Paris* **206,** 127-128.

Dawydoff, C. (1938b). Les Coeloplanides indochinoises. *Arch. Zool. Exp. Gén.* **80,** 125-162.

Dawydoff, C. (1950). La nouvelle forme de Ctenophores planarisés sessiles provenant de la Mer de Chine Méridionale (*Savangia atentaculata,* nov. gen. nov. spec.). *C. R. Acad. Sci. Paris* **231,** 814-816.

Driesch, H., and Morgan, T. H. (1895). Zur Analysis der ersten Entwickelungsstadien des Ctenophoreneies. *Arch. Entwicklungsmech. Organ.* **2**, 204-224.

Dunlap, H. L. (1966). Oogenesis in the Ctenophora. Ph.D. Thesis, Univ. of Washington, Seattle, Washington, 230 pp.

Esterly, C. O. (1914). A study of the occurrence and manner of distribution of the Ctenophora of the San Diego region. *Univ. Calif. Publ. Zool.* **13**, 21-38.

Farfaglio, G. (1963). Experiments on the formation of the ciliated plates in ctenophores. *Acta Embryol. Morphol. Exp.* **6**, 191-203.

Fawcett, D. W. (1961). Intercellular bridges. *Exp. Cell Res. Suppl.* **8**, 174-187.

Fischel, A. (1897). Experimentelle Untersuchungen am Ctenophorenei, I. *Arch. Entwicklungsmech. Organ.* **6**, 109-130.

Fischel, A. (1898). Experimentelle Untersuchungen am Ctenophorenei, II, III, IV. *Arch. Entwicklungsmech. Organ.* **7**, 557-630.

Fischel, A. (1903). Entwicklung und Organdifferenzierung. *Arch. Entwicklungsmech. Organ.* **15**, 679-750.

Fol, H. (1869). Ein Beitrag zur Anatomie und Entwicklungsgeschichte einiger Rippenquallen. Inaugural-Dissertation, Medicinischen Facultät der Friedrich-Wilhelms-Universität zu Berlin, 1-12, 4 pl.

Freeman, G. (1966). Studies on regeneration in the ctenophore *Coeloplana*. *Amer. Zoolog.* **6**, 577.

Freeman, G. (1967). Studies on regeneration in the creeping ctenophore, *Vallicula multiformis*. *J. Morphol.* **123**, 71-84.

Freeman, G., and Reynolds, G. T. (1969). The development of bioluminescence in the ctenophore *Mnemiopsis leidyi*. *Amer. Zoolog.* **9**, 1119.

Freeman, G., and Reynolds, G. T. (1972). The segregation of developmental potential during early cleavage stages in the ctenophore *Mnemiopsis leidyi*. *Biol. Bull.* **143**, 461-462.

Freeman, G., and Reynolds, G. T. (1973). The development of bioluminescence in the ctenophore *Mnemiopsis leidyi*. *Devel. Biol.* **31**, 61-100.

Freeman, G., Reynolds, G. T., and Milch, J. R. (1971). A cytoplasmic interaction during the first stages of embryogenesis that plays a necessary role in the formation of light producing cells in the ctenophore *Mnemiopsis*. *Biol. Bull.* **141**, 386-387.

Garbe, A. (1901). Untersuchungen über die Entstehung der Geschlechtsorgane bei der Ctenophoren. *Z. Wiss. Zool.* **69**, 472-491, pl. 36, 37.

Ghiselin, M. (1969). The evolution of hermaphroditism among animals. *Quart. Rev. Biol.* **44**, 189-208.

Greve, W. (1970). Cultivation experiments on North Sea ctenophores. *Helgoländer Wiss. Meeresunters.* **20**, 304-317.

Heider, K. (1927). Vom Nervensystem der Ctenophoren. *Z. Morphol. Ökol. Tiere* **9**, 638-678.

Hertwig, R. (1880). Ueber den Bau der Ctenophoren. *Jenaische Z. Naturwiss.* **14**, 313-457, pl. 15-21.

Horridge, G. A. (1964). Presumed photoreceptor cilia in a ctenophore. *Quart. J. Microsc. Sci.* **105**, 311-317.

Horridge, G. A., and Mackay, B. (1964). Neurociliary synapses in *Pleurobrachia* (Ctenophora). *Quart. J. Microsc. Sci.* **105**, 163-174.

Hyman, L. H. (1940). "The Invertebrates. I. Protozoa through Ctenophora," 726 pp. McGraw-Hill, New York.

Hyman, L. H. (1959). "The Invertebrates. V. Smaller Coelomate Groups," 783 pp. McGraw-Hill, New York.

Koch, E. A., Smith, P. A., and King, R. C. (1967). The division and differentiation of *Drosophila* cystocytes. *J. Morphol.* **121**, 55-70.

Komai, T. (1918) On ctenophores of the neighborhood of Misaki. *Annot. Zool. Japon.* **9,** 451-474, pl.7.
Komai, T. (1920). Notes on *Coeloplana bocki* n.sp. and its development. *Annot. Zool. Japon.* **9,** 575-584.
Komai, T. (1922). Studies on two aberrant ctenophores, *Coeloplana* and *Gastrodes,* 102 pp., 9 pl. Publ. by the author, Kyoto.
Komai, T. (1934). On the structure of *Ctenoplana. Mem. Coll. Sci., Kyoto Imperial Univ. Ser. B* **9,** 245-256.
Komai, T. (1942). The structure and development of the sessile ctenophore *Lyrocteis imperatoris* Komai. *Mem. Coll. Sci., Kyoto Imperial Univ., Ser. B* **17,** 1-36.
Komai, T. (1963). A note on the phylogeny of the Ctenophora. *In* "The Lower Metazoa: Comparative Biology and Phylogeny" (E. C. Dougherty, ed.), pp. 181-188. Univ. Calif. Press, Berkeley, California.
Komai, T., and Tokioka, T. (1942). Three remarkable ctenophores from the Japanese seas. *Annot. Zool. Japon.* **21,** 144-151.
Korotneff, A. (1888). *Cunoctantha* und *Gastrodes. Z. Wiss. Zool.* **47,** 650-657, pl. 40.
Korotneff, A. (1891). Zoologische paradoxen (*Cunoctantha* und *Gastrodes*). *Z. Wiss. Zool.* **51,** 613-628, pl. 30-32.
Korschelt, E. (1936). "Vergleichende Entwicklungsgeschichte der Tiere." G. Fischer, Jena, 536 pp.
Korschelt, E. and Heider, K. (1895). "Textbook of the Embryology of Invertebrates" (E. L. Mark and W. M. Woodworth, transl.) Part 1. 484 pp. Swan Sonnenschein, London.
Kowalewsky, A. (1866). Entwicklungsgeschichte der Rippenquallen. *Mém. Acad. Imp. Sci. St. Pétersbourg, sér.* 7 **10,** 1-28.
Krempf, A. (1921). *Coeloplana gonoctena:* biologie, organisation, développement. *Bull. Biol. Fr. Belg.* **54,** 252-312, pl. 5.
Krumbach, T. (1927). Ctenophora. *In* "Die Tierwelt der Nord- und Ostsee," lief VII, pp. 1-50. Grimpe und Wagler, Leipzig.
Künne, C. (1939). Die Beroë (Ctenophora) der südlichen Nordsee, *Beroë gracilis,* n. sp. *Zool. Anz.* **127,** 172-174.
La Spina, R. (1963). Development of fragments of the fertilized egg of ctenophores and their ability to form ciliated plates. *Acta Embryol. Morphol. Exp.* **6,** 204-211.
MacBride, E. W. (1914). "Text-book of Embryology," Vol. I, Invertebrata, 692 pp. Macmillan, New York.
Mayer, A. G. (1912). Ctenophores of the Atlantic Coast of North America. Carnegie Inst. of Washington publ. no. 162, 58 pp.
Metchnikoff, E. (1885). Vergleichend-embryologische Studien. 4. Über die Gastrulation und Mesodermbildung der Ctenophoren. *Z. Wiss. Zool.* **42,** 648-656, pl. 24-25.
Meyer, G. F. (1961). Interzelluläre Brücken (Fusome) im Hoden und im Ei-Nährzellverband von *Drosophila melanogaster. Z. Zellforsch.* **54,** 238-251.
Mortensen, T. (1912). Ctenophora. The Danish Ingolf-Expedition 5, no. 2. Copenhagen, 95 pp., 10 pl.
Mortensen, T. (1913). On regeneration in ctenophores. *Vidensk. Meddel. Dansk Naturhist. Foren., Copenhagen* **66,** 45-51.
Moser, F. (1909). Die Ctenophoren der Deutschen Südpolar-expedition, 1901-1903. *Deut. Süd-polar-Expedit.* **11,** Zool. 3, 117-192, pl. 20-22.
Nagano, T. (1961). The structure of cytoplasmic bridges in dividing spermatocytes of the rooster. *Anat. Rec.* **141,** 73-79.

Nelson, T. C. (1925). On the occurrence and food habits of ctenophores in New Jersey inland coastal waters. *Biol. Bull.* **48**, 92-111.

Okada, Y. K. (1931). Studies on regeneration of coelenterata. *3rd.* part: Hydromedusae, Charybdea and Coeloplana. *Mem. Coll. Sci., Kyoto Imp. Univ. Ser. B* **7**, 205-221.

Ortolani, G. (1964). Origine dell'organo apicale e di derivate mesodermici nello sviluppo embrionale di Ctenofori. *Acta Embryol. Morphol. Exp.* **7**, 191-200.

Rankin, J. J. (1956). The structure and biology of *Vallicula multiformis*, gen. et sp. nov., a platyctenid ctenophore. *J. Linn. Soc. London Zool.* **43**, 55-71.

Raven, C. P. (1961). "Oogenesis: the Storage of Developmental Information", 274 pp. Pergamon, New York.

Reverberi, G. (1957). Mitochondrial and enzymatic segregation through the embryonic development in ctenophores. *Acta Embryol. Morphol. Exp.* **1**, 134-142.

Reverberi, G. (1966). Quelques nouvelles recherches expérimentales sur le développement des cténophores. *Année Biol.* **7**, 375-390.

Reverberi, G. (1971). Ctenophores. *In:* "Experimental Embryology of Marine and Fresh-water Invertebrates" (G. Reverberi, ed.), pp. 85-103. North-Holland Publ., Amsterdam.

Reverberi, G., and Ortolani, G. (1963). On the origin of ciliated plates and of the mesoderm in the ctenophores. *Acta Embryol. Morphol. Exp.* **6**, 175-190.

Reverberi, G. and Ortolani, G. (1965). The development of the ctenophore's egg. *Riv. Biol. (Lisbon)* **58**, 113-137.

Schneider, K. (1904). Histologische Mitteilungen. 1. Die Urgenitalzellen der Ctenophoren. *Z. Wiss. Zool.* **76**, 388-399, pl. 24.

Tanaka, H. (1931a). Reorganization in regenerating pieces of *Coeloplana*. *Mem. Coll. Sci. Kyoto Imp. Univ. Ser. B* **7**, 223-246, pl. 18.

Tanaka, H. (1931b). *Coeloplana echinicola*, n. sp. *Mem. Coll. Sci. Kyoto Imp. Univ. Ser. B* **7**, 247-250, pl. 19.

Tanaka, H. (1932). Remark on the viviparous character of *Coeloplana*. *Annot. Zool. Japon.* **13**, 399-403, pl. 23.

Tokin, B. P. (1961). Regenerative ability of ctenophores. *Dokl. Biol. Sci.* **141**, 1004-1006.

Totton, A. (1954). Egg-laying in Ctenophora. *Nature (London)* **174**, 360.

von Ledenfeld, R. (1885). Über Coelenteraten der Südsee. *Z. Wiss. Zool.* **41**, 673-682, pl. 33.

Willey, A. (1896). On *Ctenoplana*. *Quart. J. Microsc. Sci.* **39**, 323-342, pl. 21.

Yatsu, N. (1911). Observations and experiments on the ctenophore egg. *Annot. Zool. Japon.* **7**, 333-346.

Yatsu, N. (1912). Observations and experiments on the ctenophore egg. *Annot. Zool. Japon.* **8**, 5-13.

Ziegler, H. E. (1898). Experimentelle Studien über die Zelltheilung. III. Die Furchungszellen von *Beroë ovata*. *Arch. Entwicklungsmech. Organ.* **7**, 34-64.

Zirpolo, G. (1930). Ricerche sui ctenofori. I processi di regolazione e die regenerazione. *Arch. Zool. Ital.* **14**, 115-156.

Chapter 5

PLATYHELMINTHES (TURBELLARIA)

Catherine Henley

5.1 Introduction

In setting forth the aims of this volume, the Editors stated, "We hope the treatise will point up areas in which additional work might profitably be done . . . ." Perhaps no other group can offer as many such possibilities as the turbellarian flatworms, because for many of the topics to be covered in the chapters of this volume, little, if any, work has been done on these forms.

There are many reasons for this neglect, not the least of which are the formidable problems involved in identifying the animals with which one is to work; the importance of accurate nomenclature was early made clear by Ball (1916) in his careful analysis and correction of a fundamental error of this sort made by Linton (1910) and Patterson (1912). For the vast majority of turbellarians, it is essential that identification be based on careful study of complete sets of serial sections of sexually mature specimens and on reference to the enormous literature on the subject, much of which is in German. Collection of living material is sometimes difficult, at least for the beginner, because some of the worms are quite small and because turbellarians live in a wide variety of habitats. These range from sand or mud bottoms, wharf-pilings (frequently in association with colonial cnidarians), under rocks, or on algae, to the gills of horseshoe crabs (*Bdelloura*), the interiors of molluscan shells inhabited also by hermit crabs (*Stylochus*) and the branchial chambers of the whelk *Busycon* (*Hoploplana*). Hyman (1944a) described a polyclad, *Taenioplana teredini*, which lives in the burrows of the shipworm *Teredo;* she suggested that as the worm is always found in burrows lacking shipworms it might be a useful enemy of these pests. The same author (1944b) reported that *Ectocotyla* was an ectoparasite of the crab *Pagurus*. [She classified the animal as an acoel in this paper, but in a later brief statement (1959a "Retrospect," p. 731) stated that it is in fact an alloeocoel of the monocelid group.] *Stylochoplana parasitica* lives in the pallial groove of a chiton (Kato, 1940), and there are many such examples known of rather bizarre associations of turbellarians with other animals. Some acoels and polyclads are pelagic as adults, and many polyclads have pelagic larval and juvenile forms (Hyman, 1951). Turbellarian worms are found at a wide variety of depths, from very shallow to very deep water and in the psammon. Thus, one must be forearmed with some knowledge of what forms are suitable for what experimental approaches, of their frequency of occurrence, and of the conditions necessary for their maintenance in the laboratory.

It should be pointed out in the latter connection that any turbellarians kept under adverse environmental conditions, or deprived of food, very rapidly begin to show retrogressive changes in the reproductive system; these changes usually begin with the gametes. Costello (unpublished observations) found that specimens of *Polychoerus carmelensis* brought in from a tidepool and left in a bucket at cool room temperatures for only 6 hours had already resorbed portions of their reproductive systems. Thus, observations should be made only on freshly collected animals, or, at the very least, on ones maintained under the best possible laboratory conditions with adequate food, temperature controls, and frequent changes of seawater.

Nearly all the information known on these and a great many other features of turbellarians is summarized in the monumental work of

Hyman (1951); see, also, her "Retrospect" (1959a). Other useful sources of information and references include Hyman (1939a), Kato (1940, 1968), Costello *et al.* (1957) and Bush (1964); revisions and updatings of the last two references may be available at this time.

It is important to emphasize that, to my knowledge, no free-living marine flatworm is readily available in the large quantities necessary for certain types of biochemical and physiological investigations which would undoubtedly yield valuable information in some of the unexplored areas of reproductive physiology of lower forms, as they already have for more plentiful animals such as the echinoderms. Similarly, the relatively inaccessible habitats of many marine turbellarians not only make them difficult to collect but also render difficult, if not impossible, the simulation of these habitats in the laboratory for experimental work. *Convoluta* is one form which has been cultured successfully away from its normal habitats (Dorey, 1965), as have *Aphanostoma, Otocelis* and *Avagina.* This success may be due largely to the fact that *Convoluta* has symbiotic algae which probably supply it certain metabolities.

On the other hand, for fields of investigation which do not require large quantities of biological material, the flatworms offer unique opportunities, advantages, and challenges. In the electron microscopy of spermatozoa, for example, pioneering work was done by Christensen (1961), Shapiro *et al.* (1961), Klima (1962), Silveira and Porter (1964), Hendelberg (1965), and Burton (1966). (See Section 5.3.4 for more recent references on this subject.) Electron microscopy might very profitably be applied also to the morphology of female gametes, and to many developmental features already known from studies using light microscopy.

The classification of the flatworms has been the subject of much debate, a great deal of which remains unresolved. Four recent principal schemes have been advanced, those of Meixner (1938), Karling (1940, 1967), Hyman (1951), and de Beauchamp (1961) (Table I). These are all modifications of a number of older systems. Meixner's and Karling's systems were rejected by Hyman (1951, 1959a), and the status of de Beauchamp's is still largely unsettled. Therefore, in this chapter I shall use the nomenclature of Hyman, with the understanding that at least some revisions of it are almost certainly necessary. This seems justified because hers is the best known to nonspecialists in the field and because it is readily available to those not fluent in German and/or French. For further discussion of this subject, see Ax (1963) and for a good example of the complexities involved, see Hyman (1959a) and Karling (1967).

TABLE I

A Comparison of the Hyman, de Beauchamps, and Karling Systems of Classification

Class Turbellaria
Hyman, 1951:
Order Acoela
Order Rhabdocoela (includes catenulids, macrostomids, and neorhabdocoels)
Order Alloeocoela
Order Tricladida
Order Polycladida
de Beauchamps, 1961:
Order Archoophora (includes acoels, catenulids, stenostomids, macrostomids and microstomids)
Order Polyclades (largely unchanged from the Hyman system)
Order Triclades (largely unchanged)
Order Protriclades (includes monocelids, otoplanids, *Bothrioplana*)
Order Eulecithophores: (includes most of Hyman's rhabdocoels and many alloeocoels: plagiostomids, dalyelliids, protovorticids, graffilids, umagillids, pterastericolids, fecampiids, mesostomids)
Order Perilecithophores: (Genera *Prorhynchus, Geocentrophora, Gnosonesima, Gnosonesimila*)
Karling, 1967:
Order Archoophora
Suborder Acoela
Suborder Nemertodermatidae
Suborder Catenulida
Suborder Macrostomida
Order Polycladida
Order Prolecithophora
Order Proseriata
Order Tricladida
Order Rhabdocoela (= Neorhabdocoela of Meixner, 1938)
Order Temnocephalida
Order Lecithoepitheliata

5.2 Asexual Reproduction

5.2.1 Occurrence and Types

Asexual reproduction, with a very few exceptions, is restricted to the freshwater rhabdocoels of the catenulid group, the microstomids (which are both marine and freshwater), some land triclads, and

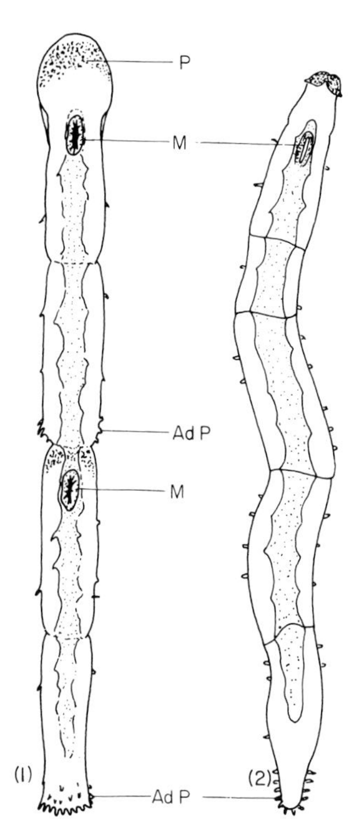

FIG. 1 AND 2. Asexual reproduction in two species of the rhabdocoel *Microstomum*. The process occurs by paratomy in *M. rubromaculatum* (Fig. 1) and by architomy in *M. papillosum* (Fig. 2). See text for further details. Ad P, adhesive papilla; M, mouth; P, anterior pigment. [Redrawn from Westblad (1952).]

freshwater planarians (Hyman, 1951). The process has been studied quite extensively in the freshwater groups (see, e.g., Curtis, 1902; Hyman, 1939b, 1951; Brønsted, 1969), less so in the marine. In all, however, the process appears to involve either transverse fission and the formation of chains of zooids (Fig. 1 and 2) or fragmentation. In the case of fission, the locus where division will occur is often marked first by an indentation and then by the appearance of ciliated pits (Hyman, 1951). During the process, the rear end of the worm is firmly attached to the substratum, while the anterior end moves away and breaks off. Both portions produce new worms. Hyman (1951) points out that both adhesion and locomotion are essential in the fission process which can be delayed or prevented by greasing the container in which the worms are kept. In the freshwater worm *Phagocata* there is fragmentation into a number of pieces which subsequently encyst; fission does not occur.

A few instances of asexual reproduction have been described for acoels, an early case being the report by Wager (1913) for a form he identified as *Convoluta roscoffensis*. His work apparently languished in obscurity until it was followed up by Marcus and Macnae (1954); the species was later identified as *C. macnaei* by Du Bois-Reymond Marcus (1957). Marcus and Macnae described their observations as representing an example of architomy (in which fission occurs before regeneration of the parts to be lost has occurred, as opposed to paratomy in which fission occurs after regeneration). Working with the same form, Ax and Schulz (1959) contended instead that paratomy occurred and they documented their findings with drawings and photographs. It is interesting to note that Marcus and Macnae (1954) found spermatozoa to be present within the parenchyma of sectioned animals, despite the absence of any trace of a penis. Wager (1913) likewise had found a penis to be lacking, but declined to commit himself on the question of whether sexual reproduction has been abandoned completely in favor of asexual. Marcus and Macnae's results seem to indicate that it had not. However, it should be pointed out in this connection that absence of a penis might mean only that the specimens studied were in a state of retrogressive degradation, known to occur for many forms. Wager suggested that the frequency with which asexual fission is observed might be correlated with the extreme fragility of *Convoluta*, which fragments very readily.

Hanson (1960) also reported the occurrence of asexual reproduction in an acoel, *Amphiscolops langerhansi*; he declined to accept the report of Marcus and Macnae (1954) as describing a case of true asexual reproduction and drew a distinction between their description of architomy (depending on "mechanical forces exterior to the organism") and true asexual reproduction (inherent in the organism). It is not clear, however, that his case of "true asexual reproduction" occurred in the absence of such external factors, so that the distinction may be an academic one. Hanson considered the possibility that the "small odd-shaped worms" appearing in his isolates might in fact be the products of sexual reproduction, but rejected it on a number of grounds, notably on differences in appearance between newly hatched products of known sexual reproduction and the forms he observed. The latter had symbiotic algae (characteristic of the adults of this form), as well as the concrements also characteristic of the adults. Sexually produced young lacked both these features but had a statocyst. Hanson's young forms lacked statocysts (which are diagnostic for the genus); he attributes this to the fact that the new worms were derived from posterior regions lacking such structures and, presumably, incapable of regenerating them.

Dörjes (1966) has also reported asexual reproduction in an acoel, *Paratomella unichaeta;* this involved paratomy and the most anterior two of the three zooids produced had recognizable sexual apparatus present.

5.2.2 Factors Influencing Asexual Reproduction

Child (1910a) found that in planarians which are in proper physiological condition for asexual reproduction, even a very slight jar of the containing vessel may be sufficient to induce fission. He also reported in the same paper that removal of the planarian's head would result in fission. [In this connection, it is interesting to point out that Giesa (1966) found amputation of the head, with subsequent regrowth of a new individual from the front portion, was the only means by which he was able to bring the alloeocoel *Monocelis* into *sexual* activity, under laboratory conditions.] Child also found that if he cut off the posterior region of planarians and treated the head end with temperature shock, there was an 80% increase in fissions in animals kept at 27°C, as compared with only 5% in those kept at 8°-10°C. In nature, an increase in temperature likewise induced asexual reproduction in normal, intact individuals. The role of all these factors, like that of water movement, as described below, remains to be investigated for those marine turbellarians undergoing asexual division.

Westblad (1952) states that the marine form *Microstomum rubromaculatum* reproduces exclusively by transverse fission, and pointed to the complete absence of sex organs in this form, as in many other microstomids. More commonly, in the freshwater forms at least, asexual reproduction appears to be the rule only until the advent of sexual maturity (often in the autumn), when fission usually (but not always) ceases and sexual reproduction occurs (Hyman, 1951). Another factor which appears to be involved in the occurrence of asexual reproduction in freshwater forms is water movement; Hyman (1939b) found that in Chicago, planarians living where they were exposed to strong wave action in Lake Michigan became sexual, whereas sexuality did not develop in the same forms living farther back along the Lake shore, where the water was quiet. She states (1939b; p. 271), "I know of no case in which sexual *D. tigrina* have been taken from ponds with still water." The role of lack of water movement in inducing asexual reproduction among marine forms has apparently not been investigated experimentally.

The role of darkness in bringing on asexual development was stressed by Hanson (1960), who stated that asexual reproduction occurred only during the absence of natural or artificial light. Only on two occasions did he observe the actual process, and both were under conditions of darkness. Experimentally, he induced asexual reproduction in worms (p. 109) "known to be able to reproduce asexually" by keeping them under constant illumination for periods of 24 or 48 hours, and then placing them in the dark. Under such conditions progeny were produced asexually within an hour or two.

5.3 Sexual Reproduction

5.3.1 Sexual Dimorphism, Hermaphroditism, and Sex Determination

Sexual dimorphism is the exception rather than the rule among Turbellaria, practically all of which are hermaphroditic. Two species of a genus constituting one of the exceptions are dioecious and do exhibit sexual dimorphism, as described by Christensen and Kanneworff (1965) for *Kronborgia amphipodicola* and by Kanneworff and Christensen (1966) for *K. caridicola*. This neorhabdocoel genus is endoparasitic, *K. amphipodicola* occurring in the hemolymph near the digestive tract of an amphipod and *K. caridicola* in the body cavity of shrimp. In *K. amphipodicola* there are marked differences in size and morphology between males and females, the former being smaller and more active than the latter. Both sexes apparently remain in their amphipod host until sexual maturity, at which time they depart, the males going first and the larger, more sluggish females going later. The process of departure appears to be quite traumatic for the host, which was described as becoming motionless and stiff as the worms emerge, and dying soon after. Furthermore, it is rendered sterile by the parasites. There are at least two factors governing the size of the worms at sexual maturity: size of host and number of parasites present. In larger hosts, parasites were larger, and if more than one parasite was present, each was smaller. Mature females were found to range in size between 10 and 40 mm, and the ratio of males to females was sometimes as high as 25:1. Christensen and Kanneworff found no evidence, in careful study of serial sections, for the existence of any protandric hermaphroditism so that this appears to be a truly dioecious condition. *K. caridicola* is much like *K. amphipodicola* according to Kanneworff and Christensen (1966), except that the worms are much larger. The authors found only females, but suggested that dwarf males exist. They found no pigment, eyes, gonopores, or other openings in the epidermis; mouth, pharynx, intestine, and excretory system were all lacking. Marked retrogression of the digestive and sensory systems of parasites is quite common, the tapeworms and *Sacculina* being classical examples.

So far as I know , there is no available information concerning the factors governing sex determination in the few cases known to be dioecious, nor is it known what determines the order in which the two sets of sexual apparatus appear in the maturing hermaphroditic animal (see Section 5.3.5.3). It is probably significant in this connection that the diploid chromosome numbers of all Turbellaria listed by Makino (1951) are even (with two exceptions, both of dubious reliability). This implies that only one type of male gamete and one type of female gamete can be produced in these primitive hermaphroditic forms, and that an X-O chromosome condition does not exist.

5.3.2 Anatomy of the Reproductive System

Nearly all turbellarians have evolved reproductive systems of such incredible variety and complexity that one would be foolhardy to attempt a comprehensive description, even in their principal permutations among the various groups. Beklemishev (1969, p. 393) points out that the flatworms, which were the "first group to develop a complex genital apparatus, display almost the greatest diversity in its structure." In another, more colorful exposition of the same theme, Beklemishev (1969) says (p. 399), "With the appearance of internal fertilization various adaptations for performing it have been poured out, as it were, from the cornucopia, and we see their independent appearance and rapid refinement in different groups of flatworms."

The literature prior to 1951 has been authoritatively, extensively and lucidly summarized in the treatise of Hyman (1951); see, also, her chapter on "Retrospect," (1959a) which cites additional references up to that date. There are a few general characteristics of turbellarian reproductive systems which can be briefly summarized here, and the primitive plan, as seen in one member of the acoel group, is presented. In addition, some of the contributions which have appeared since 1951 are indicated. However, to understand the amazing adaptations that turbellarians have evolved, in order to cope with specific reproductive problems associated with specific habitats and other factors, one must go to the original literature. Likewise, this is essential if one wishes to work with any specific given form.

5.3.2.1 Female Reproductive System

The ovary is connected more or less intimately (depending to some extent on the group) with a structure, the vitellarium, which supplies yolk, and which may be extremely diffuse, as in many acoels, or quite compact, as in some of the more advanced forms. The cells of the vitellarium are generally considered to be abortive oocytes. Evidence for this is found in the case of the alloeocoel group Lecithoepitheliata, according to Hyman (1951), where it can be observed that some oocytes become true eggs while others develop into nurse cells, surrounding each egg and providing it with yolk. It is frequently stated as a dictum that the ovary and vitellarium form a single structure, the germovitellarium, in the acoels, but this is not true in the case of some of them at least: Costello and Costello (1938) found a quite sharp demarcation between the ovarian and vitellarian regions in *Polychoerus carmelensis*. This reflects the condition prevailing in most of the higher forms (Hyman, 1951), where the two structures are entirely separate. Ax and Dörjes (1966) have suggested that the terms "vitellarium" and "germarium" should be applied only in connection with the production of complex ectolecithal eggs. They prefer to use "yolk-part" and "germ-part," respectively. Costello (unpublished data) has shown with the electron microscope that the yolk-producing cells in *P. carmelensis* have abundant rough endoplasmic reticulum and conspicuous polysomes, consistent with their protein-synthesizing activity.

The female copulatory system may involve, again, a highly variable combination of the following components. There is usually a gonopore (which may be dorsal or ventral in position, more commonly the latter) opening into a vagina of variable length. A well-defined oviduct is often lacking. Typically, there is a sac for receiving spermatozoa from another hermaphrodite in reciprocal mating. If this sac retains spermatozoa only briefly it is called a copulatory bursa, if it retains them for longer periods it is a seminal bursa, and if it stores them for more or less prolonged times it is commonly referred to as the seminal receptacle. The latter is usually merely a widened portion of the oviduct. Some Turbellaria (notably the polyclads) may retain ova within the oviducts for relatively long periods of time during which a shell may be added. There is a variety of eosinophilous glands associated with the female complex, some of which are probably the cement glands secreting material to attach eggs to substratum after they are laid. Recently Schilke (1970) has suggested that some of the eosinophilous glands emptying into the bursa of certain kalyptorhynch rhabdocoels may function in sperm capacitance (see below).

The ventral portion of the bursa of the acoel *Polychoerus* is composed of a variable number of "mouthpieces" (Mark, 1892) (Fig. 3) or nozzles, the ventral ends of which open into the lateral parenchyma. In living specimens of *P. carmelensis* Costello and Costello (1968) observed that active spermatozoa tried to enter the proximal narrow lumen of a nozzle (which is smaller than the diameter of the spermatozoon); dark, inert, threadlike material was seen at the distal portions of such nozzles. Karling (1962c) observed spermatozoa of the alloeocoel *Allostoma* within the nozzle of the bursa. We (Costello and Henley, unpublished data) have histochemical evidence that the nozzles are intensely PAS-positive (suggesting a mucoprotein content). DNA accumulates in-and-around them in rather substantial quantities, as demonstrated by the Feulgen and azure B reactions, using appropriate DNase controls. Electron microscopy reveals that the nozzles are made up of ruffled circular lamellae arranged around a minute tripartite lumen (Fig. 4).

Nozzles have been shown to occur in a number of other forms, including the alloeocoels *Pseudostomum arenarum, Reisingeria,* and *Cylindrostoma* (Westblad, 1954), *Anoplodiopsis,* a rhabdocoel parasitizing echinoderms (Westblad, 1953), two kalyptorhynch rhabdocoels, *Cicerina* and *Paracicerina* (Karling, 1952), and the alloeocoels *Pseudostomum californicum* and *Allostoma amoenum* (Karling, 1962c). Haswell (1905) likewise found a variable number of nozzles associated with the bursa of the acoel *"Heterochoerus"* (= *Amphiscolops*), whereas Hanson (1961) reported that for the acoel *Convoluta sutcliffei* only a single nozzle was associated with the bursa. Kozloff (1965) described a similar situation for the acoel *Paratocelis,* and found two nozzles per bursa in another acoel, *Diatomovora;* one of these was directed anteriorly and the other dorsally or anterodorsally. Ax and Dörjes (1966) observed 10–40 bursas, each with a nozzle, per animal in the acoel *Oligochoerus limnophilus.*

The enigmatic nature of these structures remains a fascinating problem and one which would doubtless yield findings of considerable interest, both with regard to their specific function(s) and to broad general implications concerning the physiology of reproduction in these forms.

5.3.2.2 Male Reproductive System

The testes are usually widely dispersed (except in the rhabdocoels), often in the so-called follicular configuration (although frequently such follicles are not surrounded by any true bounding membrane or other structure); they may be single (Ax, 1954), paired (most commonly), or sometimes multiple. One or more sperm ducts

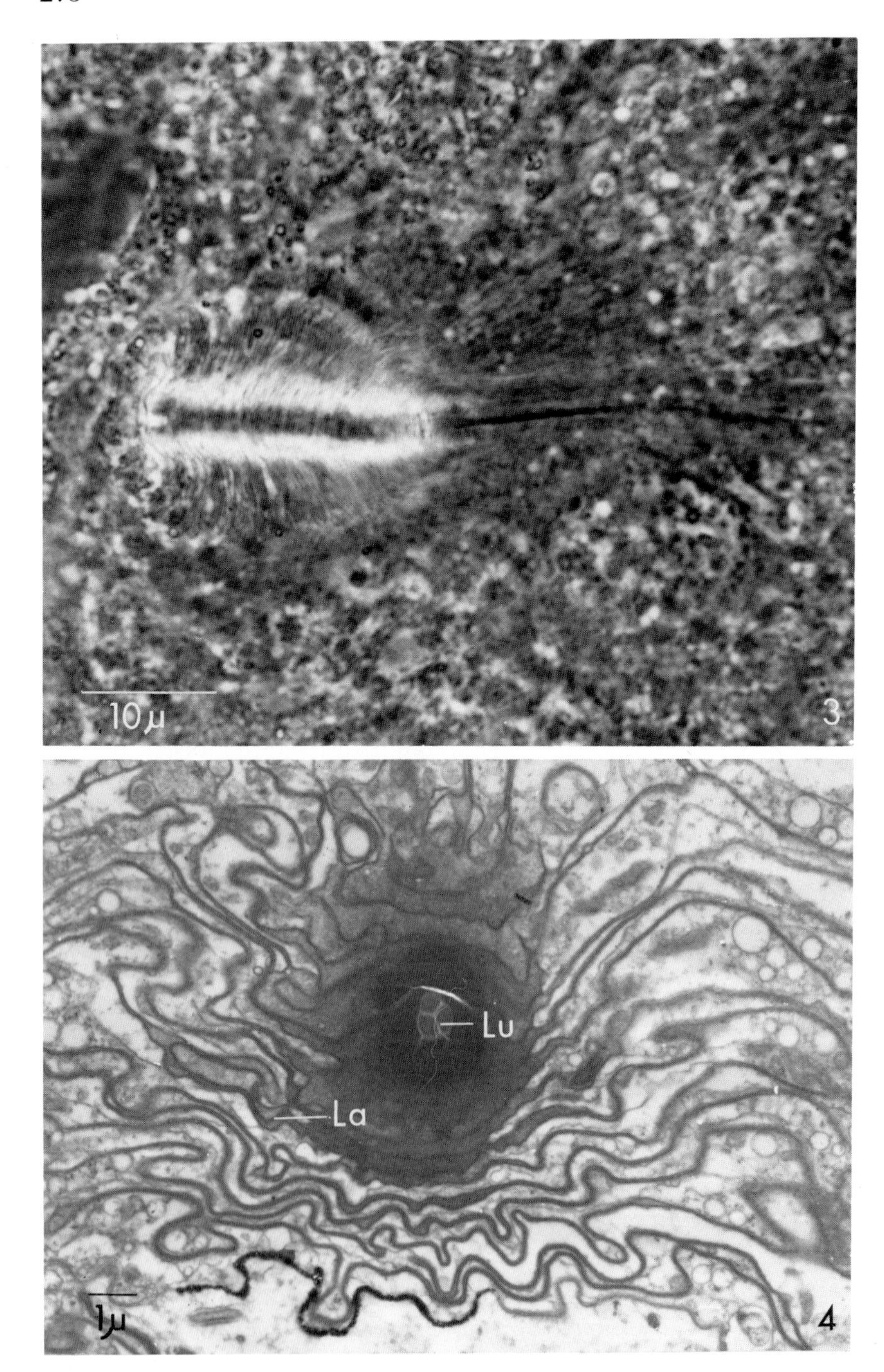
10 μ
3
Lu
La
1μ
4

convey spermatozoa from the testes to the seminal vesicle and then to the ejaculatory duct and the penis (protrusible) or cirrus (eversible). This organ is often armed with an impressive array of spines and other projections, or it may be unarmed. It is sometimes muscular and sometimes reinforced with cuticular material, and it is usually extended through the gonopore during copulation. Also associated with the male copulatory apparatus are one or more accessory gland structures, including the so-called prostatic apparatus. The seminal vesicle is variable in size, form and occurrence, but usually has a thick muscular wall, presumably involved in the process of ejaculation. The sperm ducts also may be invested with a coating of musculature, as in the case of the freshwater rhabdocoel *Microdalyellia* (Fig. 5).

Multiplication of one or more components of the male system is not uncommon, particularly among the polyclads. In the acoel *Childia*, there are two penes (Fig. 6), and another acoel, *Tetraposthia*, has four (Hyman, 1951). Ax and Dörjes (1966) found 8 pairs of ventral prostatoid organs in the acoel *Oligochoerus limnophilus*. For a review of this topic, see Ax (1957, 1966).

5.3.2.3 Reproductive System of an Acoel

One of the simple arrangements of male and female organs is to be found in the acoel *Polychoerus carmelensis*, described by Costello and Costello (1938) (Fig. 7). The female system, in addition to the seminal bursa discussed above, includes two strands of diffuse oocytes (o) which join the "vitellaria" (vi) at a point somewhat posterior to the mouth. These "vitellaria" end blindly at the level of the female aperture (♀). As the oocytes mature, they move posteriorly, toward and into the "vitellaria." From the female gonopore there extends a large vagina, which dorsally and anteriorly gives off a vaginal pocket (vp); the anterior end of this pocket is embedded in the bursa (bs) and opens into it.

The male system has the usual diffuse testes (t) whose products are emptied into the sperm ducts (vd) passing through the body parenchyma to the ejaculatory duct (ed) of the muscular penis (p). There is a single male sex opening (♂) through which the penis is protruded during sexual activity; it is located posterior to the female opening,

FIG. 3. Nozzle ("mouthpiece") associated with bursa of *Polychoerus carmelensis*. Note the dark, threadlike line (right) at the distal end of the structure; this apparently represents the remains of a spermatozoon. (Phase contrast micrograph.)

FIG. 4. Nozzle of *P. carmelensis* in cross section and seen by electron microscopy. There is a minute central tripartite lumen (Lu) and ruffled lamellae (La) surround it.

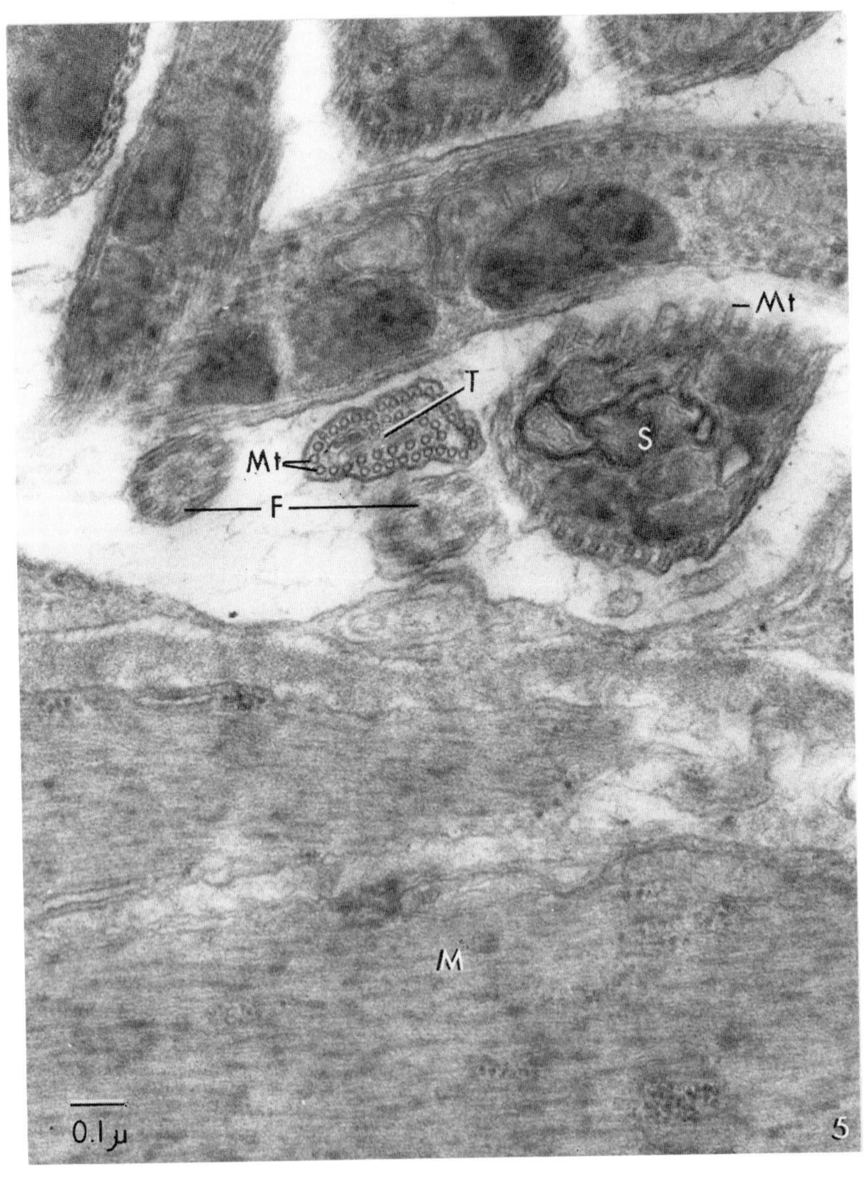

▲FIG. 5. Smooth mucle layer (M) surrounding a sperm duct of the freshwater rhabdocoel *Microdalyellia*. There are two "9 + 1" flagella (F) cut in very oblique section and spermatozoa (S) (likewise cut in glancing sections), with cortical singlet microtubules (MT). All these singlet microtubules terminate at the anteriormost tip (T) of the spermatozoon in a compact mass of microtubules. (Electron micrograph.)

▶FIG. 6. Successive sections (A, B, and C) through the paired penes (P) of the acoel *Childia*, as they approach the male gonopore (G). (Light micrographs of paraffin sections.)

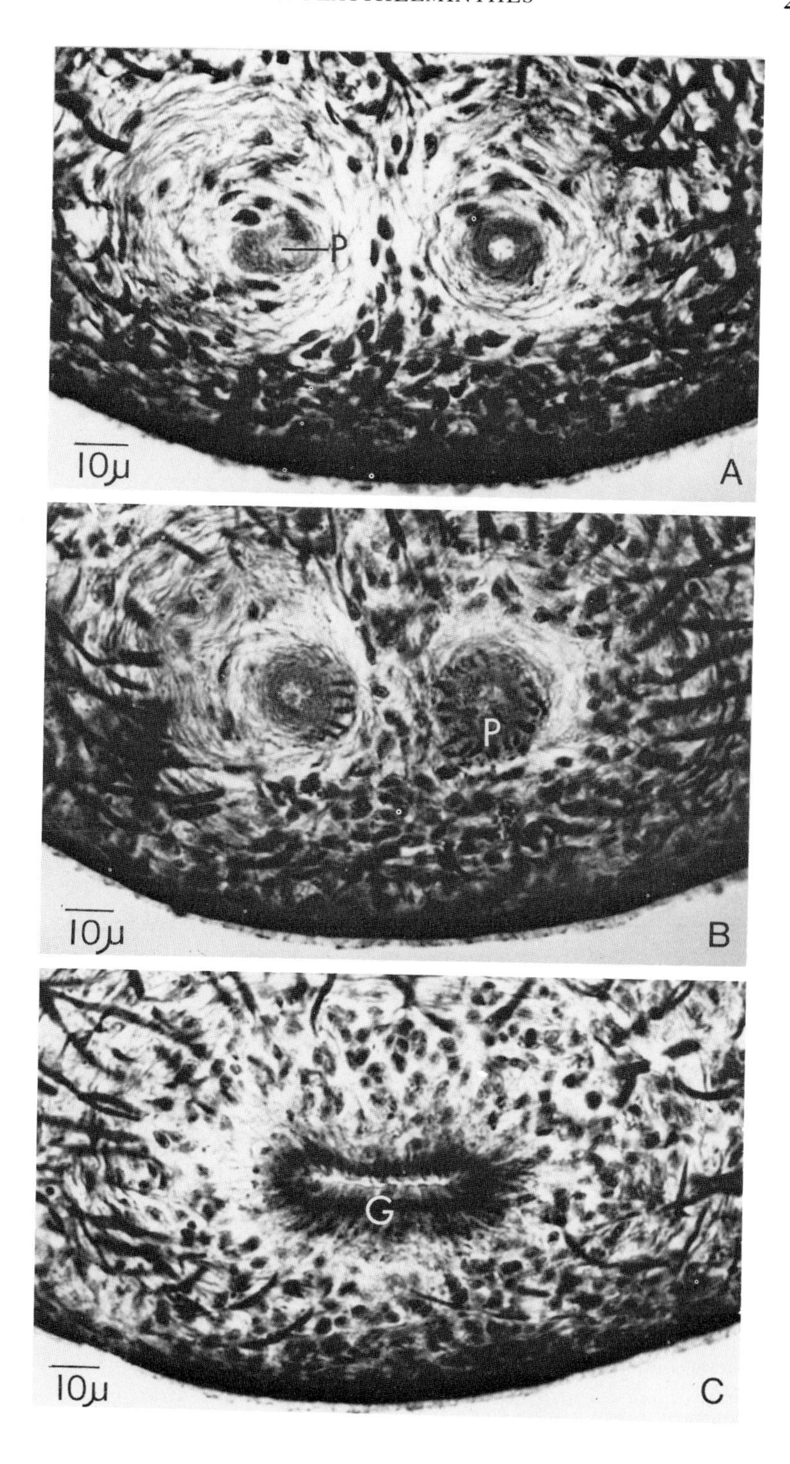
P
10μ
A
P
10μ
B
G
10μ
C

as is usual in most acoels, reversing the situation for other turbellarians.

5.3.2.4 Some Additional, More Recent, Observations

Westblad (1949a,b) described the reproductive apparatus in two aberrant turbellarians, *Meara stichopi* (which lives in the gut of a holothurian) and *Xenoturbella bocki* (which is a mud-dweller). *Meara* has no gonoducts, seminal bursa or vagina; the male and female gonads are sharply delimited from one another. In a masterly understatement Westblad says (1949a, pp. 51-52), "It is not easy to see how the sperm cells are brought into an animal lacking female supplementary organs, lacking a penis, and even lacking glands for fastening spermatophores at the surface of the body." He suggests that probably this feat is accomplished during copulation by the animals' bodies remaining in physical juxtaposition for long enough to allow spermatozoa to penetrate the epithelium of the partner. It is somewhat difficult to imagine the exact means by which this is accomplished, however. *Xenoturbella* lacks a male copulatory apparatus and Westblad (1949b) thinks that both eggs and spermatozoa leave the body via the mouth, to be attached to the surface of another animal by mouth. There are no yolk-producing cells of any sort, and the suggestion is made that the eggs obtain nutrients exclusively from the host's intestinal epithelium. [It should be noted, however, that Ax (1963) does not consider *Xenoturbella* a platyhelminth.]

Dahm (1950) described another case, in the alloeocoel *Bothrioplana*, in which the male copulatory organ, if present, is very much reduced and serves merely as a storage point for spermatozoa. He states that usually spermatozoa are not even produced, and in the event that spermatogenesis does occur, the products are incapable of fertilization. The female reproductive system consists of separate ovaries and vitellaria, and reproduction is said to be parthenogenetic (Reisinger, 1940) with one ovum from each ovary uniting and combining with cells of the vitellarium to form a cocoon. It is not clear, however, why spermatozoa should be produced and stored if they are not to be used.

Striking modification of the penis in *Microstomum spiriferum* is figured by Westblad (1952). There are strong muscles associated with it which effect rotatory movements during copulation.

Karling (1954) likewise found a complex system of musculature associated with the male copulatory complex of a kalyptorhynch rhabdocoel. The copulatory bursa of this form has a well developed

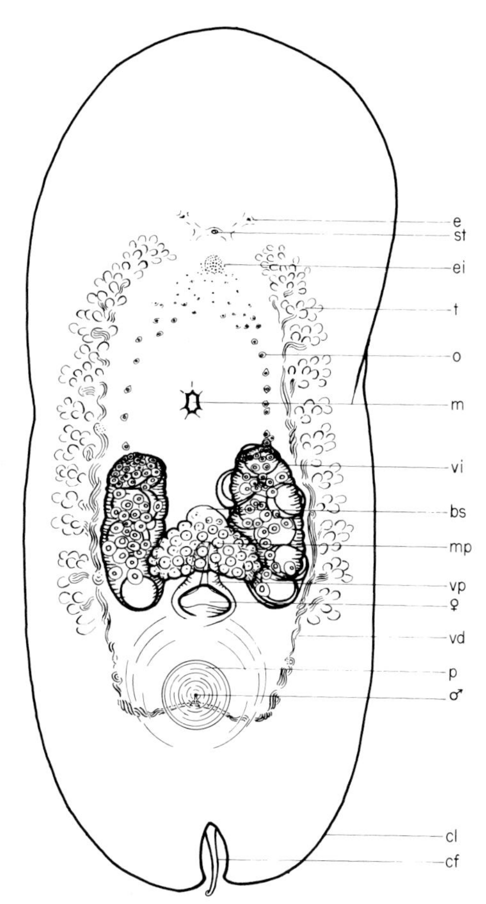

FIG. 7. Reproductive system of the acoel *Polychoerus carmelensis*. Abbreviations: bs, bursa seminalis; cf, caudal filament; cl, caudal lobe; e, eye; ei, "Eilager"; ♀, external female aperture; ♂, external male aperture; m, mouth; mp, "mouthpiece" of bursa; o, oocyte on way from "Eilager" to vitellarium; p, penis bulb; st, statocyst; t, testis follicles; vi, vitellarium containing young oocytes and mature eggs; vd, vas deferens; vp, vaginal pocket. [From Costello and Costello (1938), reproduced with permission of the authors and the Managing Editor, The Biological Bulletin.]

basal lamina, which is described as resembling cuticle in its appearance. In the alloeocoel *Serpentiplana*, Karling (1964) reported that there was complex cuticularization of the male apparatus, with modification of 2 of the approximately 50 bristles to form strongly hooked structures with denticles; these were said to support the ejaculatory duct during copulation.

A rare instance of multiplication of the female reproductive system was described by Hyman (1959b) in a tropical polyclad, *Nymphozoon*. There were 8 sets of female apparatus and 2 of the male;

Hyman had only a single sexually mature specimen of this form and was thus unable to ascertain whether the occurrence of 8 sets is a constant or a variable condition.

Karling (1962a) described and figured a single supernumerary copulatory apparatus in the alloeocoel *Multipeniata*; the components of the system were the same as those of the primary set, except smaller. Karling discussed the possible value to the organism of a reserve copulatory apparatus for use in event of loss of the primary one. In his 1962b paper, the same author noted the presence of regularly arranged hollow tubercles on the penis in three species of the alloeocoel *Plagiostomum*. These emitted a sticky eosinophilous secretion produced by glands at their bases. Karling also described a peculiar bulb, forming a sort of muscular glans penis, in *P. abbotti*; he suggested that this might function as a kind of sucker during copulation.

Ax and Ax (1967) reported a unique type of bursa in an acoel, *Pluribursaplana;* numerous small single cells in the caudal region differentiate by formation of an intracellular ciliated lumen into small bursa bladders which contact two vaginas opening into dorsolateral pores. More rostrally, the bursa bladders join to form a large bursa sac from which the ciliated sperm duct runs to the testes.

The first occurrence of supernumerary genital apparatus in a rhabdocoel was reported by Karling (1969), although, as noted above, such duplication has been known to occur regularly in a number of other turbellarians. The case reported for *Ethmorhynchus* involved a supernumerary reproductive complex consisting of a common atrium, bursa with adjacent unpaired ovary, common oviduct, and male copulatory apparatus. There was no gonopore and the seminal vesicle was empty with no connection to the testes. The entire supernumerary complex had a polarity inverse to that of the normal one. Karling suggested that it formed from body wall components as a consequence of a sort of inducing action of the ovary.

Apelt (1969) presented much detailed new information concerning the anatomy, mating habits, and egg-laying of three acoels, *Convoluta convoluta*, *Archaphanostoma agile*, and *Pseudaphanostoma psammophilum*, relating these features to the varying habitats of the animals.

It was suggested by Bashiruddin and Karling (1970) that weakly cuticularized "fingers" associated with the male copulatory apparatus of the dalyellioid rhabdocoel *Triloborhynchus* facilitate free movement of the penis stylet during copulation. They also reported the occurrence of only one ovum per animal, a rather unusual state of affairs for the flatworms.

Borkott (1970) has described the occurrence in the freshwater rhabdocoel *Stenostomum*,of what he interpreted as a bisexual gonad, the anterior portion of which is functionally a testis while the posterior is a "secondary ovary." He also stated that some ova segmented in the absence of spermatozoa and suggested that this might represent parthenogenetic differentiation.

5.3.3 Origin of Germ Cells and Gonads

This is another of the many topics for which practically no information is available for Turbellaria. Hyman (1951, p. 111) states that "The sex cells of the Turbellaria come from the free cells of the mesenchyme that migrate to appropriate locations and undergo gametogenesis."

Costello (unpublished observations) has found that in the acoel *Polychoerus* the earliest oogonia are characterized by large nuclei with prominent nucleoli; as the primary oocytes grow these nucleoli become very enlarged with two distinctly different regions.

Dörjes (1966) found that young oogonia of the acoel *Paratomella* were recognizable by their finely granular, grayish cytoplasm, but he did not give any information concerning their origin. As the primary oocytes grow, cyanophilic granules accumulate and coalesce.

Ax and Dörjes (1966) reported that the youngest oogonia of the acoel *Oligochoerus limnophilus* could be differentiated from the surrounding parenchyma by their larger nuclei, and by the fact that they were surrounded by only a thin rim of cytoplasm. There ensued a volume increase, a change in staining qualities, and the acquisition of yolk granules, in part, at least, by a process of phagocytosis of yolk cells in the nutritive portion of the ovary. The oocytes then were incorporated into the ovary to begin meiosis.

Beklemischev (1969) presented some interesting, although largely undocumented, views on the origin of copulatory organs in Turbellaria. He suggested that the male system is derived from cutaneous pyriform organs and glandular spines, but no real evidence was cited for this.

Newton (1970a) gave a detailed description of the development of oocytes of the freshwater alloeocoel, *Hydrolimax grisea*. He found that they first appeared in sections of animals fixed during the late summer, that they had clear cell boundaries and were characteristically larger than the parenchyma cells.

For another freshwater form, the rhabdocoel *Stenostomum*, Borkott (1970) reported that the subepidermal musculature appears to play a role in evoking differentiation of male components of the reproductive system; however, no such interaction was reported as being involved in development of the female system.

5.3.4 Morphology of the gametes

5.3.4.1 Ova*

The morphology of flatworm ova has received very little attention in the literature, apart from some investigations on their development after fertilization (see Section 5.4.1). In living oocytes of *Polychoerus carmelensis* observed by phase contrast microscopy (Fig. 8), a large spherical germinal vesicle is present, with a conspicuous nucleolus. The yolk is rather uniformly distributed in the cytoplasm of oocytes of this form, as in those of most acoels and many alloeocoels, whereas in the more advanced groups the nutritive materials are provided from an exogenous source, the yolk cells (see Section 5.3.2.1). Living oocytes and ova of the acoel *Childia* have rather bright brownish-orange granules uniformly distributed over their surfaces. Staining by the Feulgen method reveals that these granules are clearly Feulgen-positive, indicating the presence of DNA in quite dense small arrays (Henley, unpublished data). Treament with DNase eliminates this positive Feulgen reaction. It is quite possible that the granules may, in fact, be the nuclei of symbiotic algae rather than pigment; many acoels (e.g., *Convoluta*) are known to have similar arrays of symbiotic algae within the body (Hyman, 1951). Such symbionts are said to be picked up in the adult (Ax and Apelt, 1965) by free mesenchyme cells of the animal, and to be conveyed to the body layers beneath the superficial musculature. Thus far, however, we have been unsuccessful in demonstrating such symbiotic organisms, to our satisfaction, in sections of *Childia*, at either the light or electron microscope levels. It is possibly suggestive that the best collecting places for *Childia* in the waters around Woods Hole, Massachusetts, have very stagnant, odoriferous black mud as a substrate. If symbionts are indeed present in this organism, they might well play an important role in supplying oxygen to the worm. Further investigation of this point is in progress.

Living oocytes and ova of the turbellarians we have investigated readily change shape to accommodate to body movements; this can easily be observed in intact worms subjected to very slight pressure

*(Editors' note: Meiosis of the female gamete of platyhelminthes occurs after the fusion of the egg and sperm; Dr. Henley uses the term "ova" to refer to both full-grown oocytes and zygotes.)

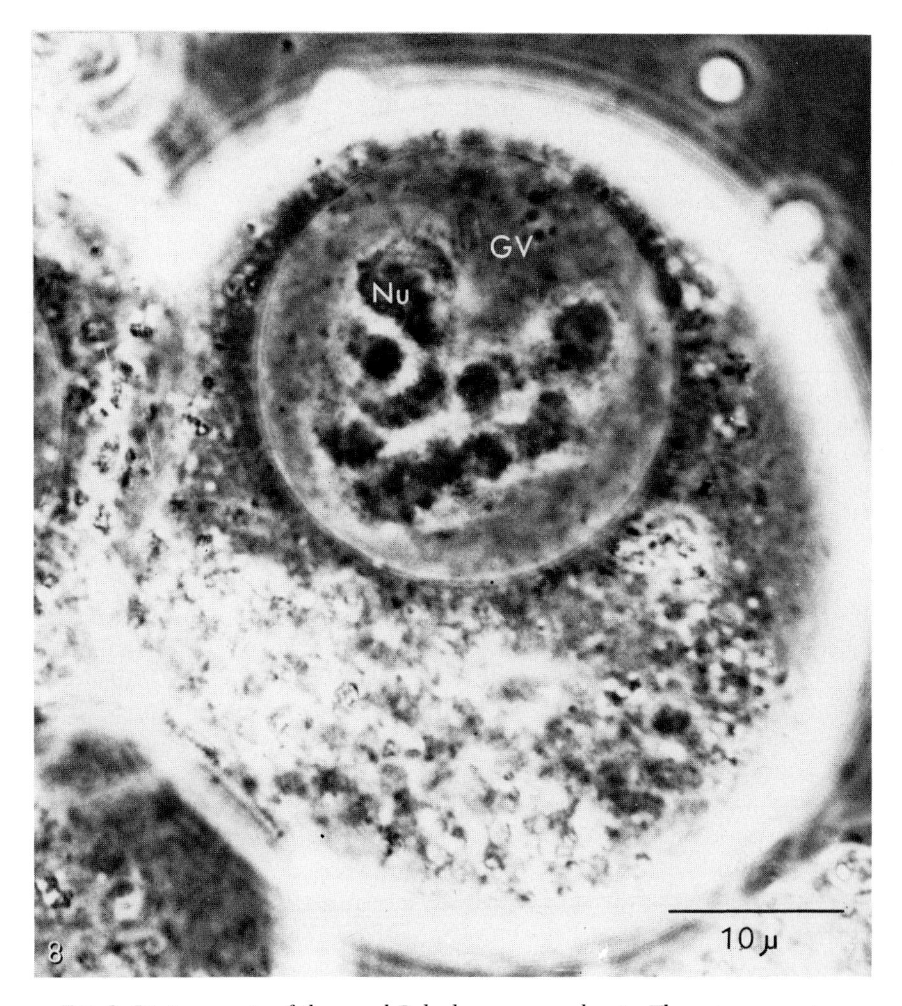

FIG. 8. Living oocyte of the acoel *Polychoerus carmelensis*. There is a conspicuous germinal vesicle (GV) containing a skeinlike nucleolus (N). (Phase contrast micrograph.)

from a coverglass. Marked deformations of shape can occur in oocytes and ova followed by a rapid return to the normal spherical form when pressure is released. A similar mobility of the cytoplasm of ripe ova of the acoel *Amphiscolops* was noted by Haswell (1905) who observed that they were sometimes wedged in the surrounding parenchyma to such an extent that they were greatly distorted and elongated.

Surface (1907), studying the development of the polyclad "*Planocera*" (=*Hoploplana*), and Papi (1953) describing the embryology of the rhabdocoel *Macrostomum*, both reported striking amoeboid surface changes in the fertilized ova preceding polar body formation and early cleavages. Thus, we have here another bit of evidence concerning the nonrigidity of the eggs of some turbellarians. It is interesting to note, in this connection, that very similar waves of surface activity have been reported by Pasteels (1950) for fertilization membranes around ova of the annelid *Chaetopterus* for comparable periods of maturation and cleavage.

In most turbellarians the youngest oocytes are located in the anteriormost portion of the animal's reproductive tract, and progressively more mature stages are found as one proceeds posteriad (Costello and Costello, 1938; Hanson, 1961). Usually oocytes occur singly or are in small groups although Ball (1916) described the oogonia of the rhabdocoel *Paravortex* as being tightly packed in rouleaux. Young oocytes of the aberrant form *Xenoturbella* were found by Westblad (1949b) to be inimately associated with the gut, and to lack a specialized membrane. These oocytes were usually isolated but sometimes were found in groups of two or three; when mature they had a firm inner membrane, which Westblad considered to be a product of the egg, and an outer envelope of vacuolated mucus. This unusually formidable covering is presumably associated with the egg's peculiar subsequent history, during which it is conveyed to the intestine of the worm and thence to the outside. Eggs lacking the protective covering are digested by intestinal cells.

Pigment granules of various types are found in the eggs of many turbellarians. Those of *Polychoerus* contain orange granules, superficially resembling the surface structures on ova of *Childia* (see above) but lacking their content of DNA, as determined by the Feulgen procedure (Costello, personal communication). Kato (1940) stated that the cytoplasm of eggs of many of the Japanese polyclads he studied had dense masses of granules in their cytoplasm, colorless or slightly pink. He also reported a dark brown pigment in ova of the polyclad *Thysanozoon;* this pigment accumulates at the animal pole after the eggs have been laid.

Newton (1970a) described the remarkable oocytes of a freshwater alloeocoel, *Hydrolimax*. These have a conspicuous yolk halo (first described by Hyman, 1938), which does not appear to play a nutritive role and is not incorporated into the definitive cocoon. Surrounding the oocyte nuclei he found a large (and variable) number of supernumerary asters and central bodies; this is of particular interest, because in these cases the germinal vesicle is still intact and entry of the spermatozoon has not yet occurred, so that the origin of the supernumerary asters remains, for the present, a puzzle.

The size of full-grown turbellarian ova varies greatly. Among the polyclads, for example, Kato (1940) studied the development of ova varying from 85 μm in diameter *(Stylochus)* to 330 μm *(Planocera)*. Some terrestrial triclads are said by Kato (1968) to have ova 20 mm in diameter! Fixed and sectioned ova of *Polychoerus carmelensis* were reported by Costello (1961) to vary between 200 and 280 μm in diameter, while those of *Hoploplana* are 100 μm in diameter (Costello *et al.*, 1957).

Only very slight beginnings have been made in the electron microscopy of turbellarian eggs. Costello and Henley (unpublished data) have found that the cytoplasm of the ripe ovum of the acoel *Polychoerus* is relatively free of cell inclusions, at both the light and electron microscope levels, while the oocyte of the acoel *Childia* (Fig. 9) has an abundance of rather dense granules (some of which are presumably yolk), vacuoles, and a fair number of mitochondria. The egg is invested with a complex lamellated outer coat which is closely applied to its surface. We have seen no evidence thus far of the surface microvilli characteristic of the ova of many other invertebrates. Study of flatworm eggs by electron microscopy is complicated by the technical difficulties associated with the relatively large size of the ova, their yolk content, and, after oviposition, by the presence of a cocoon or jelly layers.

5.3.4.2 Spermatozoa

At the level of light microscopy, there are three general forms to which the spermatozoa of most turbellarians may be referred. (1) Some are very long, slender and threadlike, with or without visible undulating membranes but lacking free flagella (Fig. 10) (e.g., the acoels *Polychoerus, Anaperus, Convoluta, Mecynostomum)*. (2) Shorter and more compact spermatozoa characterize other turbellarians (Fig. 11); sometimes bristles are present (as in the freshwater rhabdocoel *Macrostomum*), or paired undulating membranes (the acoel *Childia*) or "wings" (the alloeocoel *Plagiostomum*). Free paired flagella are again lacking. (3) The third type includes spermatozoa of moderate length, with free paired flagella, usually inserted subterminally (Fig. 12); the body of the spermatozoon is usually shorter than the flagella. Examples of this third type are to be found in spermatozoa of the triclads *Bdelloura* and *Dugesia,* the freshwater rhabdocoels *Mesostoma georgianum* and *Microdalyellia*, among others. These basic patterns may be modified in a number of ways, as for example in the case of the acoel *Paraphanastoma* (Hendelberg,

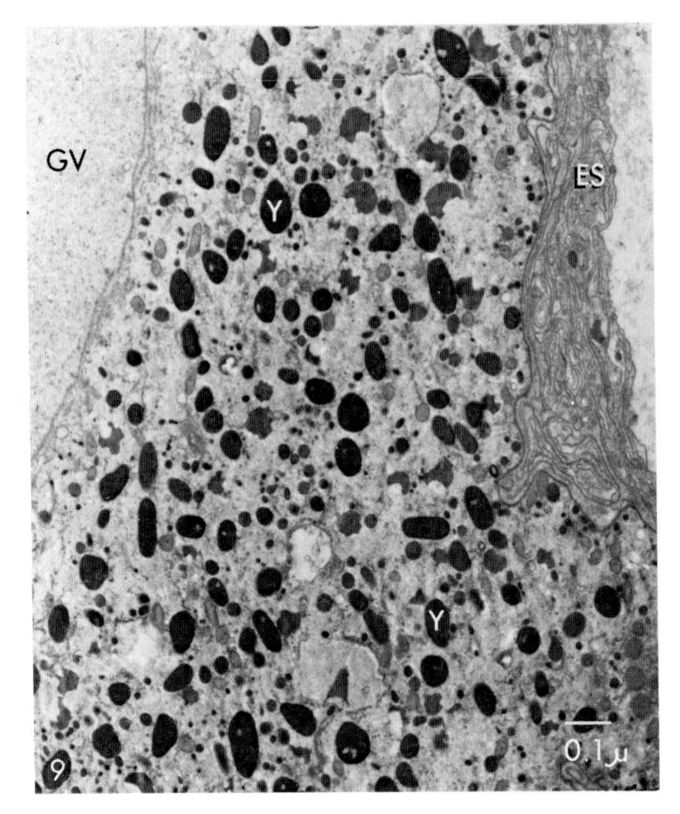

FIG. 9. Cytoplasm of an oocyte of the acoel *Childia*, with the margin of the germinal vesicle (GV) at left and the egg surface (ES) at right. The larger black granules are yolk (Y) and there are several other types of cytoplasmic inclusions.(Electron micrograph.)

1969) which has a terminal bundle of filaments. Kozloff (1965) reported that from much of the anterior half of the long threadlike spermatozoon of the acoel *Raphidophallus*, there are "delicate, cilia-like projections."

There appears to be no set pattern to the occurrence of a given type of morphology, with respect to systematic position, and even

FIG. 10. Living spermatozoon of the acoel *Anaperus*. The nucleus (N) is anterior and behind it is a double row of small refractile bodies (RB). There are also two very conspicuous refractile regions (RR) in the tail. (Phase contrast micrograph.)

FIG. 11. Compact living spermatozoa of the alloeocoel *Monoophorum*.The nucleus (N) is ovoid in shape and there are slender anterior and posterior extensions of the body of the sperm. (Phase contrast micrograph.)

FIG. 12. Living spermatozoon of the triclad *Bdelloura*. Two long free flagella (F) are inserted subterminally at the anterior end of the spermatozoon (S). (Phase contrast micrograph.)

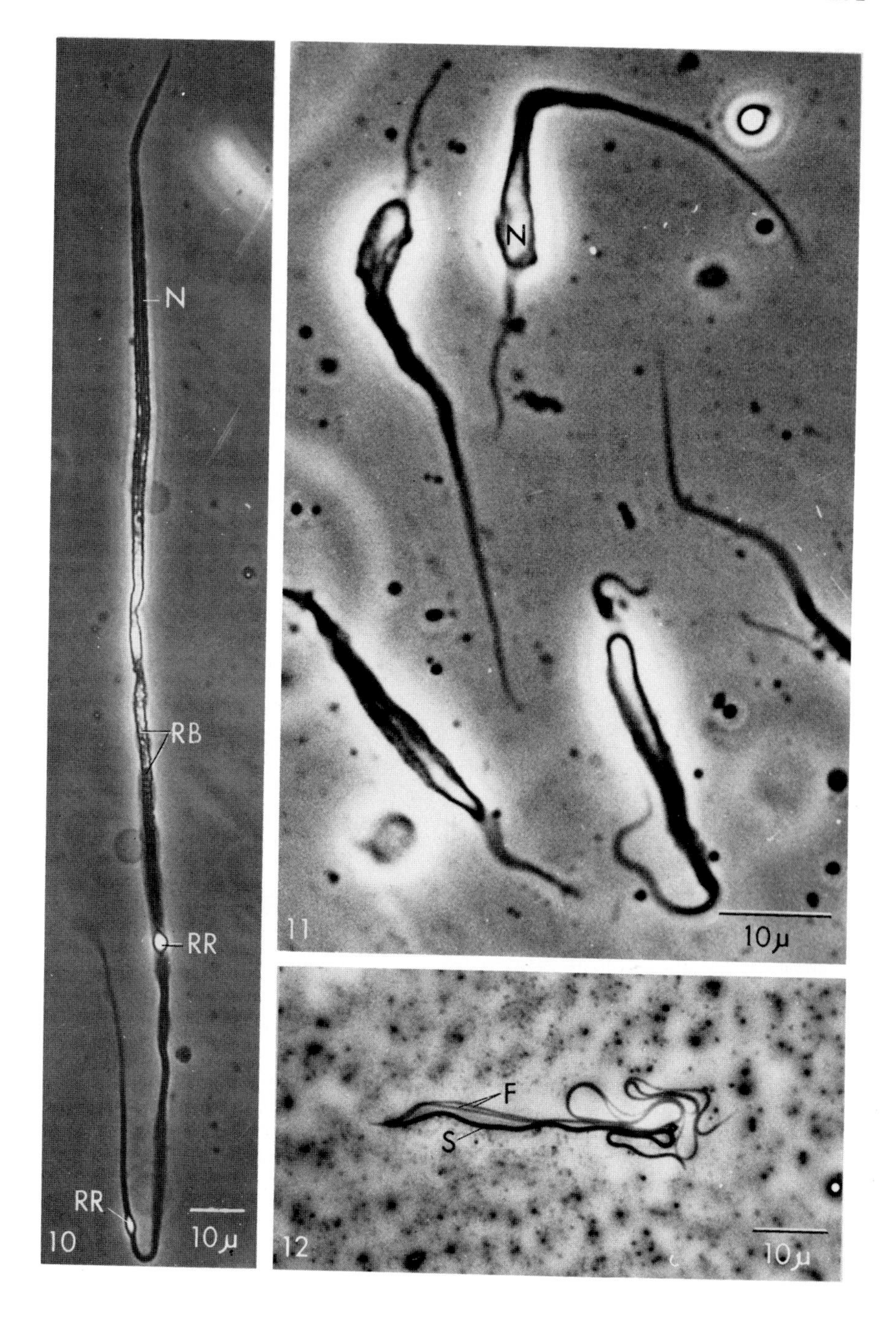
N
RB
RR
RR
10
10μ
N
11
10μ
F
S
12
10μ

within a genus quite striking differences in morphology of spermatozoa may exist. Westblad (1954), for example, noted that the spermatozoon of *Pseudostomum quadrioculatum* is only 35 μm long, while *P. gracilis* is described as having a long, thin and threadlike spermatozoon, in contrast to those of other species within the genus. As noted above, some acoels have very long slender spermatozoa, those of *Polychoerus carmelensis*, for example, being 300-400 μm long (Henley *et al.*, 1968). *Childia*, by contrast, has a rather compact spermatozoon (Fig. 13 and 14) only 50-60 μm long (von Graff, 1911; Westblad, 1945; Henley, 1968); this is readily referable to the ultrastructure of the longer and more slender spermatozoa of *Polychoerus*, being a foreshortened version of the latter, with paired undulating membranes. Peebles (1915) reported that the spermatozoon of another acoel, *Aphanostoma pulchella*, had a large rounded head and a threadlike tail. She gives no dimensions, but her Fig. 2 shows them as being moderately short. The mature spermatozoon of *A. diversicolor* is described by Hendelberg (1969) as being about 70 μm long, but threadlike. It is not clear whether this difference in morphology is a real one, or whether Peebles was, in fact, seeing spermatids and not mature spermatozoa. Kato (1940) stated that spermatozoa of the polyclad *Stylochus aomori* have easily distinguishable head, middlepiece and tail, but Thomas (1970 and unpublished data) found no such demarcation in spermatozoa of *Stylochus zebra* (Fig. 24A). Our observations indicate that as a general rule, it is rather difficult to delimit one region of a turbellarian sperm from another in the living condition; a middlepiece, as such, is lacking in all forms we have studied.

One of the most interesting features of some spermatozoa from flatworms, especially those of the acoels, is the presence of one or more rows of refractile granules in the tail. Such granules have been reported by Repiachoff (1893) for an unidentified turbellarian, by Haswell (1905) for the acoel *Amphiscolops*, by Luther (1912) for the acoel "*Palmenia*" (= *Anaperus*), by Ax and Dörjes (1966) for the

---

FIG. 13. Living late spermatid of the acoel *Childia*. The nucleus (N) has still not taken on the more oval shape characteristic of the mature spermatozoon. A double row of refractile bodies (RB) is found along the length of the tail behind the nucleus, and a short filamentous process (FP) extends forward from the tip of the cell. (Phase contrast micrograph.)

FIG. 14. Fixed and stained (toluidine blue and safranin) mature spermatozoon of *Childia*. The refractile bodies (RB) are heavily stained. The arrows indicate a region where the double undulating membranes are flattened out, and dark lines indicate the positions of the two incorporated 9 + 0 axonemes. (Light micrograph.) [From Henley (1968), reproduced with the permission of the Managing Editor, The Biological Bulletin.]

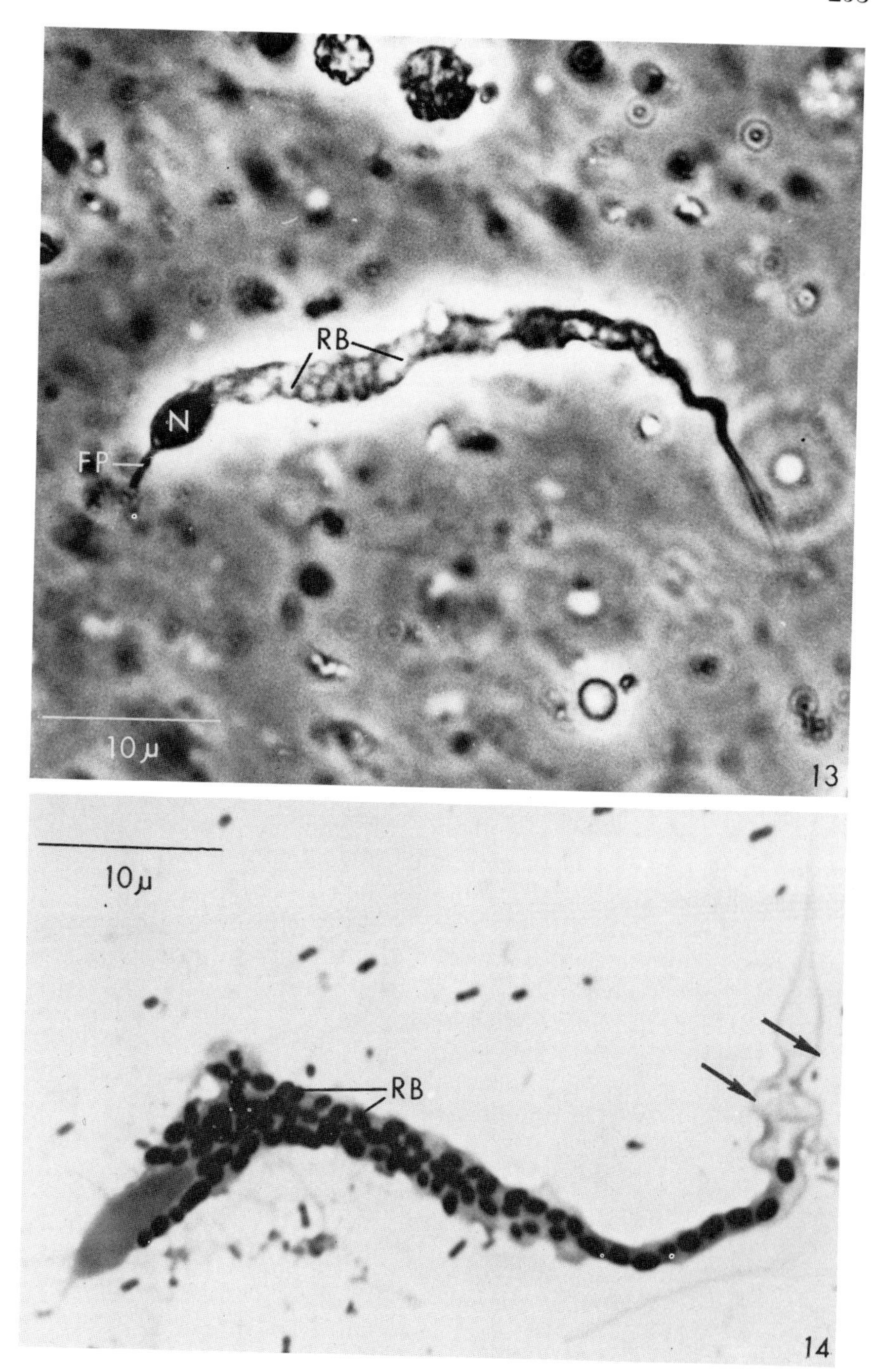
RB
N
FP
10 μ
13
10μ
RB
14

freshwater acoel *Oligochoerus*, by Henley (1968) for the acoel *Childia* (Fig. 14), and by Henley *et al*: (1968) for the acoel *Polychoerus*. It seems quite likely that such bodies, which appear superficially similar by light microscopy, may eventually be shown to have very different ultrastructural and cytochemical properties. We have already found this to be true for the refractile bodies of *Childia, Polychoerus* and *Anaperus* insofar as ultrastructure is concerned (see below); their cytochemical properties seem to be quite similar, however, At the present time nothing is really understood about their function(s), but since most spermatozoa are characteristically "stripped for action," it seems unlikely that these structures are merely going along for the ride.

Haswell (1905) described what he thought to be a case of dimorphism of the spermatozoa within one individual *Amphiscolops* (Acoela), one type being 8-10 times larger than the other and with a "contractile spiral flange." The anterior end of the shorter spermatozoon terminated in an abrupt, nearly transverse, face and there was a short filamentous portion. Both types had the granules referred to above, arranged along the length of the tail. Haswell does not appear to have considered the possibility that the smaller type might actually be a spermatid; on the basis of our observations for a number of forms on spermiogenesis in the living condition, it seems likely that this might be the explanation for the apparent dimorphism.

Hallez (1909) illustrated the spermatozoon of the rhabdocoel *Paravortex* as having a short, sickle-shaped head, with a long slender tail tapering to a fine point; a similar morphology was reported for a different species of the same genus by Ball (1916). Nuclei of several species of the alloeocoel *Plagiostomum* were shown by Karling (1962b) to be strikingly helical, whereas nuclei of spermatozoa of other species of the same genus had a more conventional cylindrical form. No matter what their shape, the nuclei of the spermatozoa of most turbellarians can be shown to be a single entity, but Karling (1940, 1962b) described an exception to this, in the case of spermatozoa of two species of the alloeocoel *Pseudostomum*. These contained a single row of granules running through the center of the rather short spermatozoon, each having a basophilic periphery; they extended for about ⅞ the length of the cell. There was also an eosinophilic strand parallelling the granules along most of their length; this was said to describe a spiral path. Similar beaded nuclei have been described for a number of freshwater forms, and for a few other marine ones. In spermatozoa of the alloeocoel *Cylindrostoma*, Karling (1962c) observed a distinct"cyanophilous" thread (presumably the nucleus) in the anterior region; it continued into the tail region, although less clearly.

There is growing evidence that motility of turbellarian spermatozoa is governed, in some unknown fashion, through a factor or factors present only in the female portion of the hermaphrodite's genital tract. As long ago as 1916, Ball reported that spermatozoa of the rhabdocoel *Paravortex* are motionless in the seminal vesicle of the male system, and suggested "Presumably the fluid inside the female passages ordinarily stimulates them to action," (p. 468). Release of spermatozoa into seawater also stimulated them to become active, although Ball thought the type of writhing shown under these conditions was not the normal mode of progress. Hyman (1951) pointed out that selffertilization usually does not occur in these hermaphrodites because spermatozoa are motionless in the male system and do not become motile until they are ejaculated during copulation. Costello and Costello (1968) made extensive studies of living intact specimens of *Polychoerus carmelensis*; they found that spermatozoa of this form were almost invariably motionless when still in testes, sperm ducts, ejaculatory ducts or (in contrast to Ball's observations discussed above) discharged into seawater. But spermatozoa contained in the seminal bursa of the female tract, or in the nondefined paths leading through the parenchyma from the bursa to the nozzles were highly active, "writhing with snake-like undulations." My own observations on spermatozoa of *Childia* are in complete accord with these latter findings, except for the fact that these cells are sometimes quite motile in seawater. This might very well be explained by the emanation of some secretion from the female tract into the small quantity of seawater surrounding the animal under a coverslip. Also, *Childia* lacks the nozzles and seminal bursa characterizing *Polychoerus*, and the role(s) these structures may play in conferring or facilitating motility is not known. Schilke (1970, p. 169) states that for kalyptorhynch (rhabdocoel) spermatozoa "Sperm capacitation probably occurs in the female ducts under the influence of substances secreted by the female genital-glands." It is not clear whether he is using the term "capacitation" in the usual sense, as applied commonly to mammalian spermatozoa, of a stripping-off of one or more superficial layers from the head of the spermatozoon prior to an acrosome reaction (Pikó, 1969); in any event his findings do implicate the female genital system in prolonging the fertilizability of spermatozoa. This entire subject obviously requires further and detailed investigation.

In addition to playing a possible role in the motility of spermatozoa acquired during copulation, the seminal bursa (in those forms possessing one) of the female tract has also been shown to be involved in the destruction of excess spermatozoa, over and above those relatively few required to fertilize the eggs. Mark (1892) postulated that the granular secretion within the bursa of the acoel *Polychoerus caudatus* functions "in consumption of the spermatozoa acquired at copulation" (p. 309). Cernosvitov (1932) suggested that in turbellarians, sperm resorption is an oft-repeated process, and that it plays a specific physiological role in the life of the organism in supplying nutritive material. Thus, in his view spermatozoa might be considered to recycle in the metabolic economy of the animal. Karling (1954) and Schilke (1970) both suggested that the resorptive portion of the bursa or the gut of kalyptorhynchs might be a repository for excess spermatozoa. Jennings (1968), studying two freshwater temnocephalid rhabdocoels by use of histochemical techniques, also concluded that a portion of the seminal bursa functions in resorption of spermatozoa and yolk globules, affirming the earlier suggestion by Haswell (1909) for other temnocephalids. Jennings found that the walls of the bursa and the duct leading to it had a strong positive reaction for acid phosphatase and certain proteolytic enzymes, while the contents of the bursa showed only a very slight positive reaction. The surrounding parenchyma also showed a strong positive reaction for acid phosphatase, suggesting that it may also function in a similar capacity. Since the contents of the bursa and duct showed only slightly positive reactions, Jennings suggested that destruction of excess spermatozoa and yolk globules may be accomplished by gradual histolysis, presumably followed by resorption of soluble materials. The source of the enzymes is not known, but he pointed out that since the bursa is in close association with the gastrodermis enzymes may originate in phagocytes there and merely diffuse to the bursa. Costello and Costello (1968) postulated that the nozzles, near the bursa of the acoel *Polychoerus carmelensis*, may act as a device for hindering and digesting excess sperm. The likelihood that the bursa may act in this fashion gains credence from the fact that, as noted above in Section 5.3.2.1, the diameter of the nozzle lumen is narrower than that of the spermatozoon.

The spermatozoa of many invertebrate and vertebrate species are characterized by the presence of an acrosome at the anteriormost tip; such a structure has thus far not been reported for any members of the Platyhelminthes (Dan, 1967) on the basis of ultrastructural studies. Some, but not all, the spermatozoa in living (Fig. 13) or fixed squash preparations of *Childia* have a filament extending forward from the tip of the cell (Westblad, 1945; Henley, 1968), but I have been unsuccessful in demonstrating by electron microscopy any structure which might account for this possible acrosome reaction. Hallez (1909) stated that a small acrosome was visible by light microscopy at the tip of the spermatozoon of the rhabdocoel *Paravortex*. The only other account I know of which even suggests the presence of an acrosome in spermatozoa of turbellarians is that of Karling (1962c, p. 203) who makes the following statement concerning the alloeocoel *Allostoma:* "In the 'nozzle' of the bursa I once saw three heavily stained arrows, which were first regarded as cuticular bursal structures, but were later discovered to be a kind of acrosomes belonging to sperms on the way towards the ovary." No further amplification of this statement is given.

Electron microscopy has yielded considerable amounts of new information on the morphology of turbellarian spermatozoa beyond the findings summarized above and the somewhat stark statements by Hyman (1951, pp. 117-118): "The sperm of the Turbellaria resemble those of other animals. They are usually long and filamentous and are provided with two flagella or two bristles. . . , or tapering at one or both ends to a flagella-like filament. The alloeocoel genus *Plagiostomum* has characteristic broad sperm with wing-like lateral extensions." One of the more important generalizations revealed by electron microscopy concerning the ultrastructure of spermatozoa from flatworms has been the finding that if flagellar axonemes are present at all, they are invariably paired. Hendelberg (1969) has discussed this point, and its possible phylogenetic significance, at some length.

One of the early pieces of work using electron microscopy for the study of flatworm spermatozoa was that of Klima (1962), who described the paired free flagella of freshwater triclad *(Planaria* and *Dendrocoelum)* spermatozoa as having the 9 peripheral doublet microtubules arranged in a circle, an arrangement already known to occur in a number of varieties of cilia and flagella (Manton, 1952; Fawcett and Porter, 1954). Instead of the usual two central singlet microtubules of the 9 + 2 pattern, however, Klima found a single cylindrical structure with 9 spokelike radiations passing from it to the peripheral doublets. Such an axial unit is now said to have the "9 + 1" pattern of microtubules (for a discussion of the reasons for placing quotation marks around the designation, see Henley *et al.*, 1969). The "9 + 1" pattern of microtubules has now been shown to occur in spermatozoa of a number of free-living and parasitic flatworms, some marine and some freshwater in habitat. See, for example, the papers by Shapiro *et al.*, (1961), Hendelberg (1965, 1969), Silveira and Porter (1964), Burton (1966, 1967, 1968), Rosario (1967), Tulloch and Hershenov (1967), Henley *et al.*, (1969), Silveira (1969), and Thomas

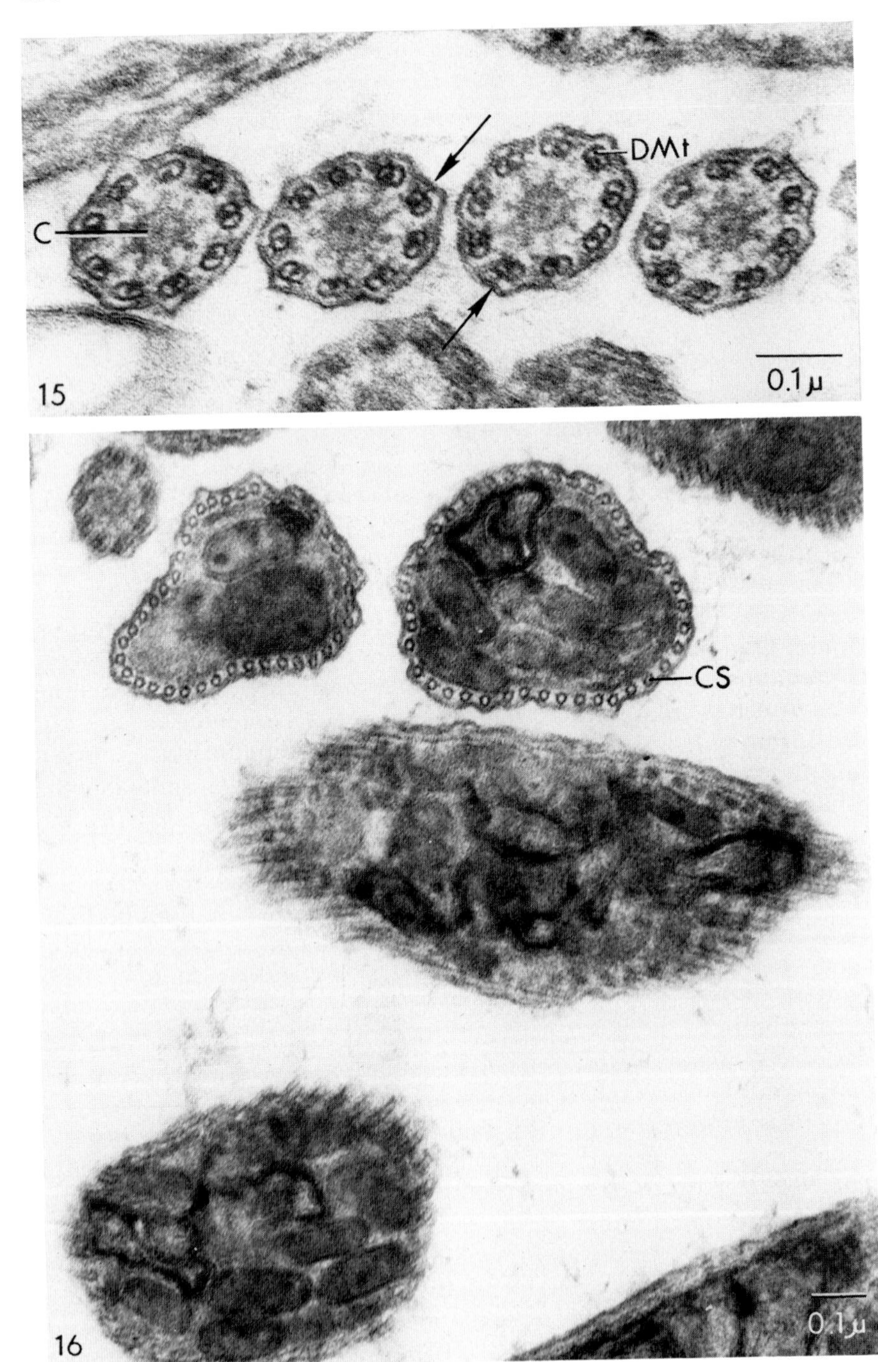
DMt
C
0.1 μ
15
CS
16
0.1 μ

(1970). Spermatozoa of the freshwater rhabdocoel *Microdalyellia* exemplify this arrangement of flagellar microtubules in spermatozoa (Fig. 15). In addition to the doublet microtubules and the central core found in the flagella, there is also a system of cortical singlet microtubules around the periphery of the spermatozoon (Fig. 16); these are often found to be wound in a steeply pitched spiral (Fig. 20) and presumably contribute to the characteristic undulatory movements of the cell. Irregularly arranged cross bridges occur between these cortical singlet microtubules; these are only at certain levels, so that a given section will show such cross connections between some, but not all, singlets.

Some early workers went so far as to suggest that the "9 + 1" pattern might be a normal one for spermatozoa of the flatworms, but there is a growing body of evidence that this is not the case. Christensen (1961), in an unillustrated abstract, described a "pellicle" of spirally wound singlet microtubules beneath the plasma membrane of the spermatozoon of the marine alloeocoel *Plagiostomum*. No evidence of free or incorporated flagella was found, so that the motility of this spermatozoon was attributed solely to the system of cortical microtubules. A similar situation exists in the spermatozoon of the freshwater rhabdocoel *Macrostomum* (Fig. 17); see also, Thomas and Henley (1971).

Complicating the picture even more were the reports of Henley *et al.* (1968), Costello *et al.* (1969) and Henley and Costello (1969), which described an axonemal pattern of microtubules within paired undulating membranes, in spermatozoa of the acoels *Childia, Polychoerus carmelensis, P. caudatus,* and *Anaperus gardineri.* These incorporated axonemes had the usual 9 peripheral doublets but lacked any central elements at all, the 9 + 0 pattern (Fig. 18). Furthermore, the spermatozoa of all were motile, under certain conditions at least (see above), thus upsetting the earlier prediction by Inoué (1959) and others that 9 + 0 patterns occurred only in nonmotile and sensory cilia and flagella. Hendelberg (1969) found that spermatozoa of the acoels *Mecynostomum* and *Convoluta* likewise had incorporated axial units, similar to those we reported except that his material had the 9 + 2 pattern. Once again, we have evidence here

FIG. 15. Four free flagella of spermatoza of the rhabdocoel *Microdalyellia*, sectioned within the body of the animal. There is a central core (C) in each flagellum, from which spokes radiate out to the 9 peripheral pairs of microtubules (DMT). Arrows indicate cross connections between doublets and plasma membrane of the flagellum. (Electron micrograph.)

FIG. 16. Spermatozoa of *Microdalyellia* cut in transverse (upper) and longitudinal (center) sections, showing the cortical singlet microtubules (CS). (Electron micrograph.)

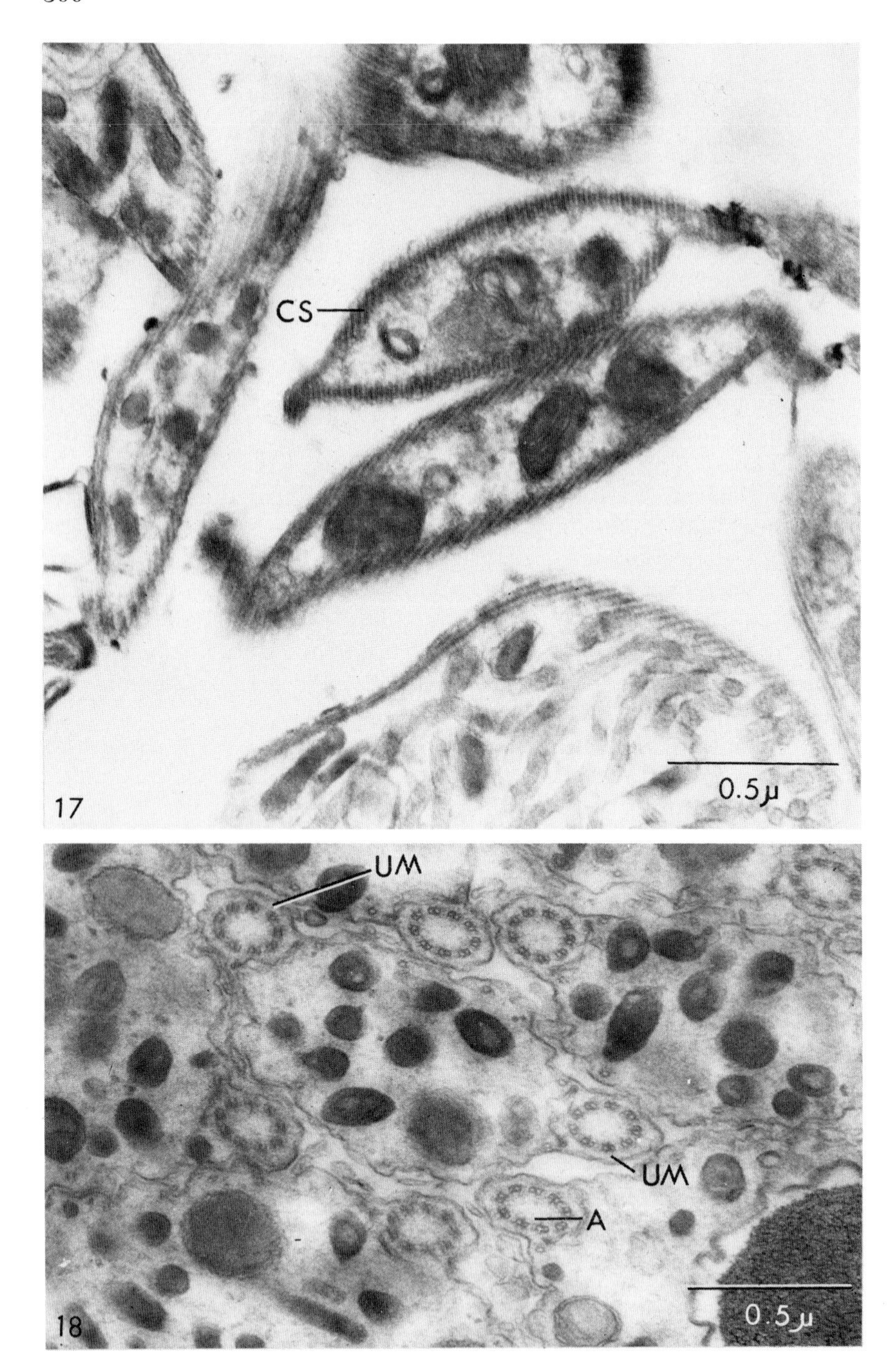
CS
0.5μ
17
UM
UM
A
0.5μ
18

for dissimilar morphology in the gametes of rather closely allied groups.

In spermatozoa of the polyclad *Stylochus* (Thomas, 1970 and unpublished results), there are two "9 + 1" axonemes, but these are neither free, as in the cases described above, nor completely incorporated into the cytoplasm of the sperm tail, as in spermatozoa of the acoels. Instead they are lodged in shallow depressions outside the plasma membrane (Fig. 19); fine fibrous elements, rather uniformly spaced, connect the axonemes to the outside of the plasma membrane. A similar situation obtains in the spermatozoon of the polyclad *Notoplana* (Henley, unpublished observations). Both types of spermatozoa have cortical singlet microtubules which, under certain conditions, can be shown to have a strikingly helical pattern in their walls (Fig. 24C). Again, as in the case of the spermatozoon of *Microdalyellia,* there are cross bridges between neighboring singlets at certain intervals and the microtubules can be seen, in favorable sections, to be attached by short stalklike segments to the plasma membrane of the spermatozoon. Similar stalklike connections have been demonstrated by Silveira and Porter (1964) for cortical singlets in spermatozoa of *Dugesia* and *Bdelloura.* In addition to these connections and the cross bridges between adjacent singlet microtubules already described, there are other modes of attachment of microtubules as well. Henley *et al.* (1969) found short connections between peripheral doublets and the flagellar membrane, and between doublets and the core of flagella having the "9 + 1" pattern. The relationship of all these complex interconnections to motility is still obscure but undoubtedly of great significance.

It should be pointed out that despite the variations in arrangement of the component microtubules in spermatozoa, cilia (epithelial and flame cell) in all the forms we have studied have the 9 + 2 pattern.

There is variability in the morphology of the nucleus in turbellarian spermatozoa studied by electron microscopy; in some forms the nucleus is uniformly lamellar during early stages of spermiogenesis (Silveira and Porter, 1964; Silveira, 1970) but later becomes regionally condensed into a cross-shaped area embedded in a lamellar portion. For other forms, such as the freshwater rhabdocoel

FIG. 17. Transverse (center) and longitudinal (left) sections through the spermatozoon of the rhabdocoel *Macrostomum.* This cell has no axial filament complexes associated with it, and there are only cortical singlet microtubules (CS). (Electron micrograph.)

FIG. 18. Transverse section through very late spermatids of the acoel *Childia.* Reincorporation (see Section 5.3.5.2) of the 9 + 0 axonemes (A) is almost complete, within the paired undulating membranes (UM), one at either lateral surface of the tail. (Electron micrograph.)

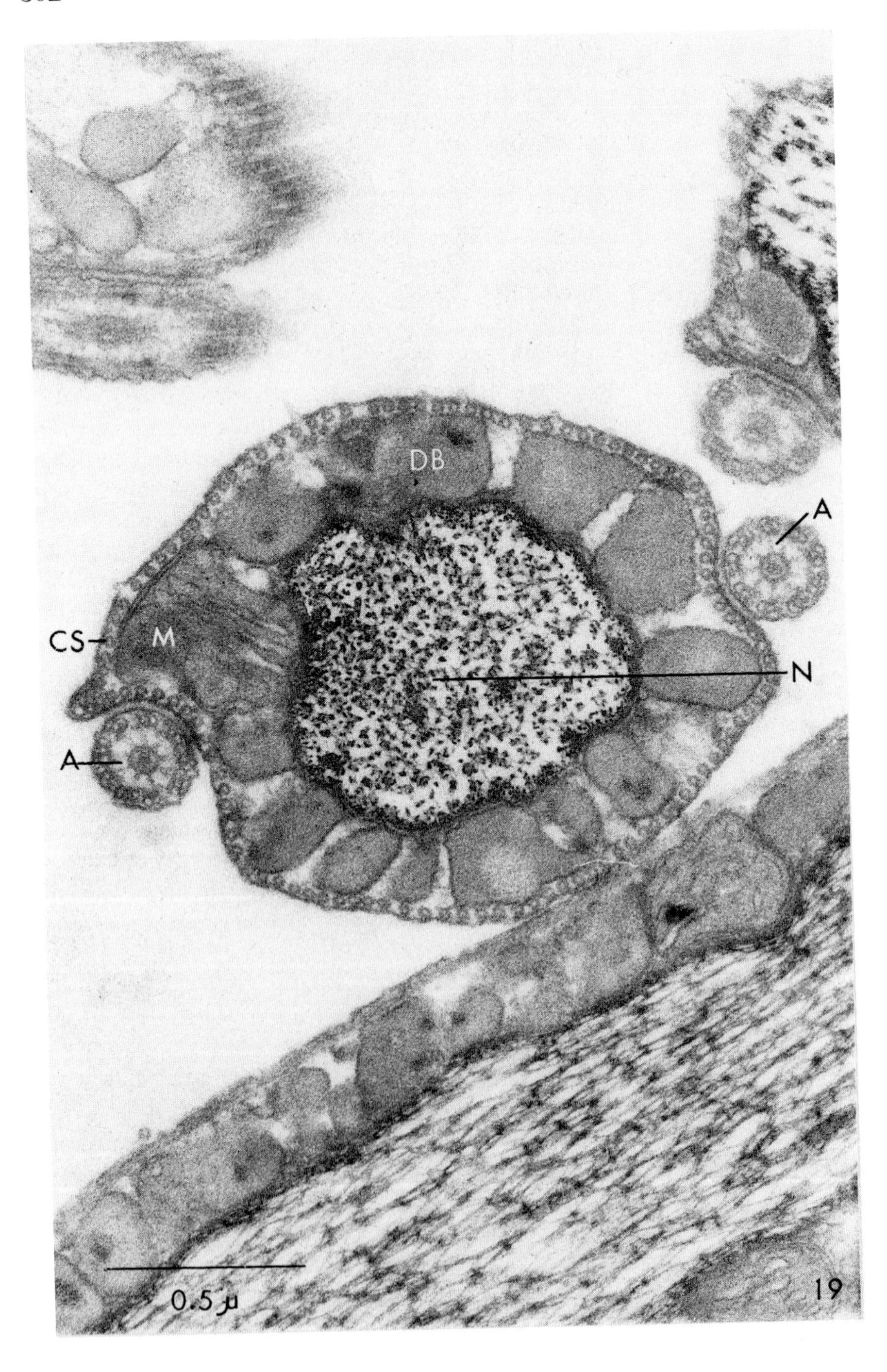
DB
A
CS
M
N
A
0.5 μ
19

Mesostoma, the lamellated condition found in spermatids (Fig. 20) is changed to a uniformly condensed state in the mature spermatozoon, while in still others (the polyclads *Stylochus* and *Notoplana*, for example) the nucleus remains diffuse and somewhat lamellar in the mature state (Fig. 19). There is little change in the nuclear morphology of spermatozoa of *Childia*, *Polychoerus*, and *Anaperus* during development. In the case of *Childia*, the nucleus is quite dense from the onset of spermiogenesis, while in spermatozoa of *Polychoerus* and *Anaperus*, it is considerably more diffuse. Spermatids and spermatozoa of *Polychoerus* tend to have a more electron-dense nuclear periphery (Fig. 21).

The spermatozoa of many Turbellaria are characterized by the presence of discrete bodies of varying morphology and size, some visible by light microscopy and some discernible only by electron microscopy. The refractile granules in the spermatids and spermatozoa of *Childia*, for example, are quite large and easily seen in the living condition (Fig. 13); electron microscopy reveals that they have a paracrystalline structure (Fig. 22) and are surrounded by a conspicuous ruffled membrane. Some of their histochemical properties have been described (Henley, 1968), and their origin, in association with the Golgi apparatus, was documented by Silveira (1967). Although similar histochemical reactions can be demonstrated for the refractile bodies in the spermatozoon of *Polychoerus*, their morphology, as seen by electron microscopy, is very different: they are much smaller than those of spermatozoa of *Childia* and are not paracrystalline, but rather homogeneous, electron-dense spheres. If a bounding membrane is present, it is so closely applied as to be indiscernible. In spermatozoa of the freshwater rhabdocoels *Mesostoma* and *Microdalyellia* and of the marine polyclad *Stylochus*, there are also regularly arranged discrete bodies, but these are not readily visible by light microscopy, and they differ somewhat from one another in their morphology as revealed by electron microscopy. In spermatozoa of *Microdalyellia*, for example (Fig. 23), many of these structures are very electron-dense throughout, while others show a less dense region. The spermatozoa of the polyclad *Stylochus* (Fig. 19) have a variable number of rather dense structures without cristae,

FIG. 19. Spermatozoa of the polyclad *Stylochus*, cut in transverse (center) and longitudinal (lower right) sections. Two "9 + 1" axonemes (A) lie just outside the limiting membrane of the spermatozoon, attached to it by very fine filamentous connections. Cortical singlet microtubules (CS) are just beneath the surface and there are large dense bodies (DB) between them and the nucleus (N). One mitochondrion (M) with cristae is interspersed among these dense bodies. The lamellated nature of the nucleus is clearly apparent in the cell cut in longitudinal section at lower right. (Electron micrograph.)

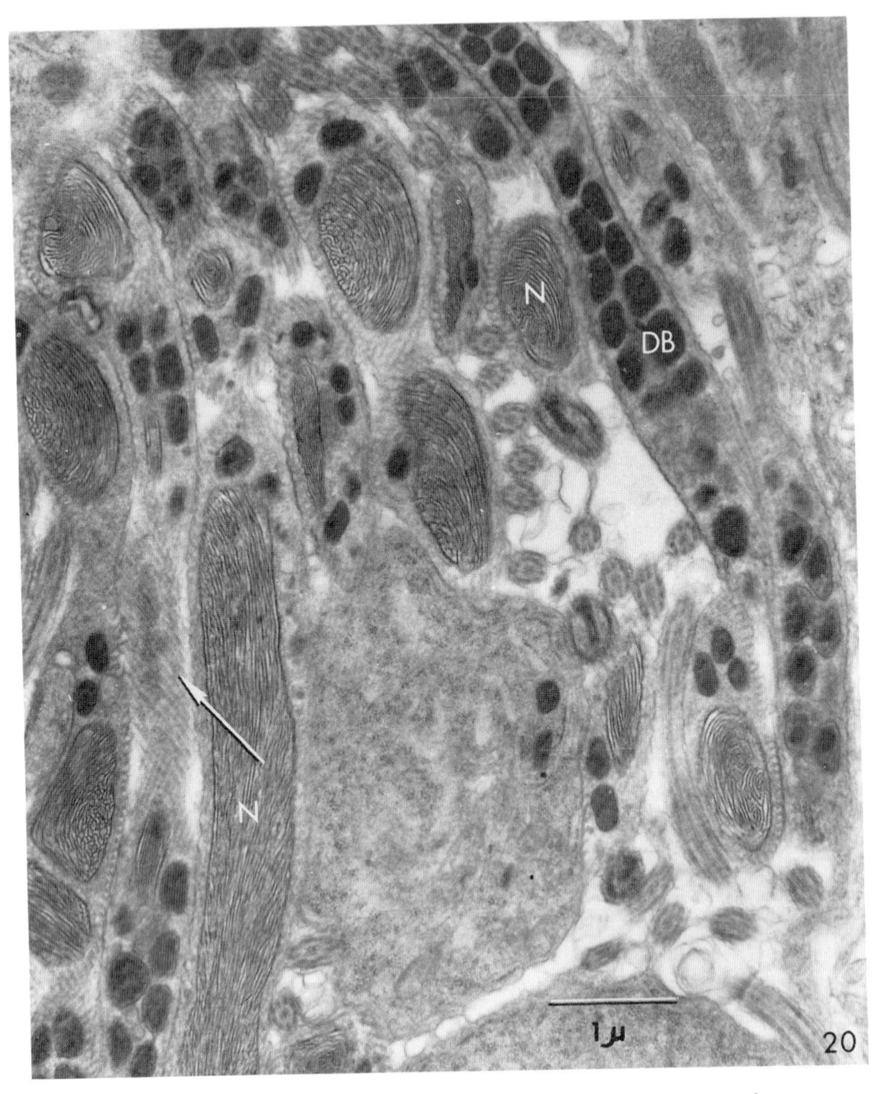

FIG. 20. Late spermatids of *Mesostoma*, cut at a number of levels and in various planes of section. The lamellar nucleus (N) found at this stage later condenses to a more compact form. In the longitudinal sections through the tail there are conspicuous dark structures (DB). The arrow indicates the spiral path described by the cortical singlet microtubules. (Electron micrograph.)

each surrounded by a membrane; these are arranged around the periphery of the nucleus and continue down the length of the tail after termination of the nucleus. A given section shows, also, a variable number (0-3) of large recognizable mitochondria with cristae. In the

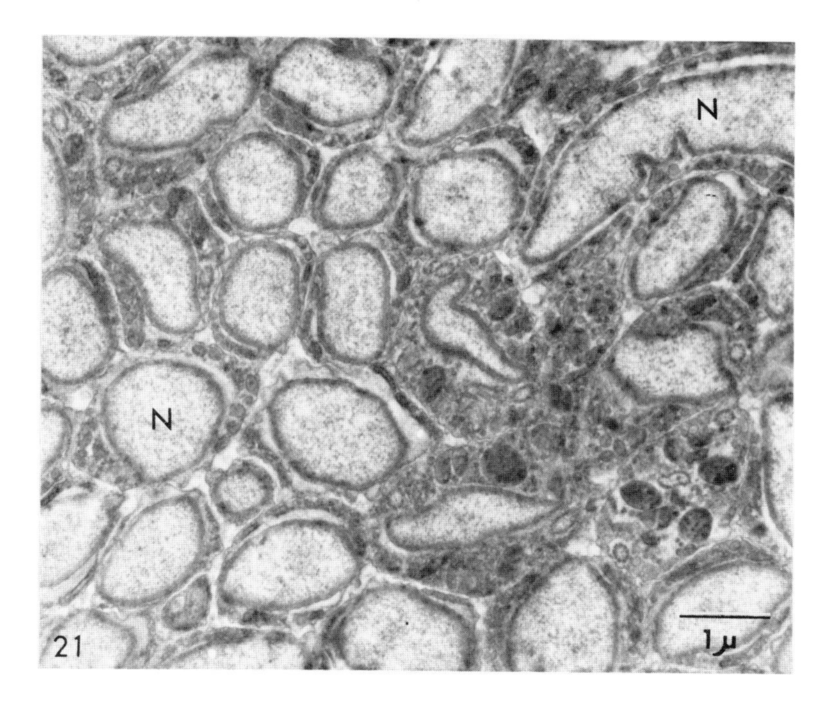

FIG. 21. Transverse sections though the nuclear region (N) of spermatozoa of *Polychoerus carmelensis*. There is an electron-dense periphery to each nucleus but no trace of lamellation. (Electron micrograph.)

spermatozoon of the polyclad *Notoplana* there again is a variable number of organelles in the tail of rather uniform size and appearance (all have definite cristae within). The evidence suggests, therefore, that there may exist a series of modifications of the mitochondria in spermatozoa of turbellarians, leading from a condition where unmodified mitochondria are present *(Notoplana)*, to ones where a small number of the organelles remains unmodified *(Stylochus)* or where transitional states exist between the modified and unmodified conditions *(Microdalyellia)*. The functional significance of such modifications is not clear (if indeed they actually exist—the evidence on this point obviously is only suggestive and not conclusive). Further comparative studies to clarify the matter are urgently needed and are in progress.

All the findings on ultrastructure discussed above were made principally with the aid of conventional thin-sectioning techniques. Another useful method is that of negative staining, first described by Brenner and Horne (1959) for the study of viruses, and later extended by a number of investigators to observations on microtubules.

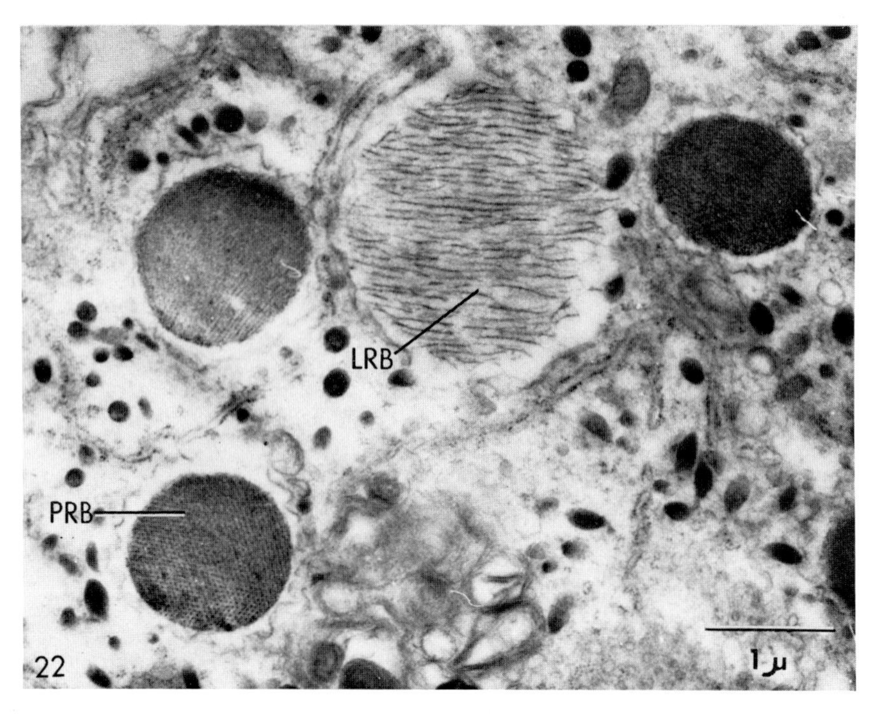

FIG. 22. Four refractile bodies in various planes of section and in two stages of condensation from a late spermatid of *Childia*. These initially have a loosely lamellar organization (LRB) but this eventually condenses to the paracrystalline condition (PRB) characteristic of the completed organelle. (Electron micrograph.)

Essentially, the technique involves the deposition of amorphous, highly electron-dense material within the cavities of microtubules and around their surfaces so that the biological material appears white against the dark negative "stain." Sodium or potassium phosphotungstate (PTA) at pH 6.8 is a commonly used negative stain. In addition to outlining microtubules it appears to have considerable macerating action on proteinaceous material so that in the spermatozoa where axonemes are enclosed within undulating membranes the microtubules are released. In such unsectioned material one can, by judicious manipulation of the duration of treatment with PTA, obtain complete complements of doublet and cortical singlet microtubules and core structures in forms having the "9 + 1" pattern of microtubules (Fig. 24 and 25). Similarly, the 9 doublets of the 9 + 0 axonemes can readily be demonstrated, often still attached to the basal plate (Fig. 26).

Another application of negative staining is in demonstrating mitochondria, both the conventional variety as found in spermatozoa of

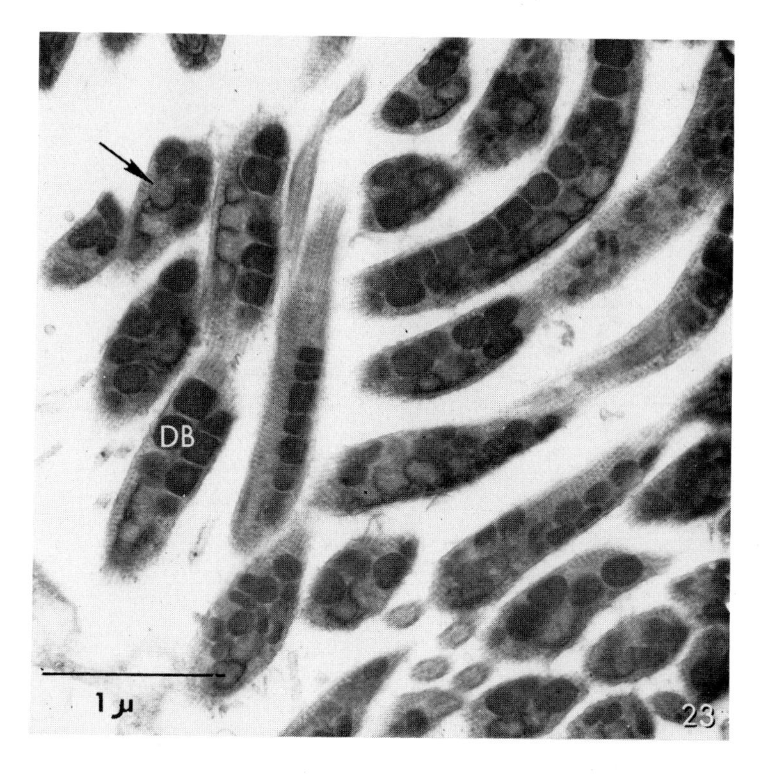

FIG. 23. Longitudinal sections through the tails of spermatozoa of *Microdalyellia*. There are rather chunky structures (DB) arranged in pairs along the length of the tail; some of these are more electron-dense than others. The less electron-dense structures (arrow) can be seen in higher magnification micrographs to have recognizable cristae within them. (Electron micrograph.)

Notoplana (Fig. 27) and those structures which may represent transitional or modified mitochondria in spermatozoa of *Mesostoma* (Fig. 28 and 29). The mitochondria of spermatozoa of *Notoplana* consistently appear in our material as dense black structures often scattered at some distance from the spermatozoon; it is only those organelles which remain "entrapped" in the cortical singlet microtubules (Fig. 27B) that have the more usual light appearance of comparable structures in spermatozoa of *Mesostoma* and *Microdalyellia*.

5.3.5 Gametogenesis

The meiotic events leading to the formation of haploid gametes in turbellarians appear to be entirely comparable to similar processes in other Metazoa, except for the fact that in the flatworms, gonial divisions (and later ones as well in some forms) often take place without

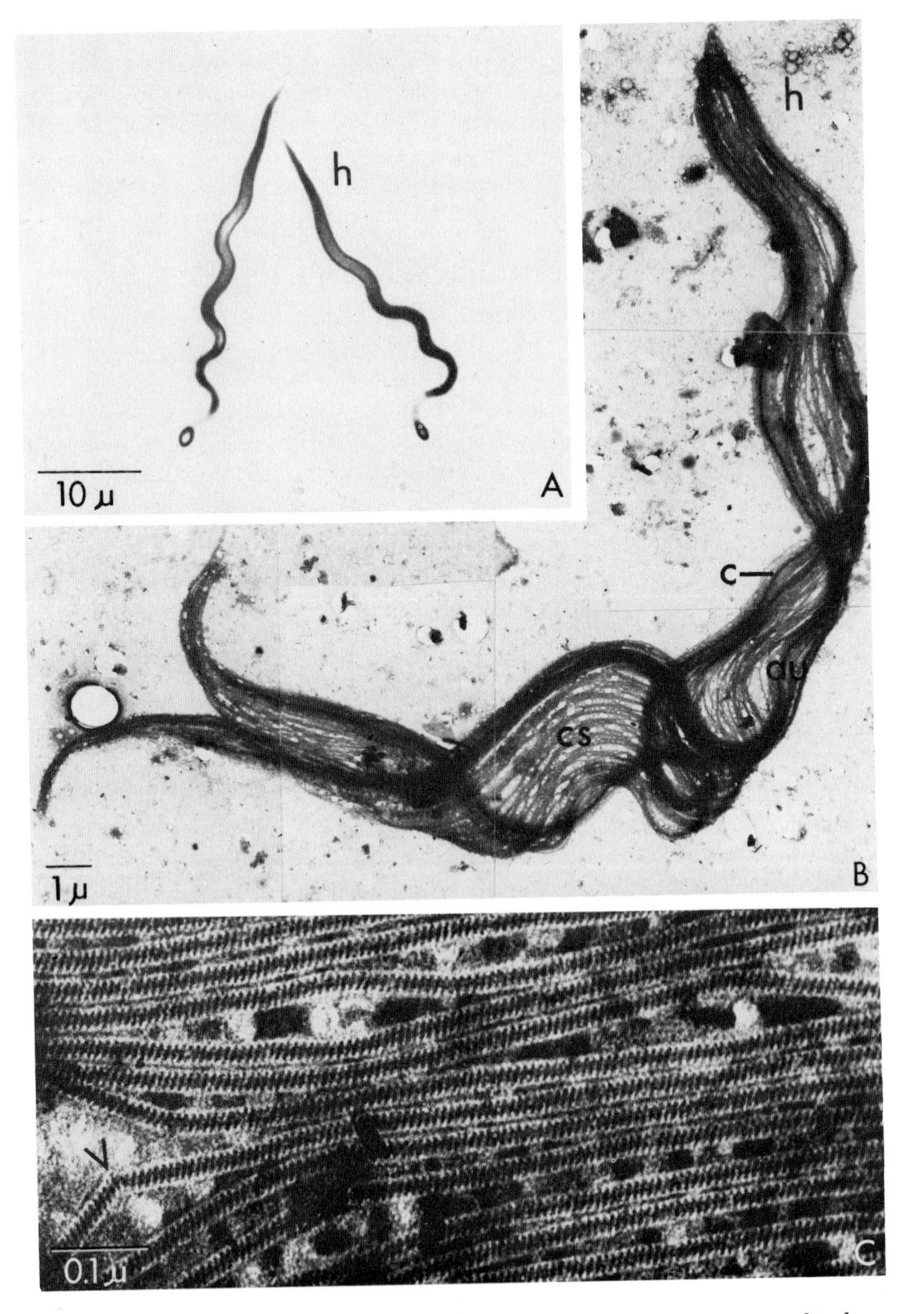

FIG. 24. (A) Living spermatozoon of the polyclad *Stylochus;* the anterior end is designated by h. (Phase contrast micrograph.)

cytokinesis. Thus the nuclei are embedded in a common syncytial mass of cytoplasm which is divided to form the definitive gametes at varying periods.

5.3.5.1 OOGENESIS

Patterson (1912) gave an extensive but unfortunately largely incorrect account of oogenesis in the rhabdocoel *Paravortex*. His errors in observation were corrected by Ball (1916), studying the same form. Ball extended his observations further, including an account of the history of the so-called "yolk nucleus," actually a mass of mitochondria. As is true for oocytes of many other turbellarians, meiotic divisions in *Paravortex* do not commence until after the cells have left the ovary, have been accessible to spermatozoa, and have become surrounded by yolk cells.

A departure from the more or less conventional processes of meiosis characteristic of most flatworms was described by Reisinger (1940) for the alloeocoel *Bothrioplana*. This peculiar form is said to develop by a process of dioogony, the obligatory development of a single embryo from two cells; associated with this is the complicating factor of parthenogenesis. Reisinger traced the chromosome numbers for the two oocytes which gave rise to a single embryo and found that from the initial diploid condition a tetraploid chromosome number developed. This was then reduced back to the diploid number in the eight meiotic products, all of which contributed to the diploid, parthenogenetically activated embryonic blastema. He compared this process with the normal changes in chromosome number from diploid to haploid in forms where two oocytes, enclosed within a common capsule, underwent the usual meiotic divisions and were then fertilized by haploid spermatozoa.

Costello (1961)* studied the meiotic divisions in oocytes of the acoel *Polychoerus carmelensis*. He found that the first maturation

(B) Montage of adjacent low power electron micrographs of the entire complement of microtubules from the spermatozoon of *Stylochus:* cortical singlets (cs), axonemes (au) composed of peripheral doublets and the core of the "9 + 1" pattern. (Negatively stained with phosphotungstic acid.)

(C) Cortical singlet microtubules from the spermatozoon of *Stylochus*, at higher magnification. Note the strikingly helical pattern in the walls of the tubules, and the sharp bend designated by V. [All figures in this plate are from Thomas (1970), reproduced with the permission of the author and the Managing Editor of The Biological Bulletin.]

*Dr. Costello informs me that the description of the centrioles of the second maturation spindles given in his 1961 paper is incorrect. It was based upon atypical cases and will be redescribed by him in a forthcoming paper.

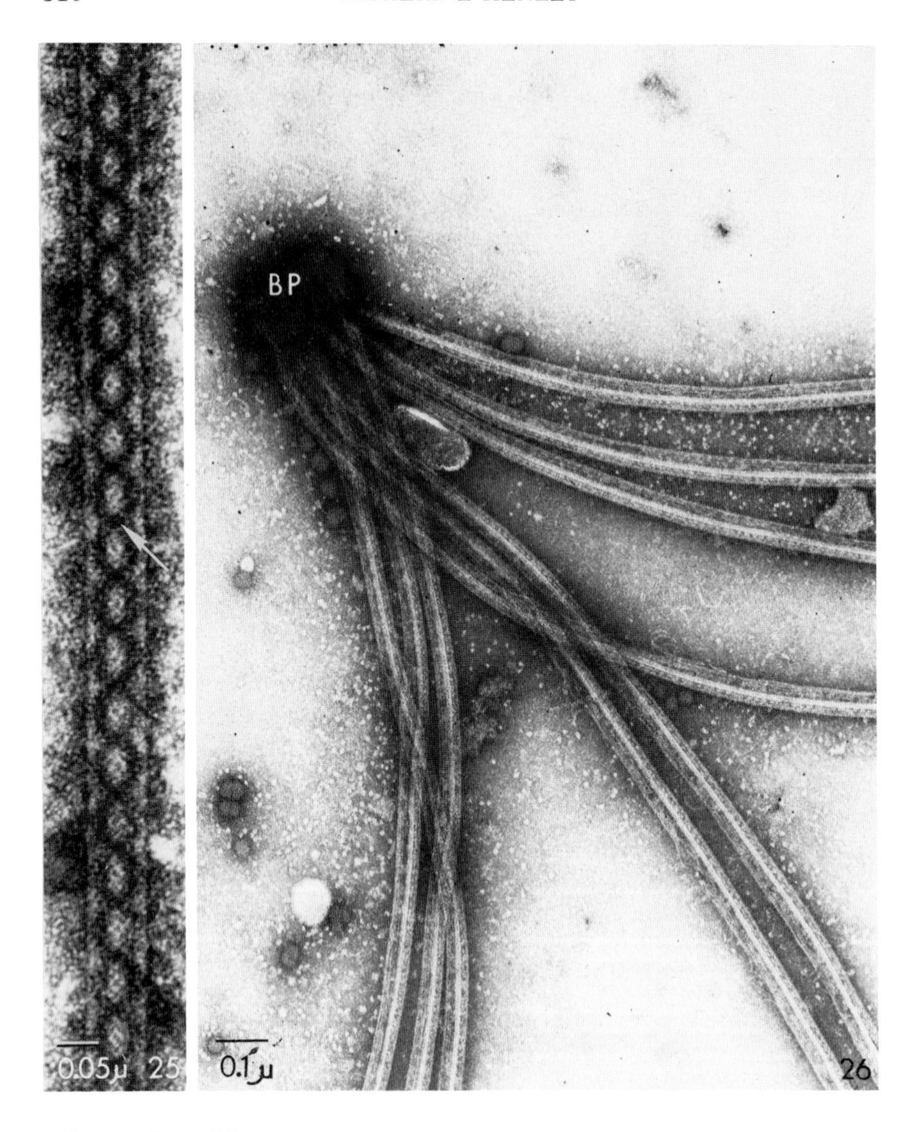

FIG. 25. Core of the negatively stained "9 + 1" axoneme of spermatozoa of *Dugesia;* this is the region designated "C" in the micrograph of sectioned flagella (Fig. 15) having the same pattern of microtubules. The darker areas (arrow) are more electron-dense because they are apparently hollow and contain the negative stain (PTA).

FIG. 26. Negatively stained 9 + 0 axoneme from the spermatozoon of *Childia.* The 9 doublet microtubules (each with a longitudinal white line representing the shared common wall of the two components) are still attached to the basal plate (BP). Protofibrillar substructure is apparent along the lengths of the doublet microtubules. (Electron micrograph.)

spindle was considerably longer and more slender than that of the second meiotic division (75 μm long, as opposed to about 38 μm). In contrast to the cleavage spindles of this form, which are exclusively central with all chromosomes arranged peripherally, the first meiotic spindle was found to have at least three chromosomes located within the spindle, the others being peripheral. There were two rather inconspicuous asters, with small bivalent centrioles; the spindle had an extranuclear origin near the germinal vesicle and was generated by two centrioles moving away from the nucleus. The first maturation spindles of the oocytes of *Polychoerus* have 17 typical tetrad chromosomes. These condense from 17 slender tetrapartite strands resulting from synapsis. Synapsis occurs after the germinal vesicles have grown to a large size as the skeinlike nucleolus disappears and before there is an appreciable accumulation of yolk and other inclusions by the oocyte. The second maturation spindles show 17 dyads. Both polar bodies are characteristically internal in oocytes of *Polychoerus;* they lie for a time in the cytoplasm and eventually disintegrate.

The spermatozoon enters the oocyte before germinal vesicle breakdown and lies dormant until the completion of the maturation divisions. Then, while the 17 haploid chromosomes of the egg transform into 17 karyomeres (chromosome vesicles; see Section 5.4.1.1) the nucleus of the spermatozoon undergoes a series of changes to produce 17 paternal karyomeres. A pair of large centrioles, presumably of paternal origin, appears alongside or between the two groups of karyomeres. These centrioles are 5 μm long and about ¼ μm in diameter and constitute the cleavage centers. They are oriented precisely at right angles to one another and maintain this orientation as they move apart generating the first cleavage spindle between them. Meanwhile, the two sets of 17 karyomeres have been transforming into the definitive chromosomes and are being aligned at the periphery of the spindle. The 34 chromosomes (all J- or V-shaped) are attached to mantle fibers of the spindle with their free ends extending away from the spindle (Costello, 1970). The centrioles of the first cleavage spindle have maintained their right angle orientation with respect to one another. By metaphase of the first cleavage, however, each centriole has a tiny daughter bud at the exact middle of each mother centriole. This bud is at right angles to the mother centriole but oblique to the cleavage axis. This orientation of centrioles, and their mode of replication, have been suggested by Costello (1961) as being the mechanical basis for the alternating planes of oblique cleavage (see Section 5.4.1.2).

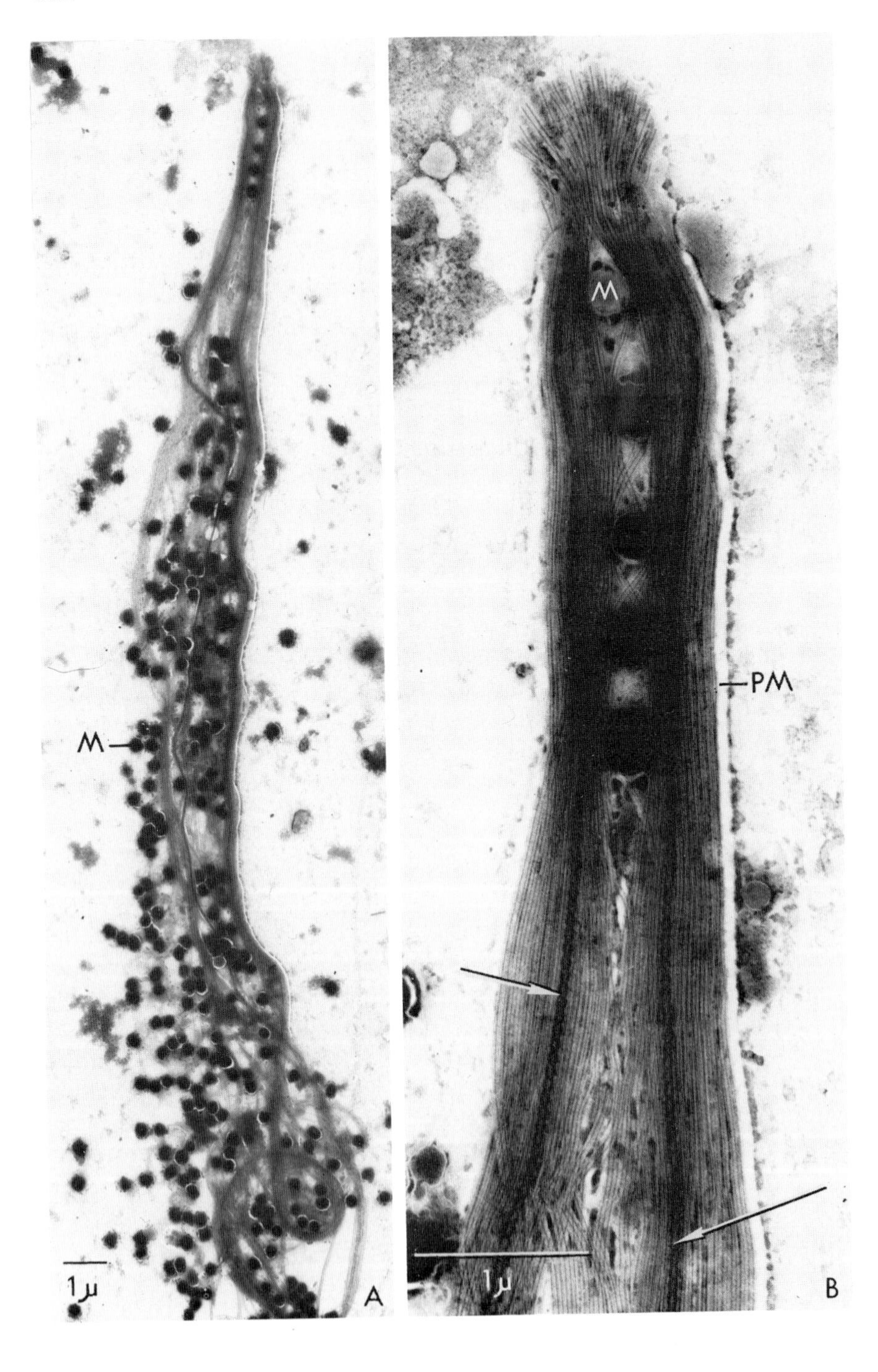
M
1μ
A
M
PM
1μ
B

Newton (1970a) has described oogenesis in the freshwater alloeocoel *Hydrolimax*. In oocytes of this form an accessory parenchymal cell becomes very intimately associated with the growing oocytes and remains thus associated until cocoon formation begins. Its function is not clearly evident but the author suggested that it may serve as a temporary source of nutritive materials; later it gives rise to the yolk halo characteristic of this form which is apparently not nutritive in function (see Section 5.3.4.1). Newton is of the opinion that formation of the meiotic spindles is entirely independent from the process or consequences of sperm penetration.

The meiotic condition of the egg at the time of oviposition varies, being at the metaphase of the first meiotic division for the rhabdocoel *Macrostomum* (Papi, 1953) and for some of the Japanese polyclads studied by Kato (1940). As noted above, the zygote of the acoel *Polychoerus* is arrested at the metaphase of the first cleavage division until it is deposited in seawater: the presence within the bodies of these hermaphroditic individuals of a number of zygotes with large first cleavage spindles is correlated with this delay in egg-laying. The very large size of the amphiasters, visible through the body wall of the worm, once prompted the early embryologists to refer to them as "polar suns." This resting stage is of considerable duration compared with the transitory, fleeting moment characteristic of the metaphase stage of many other marine invertebrates.

5.3.5.2 Spermatogenesis

Hallez (1909) studied spermatogenesis in the rhabdocoel *Paravortex* and found that it proceeded quite rapidly. The four haploid nuclei resulting from the meiotic divisions were often found to be embedded in a common mass of cytoplasm. In spermiogenesis the nuclei of the spermatids condensed and became excentric in position and the cytoplasm was more hyaline than at earlier stages. One axial filament was borne by each centrosome (presumably a single one for each spermatid, although this is not made clear); this axial filament,

---

Fig. 27. (A) Low power electron micrograph of the negatively stained microtubular complement from spermatozoa of *Notoplana*. The dense black granules (M) correspond to structures which, in sectioned material, have cristae and appear to be typical mitochondria.

(B) Uppermost portion of the micrograph shown in (A) at higher magnification. Doublet and singlet microtubules and the two cores (arrows) of the axonemes terminate here, but their precise relationships are obscure. The structures thought to be mitochondria (M) are light, presumably because the electron-dense negative stain has not penetrated them. The white line running longitudinally at each side (PM) is the plasma membrane, relatively unaffected by the negative stain.

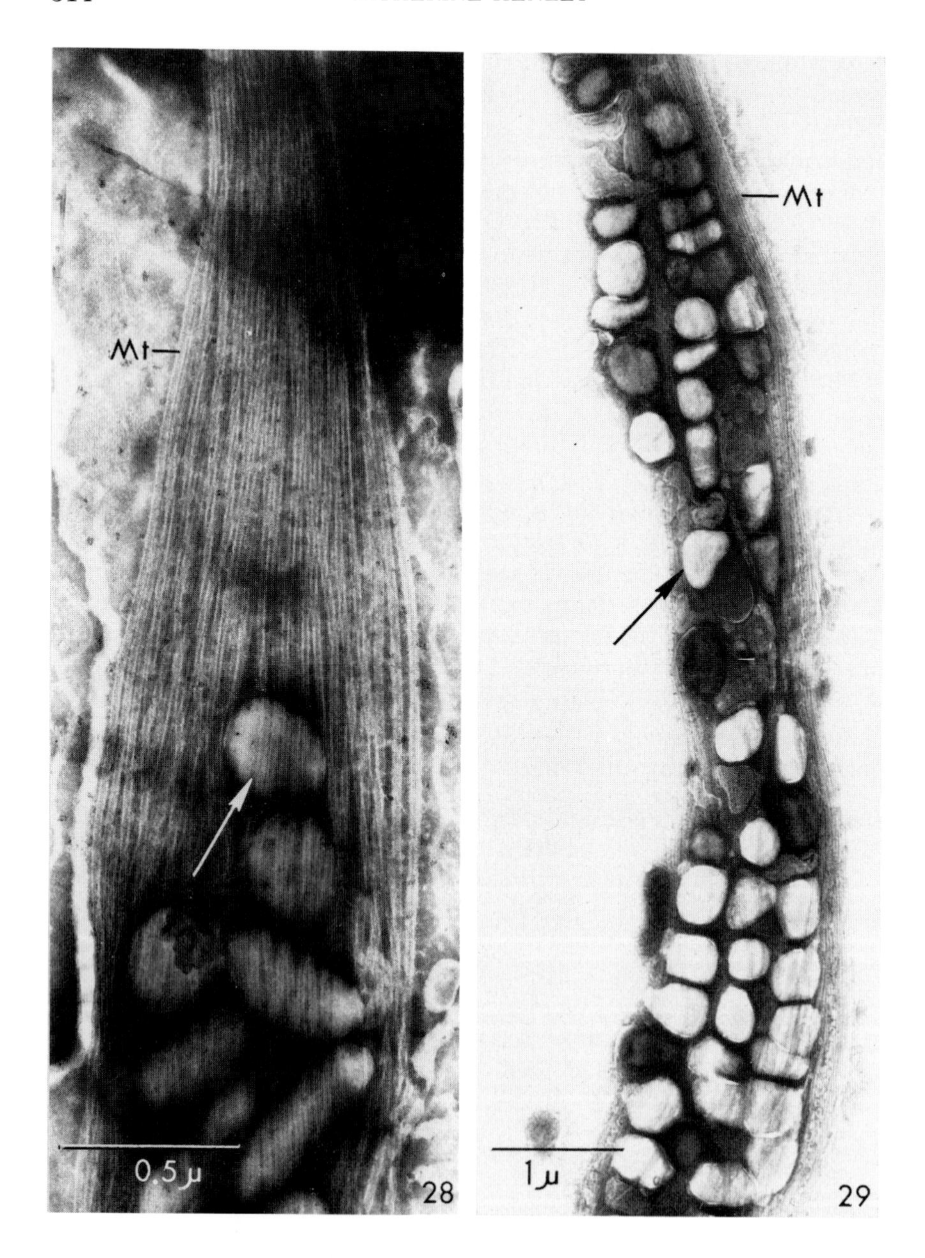

FIGS. 28 AND 29. Two negatively stained preparations of spermatozoa of *Mesostoma*, showing microtubules (Mt) and predominantly light structures (arrows) which, for the most part, are resistant to the action of the negative stain. In sectioned material, no cristae are visible within them. (Electron micrographs.)

or flagellum, was sometimes free and sometimes embedded in cytoplasm. Subsequent events included elongation of the flagellum and of the nucleus to an S-shape; during this period the flagellum was observed to have a spiral configuration in living spermatids.

These observations take on new meaning when considered in conjunction with the studies of Hendelberg (1965, 1969) on spermiogenesis in a number of turbellarians, and with our unpublished findings for the acoels *Polychoerus* and *Childia*. Both light and electron microscopy were used by Hendelberg and by us. There has been no evidence, at the level of electron microscopy, for spermatids of flatworms giving off a single flagellum during spermiogenesis. Invariably, even in the case of those spermatozoa having a threadlike shape in the mature condition, two flagella are given off; these are free, in early stages of spermiogenesis, but later become incorporated into a long tongue of cytoplasm, the spermatid shaft (Hendelberg, 1969) (Fig. 30).

A number of aspects of the spermatogenesis of *Polychoerus* have been worked out (Costello, unpublished data). These observations were based upon extensive phase contrast study of the living, developing gametes, upon light microscopical examination of fixed and stained material embedded in plastic and in paraffin, and upon electron microscopy of glutaraldehyde-fixed material. The meiotic cycle is similar to that found in other Metazoa, with 17 tetrads in the primary spermatocytes and 17 dyads in the secondary spermatocytes. Four primary spermatocytes are associated as a group so that after two meiotic divisions a bundle of 16 spermatids results; this metamorphoses into a bundle of 16 spermatozoa temporarily held together by a residual cytoplasmic mass at their anterior ends. The centriolar arrangements are more complex than in uniflagellated spermatozoa since two axonemes of microtubules and a tail "keel" structure of microrods are centriolar derivatives. There are numerous refractile bodies formed in the spermatocytes, presumably through the activity of numerous dictyosomes (Golgi apparatus) demonstrable by electron microscopy. Mitochondria, small dense bodies, and a very extensive endoplasmic reticulum are among the organelles of these cells.

In the case of spermatids which are to develop into the threadlike form exemplified by *Polychoerus*, two free flagella, each with a plasma membrane of the usual sort, are given off by the spermatid at a stage when it is still more or less round and compact (Fig. 30H). There follows a process of cytoplasmic and nuclear elongation, during which the spermatid shaft incorporates the free flagella, one

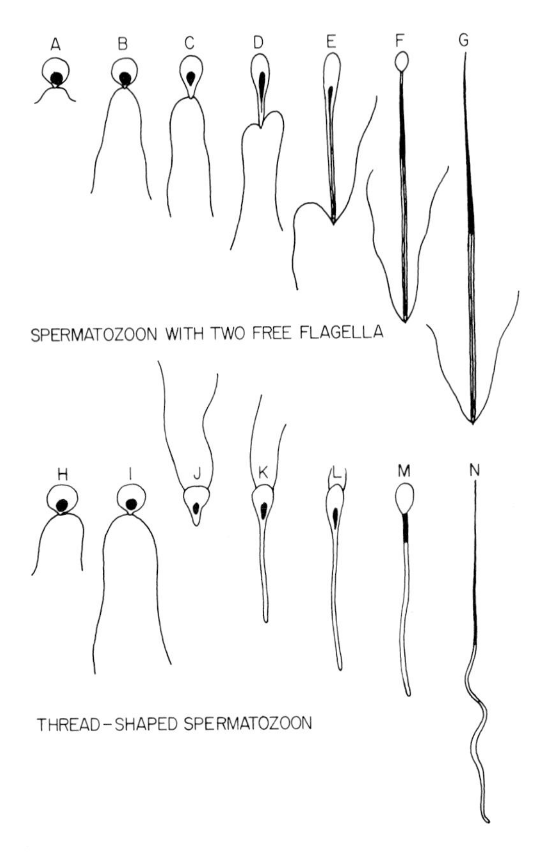

FIG. 30. Diagrams illustrating the possible steps in spermiogenesis of spermatozoa with free flagella and with incorporated axonemes. For further explanation, see text. [Redrawn and modified from Hendelberg (1969).]

on each side of the cell. According to Hendelberg's interpretation (based largely on light microscopy) the direction of cytoplasmic elongation is toward what will become the anterior end of the mature spermatozoon, the nucleus thus lying in the more slender portion of the cell (at top, Fig. 30N). In the case of spermatozoa which, in the mature condition, have two free flagella (Fig. 30A-G) Hendelberg believes the process of spermiogenesis begins in the same fashion as that just described for threadlike spermatozoa. Two free flagella are given off by the young spermatid; these elongate but are not incorporated into the spermatid shaft. The shaft, including the nucleus, elongates anteriad; it is not clear from his description whether Hendelberg considers that a long slender portion of the nucleus is extended into the region where the two flagella are inserted, but his figures imply that such is the case. In both the threadlike spermatozoa and those with free flagella a residual mass of cytoplasm is cast off at the conclusion of spermiogenesis (Fig. 30F, M), and in both, slender cytoplasmic bridges persist between developing spermatids for variable periods of time. The end result of spermiogenesis in both cases, according to Hendelberg's views, would place the nucleus at the slender end of the spermatozoon away from the axonemes, whether these be free or incorporated. We are not convinced that his interpretation is entirely correct, but it does offer some useful clues in our study of spermiogenesis in *Polychoerus* and *Childia*.

One stage of the flagellar incorporation process is shown in Fig. 31 for the spermatid of *Polychoerus*. The two 9 + 0 axonemes are each provided with a surrounding membrane at this time, shortly before their incorporation into the cytoplasmic mass. In sections through mature spermatozoa, there is no sign of this surrounding membrane, the microtubules of the axial filament complex lying naked in the cytoplasm. In the case of *Polychoerus*, the flagella are apparently extruded from the spermatid as rapidly as they are formed, but in another acoel, *Childia* (Fig. 32), the axonemes elongate very markedly within the cytoplasm of the spermatid and become coiled around and around within the cell. A typical section shows numerous transverse, oblique, and longitudinal sections of the axonemal coils before they are extruded. There is no membrane around the axonemes while they are coiled, but in favorable sections, one can readily count the 9 + 0 complement of microtubules shown to be characteristic of this form by Costello *et al.* (1969). Sometimes these are quite symmetrically rounded, but in other cases it appears that some flattening and deformation of the axonemes occurs as their component microtubules are synthesized (Fig. 32, arrows). This is probably a consequence of pressure exerted by the cytoplasm, which is in turn impinged upon by neighboring spermatids and somatic cells. Eventually, the axonemes acquire membranes, as they are being extruded from the spermatid; they attain a maximum length, while free, of about 50 μm (Hendelberg, 1969). Incorporation of the flagella into the spermatid shaft ensues by a process quite similar to that seen in spermatids of *Polychoerus*. The mature spermatozoon then has two 9 + 0 axonemes incorporated into the lateral surfaces of paired undulating membranes. The uncoiling process, preceding flagellar extrusion, must present some interesting mechanical problems.

We believe that the striking difference in the process of extrusion of the flagella in spermatids of *Polychoerus* and *Childia* can readily be explained if one considers the relative length of the mature sper-

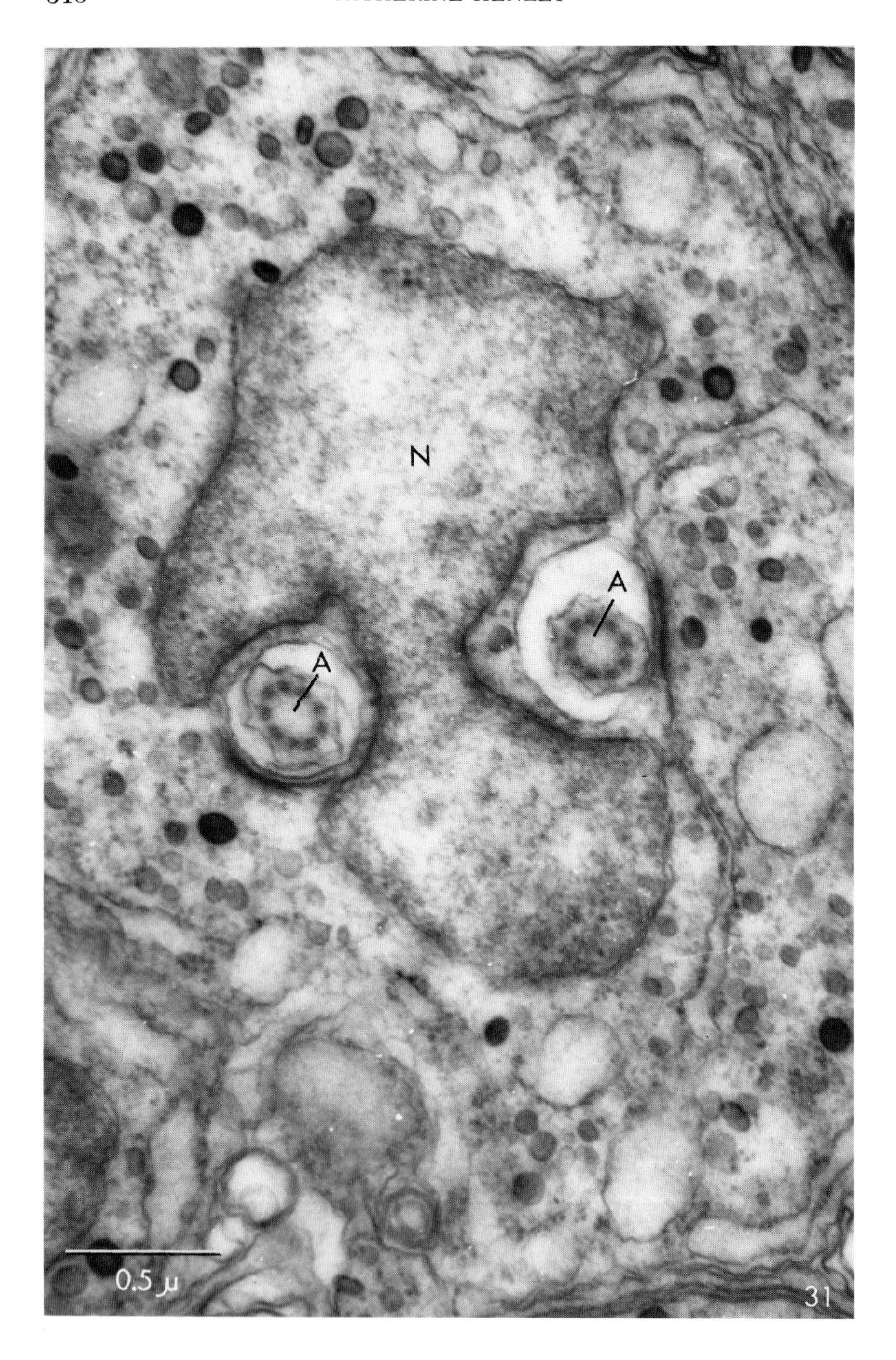
N
A
A
0.5 μ
31

matozoon of each (300 μm vs. 50-60μm). It is known that synthesis of axonemal microtubules can occur rapidly, but the process of elongation of the spermatid shaft and the formation of the undulating membranes is a relatively slow one. In the case of *Polychoerus* spermatids the axonemes are much longer than they are in comparable cells of *Childia* and there would be neither sufficient time nor enough space to coil such long axonemes inside the spermatid. The problem of unwinding would likewise be much more complex for very long axial filament complexes.

5.3.5.3 Sequence of Sexual Development

In the predominantly hermaphroditic turbellarians, development of the male portion of the reproductive system has frequently been observed to precede formation of the female organs. In some instances there is protandric hermaphroditism in which the male system forms first and then degenerates, followed by the gradual development of the female system. This situation was found by Hallez (1909) to occur in the rhabdocoel *Paravortex,* but for the same genus Ball (1916) reported that degeneration of the male system was never complete, although retrogression of some components did occur after the onset of formation of the female organs. In the freshwater rhabdocoel *Mesostoma georgianum,* Darlington (1959) observed that development of both sets of sexual apparatus was almost simultaneous; the penis, testes, ovary, and copulatory bursa being formed first and the other components soon afterward. In the case of the acoel *Polychoerus carmelensis* (at Pacific Grove, California) Costello (personal communication) found no trace of ripe gonads in animals examined early in March. The testes and associated male structures developed rapidly thereafter and mature spermatozoa were present by the middle of April, but very few mature ova were observed until the middle of May. Under conditions of starvation the first visible signs of degeneration were in the nuclear membranes of spermatids and spermatozoa, but very soon all female components of the system likewise began to regress and disappearance of both sets of genital apparatus was more or less simultaneous.

The role of sequential development of reproductive structures in

FIG. 31. Electron micrograph of a spermatid of *Polychoerus*, at a stage comparable to Fig. 30K, L. Free flagella have earlier been given off (Fig. 30I) and are now being incorporated into a slender mass of cytoplasm. Two cup-shaped indentations in the nucleus (N) surround the two 9 + 0 axonemes (A), each of which has a typical flagellar membrane; these membranes will later disappear, leaving the microtubules naked in the cytoplasm of the undulating membranes.

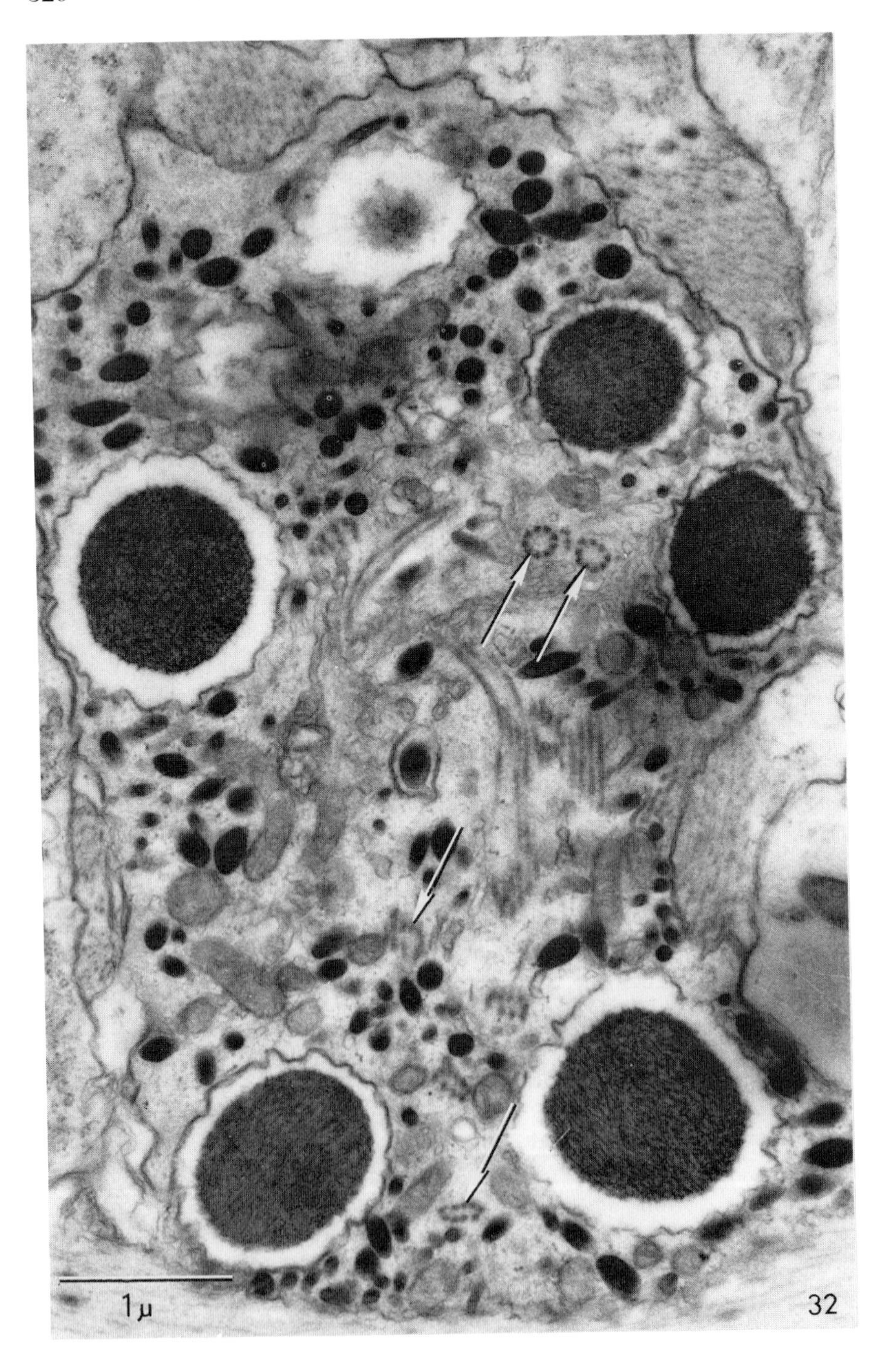
1μ
32

animals where great reliance is placed on such structures in classification has been pointed out by Bush (personal communication). She has observed that one can obtain a graded series, in many turbellarians, from juveniles to fully mature adults, reflecting what she believes may be a rather strict order in which the various sexual components appear. This emphasizes, once again, the critical importance of being certain that one bases identification of forms on fully mature animals.

In many cases, at least, it appears that complete retrogression of the reproductive system occurs in the early autumn to be followed by its reappearance in the spring.

5.3.6 Factors Affecting Gametogenesis

This is one of many almost totally unexplored areas of the reproductive biology of turbellarians. The recorded observations are nearly always fragmentary and usually incidental to some principal line of investigation.

Dorey (1965) found that occasional 12-hour periods of darkness notably improved the reproductive performance (by implication, favoring gametogenesis) of several acoels (*Convoluta, Aphanostoma, Otocelis,* and *Avagina*). He stated "They [newly hatched juveniles] should therefore be given several 12-hour periods of darkness, in the presence of a moderate population of the *Carteria* flagellates, during the first week after hatching. Using such methods, and without adding any special organic food-materials, a culture has been maintained for ten months, through four generations" (p. 150). No sensory structures definitely mediating such an effect have been demonstrated, although light perception is known to occur via the photoreceptive ocelli (Hyman, 1951).

As noted in Section 5.1 we have found that adequate nutrition is a *sine qua non* for obtaining normal gametes in all the forms studied. This presents a problem for maintaining most turbellarians in the laboratory. An exception is to be found in the case of the freshwater rhabdocoel *Mesostoma georgianum.* Newton (unpublished data) has observed a striking response of these forms to any vibration of the dish in which they are kept and more specifically to perturbations in the surface film. This is associated with the fact that the worms feed

FIG. 32. Electron micrograph of a spermatid of *Childia.* In this form, growth and elongation of two axonemes occur within the cytoplasm of the spermatid and the microtubules are wound, as two 9 + 0 units (arrows), through the substance of the cell. Eventually these long axonemes are given off as free flagella, and then reincorporated into the cytoplasm by a process comparable to that seen in spermatids of *Polychoerus.* This section does not pass through the level where the nucleus is present.

on insects and other small animals which fall into the temporary pools on stone outcroppings where they live (Darlington, 1959). If one drops a living fleshfly into a culture dish containing *Mesostoma* the animals will immediately converge on it and devour the fly within minutes, leaving only the chitinous parts.By this simple method it is possible to maintain this species in the laboratory, in excellent condition, for prolonged periods. Unfortunately this is not the case for most marine forms. Adequate nutrition is very likely one of the most important of the many factors that might affect gametogenesis.

A few observations have been made concerning the effects of parasitism on gametogenesis. Hyman (1951) stated that the rhabdocoel *Oekicolax* parasitizes the alloeocoel *Plagiostomum* causing ovarian degeneration and that the freshwater triclad *Dendrocoelum* is sometimes infected with a trypanosomelike flagellate in its copulatory bursa; no mention is made of possible deleterious effects in the latter case. In a 1955 paper Hyman described the occurrence of a eugregarine parasite in the vagina of a land planarian, *Rhynchodemus;* the protozoan was said to have destroyed part of the vaginal wall (and, of incidental interest, to feed on the turbellarian's yolk cells), but, again, no specific statement is made concerning any effects on gametogenesis. Holmquist (1967) reported that heavy parasitism of freshwater triclads *(Dendrocoelopsis)* resulted in destruction of the testes and copulatory organ, while animals with few or no parasites had fully developed genital organs. Structures which he interpreted to be bacteria of host origin were seen by Karling (1970) in the gonads of *Pterastericola,* a rhabdocoel commensal of seastars. The effects on the worms of such infections were not given.

5.3.7 Reproductive Behavior

5.3.7.1 Mating

In many cases, selffertilization of these hermaphroditic animals is anatomically possible, but apparently it rarely occurs. There are said to be two general methods of cross insemination, hypodermic impregnation and true copulation (Hyman, 1951). The first method was formerly considered to prevail among the acoels, but more recent observations (see below) have shown that this is not the case. It does seem to occur in a few acoels, rhabdocoels and polyclads: the spermatozoa are injected, via the male copulatory apparatus through the surface of the partner, apparently at any point on the body; they then wander through the parenchyma to the oocytes. Reciprocal copulation, often with deposition of spermatozoa into the partner's copulatory or seminal bursa or into the parenchyma, is more commonly the rule, particularly among the triclads. Rather complex patterns of courtship behavior often precede the event.

Hallez (1909) never observed the process of mating in the rhabdocoel *Paravortex*, but suggested that it occurred soon after the young worm emerged viviparously from its mother and enjoyed a brief period of freedom in the intestine of the molluscan host. The first description of copulation in an acoelous turbellarian was apparently that of Peebles (1915), who gives a brief account of the process as seen in *Monochoerus lineatus*. Hyman (1937) gave a much more detailed account for *Amphiscolops;* mating was preceded by a kind of courtship in which animals wandering along the wall of the vessel gave one another what were described as quick little nips with the anterior end. Copulants then rolled into a ball with the posterior end of the outermost individual curved so that its ventral surface was in contact with the posterior ventral surface of its partner. After a very short interval of approximately 30 seconds, the two unrolled but were still firmly attached to one another via their respective penes. They then became quiescent against the glass wall of the container with ventral surfaces down and heads pointed away from one another; this position was maintained for quite prolonged periods, usually about 50 minutes. "The animals then quickly separate, each gives a comical little shake, as if settling its viscera into place, and proceeds about its business" (Hyman, 1937; p. 324). Histologically, the penis of each partner was shown to be interlocked in such a fashion that it directed ejaculated spermatozoa over the anterior edge of the partner's penis and thence into the bursa. The penes themselves were not inserted into the bursa, and Hyman noted that, in point of fact, the female genital pore is too small to receive the penis of the partner. Costello and Costello (1938), studying copulation in *Polychoerus carmelensis*, found, in contrast to Hyman's observations, that spermatozoa were deposited directly by the penis of each partner into the dorsal portion of the bursa of its mate. They likewise observed a form of courtship behavior in which prospective copulants exchanged head contacts, then faced each other with their head ends elevated and their ventral surfaces in contact. The pair then swayed slowly from side to side in opposite directions; one partner would then bend its anterior half down under its partner, the ventral surfaces of the two animals being in contact. Eventually the two formed a ball and remained in contact with one another for only 40-50 seconds, after which one partner would move away; the entire process, including courtship, required only about 1-1½ minutes, in contrast to the prolonged period described by Hyman. Hanson (1961) reported comparable mating behavior for the acoel *Convoluta*, with the process lasting less than a minute. Pearse and Wharton (1938) gave a brief description of copulation in the polyclad *Stylochus inimicus*, and included a statement as to the sexual prowess of this form (p. 613): "One pair copulated for five hours on one day, separated, and began again the next day. Several individuals copulated more than once; some at least four times."

Kato (1940) also studied mating in a number of polyclads, including *Notoplana*, *Pseudostylochus*, *Stylochus*, and *Prosthiostomum*. In all cases he found that pairing occurred during the day as well as at night; this is in interesting contrast to the observations of Costello and Costello (1938), cited above, who found that *Polychoerus* mated most frequently during the first 1-1½ hours after dawn. Other reports describing mating behavior include those of Surface (1907) for the polyclad *Hoploplana*, Darlington (1959) for the freshwater rhabdocoel *Mesostoma georgianum*, and Giesa (1966) for the alloeocoel *Monocelis*. The general pattern was comparable for all and the duration of the process was usually short.

See, also, the important paper by Apelt (1969).

5.3.7.2 Oviposition

Fertilization of most turbellarian ova occurs internally, and the eggs are then encased in delicate capsules which are embedded in a jelly mass (as for most acoels and polyclads), or in a shelled cocoon. In the latter event there may be one to several ova within a single shell. A few cases are known where ova develop within the body of the parent (e.g., the rhabdocoels *Paravortex* and *Bresslauilla*). Cement glands, usually found in proximity to the gonopore, produce an adhesive material which in many cases attaches the deposited ova to the substratum; some forms have stalked cocoons, others do not. Giesa (1968) has pointed out the correlation between the morphology of turbellarian egg cocoons and the habitat of the animal; where the water is calm there is no attachment to the substratum, and in muddy areas the cocoons are deposited on and attached to solid objects. There is attachment of the cocoon by an elongated elastic stalk to sand grains in beach surf and in even more violent environs, additional protective coats may be present in the capsule. In a 1966 paper the same author reported the occasional occurrence of encysted specimens of the alloeocoel *Monocelis*; these were found to be rolled up and encased in a sticky slime case, presumably in response to adverse environmental conditions. He found the temperature requirements for egg-deposition and development of the young of *Monocelis* to be rather narrow, 19-21°C.

Surface (1907) reported that the egg capsules of the polyclad *Hoploplana* were remarkably insensitive to temperature, since they developed normally even in direct sunlight. By contrast, the adults were very difficult to maintain in the laboratory. Surface observed that his animals always laid soon after collection and surmised that seawater might be the stimulus bringing this on. If such were the case in nature it would mean that the adults would have to leave their whelk-host, deposit ova, and then either return to the original host or find a new one; this seems a risky situation.

Pearse and Wharton (1938) made an attempt to ascertain the number of ova laid down by a single specimen of the polyclad *Stylochus inimicus* which preys on oysters. Even if the worms were not fed (contradicting the general observation made above concerning the importance of nutrition), they laid totals of 7000-21,000 eggs per individual during one month. Brooding behavior was also reported, a single individual placing its body over several egg masses.

Kanneworff and Christensen (1966) reported that a very long tubiform egg cocoon was produced by *Kronborgia caridicola;* this was dextrally coiled, 46-71 cm long, with closely packed egg capsules within. One end (which presumably had been anchored in the mud substratum) was irregularly thickened and twisted, while the other had a funnel-shaped opening with a diameter of about 1 mm, through which the hatched young emerged.

There is a fair unanimity of opinion recorded in the literature concerning the time of day at which oviposition takes place, in those forms for which the process has been studied in the laboratory. Haswell (1905), Hyman (1937), Costello and Costello (1939), and Hanson (1961), all studying acoels, found that the process took place at night and the same finding was reported by Kato (1940) for a number of Japanese polyclads. In contrast, Giesa (1966) observed that for the alloeocoel *Monocelis* deposition of ova did not appear to be related to any orderly daily periodicity. He also reported that deposition of one egg-capsule seemed to cause other animals to congregate in the vicinity, and, in the ensuing burst of egg-laying activity, to deposit theirs in the same place.

The process of egg-deposition varies with the group. In those forms lacking an oviduct (notably some of the acoels) von Graff (1911), Peebles (1915), and Bresslau (1933) concluded that fertilized ova must be deposited either via the mouth or through a rupture of the body wall. Hyman (1937) concluded that ova of the acoel *Amphiscolops* were extruded through the mouth, although she did not actually observe the process. Costello and Costello (1939), in a careful study of the process in the acoel *Polychoerus*, using both living and sectioned material, found no evidence whatever that eggs were extruded through the mouth; they were able to demonstrate clearly in their sectioned animals that the ova were extruded through an extensive rupture of the body wall. They point out that because of the marked regenerative powers of this form, such a seemingly drastic mode of oviposition is actually a natural and direct method. In turbellarians possessing an oviduct, oviposition is usually accomplished via that structure and the gonopore.

At one time it was thought that the so-called shell glands were the source of the shell around those ova having one, but more recent work indicates that the shell instead is often derived from surface granules on the egg itself. Kato (1940) showed that in oocytes of polyclads, intensely eosinophilic granules appeared in the cytoplasm at about the time when the large germinal vesicle was beginning to disappear. These coalesced when the egg was deposited in seawater and the continuum took up water, separated from the surface of the oocyte, and became the shell. Giesa (1966), however, found that the shell glands of the alloeocoel *Monocelis* did participate in laying down the capsule around the egg, together with contributions from the yolk cells. Hyman (1951) stated that most of the so-called shell glands were in reality the source of cementing substances for attachment of the egg after deposition.

5.3.7.3 Breeding Seasons

When there are marked seasonal differences (as in the temperate zones either north or south of the equator), it is commonly noted that turbellarians reach the height of reproductive activity during the summer months, as noted by Haswell (1905) for the acoel *Amphiscolops*, Costello and Costello (1939) for the acoel *Polychoerus carmelensis*, Costello *et al.* (1957) for the acoel *Polychoerus caudatus* and the polyclad *Hoploplana*, and Kanneworff and Christensen (1966) for the rhabdocoel *Kronborgia caridicola*. Hallez (1909), in contrast, stated that *Paravortex*, another rhabdocoel, was in reproductive condition throughout the year, and Darlington (1959) reported that the freshwater rhabdocoel *Mesostoma georgianum* produced thick-shelled dormant eggs from November through April, with a production peak occurring in January and February. He was unable to state whether these embryos hatched the same season they were produced, or whether a period of dormancy was required as is the case for related species. Kato (1940) reported that the Japanese species of polyclads he studied all appeared to have rather extended breeding seasons, usually from early spring to late summer. The breeding season for *Coelogynopora* (an alloeocoel) was stated by den Hartog (1964) to be at its height from January to April.

Hyman (1951) drew some generalizations on this subject, and suggested that after attaining sexual maturity most acoels probably remain permanently in a sexual state and breed throughout the year in warmer waters, seasonally in colder ones. Some rhabdocoels likewise probably continue to reproduce throughout their lives after attaining sexual maturity, but others (principally freshwater forms) have definite cycles related to habitat. This is usually associated with a habitat which is more or less temporary, as in the various species of *Mesostoma*. Animals hatch in the spring from over-wintering eggs, become sexually mature, and produce thick-walled dormant eggs which can survive drying and/or freezing. Most polyclads breed in the summer. M. B. Thomas (unpublished data) has observed that although ova of the polyclad *Stylochus* (from Woods Hole) are found only during the summer months, spermatozoa can be obtained all the year around.

5.4 Development

5.4.1 Embryonic Development

The embryology of turbellarian eggs has not received as much attention as has been devoted to those forms such as the echinoderms which are more amenable to study in the living condition and to experimental manipulation. This neglect has been due in part to the complications posed by an extensive mass of yolk (particularly in the ectolecithal forms) and by the frequent presence of an impervious and often opaque capsule. Those studies which have been done, however, open up some fascinating vistas for future work, particularly in the utilization of electron microscopy; thus far this tool has had almost no application.

One of the more interesting and unusual features of development in the group is the frequent incidence of gonomery. This is a condition where the maternal and paternal chromosomes do not mix but remain in two separate groups in the zygote and its products. Commonly, the haploid sets of chromosomes contributed by the gametes do not form typical pronuclei, each with a nuclear envelope. Instead, each chromosome forms a separate chromosome vesicle, or karyomere, with its own envelope. Each "pronucleus" is thus a morula of chromosome vesicles. When the first cleavage spindle is formed, the chromosomes condensing from the vesicles become arranged on the spindle in separate groups. This condition may persist through several cleavage cycles, the best known example being that described by Conklin (1901) for the mollusc *Crepidula*. Its occurrence in developing flatworm embryos is unrelated to systematic position. Recently, Costello (1970) has reported a special case of gonomery in the fertilized ovum of the acoel *Polychoerus carmelensis*. In an exceptionally well oriented resting first-cleavage metaphase in a sectioned egg, all 34 (2n) chromosomes were present in a single 8 μm section, and it was possible to identify the homologous chromosomes derived from the two parent gametes. They occurred in matching groups, in an arrangement which could be most simply explained only if the 17 chromosomes contributed by the ovum had been in precisely the same linear order as those contributed by the spermatozoon.

5.4.1.1 Fertilization, Zygotes

In the development of many turbellarians sperm entry into the egg occurs within the worm's body before completion of the maturation divisions. The spermatozoon remains coiled dormant in the cytoplasm until after completion of the female meiotic divisions. Association of maternal and paternal genetic material is often followed by an advance to the metaphase of the first cleavage, at which point mitosis is arrested until deposition of the zygotes has occurred (Costello, 1961). The reason for this phenomenon of mitotic arrest is not known. In the case of those forms having a capsule or cocoon surrounding the ovum, the spermatozoon is incorporated within the capsule before its deposition is completed (Ball, 1916; Papi, 1953; Giesa, 1966). Ball (1916) described and illustrated an extreme example of this for the viviparous rhabdocoel *Paravortex;* a large mass of supernumerary spermatozoa became enclosed within a vacuole in the cytoplasm of an oocyte, even though it had already been fertilized by another spermatozoon. In the more usual (and presumably normal) condition prevailing for this form the spermatozoon was found to enter the vegetal pole of the egg of *Paravortex* by Hallez (1909), Patterson (1912), and Ball (1916); Ball stated that at the time of the second maturation division of the oocyte, the spermatozoon lost its tail, became enlarged and lobulated forming karyomeres, and moved toward the center of the oocyte. After the second polar body was given off, the female pronucleus likewise became lobulated. Neither Hallez (1909) nor Ball (1916) ever observed actual fusion of the pronuclei in zygotes of this form; Patterson stated that copulation of the pronuclei occurred near the animal pole and showed a rather vague sketch illustrating two pronuclei in close association. Kato (1940), studying the development of polyclad eggs, noted that the nucleus of the spermatozoon remained rodlike in the cytoplasm of the oocyte until after completion of the first maturation division and the beginning of the second, at which time it enlarged rapidly and a sperm aster appeared. His account of subsequent events is somewhat uninformative (p. 547): "The sperm nucleus gradually approaches the egg nucleus to come into contact with it; after a while the first segmentation of egg occurs" (p. 547). Giesa (1966) reported that the threadlike nucleus of the spermatozoon of the alloeocoel *Monocelis* did not change its form while in the cytoplasm of the immature oocyte; completion of the meiotic divisions resulted in the formation of three female chromosome vesicles in the condition of gonomery referred to above. The paternal complement likewise formed karyomeres, and gonomery was observed to persist until the two-cell stage.

5.4.1.2 Cleavage, Blastulation, Gastrulation, Organogenesis

Cleavage is usually distorted and difficult to follow in those ectolecithal eggs having large amounts of yolk (eggs of nearly all turbellarians except acoels and polyclads). It has been studied in some detail, however, in the entolecithal eggs of acoels and polyclads, and there is found to be a modified form of spiral cleavage (probably better described as "oblique"; see Costello, 1955). This orderly pattern of cell division is characteristic also of annelidan and molluscan eggs. There is one notable difference, however; spiral cleavage in the higher forms is by quartets, both the first two divisions of the embryo being meridional and giving rise to four large macromeres, from which micromeres are subsequently budded off. In the embryos of acoels, by contrast, only the first cleavage is meridional, so that only two macromeres are produced; in these forms, spiral cleavage occurs by duets (Fig. 33). Most polyclad eggs exhibit the more usual spiral cleavage by quartets, but some divide by duets and, conversely, there are a few acoels (notably, *Convoluta borealis*) which occasionally cleave spirally by quartets (Beklemishev, 1969). There has been much speculation as to the possible phylogenetic significance of this type of cell division. Steinböck (1963a), who favors the view that acoels are directly evolved from ciliates, discounts the significance of spiral cleavage and asserts that it is merely a device for the best possible use of limited space within a hard-shelled egg;

however, the pattern is retained even in forms where the blastomeres do not lie against a shell wall. Ax (1963, p. 206), on the other hand, maintains that spiral cleavage, as seen in embryos of turbellarians, "must represent a highly specific state of homology; this forces the acceptance of a natural relationship between the Turbellaria and the other 'Spiralia.' "

Closely associated with spiral cleavage in embryos of annelids and molluscs are cell lineage studies which trace the developmental course of individual blastomeres to their respective contributions to the definitive larva and/or adult. Such studies have been carried out for a number of spirally cleaving eggs of turbellarians, and the results are in general harmony with the findings for other "determinate" or "mosaic" eggs also exhibiting spiral cleavage. It is important to bear in mind the fact that these terms are purely relative, as Costello (1955) pointed out. The blastomeres of turbellarian embryos have a general tendency to be rather widely separated from one another, and this renders difficult the study of cell lineage in many forms (see Ball, 1916; Kato, 1968).

The relatively uncomplicated segmentation pattern in the developing embryo of an acoel, *Polychoerus carmelensis,* has been described by Costello (1948, 1961); Gardiner (1895) likewise studied the development of the related species, *P. caudatus.* At first cleavage a meridional division separates the zygote into two cells of approximately equal size (Fig. 33). The second division cuts off two smaller cells (micromeres 1a and 1b), nearer the animal pole, from two macromeres (1A and 1B). The two spindles involved in this cleavage are oblique, with respect to the polar axis; the upper ends lie to the left of the lower ends, to an observer at the animal pole viewing the division figure in question. This cleavage is therefore designated as leiotropic, left-handed, or counterclockwise. The first duet micromeres (1a and 1b) next divide by a dexiotropic (right-handed or clockwise) division, producing $1a^1$ and $1a^2$, and $1b^1$ and $1b^2$. At about the same time the first generation macromeres (1A and 1B) divide again, this time almost bilaterally but somewhat dexiotropically, to produce a pair of second quartet micromeres (2a and 2b) and a pair of second generation macromeres (2A and 2B). This regular alternation of counterclockwise with clockwise cleavages is typical of spiral cleavage.

In spiral cleavage, the even-numbered division cycles are almost invariably leiotropic, the odd-numbered divisions dexiotropic. The only exceptions occur in certain snails with reversed symmetry; here the leiotropic and dexiotropic sequence is reversed, also. The generation number of the blastomeres (both micromeres and macromeres)

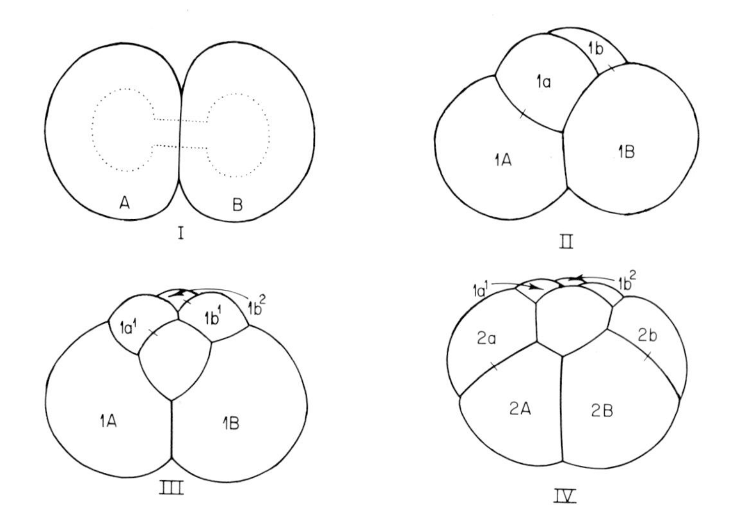

FIG. 33. Diagram of spiral cleavage by duets, as observed in ova of the acoel *Polychoerus.* (I) 2-cell stage; (II) 4-cell stage; (III) 6-cell stage; (IV) 8-cell stage. For further explanation, see text. [From Costello (1961), with permission of the author and the Managing Editor, The Biological Bulletin.]

is given by the prefix; the exponents designate the micromere products, with the lower numeral assigned to that daughter cell lying nearer the animal pole.

At the 8-cell stage, there are, then, two second-generation macromeres (2A and 2B), two second duet micromeres (2a and 2b) and four products of the first duet micromeres ($1a^1$, $1b^1$, $1a^2$, $1b^2$). A third duet of micromeres is then given off by a division which is almost bilateral, with only a slight tendency towards being counterclockwise or leiotropic. The spiral (oblique) nature of further cleavages is not clearly apparent. However, in forms which show the quartet type of spiral cleavage there is, sooner or later, a transition to bilateral cleavages. This transition foreshadows the appearance of bilaterally symmetrical larvae, or of certain bilaterally arranged structures.

The eventual fates of the cells of acoel embryos are as follows (Bresslau, 1933; Hyman, 1951). The first duet of micromeres will give rise to ectodermal structures including the nervous system; the second, and possibly also the third, duet will form ectomesoderm (epidermis and peripheral parenchyma). Part of the interior mass of the embryo will be formed by progeny of the fourth duet of micromeres. The fourth-generation macromeres and their descendants also contribute to this interior mass which, in the gastrula stage, is not clearly separable into mesoderm proper and mesendoderm.

Apelt (1969) reported extensive comparative studies on the cleavage patterns of a number of acoels. His results confirmed the general occurrence of spiral cleavage by duets in this group.

Recently, Boyer (1971) described experiments in which various blastomeres, singly and in combination, were "deleted" from the developing embryo of the acoel *Childia*. She interpreted her results as indicating that these spirally cleaving eggs are capable of considerable regulation, unlike most "mosaic" forms. It is not clear, however, what role her operative procedures may have played since she merely cytolyzed the blastomeres by puncturing them and left them in contact with the other cells of the embryo. There is thus no way of knowing what diffusible substances may have been present.

Cleavage in embryos of rhabdocoels and alloeocoels is obscured by the presence of yolk cells and is difficult to follow and to characterize. Some observers have reported indications of spiral cleavage (Papi, 1953; Giesa, 1966), while others find no clear evidence for this type of cell division. In addition to the papers cited, by Papi and by Giesa, there are several earlier papers (for references see Hyman, 1951), and the report by Seilern-Aspang (1957) for the rhabdocoel *Macrostomum*. The pattern of cell division in triclads is likewise difficult to follow; for a general account and further references, see Hyman (1951).

Nearly all polyclads exhibit a well-defined and typical spiral cleavage by quartets, with four macromeres being formed by two meridional divisions. An early study was that of Surface (1907) and, more recently, Kato (1940, 1968) has studied cleavage and later development in a number of Japanese forms. Sometimes there are notable size differences among the first four cleavage products, and in other cases such differences are not apparent. Similarly, there are considerable degrees of variation in the synchrony with which blastomeres divide; Kato (1940) found, for example, that only rarely were there marked size differences among the blastomeres, and that at the third cleavage the D macromere divided first, followed by B, C, and A in that order. Surface (1907) reported that in embryos of *Hoploplana*, the D macromere was slightly larger, and that the sequence in asynchronous division was D, B, A, and C. Characteristically, at the 32- to 36-cell stage, the third quartet macromeres (3A-3D) divide dexiotropically to produce four very small and abortive cells at the vegetal pole, 4A-4D. Although these are much smaller than their sister cells, they are called macromeres because of their position at the vegetal pole. Eventually they become incorporated within the endoderm, but play no other role in the development of the embryo.

The first quartet of micromeres will give rise to the anterior dorsal ectoderm, to the pigment of the eyes, and to the ganglia of the nervous system. In some forms, there is an invagination of these cells to form the nervous system, together with the differentiation of frontal glands to give rise to long apical sensory "hairs" which are temporary. They have been compared with similar apical sensory cilia found in larvae of annelids and molluscs (Hyman, 1951). The second quartet of micromeres also contributes to the surface epidermis and some of their progeny pass into the interior of the embryo as ectomesoderm which will form muscular and mesenchymal components of the pharynx. The third quartet micromeres have a subsequent history similar to that of the second. Three (4a-4c) of the larger sister cells constituting the fourth quartet of micromeres are nutritive and do not divide further; the fourth cell (4d) will eventually give rise to mesoderm and endoderm and is therefore entirely comparable to the mesentoblast cell of the molluscan embryo (Kato, 1940). It divides by a bilateral cleavage into $4d^1$ and $4d^2$, and the products sink into the interior of the embryo. Progeny of $4d^1$ give rise to the intestine, those of $4d^2$ to mesenchyme, glands, and the reproductive system.

A true blastula stage, with recognizable blastocoele, is not found in the development of all turbellarian embryos. Giesa (1966) observed that the coeloblastula of the alloeocoel *Monocelis* did have an extensive blastocoele and was definitely polarized with smaller blastomeres at the animal pole and larger ones at the vegetal.

Gastrulation is typically by a process of epiboly or overgrowth (found also in the development of many annelids and molluscs). The events leading to formation of a two-layered embryo were described by Giesa (1966) as involving five quite clearly delimited steps, in the case of the alloeocoel *Monocelis.* Newton (1970b) has given a detailed account of gastrulation in the embryos of another alloeocoel, the freshwater form *Hydrolimax*.

Organogenesis has been studied in most detail among polyclad embryos (see, for example, the papers by Surface, 1907; Kato, 1940, 1968), although some information is available for the other groups as well. The processes appear for the most part to differ only in details. Ball (1916) described the formation of the pharynx in embryos of the viviparous rhabdocoel *Paravortex;* a cord of cells grows out from the anterior end and acquires a lumen shortly before the embryo leaves its parent; the external layers of musculature are laid down by other cells which are the progeny of the second and third generations of micromeres. Similarly, the intestine of this form arises from a loose cord of cells (derived from the 4d micromere) proliferated just posterior to the pharynx; it acquires a lumen initially at the anterior end and only later in the posterior region. The mouth is usually formed by an invagination of ectoderm near the site of the closed blastopore and frequently shifts its position posteriad during axial changes as the embryo grows. Kato (1968) described an essentially similar series of events leading to formation of the alimentary tract in polyclads as well.

According to Kato (1940) the brain of polyclad embryos arises from a single invagination of the ectoderm at the animal pole at a stage when the embryo has developed two eyespots. The invaginated cells of the brain Anlage were described as forming a small compact group, readily distinguishable from neighboring cells by their larger size. As a consequence of mitosis, their number increases and eventually the brain is a bilobed mass lying deep within the mesenchyme in the midline. Ball (1916) described the origin of posterior nerves from the corners of the main brain mass in embryos of the rhabdocoel *Paravortex;* fibers formed which pushed backward through the parenchyma. A similar origin for the optic nerves is probable, although Ball suggested the possibility that they might arise as fibers going back to the brain from two cells near the eyes. In the alloeocoel *Monocelis* Giesa (1966) found that the brain originated from paired masses of small cells, between which a commissure developed quite early.

A characteristic structure present in the larvae of many polyclads is the frontal organ which disappears one or two days after hatching. It arises from a portion of the mass of invaginated cells giving rise also to the brain and in some forms is provided with long cilia presumed to be sensory in function (Kato, 1940). The electron microscopy of this organ is of obvious interest, particularly with reference to the pattern of microtubules in the cilia; it would be interesting to know whether a sensory cilium in primitive forms such as these would have the 9 + 0 pattern characteristic of sensory structures in higher forms, or the 9 + 2 arrangement found in most nonspecialized cilia and flagella, and (as noted in Section 5.3.4.2) in the cilia of all turbellarians we have studied thus far.

5.4.1.3 Regeneration

The capacity to replace missing parts varies considerably among the various Turbellaria, according to Hyman (1951) and Steinböck (1963b). The most extensive studies of the process have utilized freshwater forms (chiefly *Dugesia*); for a general review of this topic, see Child (1941) and Brøndsted (1969). Some studies have been made for marine turbellarians, notably those of Stevens and Boring (1905), Child (1907), Peebles (1913), Keil (1929), and Steinböck (1963b) for acoels; Lloyd (1914a, b), and Steinmann (1908) for marine triclads; and Child (1904) for polyclads. In general, all these findings implicate the cerebral ganglia as being essential for regeneration, and in most cases the process is not complete.

The fact that regeneration can occur at all in acoels is remarkable because somatic mitosis is apparently very rare, if not nonexistent in members of this group. Hanson (1961) illustrates rather obscure light micrographs of what he interpreted to be mitotic figures in the parenchyma of young worms (the acoel *Convoluta sutcliffei*) and regenerating adults, but to my knowledge there are almost no other reports in the literature of division in somatic cells of acoels. However, Dorey (1965) states that mitosis does occur in *Convoluta roscoffensis*, but gives no further details or illustrations. Regeneration appears to take place by what Morgan (1901) designated morphallaxis, or a reworking of existing materials and migrating cells at the site of injury to replace the missing parts.

Dorey (1965) suggested that the pulsatile bodies (Ersatzzellen of Luther, 1912) found in some acoels are actually replacement epidermal cells, utilized after injury to the epithelium. He admitted the possibility that these peculiar structures, consisting of tightly packed cilia within what he described as a vacuole, might instead be the products of degeneration of epithelium but felt this to be an unlikely explanation. We have found structures in the parenchyma and near the surface of the acoels *Childia* and *Polychoerus* which are very similar to those figured by Dorey, but we tend to the belief that these are products of degeneration and not replacement cells. Another possibility is that they may be portions of the surfaces of other worms which have been lost as a consequence of the abrasive action of sand or other materials and taken in as food.

Steinböck (1963b) studied the regenerative capacity of the acoel *Amphiscolops*, and found that even after very drastic cutting experiments, complete or nearly complete regeneration by morphallaxis occurred in as short a time as approximately 30 hours. The same author (1967) rejected the idea that Ersatzzellen were involved in the regeneration of *Hofstenia* (although he concedes that such cells are at least present in some acoels, notably *Childia* and *Otocelis*, among others). In support of his view, cited in Section 5.3.8.2, that a close relationship exists between ciliates and acoels, he stated that "The regeneration speed is in *H. giselae* . . . much greater than in

Planarians, and absolutely comparable with that of the Ciliata. One can say that what the Planarians achieve in days the Acoela do in hours" (p. 453).

5.4.2 Larvae

Most turbellarian flatworms have direct development, the juveniles hatching as miniature adults; only among some of the polyclads (with the exception of all cotyleans, according to Kato, 1940) do free-swimming larvae occur. Kato (1940) has suggested that hatching in those forms having direct development is probably effected by secretions from the frontal organs. The free-swimming juveniles are often strongly positively phototropic and have a tendency to crawl out of the water on the lighted side of a culture vessel, so that if one wishes to rear them to maturity, dishes should be kept in the dark.

Two principal types of free-swimming larvae occur in the indirect development of polyclads, Götte's larva (having four ciliated processes) and Müller's larva with eight ciliated processes (Kato, 1940, 1968). Lang (1884) considered the former to be a transitional stage en route to the Müller's larva, but Kato (1940) pointed out that in his extensive researches he had never observed the necessary multiplication in the number of ciliated processes after hatching, from 4 to 8, which would confirm this. Both types have rather long anterior and posterior sensory tufts, one or more eyespots, a brain, frontal organ, and heavy ciliation. They somewhat resemble the trochophore larvae of annelids and molluscs, but are more like the pilidium and Desor's larva of the rhynchocoels in having only one ciliated band and no anus (Kato, 1968). At metamorphosis, the processes degenerate and are resorbed almost simultaneously; the larva is no longer a rounded triangle in shape, but becomes oval or spherical, and flattens. The sensory tufts at anterior and posterior ends are lost (Kato, 1940) and formation of the definitive adult pharynx takes place (Kato, 1968).

An intracapsular Müller's larva has been described by Kato (1940) in the development of the polyclad *Planocera reticulata;* this was said to differ from the free-swimming Müller's larva only in details and hatched as a juvenile. Intracapsular cannibalism was observed to occur in cases where 6 or 7 embryos occupied the same capsule and one or more died. This was sometimes carried to the extreme where a single giant larva was produced, the defunct siblings having contributed to its nutrition. During metamorphosis of the intracapsular Müller's larva, the body became elongated, eyespots appeared, the ciliated processes were resorbed, and the digestive system had assumed almost its adult form. Upon hatching the miniature worm sank to the bottom and crept about on the substratum, never becoming free-swimming. Further development involved merely growth and some minor remodelling.

There appears to be no available information concerning the factors affecting metamorphosis of turbellarians having indirect development.

Acknowledgments

Aided by a grant from the National Institutes of Health, GM 15311. I am deeply indebted to Dr. Donald P. Costello and Dr. Mary Beth Thomas for much valuable advice and assistance, for aid in preparation of the illustrations, and for permission to use some of their micrographs (Dr. Costello for Fig. 3, 4, 6, 7, 8, 9, 22 and 31; Dr. Thomas for Fig. 19 and 24).

5.5 References

Apelt, G. (1969). Fortpflanzungsbiologie, Entwicklungszyklen und vergleichende Früentwicklung acoeler Turbellarien. *Mar. Biol.* **4**, 267-325.

Ax, P. (1954). Die Turbellarienfauna des Küstengrundwassers am Finnischen Meerbusen. *Acta Zool. Fennica* #81, 1-54.

Ax, P. (1957). Verfielfachung des männlichen Kopulationsapparates bei Turbellarien. *Verh. Deut. Zool. Ges. Graz.* 227-249.

Ax, P. (1963). Relationships and phylogeny of the Turbellaria. *In* "The Lower Metazoa" (E. C. Dougherty, ed.), pp. 191-223. Univ. of California Press, Berkeley, California.

Ax, P. (1966). Das choroide Gewebe als histologisches Lebensformmerkmal der Sandlückenfauna des Meeres. *Naturwiss. Rundsch.* **19**, 282-289.

Ax, P., and Apelt, G. (1965). Die "Zooxanthellen" von *Convoluta convoluta* (Turbellaria Acoela) entstehen aus Diatomeen. *Naturwissenschaften* **15**, 444-446.

Ax, P., and Ax, R. (1967). Turbellaria Proseriata von der Pazifikküste der USA (Washington). I. Otoplanidae. *Z. Morph. Tiere* **61**, 215-254.

Ax, P., and Dörjes, J. (1966). *Oligochoerus limnophilus* nov. spec., ein kaspisches Faunenelement als erster Süsswasservertreter der Turbellaria Acoela in Flüssen Mitteleuropas. *Int. Rev. Ges. Hydrobiol.* **51**, 15-44.

Ax, P., and Schulz, E. (1959). Ungeschlechtliche Fortpflanzung durch Paratomie bei acoelen Turbellarien. *Biol. Zentralbl.* **78**, 615-622.

Ball, S. J. (1916). Development of *Paravortex gemellipara*. *J. Morphol.* **27**, 453-558.

Bashiruddin, M., and Karling, T. G. (1970). A new entocommensal Turbellarian (Fam. Pterastericolidae) from the sea star *Astropecten irregularis*. *Z. Morphol. Tiere* **67**, 16-28.

de Beauchamp, P. (1961). Classe des turbellariés. *Traité Zool.* **4** (1), 35-212.

Beklemishev, W. N. (1969). "Principles of Comparative Anatomy of Invertebrates" (Translated by J. M. MacLennan; Z. Kabata, ed.), Vol. 2, pp. 393-409. Univ. of Chicago Press, Chicago, Illinois.

Borkott, H. (1970). Geschlechtliche Organisation, Fortpflanzungsverhalten und Ursachen der sexuellen Vermehrung von *Stenostomum sthenum* nov. sp. (Turbellaria, Catenulida), mit Beschreibung von 3 neuen *Stenostomum*-Arten. *Z. Morphol. Tiere* **67**, 183-262.

Boyer, B. C. (1971). Regulative development in a spiralian embryo as shown by cell deletion experiments on the acoel, *Childia. J. Exp. Zool.* **176**, 97-105.

Brenner, S., and Horne, R. W. (1959). A negative staining method for high resolution electron microscopy of viruses. *Biochim. Biophys. Acta* **34**, 103-110.

Bresslau, E. (1933). Erste Klasse des Cladus Plathelminthes. *In* "Handbuch der Zoologie" (W. Kükenthal and T. Krumbach, eds.), vol. 2 (1), pp. 52-320. Gruyter, Berlin.

Brøndsted, H. V. (1969). "Planarian Regeneration," pp. 1-276. Pergamon, Oxford.

Burton, P. R. (1966). Substructure of certain cytoplasmic microtubules: An electron microscope study. *Science* **154**, 903-905.

Burton, P. R. (1967). Fine structure of the unique central region of the axial unit of lung-fluke spermatozoa. *J. Ultrastruct. Res.* **19**, 166-172.

Burton, P. R. (1968). Effects of various treatments on microtubules and axial units of lung-fluke spermatozoa. *Z. Zellforsch.* **87**, 226-248.

Bush, L. (1964). Phylum Platyhelminthes, Class Turbellaria. *In* "Key to Marine Invertebrates of the Woods Hole Region" (R. I. Smith, ed.), pp. 30-39. Mar. Biolog. Lab., Woods Hole, Mass.

Cernosvitov, L. (1932). Studien über die Spermaresorption. IV. Verbreitung der Samenresorption bei den Turbellarien. *Zool. Jahrb., Abt. Anat.* **55**, 137-172.

Child, C. M. (1904). Studies on regeneration. *J. Exp. Zool.* **1**, 95-133, 463-512, 513-557.

Child, C. M. (1907). The localization of different methods of form-regulation in *Polychoerus caudatus. Arch. Entw.* **23**, 227-248.

Child, C. M. (1910a). Physiological isolation of parts and fission in Planaria. *Arch. Entw.* **30**, 159-205.

Child, C. M. (1910b). The central nervous system as a factor in the regeneration of polyclad Turbellaria. *Biol. Bull.* **19**, 333-338.

Child, C. M. (1941). "Patterns and Problems of Development, " 811 pp. Univ. of Chicago Press, Chicago, Illinois.

Christensen, A. K. (1961). Fine structure of an unusual spermatozoan in the flatworm *Plagiostomum. Biol. Bull.* **121**, 416.

Christensen, A. M., and Kanneworff, B. (1965). Life history and biology of *Kronborgia amphipodicola* Christensen and Kanneworff (Turbellaria, Neorhabdocoela). *Ophelia* **2**, 237-251.

Conklin, E. G. (1901). The individuality of the germ nuclei during the cleavage of the egg of *Crepidula. Biol. Bull.* **2**, 257-265.

Costello, D. P. (1948). Spiral cleavage. *Biol. Bull.* **95**, 265.

Costello, D. P. (1955). Cleavage, blastulation and gastrulation. *In* "Analysis of Development" (B. H. Willier, P. A. Weiss, and V. Hamburger, eds.), pp. 213-229. Saunders, Philadelphia, Pennsylvania.

Costello, D. P. (1961). On the orientation of centrioles in dividing cells and its significance: A new contribution to spindle mechanics. *Biol. Bull.* **120**, 285-312.

Costello, D. P. (1970). Identical linear order of chromosomes in both gametes of the acoel turbellarian *Polychoerus carmelensis:* A preliminary note. *Proc. Nat. Acad. Sci. USA* **67**, 1951-1958.

Costello, D. P., and Costello, H. M. (1968). Immotility and motility of acoel turbellarian spermatozoa, with special reference to *Polychoerus carmelensis*. *Biol. Bull.* **135**, 417.

Costello, D. P., Henley, C., and Ault, C. R. (1969). Microtubules in spermatozoa of *Childia* (Turbellaria Acoela) revealed by negative staining. *Science* **163**, 678-679.

Costello, D. P., Davidson, M. E., Eggers, A., Fox, M. H., and Henley, C. (1957)."Methods for Obtaining and Handling Marine Eggs and Embryos," pp. 40-43. Mar. Biolog. Lab., Woods Hole, Massachusetts.

Costello, H. M., and Costello, D. P. (1938). Copulation in the acoelous turbellarian *Polychoerus carmelensis*. *Biol. Bull.* **75**, 85-98.

Costello, H. M., and Costello, D. P. (1939). Egg laying in the acoelous turbellarian *Polychoerus carmelensis*. *Biol. Bull.* **76**, 80-89.

Curtis, W. C. (1902). The life history, the normal fission and the reproductive organs of *Planaria maculata*. *Proc. Boston Soc. Natur. Hist.* **30**, 515-559.

Dahm, G. (1950). On *Bothrioplana semperi* M. Braun (Turbellaria Alloeocoela Cyclocoela). The taxonomy of the "genera" and of the known "species" of the family Bothrioplanidae. *Ark. Zool.* **1**, 503-510.

Dan, Jean Clark (1967). Acrosome reaction and lysins. *In* "Fertilization" (C. B. Metz and A. Monroy, eds.), Vol. I., pp. 237-295. Academic Press, New York.

Darlington, J. D. (1959). The Turbellaria of two granite outcrops in Georgia. *Amer. Midl. Natur.* **61**, 257-294.

Dorey, A. E. (1965). The organisation and replacement of the epidermis in acoelous turbellarians. *Quart. J. Microsc. Sci.* **106**, 147-172.

Dörjes, J. (1966). *Paratomella unichaeta* nov. gen. nov. spec., Vertreter einer neuen Familie der Turbellaria Acoela mit asexueller Fortpflanzung durch Paratomie. *Veroff. Inst. Meeresforsch. Bremerhaven* **2**, 187-200.

Du Bois-Reymond Marcus, E. (1957). On Turbellaria. *An. Acad. Brasil. Ciènc.* **29**, 153-191. (cited from Ax and Schulz, 1959).

Fawcett, D. W., and Porter, K. R. (1954). A study of the fine structure of ciliated epithelia. *J. Morphol.* **94**, 221-281.

Gardiner, E. G. (1895). Early development of *Polychoerus caudatus* Mark. *J. Morphol.* **11**, 155-176.

Giesa, S. (1966). Die Embryonalentwicklung von *Monocelis fusca* Oersted (Turbellaria, Proseriata). *Z. Morphol. Tiere* **57**, 137-230.

Giesa, S. (1968). Die Eikapseln der Proseriaten (Turbellaria, Neoophora). *Z. Morphol. Tiere* **61**, 338-346.

von Graff, L. (1911). Acoela, Rhabdocoela und Alloeocoela des Ostens der Vereinigten Staaten von Amerika mit Nachträgen zu den "marinen Turbellarien un der Küsten Europas." *Arb. Zool. Inst. Graz* **9**, 321-428.

Hallez, P. (1909). Biologie, organisation, histologie et embryologie d'un rhabdocoele parasite du *Cardium edule* L. *Paravortex cardii* n. sp. *Arch. Zool. Exp. Gen. Ser. 4* **9**, 429-544.

Hanson, E. D. (1960). Asexual reproduction in acoelous Turbellaria. *Yale J. Biol. Med.* **33**, 107-111.

Hanson, E. D. (1961). *Convoluta sutcliffei*, a new species of acoelous Turbellaria. *Trans. Amer. Microsc. Soc.* **80**, 423-433.

den Hartog, C. (1964). Proseriate flatworms from the deltaic area of the rivers Rhine, Meuse and Scheldt I. *Proc. kon. ned. Akad. Wetensch. Sect. C* **67**, 10-34.

Haswell, W. A. (1905). Studies on the Turbellaria. *Quart. J. Microsc. Sci.* **49**, 425-467.

Haswell, W. A. (1909). The development of the Temnocephaleae. Part I. *Quart. J. Microsc. Sci.* **54**, 415-441.

Hendelberg, J. (1965). On different types of spermatozoa in Polycladida, Turbellaria. *Ark. Zool.* **18**, 267-304.

Hendelberg, J. (1969). On the development of different types of spermatozoa from spermatids with two flagella in the Turbellaria with remarks on the ultrastructure of the flagella. *Zool. Bidrag. Uppsala* **38**, 1-52.

Henley, C. (1968). Refractile bodies in the developing and mature spermatozoa of *Childia groenlandica* (Turbellaria: Acoela) and their possible significance. *Biol. Bull.* **134**, 382-397.

Henley, C., and Costello, D. P. (1969). Microtubules in spermatozoa of some turbellarian flatworms., *Biol. Bull.* **137**, 403.

Henley, C., Costello, D. P., and Ault, C. R. (1968). Microtubules in the axial filament complexes of acoel turbellarian spermatozoa, as revealed by negative staining. *Biol. Bull.* **135**, 422-423.

Henley, C., Costello, D. P., Thomas, M. B., and Newton, W. D. (1969). The "9 + 1" pattern of microtubules in spermatozoa of *Mesostoma* (Platyhelminthes, Turbellaria). *Proc. Nat. Acad. Sci. USA* **64**, 849-856.

Holmquist, C. (1967). *Dendrocoelopsis piriformis* (Turbellaria Tricladida) and its parasites from Northern Alaska. *Arch. Hydrobiol.* **62**, 453-466.

Hyman, L. H. (1937). Reproductive system and copulation in *Amphiscolops langerhansi* (Turbellaria Acoela). *Biol. Bull.* **72**, 319-326.

Hyman, L. H. (1938). North American Rhabdocoela and Alloeocoela. II. Rediscovery of *Hydrolimax grisea* Haldeman. *Amer. Mus. Nov.* **1004**, 1-19.

Hyman, L. H. (1939a). Some polyclads of the New England coast, especially of the Woods Hole region. *Biol. Bull.* **76**, 127-152.

Hyman, L. H. (1939b). North American triclad Turbellaria: IX. The priority of *Dugesia* Girard 1850 over *Euplanaria* Hesse 1897 with notes on American species of *Dugesia*. *Trans. Amer. Microsc. Soc.* **58**, 264-275.

Hyman, L. H. (1944a). A new Hawaiian polyclad flatworm associated with *Teredo*. *Occas. Pap. Bernice P. Bishop Mus., Honolulu* **18**, 73-75.

Hyman, L. H. (1944b). Marine Turbellaria from the Atlantic coast of North America. *Amer. Mus. Nov.* #1266, 1-15.

Hyman, L. H. (1951). "The Invertebrates: Platyhelminthes and Rhynchocoela; the Acoelomate Bilateria," pp. 1-531. McGraw-Hill, New York.

Hyman, L. H. (1955). Miscellaneous marine and terrestrial flatworms from South America. *Amer. Mus. Nov.* #1742, 1-33.

Hyman, L. H. (1959a). "The Invertebrates: Smaller Coelomate Groups," pp. 731-767. McGraw-Hill, New York.

Hyman, L. H. (1959b). A further study of Micronesian polyclad flatworms. *Proc. U. S. Nat. Mus.* **108**, 543-597.

Inoué, S. (1959) Motility of cilia and the mechanism of mitosis. *In* "Biophysical Sciences—A Study Program" (J. L. Oncley, ed.), pp. 402-408. Wiley, New York.

Jennings, J. B. (1968). Feeding, digestion and food storage in two species of temnocephalid flatworms (Turbellaria: Rhabdocoela). *J. Zool. London* **156**, 1-8.

Kanneworff, B., and Christensen, A. M. (1966). *Kronborgia caridicola* sp. nov., an endoparasitic turbellarian from North Atlantic shrimps. *Ophelia* **3**, 65-80.

Karling, T. G. (1940). Zur Morphologie und Systematik der Alloeocoela cumulata und Rhabdocoela lecithophora. *Acta Zool. Fennica* #26, 1-260.

Karling, T. G. (1952). *Cytocystis clitellatus* n. gen., n. sp., ein neuer Eukalyptorhinchien-Typus (Turbellaria). *Ark. Zool.* **4**, 493-504.

Karling, T. G. (1954). Einige marine Vertreter der Kalyptorhynchien-Familie Koinocystididae. *Ark. Zool.* **7**, 165-183.

Karling, T. G. (1962a). On a species of the genus *Multipeniata* Nasanov (Turbellaria) from Burma. *Ark. Zool.* **15**, 105-111.

Karling, T. G. (1962b). Marine Turbellaria from the Pacific Coast of North America. I. Plagiostomidae. *Ark. Zool.* **15,** 113-141.

Karling, T. G. (1962c). Marine Turbellaria from the Pacific Coast of North America. II. Pseudostomidae and Cylindrostomidae. *Ark. Zool.* **15,** 181-209.

Karling, T. G. (1964). Marine Turbellaria from the Pacific Coast of North America. III. Otoplanidae. *Ark. Zool.* **16,** 527-541.

Karling, T. G. (1967). Zur Frage von den systematischen Wert der Kategorien Archoophora und Neoophora (Turbellaria). *Commentati. Biolog. Soc. Scieniarum Fennica* **30** (3), 1-11.

Karling, T. G. (1969). Ein Überzähliger Genitalapparat bei einem rhabdocölen Turbellar. *Z. Morphol. Tiere* **65,** 202-208.

Karling, T. G. (1970). On *Pterastericola fedotovi* (Turbellaria) commensal in sea stars. *Z. Morphol. Tiere* **67,** 29-39.

Kato, K. (1940). On the development of some Japanese polyclads. *Jap. J. Zool.* **8,** 537-573.

Kato, K. (1968). Platyhelminthes. *In* "Invertebrate Embryology" (M. Kumé and K. Dan, eds.; translated by J. C. Dan)., pp. 125-143. NOLIT Publ. House, Belgrade, Yugoslavia.

Keil, E. M. (1929). Regeneration in *Polychoerus caudatus* Mark. *Biol. Bull.* **57,** 225-244.

Klima, J. (1962). Elektronenmikroskopische Studien über die Feinstruktur der Tricladen (Turbellaria). *Protoplasma* **54,** 101-162.

Kozloff, E. N. (1965). New species of acoel Turbellaria from the Pacific Coast. *Biol. Bull.* **129,** 151-166.

Lang, A. (1884). Die polycladen des Golfes von Neapel. *Flora Fauna Golfes von Neapel* **11,** 1-688.

Linton, E. (1910). On a new Rhabdocoele commensal with *Modiolus plicatus. J. Exp. Zool.* **9,** 371-384.

Lloyd, D., (1914a). The influence of the position of the cut upon regeneration in *Gunda ulvae. Proc. Roy. Soc. London Ser. B* **87,** 355-366.

Lloyd, D. (1914b). The influence of osmotic pressure upon regeneration of *Gunda ulvae. Proc. Roy Soc. London Ser. B* **88,** 1-19.

Luther, A., (1912). Studien über acöle Turbellarien aus dem Finnischen Meerbusen. *Acta Soc. Fauna Flora Fennica* **36** (5), 1-59.

Makino, S. (1951). "An Atlas of the Chromosome Numbers in Animals," pp. 5-10. Iowa State College Press, Ames, Iowa.

Manton, I. (1952). The fine structure of plant cilia. *Symp. Soc. Exp. Biol.* **6,** 306-319.

Marcus, E., and Macnae, W. (1954). Architomy in a species of *Convoluta. Nature (London)* **173,** 130.

Mark, E. L. (1892). *Polychoerus caudatus* nov. gen. et nov. spec. Festschr. siebn. Geburtstage Rudolf Leuckarts, pp. 298-309.

Meixner, J. (1938). Turbellaria (Strudelwurmer). *In* "Die Tierwelt der Nord- und Ostsee," Teil IVb, Lief. (G. Grimpe and E. Wagler, eds.), pp. 1-146. Akademische Verlagsgesellschaft, Becker & Erler.

Morgan, T. H. (1901). "Regeneration," pp. 1-316. Columbia Univ. Biol. Ser., #7, New York.

Newton, W. D. (1970a). Oogenesis of the freshwater turbellarian *Hydrolimax grisea* (Platyhelminthes: Plagiostomidae) with special reference to the history of the supernumerary asters and central bodies. *J. Morphol.* **132,** 27-46.

Newton, W. D. (1970b). Gastrulation in the turbellarian *Hydrolimax grisea* (Platyhelminthes: Plagiostomidae): Formation of the epidermal cavity, inversion and epiboly. *Biol. Bull.* **139,** 539-548.

Papi, F. (1953). Beiträge zur Kenntnis der Macrostomiden (Turbellarien). *Acta Zool.*

Fennica **#78,** 1-32.

Pasteels, J. (1950). Mouvements localisés et rythmiques de la membrane de fécondation chez des oeufs fecondés ou activés *(Chaetopterus, Mactra, Nereis) Arch. Biol.* **61,** 197-220.

Patterson, J. T. (1912). Early development of *Graffilla gemellipara*—a supposed case of polyembryony. *Biol. Bull.* **22,** 173-204.

Pearse, A. S., and Wharton, G. W. (1938). The oyster "leech," *Stylochus inimicus* Palombi associated with oysters on the coasts of Florida. *Ecol. Monogr.* **8,** 605-655.

Peebles, F. (1913). Regeneration acöler Plattwürmer. *Bull. Inst. Oceanogr. Monaco* **#263.**

Peebles, F. (1915). A description of three Acoela from the Gulf of Naples. *Mitt. Zool. Stat. Neapel* **22,** 291-311.

Pikó, L. (1969). Gamete structure and sperm entry in mammals. *In* "Fertilization" (C. B. Metz and A. Monroy, eds.), Vol. 2, pp. 325-403. Academic Press, New York.

Reisinger, E. (1940). Die cytologische Grundlage der parthenogenetischen Dioogonie. *Chromosoma* **1,** 531-553.

Repiachoff, W. (1893). Zur Spermatologie der Turbellarien. *Z. wiss. Zool.* **65,** 117-137.

Rosario, B. (1967). An electron microscope study of spermatogenesis in cestodes. *J. Ultrastruct. Res.* **11,** 412-427.

Schilke, K. (1970). Zur Morphologie und Phylogenie der Schizorhynchia (Turbellaria, Kalyptorhynchia). *Z. Morphol. Tiere* **67,** 118-171.

Seilern-Aspang, F. (1957). Die Entwicklung von *Macrostomum appendiculatum* (Fabricius). *Zool. Jahrb. Abt. Anat.* **76,** 311-330.

Shapiro, J. E., Hershenov, B. R. and Tulloch, G. S. (1961). The fine structure of *Haematoloechus* spermatozoon tail. *J. Biophys. Biochem. Cytol.* **9,** 211-217.

Silveira, M. (1967). Formation of structured secretory granules within the Golgi complex in an acoel turbellarian. *J. Microsc.* **6,** 95-100.

Silveira, M. (1969). Ultrastructural studies on a "nine plus one" flagellum. *J. Ultrastruct. Res.* **26,** 274-288.

Silveira M. (1970). Characterization of an unusual nucleus by electron microscopy. *J. Submicrosc. Cytol.* **2,** 13-24.

Silveira, M., and Porter, K. R. (1964). The spermatozoids of flatworms and their microtubular systems. *Protoplasma* **59,** 240-265.

Steinböck, O. (1963a). Origin and affinities of the lower Metazoa: The "aceloid" ancestry of the lower Metazoa. *In* "The Lower Metazoa. Comparative Biology and Phylogeny" (E. C. Dougherty, ed.) pp. 40-54. Univ. of California Press, Berkeley, California.

Steinböck, O. (1963b). Regeneration experiments and phylogeny. *In* "The Lower Metazoa. Comparative Biology and Phylogeny" (E. C. Dougherty, ed.) pp. 108-112. Univ. of California Press, Berkeley, California.

Steinböck, O. (1967). Regenerationsversuche mit *Hofstenia giselae* Steinb. (Turbellaria Acoela). *Arch. Entw.* **158,** 394-458.

Steinmann, P. (1908). Untersuchungen ueber das Verhalten des Verdaungssystems bei der Regeneration der Tricladen. *Arch. Entw.* **25,** 523-568.

Stevens, N. M., and Boring, A. M. 1905. Regeneration in *Polychoerus caudatus. J. Exp. Zool.* **2,** 335-346.

Surface, F. M. (1907). The early development of a polyclad, *Planocera inquilina. Proc. Acad. Nat. Sci. Philadelphia* **59,** 514-559.

Thomas, M. B. (1970). Transitions between helical and protofibrillar configurations in doublet and singlet microtubules in spermatozoa of *Stylochus zebra* (Turbellaria, Polycladida). *Biol. Bull.* **138,** 219-234.

Thomas, M. B., and Henley, C. (1971). Substructure of the cortical singlet microtubules in spermatozoa of *Macrostomum* (Platyhelminthes, Turbellaria) as revealed by negative staining. *Biol. Bull.* **141**, 592-601.

Tulloch, G. S., and Hershenov, B. R. (1967). Fine structure of platyhelminth sperm tails. *Nature (London)* **213**, 299-300.

Wager, A. (1913). Some observations on *Convoluta. Rep. 10th Ann. Meeting South African Ass. Advan. Sci.*

Westblad, E. (1945). Studien über skandinavische Turbellaria Acoela. III. *Ark. Zool.* **36A**, 1-56.

Westblad, E. (1949a). On *Meara stichopi* (Bock) Westblad, a new representative of Turbellaria Archoophora. *Ark. Zool.* **1**, 43-57.

Westblad, E. (1949b). *Xenoturbella bocki* n. g., n. sp. a peculiar primitive turbellarian type. *Ark. Zool.* **3**, 11-29.

Westblad, E. (1952). Marine Macrostomida (Turbellaria) from Scandinavia and England. *Ark. Zool.* **4**, 391-408.

Westblad, E. (1953). New Turbellaria parasites in echinoderms. *Ark. Zool.* **5**, 269-288.

Westblad, E. (1954). Marine "alloeocoels" (Turbellaria) from North Atlantic and Mediterranean coasts I. *Ark. Zool.* **7**, 491-526.

Chapter 6

GNATHOSTOMULIDA

Wolfgang Sterrer

6.1 Introduction

Gnathostomulida, microscopic marine worms, were first described in 1956 by Ax and since then they have been given the rank of a class or even a phylum. This distinction from the Turbellaria (where they were first placed) is mainly due to two characters: a monociliated epidermis (each cell carries only one cilium), and cuticularized mouth parts in a highly specialized pharynx. By now Gnathostomulida are known from all over the world with more than 80 species and 18 genera. A first approach to their systematics has recently appeared (Sterrer, 1972). It distinguishes the two orders Filospermoidea (with filiform sperm, e.g., *Haplognathia*,Fig. 1A) and Bursovaginoidea, the latter being divided into the suborders Scleroperalia (with cuticularized bursa, e.g., *Gnathostomula*, Fig. 1B) and Conophoralia (with conulus-type sperm, e.g., *Austrognatharia*, Fig. 1C).

Late discovery, together with high diversity, meant that gnathostomulid research has been dominated by morphological questions. Furthermore, it has not yet been possible to keep and rear gnathostomulids under laboratory conditions. Consequently, our knowledge of reproduction in this group is still at the stage of gross anatomical features and deductions therefrom, supplemented by isolated observations of actual processes (Riedl, 1969, 1971; Sterrer, 1969, 1971).

6.2 Asexual Reproduction

Asexual reproduction as such is not positively known, but so-called "body fragmentation" occurs (Sterrer, 1969). About 80% of the more than 500 specimens of *Haplognathia* and *Pterognathia* observed on the Swedish West Coast (Sterrer, 1969) were anterior fragments. Only a few posterior fragments were found. The same was true for a population of *Gnathostomaria lutheri* collected at the type locality at Banyuls-sur-Mer (southern France) in October 1966: all 259 specimens observed were anterior fragments (Sterrer, 1971; see also Ax, 1964). Whereas it is improbable that both the anterior and posterior end regenerate to a complete specimen (this may be the case for the anterior end only), fragmentation seems to be a biological norm in very elongated genera and probably happens periodically in the fall, possibly in connection with oviposition.

6.3 Sexual Reproduction

All known gnathostomulids are hermaphrodites. There are, however, observations that point to an alternation of phases with male and female emphasis, possibly with a tendency towards protandry (Riedl, 1969).

6.3.1 Anatomy of Reproductive System

The anatomy of the reproductive system is fairly simple (Fig. 1), and on about the same organizational level as that of acoel turbellarians. Two puzzling features, however, are the high diversity in sperm morphology (Section 6.3.2.1) and the complex bursa cycles (Section 6.3.3). The origin of germ cells and gonads is not known.

6.3.1.1 Male Organs

The male organs consist of testes and a more or less developed copulatory organ ending in a subterminal ventral male pore.

Paired lateral testes are situated in the posterior part of the body in *Haplognathia* (Fig. 1A) and all Scleroperalia (Fig. 1B), whereas only one dorsal testis, resulting from the fusion of paired anlagen, is found in *Pterognathia* and Conophoralia (Fig. 1C).

The copulatory organ has two main forms: the simple gland-sur-

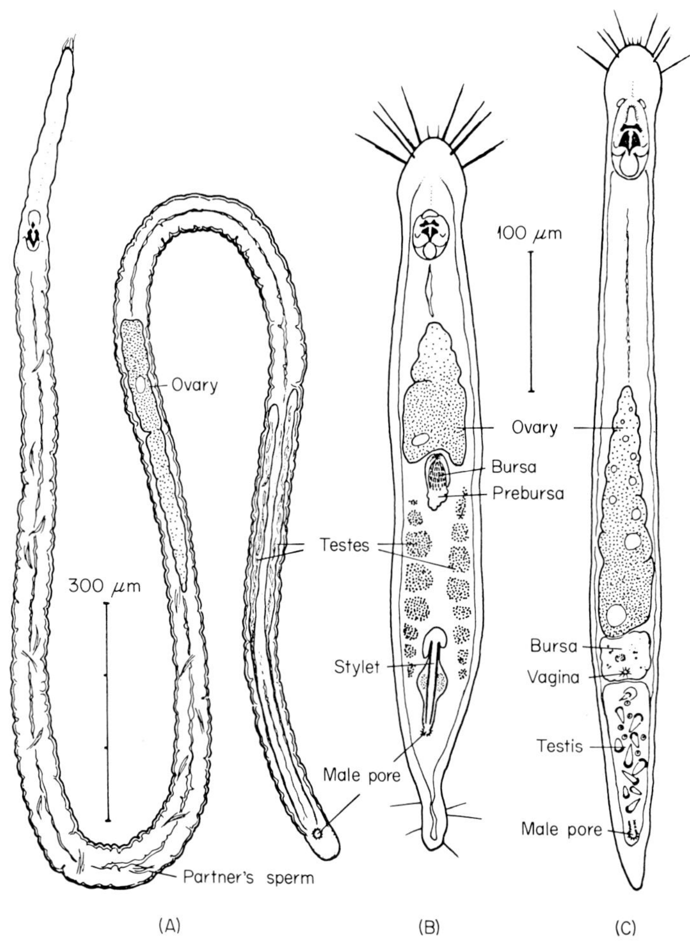

FIG. 1. Organization of Gnathostomulida, dorsal view. (A) *Haplognathia simplex* (Sterrer) (Redrawn from Sterrer, 1969). (B) *Gnathostomula jenneri* Riedl (Redrawn from Riedl, 1971). (C) *Austrognatharia kirsteueri* Sterrer (Redrawn from Sterrer, 1970). (B) and (C) in the same scale.

rounded male opening (which can hardly be called a penis) of the Filospermoidea (Fig. 1A), and the glandular and muscular penis, provided with a stylet in the Scleroperalia (Fig. 1B) and without a stylet in the Conophoralia (Fig. 1C). The stylet, usually straight, is typically composed of a dozen cuticular rods forming a tube around the male canal. It is more elongated, slightly bent and only faintly cuticularized in Gnathostomariidae. Two types of granular gland secretion accompany the copulatory organ throughout the gnathostomulids: a fine granular type which sheathes the copulatory canal or stylet, leaving only the proximal part uncovered, and a coarse granular type often forming a ring around the proximal part. A more or less muscular vesicula seminalis is often present in Scleroperalia.

6.3.1.2 Female Organs

The most consistent organ throughout the group is the ovary (Fig. 1). It is always unpaired and situated mediodorsally in the midbody region. It is shaped like an elongated pear, as the thick caudal section usually contains one mature egg.

Two accessory female organs, a vagina and a bursa system, are present in the order Bursovaginoidea. A permanent vagina can be found in Conophoralia (Fig. 1C), whereas it may be temporary or reduced in Scleroperalia. The vagina is a simple, occasionally thick-lipped dorsal pore above or slightly behind the bursa, into which it opens through a short, wide canal.

The "bursa system" derives its name from functional similarities (sperm storage) to bursal organs in Turbellaria, although a homology has yet to be proven. It is invariably situated dorsally, immediately behind the ovary. Two distinct types are known. In the Conophoralia, the bursa system consists of one soft-walled sac without a distinct form, i.e., it simply takes up the space between ovary, testis, intestine, and body wall (Fig. 1C).

In Scleroperalia, the "bursa system" is more complicated (Fig. 1B), and only recently has a model been proposed that explains both anatomical and functional peculiarities (Riedl, 1971). The following description refers to the genus *Gnathostomula* but is probably true for the whole Scleroperalia (Fig. 4A).

The "bursa system" consists of an anterior bursa and a posterior "prebursa." Both seem to be derived at least in part from the tissue injected by the partner in copulation (see Section 6.3.3). The prebursa, in fact, is the sperm mass that was injected by the partner during copulation. It is a spherical ball of hyaline appearance and may be separate from the bursa; in most cases, however, it is connected with the bursa, or finally forms a short bursa appendage with walls of variable shape.

The bell-shaped bursa itself consists of more stable walls whose thickness increases at the anterior end. A cross section shows that each layer of the bursa wall consists of three separate parts (cells or cell lobes) which meet in three longitudinal lines. These thickened lines appear in the light microscope as "cristae," running from the mouthpiece toward the end of the bursa wall. The anterior end of the bursa always forms a "mouthpiece," again an expression derived from turbellarian anatomy, although not even a functional analogy is proven in this case. It shows a lamellate structure as if it were made out of a series of discs. These "discs" are in fact the cuticularized and thickened edges of the cells forming the walls of the bursa itself. Although a central canal in the mouthpiece appears to be present in phase contrast studies, this has yet to be proven with electron microscopy.

6.3.2 Gametogenesis

Except for sperm morphology and isolated data on spermatogenesis, little is known. We have no information yet on gametogenic cycles and their determining factors.

6.3.2.1 Sperm Morphology and Spermatogenesis

The sperm is represented by three major types which seem to be structurally very distinct from each other.

1. The filiform type, (Fig. 2F, G) typically consisting of a spiral head, a middle piece, and a tail (Sterrer, 1969), has given its name to the Filospermoidea and is not found anywhere else within the Gnathostomulida. The microtubular pattern of the sperm tail in Filospermoidea is 9 + 2 (Rieger and Sterrer, unpublished).

2. The dwarf type occurs in the Scleroperalia and can be divided into three subtypes: (a) the small round type, found in *Agnathiella* (Fig. 2C), Mesognathariidae and in a *Gnathostomula* species (Fig. 2B); (b) the large round type, which is an exclusive feature of Gnathostomariidae (Fig. 2A), and functionally connected with the very reduced stylet of this family; and (c) the droplet type, found in Onychognathiidae (Fig. 2D) and—with the above exception—in Gnathostomulidae (Fig. 2E). Light as well as electron microscopy (Riedl, 1969) have shown that these three subtypes basically follow the same structural pattern: a round or more or less elongated sperm body which bears a row or bunch of short, nonciliary filaments.

3. The "conulus" type (Fig. 2H), typically consisting of hat, cingulum, body, and matrix (Sterrer, 1970) is still the most puzzling. Whereas there is much evidence for the conulus (which can be up to 45 μm long) being a single giant sperm, it cannot be excluded as yet that it represents a spermatophore.

Spermatogenesis of the dwarf type (Fig. 3) is best known, thanks to ultrastructural analyses by Riedl (1969) and Graebner (1968, 1969) on representatives of the genus *Gnathostomula*. In the spermatocytes, vesicles originating from the Golgi apparatus are disposed in a single

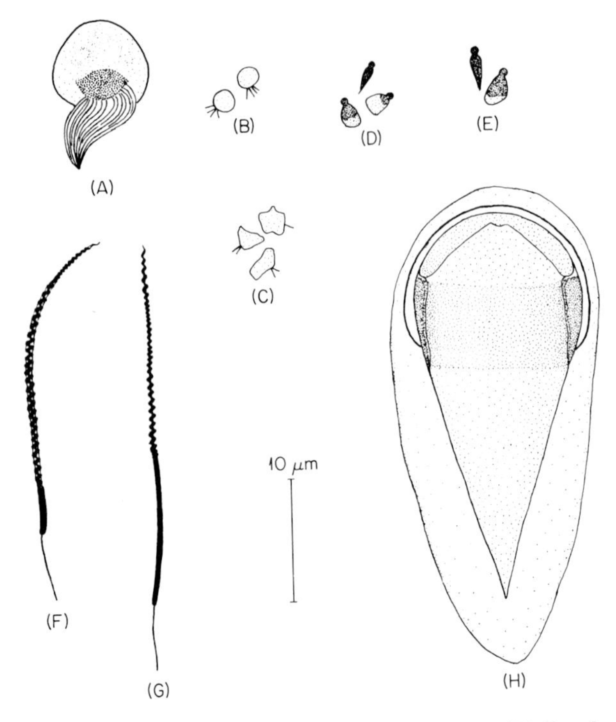

FIG. 2. Sperm morphology. (A) *Gnathostomaria nov. spec.* (B) *Gnathostomula jenneri* Riedl, (C) *Agnathiella beckeri* Sterrer, (D) *Onychognathia filifera* Riedl, (E) *Gnathostomula microstyla* Riedl, (F) *Pterognathia swedmarki* Sterrer, (G) *Haplognathia simplex* (Sterrer), (H) *Austrognathia riedli* Sterrer. All in the same scale.

array around the nucleus. There is a temporary contact between the vesicles and the nucleus. In the spermatids, the vesicles migrate toward the plasma membrane and finally come to lie in regularly spaced outpocketings of the membrane (Fig. 3, VP). The arrangement of these protrusions in sperms varies from species to species—in clusters along one side (*G. paradoxa,* see above: "droplet type"), or in a single cluster at the base of the sperm (*G. jenneri,* see above: "small round type").

Spermatogenesis of the filiform type is poorly known, and that of the conulus not at all. The filiform sperm, in *Haplognathia*, develops from rounded spermatids which, in groups of four, get more and more condensed. Finally, the nucleus, now about 1-2 μm in diameter and optically very dense, stretches, and a tail develops. This is also the stage when the head elongates and develops a spiral. It is interesting to note that tails in spermatids are usually longer than in the mature sperm.

In Conophoralia, conuli develop from cells with large nuclei. However, the identity of parts of the spermatocyte with parts of the conulus (hat, cingulum, body, matrix) has not been ascertained. The fact that groups of 4 identical conuli can be found occasionally in the testis suggests that they are sperms rather than spermatophores.

No overall direction of spermatogenesis with regard to body axis could be ascertained in Filospermoidea and Scleroperalia, except that earlier stages seem closer to the testis wall, and spermatozoa are produced toward the lumen. In Conophoralia, however, spermatogenesis appears to start at the caudodorsal end of the testis, proceeds frontally, and then turns ventrally at the anterior end of the testis. At this turning point appear the first conuli, which then gain in size toward the posterior end of the animal. Usually one or two mature conuli can be found in the lumen of the copulatory organ.

6.3.2.2 Oogenesis

Hardly anything is known on oogenesis. Only one egg at a time reaches maturity and then occupies the caudal part of the ovary. Its length (up to 300 μm) may correspond to about 1/10 of the body length. The mature egg seems fairly rich in yolk material.

6.3.3 Reproductive Behavior and Mating

As in many representatives of the interstitial fauna, gregariousness has been observed in Gnathostomulida. Its importance as a factor increasing the chances of meeting a partner is obvious. No other behavioral data are available yet.

In spite of the fact that approximately 15,000 specimens of gnathostomulids have been documented so far, mating has never actually been observed. It should be added, though, that it has been impossible to keep animals alive in the laboratory for any prolonged period of time. Our conception of mating, therefore, is based on anatomical facts rather than direct observations.

Three types of copulation must be postulated: (1) sperm penetration into the partner's body, (2) hypodermic impregnation, and (3) injection into the vagina. It is interesting to note the correlation between mobility of the sperm and differentiation of a copulatory organ. In Filospermoidea, the sperm has strong locomotory ability (by rotating and writhing), whereas the male pore is devoid of any musculature or stylet; the immobility of the sperm in Bursovaginoidea, on the other hand, is compensated by a muscular or stylet-armed injectory penis.

Whereas simultaneous cross fertilization seems improbable in Bur-

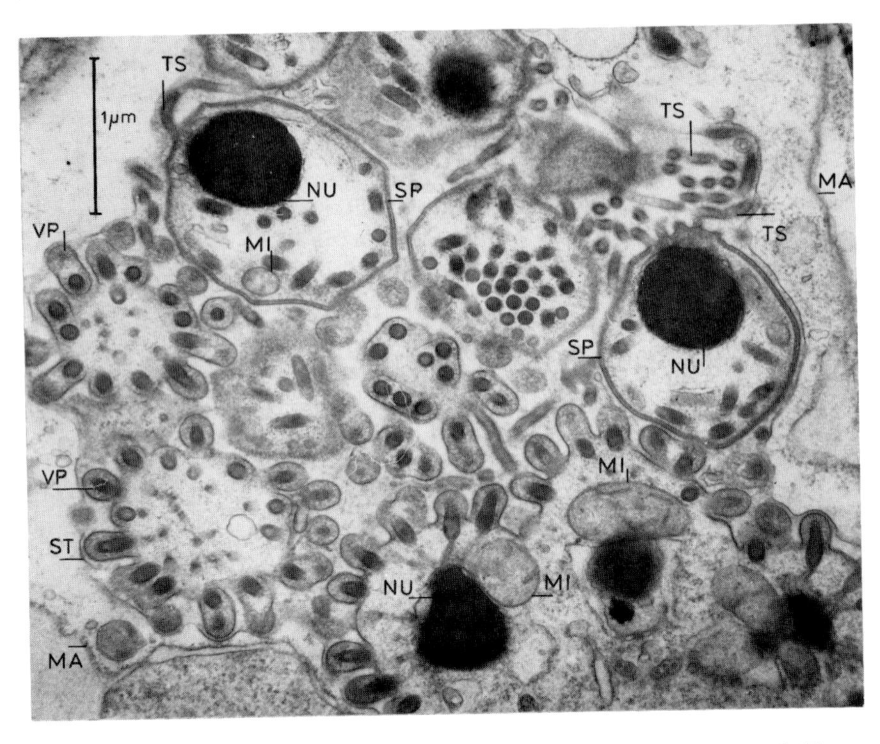

FIG. 3. Spermatogenesis of *Gnathostomula jenneri* Riedl, electron micrograph (From Riedl, 1969). Below spermatids, above sperm. (MA) matrix of testis follicle, (MI) mitochondrion, (NU) nucleus, (SP) sperm, (ST) spermatid, (TS) taillike structure, and (VP) vesicular peduncle. [Copyright 1969 by the American Association for the Advancement of Science. Reproduced with permission.]

sovaginoidea for anatomical reasons (male pore ventrally, female pore dorsally situated), it may occur in Filospermoidea. A further observation is that the receiving partner can be a juvenile, and sperm is then stored until female maturity is reached.

Sperm penetration through the partner's body takes place in Filospermoidea. The glands surrounding the male opening are everted and stick to the partner's skin (Sterrer, 1969). It is not known whether a particular region of the partner's body is preferred. Sperm is then injected, or rather bores its way into the partner by corkscrewlike motion. It can be found stored in the gut epithelium of the partner, mostly in the midbody region, but even in the parenchyma of the head. It is not known how sperm are dispersed in the partner's body nor how the mature egg is fertilized.

Hypodermic injection must occur in those Bursovaginoidea that lack a vagina. However, injection is here restricted to the dorsal body region behind the bursa. In Bursovaginoidea possessing both vagina and bursa, sperm is injected into the vagina, and then stored in the bursa.

In Conophoralia, the bursa usually contains only one conulus which differs characteristically from conuli encountered in the testis. It appears hollow, often with a hole on the tip, the conus covered with droplets (Sterrer, 1970). Occasionally, the bursa carries two conuli tied together like Siamese twins, both displaying the typical transformation. It may be assumed that the changes that occur to the conuli in the bursa have to do with fertilization.

In Scleroperalia, the processes following copulation seem still more complicated (Fig. 4). In their recent studies on *Gnathostomula*, Riedl (1971) and Müller and Ax (1971) come to the following conclusions: A mature testis follicle with sperm is injected under the partner's skin in the midbody region where it forms the first prebursa. Subsequently it develops into the first bursa by differentiating typical layers and a mouthpiece (see Section 6.3.1.2). It is not known whether the bursa is formed by injected ("male") or "female" tissue, or both. If there is already a bursa (i.e., from a foregoing copulation), then the prebursa joins and refills the bursa. The prebursa, therefore, forms or contributes to the formation of the new bursa as well as refilling an already existing bursa. In both cases it becomes, after several stages, completely integrated in the new, or newly filled bursa. The bursa thus goes through a complicated cycle in which Riedl (1971) distinguishes 11 "stages of formation" (e.g., early, full, senile, decay) and 3 "methods of filling" (primary, refill, secondary). Whether the

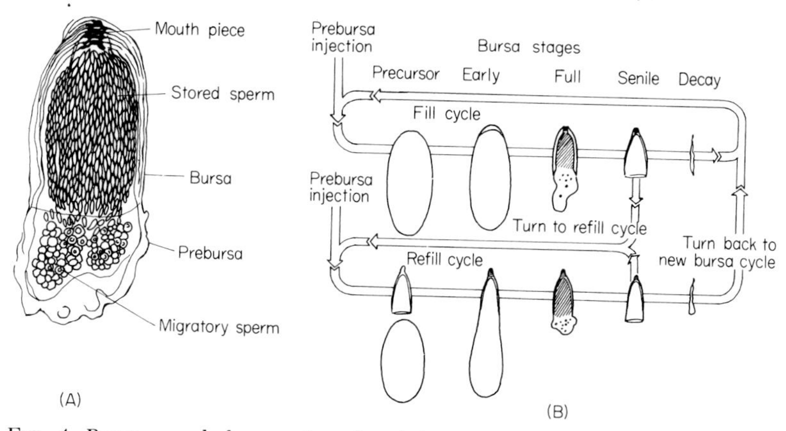

FIG. 4. Bursa morphology and cycles. (A) Bursa and prebursa of *Gnathostomula jenneri* Riedl. (B) Bursa formation and cycles in *Gnathostomula*. [Both figures modified from Riedl, 1971.]

bursa mouthpiece actually acts as a device to channel restricted numbers of spermatozoa to the mature egg is not known.

6.3.4 Oviposition

Oviposition has been observed only in *Gnathostomula jenneri* (Riedl, 1969). The animal attaches itself to the substratum by its dorsal epithelium and starts performing peristaltic movements. The zygote is thereby forced backwards between bursa and gut and finally breaks through the dorsal epithelium behind the bursa. This peculiar route the zygote takes (one would rather expect it to break through where it ripened, i.e., anterior to the bursa) seems to indicate the presence of a temporary vagina. Occasionally zygotes (ova?) are accidentally pressed into the gut where they are presumably digested.

After oviposition the zygote (having a diameter of 55 μm in *Gnathostomula jenneri*) becomes spherical and sticks to the substratum.

Both sexually mature and juvenile specimens are encountered all year round, even in temperate climates; for *Gnathostomula paradoxa*, however, Müller and Ax (1971) report increased reproduction in the spring leading to a higher representation of juveniles in the fall.

6.4 Development

Development is direct, as in most representatives of interstitial fauna. The little information we have is from Riedl's (1969) report on *Gnathostomula jenneri* (Fig. 5) and my unpublished observations on a hatching juvenile of *Haplognathia nov. spec.* According to Riedl (1969, p. 449)

> Cleavage begins a few hours after oviposition (temperature was maintained at 22°C). The first and second divisions are meridional,. . .nearly equal, and holoblastic. . .the second division sometimes may be nonsynchronous with the C-D blastomere (cleavage cell) being delayed. Beginning with the third division, eggs exhibit a spiral plan with alternation between dexio- and leiotropic planes of cleavage. The first equatorial division is unequal, with dexiotropic or clockwise displacement of the first quartet of micromeres. Further cleavages, after 20 hours, result in a mass of micromeres covering the animal hemisphere. On the third day, "tip" cells of the micromere cross appear (similar to the annelid type of cleavage). On the fourth day, epibolic gastrulation is completed in accord with typical mesolecithal cleavage and without previous formation of a blastocoel or blastopore. A pair of larger blastomeres, presumably $4d^1$ and $4d^2$, still remains visible for another day. Postgastrula

development is slow. Yolk resorption takes place, and a belt of long, unmovable fibers is formed on the outer surface of the egg capsule.

Prior to hatching the embryo of *Haplognathia nov. spec.* can be seen rotating within the egg capsule by vigorous ciliary motion (Fig. 6A). Hatching takes place by rupture of the capsule (Fig. 6B). A juvenile of *Gnathostomula jenneri* is about 100 μm long and 40 μm in diameter. It is completely monociliated and bears a short tail with stiff sensory cilia. The anterior sensorium is missing, as well as the jaws. There is, however, a rudiment of the pharynx, with the central part of the basal plate being distinguishable. This latter observation was verified also in a new species of *Haplognathia*. It is interesting to note that the egg capsule containing the juvenile of *Haplognathia*

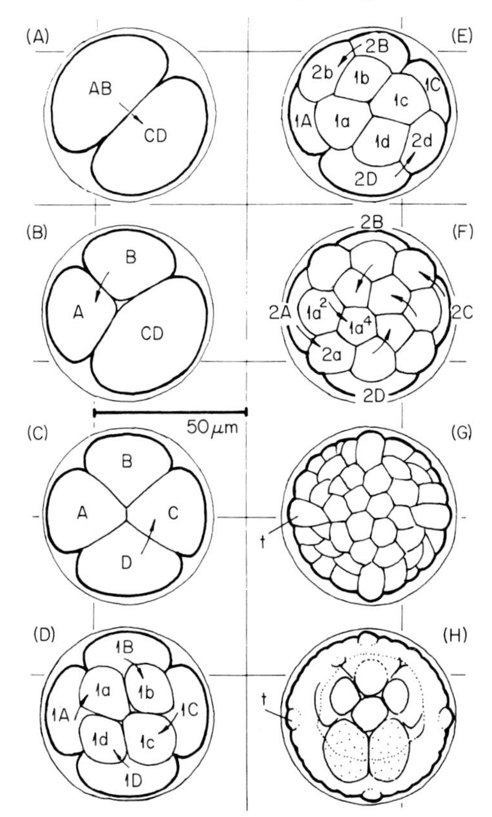

FIG. 5. Early embryology of *Gnathostomula jenneri* (From Riedl, 1969). (A)-(G) drawn from the animal pole, (H) from the vegetative pole of the embryo. The "tip" cells of the micromere cross (t) appear in (G) and (H); in (H) the edge of epibolic gastrulation (broken line) and the two possible mesoblasts (dotted) are marked. [Copyright 1969 by the American Association for the Advancement of Science. Reproduced with permission.]

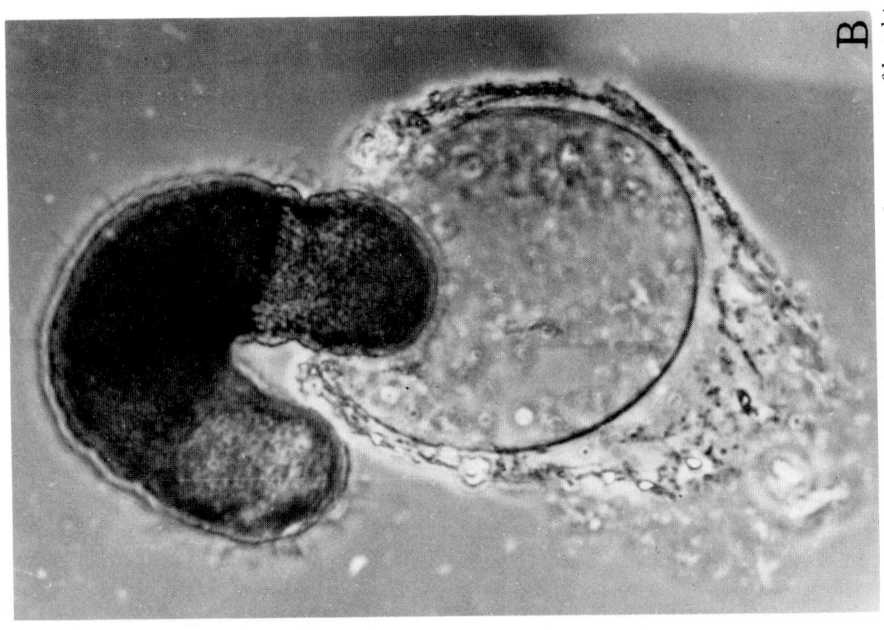

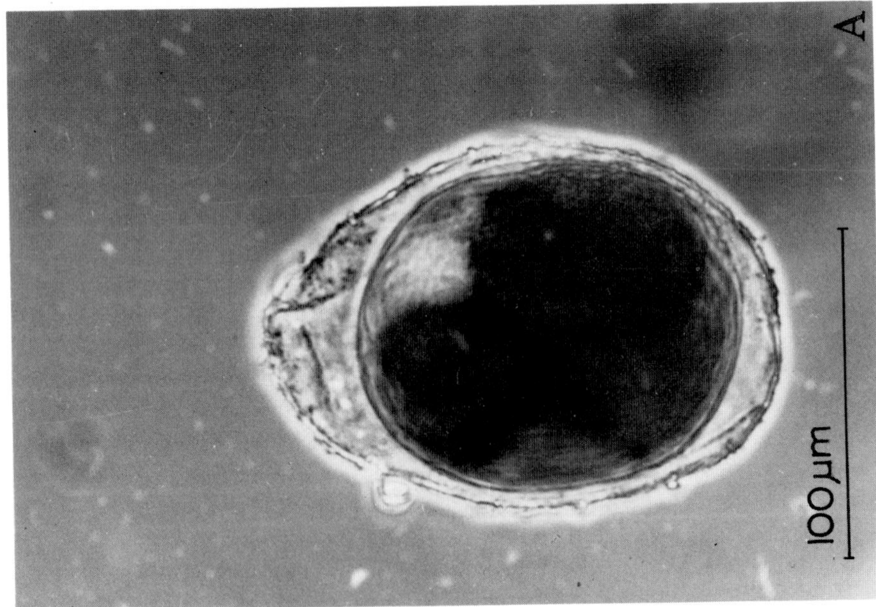

FIG. 6. Hatching of a *Haplognathia nov. spec.* In (A) the embryo is rotating in the egg capsule; in (B) it is in the process of hatching.

nov. spec. (Fig. 6) came from a "black" sand layer, or "sulfide system" (which is the typical biotope for Gnathostomulida as a whole). This suggests that development takes place in complete anoxybiosis (Fenchel and Riedl, 1970).

In *Gnathostomula paradoxa,* which has a maximum body length of 1000 μm, the copulatory organ develops in specimens 600 μm long, followed by the bursa, whereas both the ovary and the testes appear first in specimens 600-700 μm long (Müller and Ax, 1971).

6.5 References

Ax, P. (1956). Die Gnathostomulida, eine rätselhafte Wurmgruppe aus dem Meeressand. *Abh. Akad. Wiss. Lit. Mainz Math.-Nat. Kl.* 8, 1-32.

Ax, P. (1964). Die Kieferapparatur von *Gnathostomaria lutheri* Ax (Gnathostomulida). *Zool. Anz.* **173**, 174-181.

Fenchel, T. M., and Riedl, R. J. (1970). The sulfide system: a new biotic community underneath the oxidized layer of marine sand bottoms. *Mar. Biol.* **7**, 255-268.

Graebner, I. (1968). Ergebnisse einer elektronenmikroskopischen Untersuchung von Gnathostomuliden. Verhandlungen der Deutschen Zoologischen Gesellschaft in Innsbruck, p. 580-599.

Graebner, I. (1969). Vergleichende elektronenmikroskopische Untersuchung der Spermienmorphologie und Spermiogenese einiger Gnathostomula-Arten: *Gnathostomula paradoxa* (Ax, 1956), *Gnathostomula axi* (Kirsteuer, 1964), *Gnathostomula jenneri* (Riedl, 1969). *Mikroskopie* **24**, 131-160.

Müller, U. und Ax, P. (1971). Gnathostomulida von der Nordseeinsel Sylt mit Beobachtungen zur Lebensweise und Entwicklung von *Gnathostomula paradoxa* Ax. Mikrofauna d. Meeresbodens, *Akad. Wiss. Lit. Mainz* **9**, 1-41.

Riedl, R. J. (1969). Gnathostomulida from America. First record of the new phylum from North America. *Science* **163**, 445-452.

Riedl, R. J. (1971). On the genus *Gnathostomula* (Gnathostomulida). *Int. Rev. gesamten Hydrobiol.* **56**, 385-496.

Sterrer, W. (1969). Beiträge zur Kenntnis der Gnathostomulida. I. Anatomie und Morphologie des Genus *Pterognathia* Sterrer. *Ark. Zool.* **22**, 1-125.

Sterrer, W. (1970). On some species of *Austrognatharia, Pterognathia* and *Haplognathia* nov. gen. from the North Carolina coast (Gnathostomulida). *Int. Rev. gesamten Hydrobiol.* **55**, 371-385.

Sterrer, W. (1971). On the biology of Gnathostomulida. *Vie Milieu,* Suppl. 22, 493-508.

Sterrer, W. (1972). Systematics and evolution within the Gnathostomulida. *Syst. Zool.* **21**, 151-173.

Chapter 7

NEMERTINEA

Nathan W. Riser

7.1 Introduction

Sabatier's "law" of cellular differentiation in animals which was propounded in 1883 was based upon his studies of the nemertinean *Tetrastemma flavidum* Ehrenberg 1831. The primary points of the "law" were that all cells are polarized and that if both the central (containing the nucleus) and the peripheral portion were maintained a neuter cell would result. If the central portion disappeared, the cell would be male but if the periphery degenerated or disappeared the cell would be female. Lee (1887) used a large number of species in his investigation of spermatogenesis and spermiogenesis in nemertineans, and in the process rendered the theory of Sabatier untenable. The latter had made many of his observations on worms flattened under pressure and apparently had confused gonads with sex cells.

The phylum Nemertinea (also called Nemertina, Nemertea, Nemertini, Rhynchocoela) is known to many biologists because of the wide use of a few species for embryological and regeneration studies. Asexual reproduction of nemertineans by fragmentation was used by Hubrecht (1887) as a major support for his theory of the origin of segmentation. Oddly, investigations utilizing representatives of this little known group have contributed an unusually large amount of information in the areas of cytomorphology, oogenesis, spermatogenesis, spermiogenesis, and fertilization in addition to embryology and regeneration. Except for its ecology, very little is known about this group of animals, in part because it is difficult to identify species. This is further complicated by the fact that preserved worms lose their color and, when sent to a specialist, must be serially sectioned; even then they can often be identified only to family or genus. Internal morphology is poorly known and most species have been described either on the basis of color patterns of the living worms or upon internal morphology of preserved specimens, but rarely on a combination of the two. A large percentage of the species are burrowers and are infrequently encountered. The majority of intertidal species are seasonal in occurrence with a life-history extending through only one breeding cycle. Some species which can be collected in large numbers such as *Amphiporus angulatus* (Fabr., 1774) and *A. lactifloreus* (Johnston, 1828) from New England waters live poorly in the laboratory. However, McIntosh (1873) gives the impression that laboratory rearing can be quite simple and the latter species has been used extensively for laboratory studies by European workers. Thus, most laboratory investigations have been restricted to a few forms which live well in captivity.

7.2 Asexual Reproduction and Regeneration

Nemertineans are capable of wound healing and most species which have been studied can regenerate regions behind the foregut. Species capable of anterior regeneration are the only ones reported to undergo asexual reproduction. Asexual reproduction has been described by Coe (1930a, 1931) for *Lineus socialis* (Leidy, 1855), and *L. vegetus* Coe, 1931, both species in which the fragments tend to encyst while undergoing reorganization, and by Gontcharoff (1951) for *Lineus sanguineus* (Rathke, 1799), and *L. pseudo-lacteus* Gontcharoff, 1951, for neither of which encystment is mentioned. Monastero (1928) described fragmentation, followed by encystment of the fragments and regeneration in *Lineus nigricans* Bürger, 1892 but made no mention of sexual reproduction. The phylum exhibits a short series of reproductive patterns ranging from strictly asexual in *L. sanguineus*, alternation of sexual with asexual in *L. socialis* and *L. vegetus*, and strictly sexual in most other species.

Coe (1929a, b, 1930b, 1932, 1934a, b), Dawydoff (1909, 1910), Nusbaum and Oxner (1910a, b, c, 1911a, b, 1912a, b) and Oxner (1909, 1910a, b, 1912) published the basic information upon which our knowledge of nemertinean regeneration is based. The primary features of regeneration were detailed and added to by Gontcharoff in 1951 and were reviewed by Gontcharoff (1961) and Hyman (1951). The inability of pieces anterior to the brain to regenerate and the tendency for posterior regeneration have demonstrated an axial gradient which is expressed to different degrees in different species. A significant fact recorded by Coe (1929b) is that in *Lineus socialis* and *L. vegetus*, species capable of both anterior and posterior regeneration, wound healing does not occur if the fragments are less than half as long as the body width, therefore regeneration cannot occur. He further stressed that regeneration of fragments follows axial gradients, *i.e.*, the nearer to the head the more rapid the regeneration of the piece. Furthermore, a part of the brain or nerve cord is necessary for regeneration. The inhibition of anterior regeneration in *L. vegetus* by homogenates of regions anterior to a fragment, and the inhibition of posterior regeneration by homogenates from regions posterior to a fragment were interpreted by Tucker (1959) to indicate an anterior-posterior "flow of inhibiting information."

Dawydoff (1909, 1942) described wound healing in *Lineus lacteus* (Rathke, 1843) as beginning with dedifferentiation of cells. Bierne's (1962b) studies on regeneration of the rhynchodaeum, proboscis, and rhynchocoel of *Lineus ruber* (Müller, 1771), *L. sanguineus*, and *Amphiporus lactifloreus* utilizing modern techniques supported this concept. Coe (1934b) had described a process involving migration of cells in *Lineus socialis*, not only cells in the vicinity of the damage, but also "dormant cells situated in the parenchyma" of "all parts of the body." The incomplete evidence on *L. sanguineus* by Bierne is important because this species reproduces asexually as does *L. socialis*, and there is reason on the basis of comparative studies on sponges, turbellarians, annelids, etc. to assume that nemertinean species which undergo asexual reproduction have a reserve of undifferentiated cells in contrast to species incapable of such reproduction. Before an understanding of asexual reproduction can be attained, the problem of differentiation vs. dedifferentiation in the asexually reproducing form must be resolved. Information obtained from studies of posterior regeneration or the regeneration of lost parts is not *a priori* evidence of what happens in asexual reproduction.

The role of the nervous system in regeneration and asexual reproduction is not clear. Complete anterior regeneration of *Prostoma graecense* (Böhmig, 1898) according to Kipke (1932) occurs if the animal is cut immediately behind the brain so that the posterior piece contains all of the rhynchocoel, proboscis, intestine, and lateral nerve cords. The lateral nerve cords of nemertineans each consist of a fibrous core with nerve cell-bodies wrapped around it or with the cell-bodies in dorsal and ventral tracts applied to it, but no ganglia. Thus, the rate of regeneration of a fragment of an asexually reproducing species cannot be influenced by fibers originating in the brain, and in the absence of ganglia, the cord, in spite of its necessary presence for regeneration, cannot be responsible for the gradient. The inducing and inhibiting factors in nemertinean morphogenesis and regeneration are beginning to be uncovered. The antagonism between these factors controls the process of regeneration and is responsible for axial gradients. Posterior regeneration of nemertineans has been the most valuable tool in attacking the problems of sexual reproduction. (The publications of Oxner, and Nusbaum and Oxner on *L. ruber* do not distinguish between the four species that constitute the "*L. ruber complex*" of *L. ruber, L. viridis, L. sanguineus,* and *L. pseudo-lacteus.*) McIntosh (1874) described posterior regeneration of a male *Lineus marinus* (Montagu, 1804) and the shedding of sex products by the headless fragments 5 months later, but made no reference to the sexual condition of the regenerator. Oxner (1911) described the regeneration of the posterior region of an immature *Lineus lacteus* and the development of sex products in the regenerator. The regeneration of gonads of *Lineus ruber* following ablation immediately behind the mouth was investigated by Bierne (1962a) using animals in which active gametogenesis was occurring as well as ripe individuals ready to spawn. The development of gonads and sex products in these instances implies the existence of undifferentiated toti-potent cells which are controlled by inhibitors and inducers. Posterior regeneration by heterospecific transplants was studied by Bierne (1967a). In these studies, chimaeras of *L. ruber-L. viridis, L. sanguineus-L. lacteus, L. sanguineus-L. pseudo-lacteus* were utilized, the restriction resulting from the arrangement of muscles, i.e., *L. ruber* and *L. viridis* have a body musculature that allows contraction but not coiling, while the other three species coil. In posterior regeneration, the regenerating part is derived from the region immediately anterior to the line of section and retains the specific qualities of that portion of the chimaera. Mixed regenerators of *L. sanguineus-L. pseudo-lacteus* following this pattern developed gonads in the *L. pseudo-lacteus* side, but not in the other, and anterior regeneration of the *L. sanguineus* side produced a new head for the chimaeras according to Bierne (1967c). In this latter study, chimaeras consisting of a head of one species and postoesophageal region of another were cut into two parts through the intestinal region. The cephalic region of these heterospecific grafts had no influence upon phenotypic regeneration. A posterior fragment of a *L. lacteus* base did not undergo anterior regeneration while a comparable fragment from a *L. sanguineus* base did. The anterior fragments regenerated according to the phenotype of the postoesophageal region. Thus Bierne concluded that the cells involved in posterior regeneration in the family Lineidae were derived from the region immediately in front of the severed edge.

7.3 Sexual Reproduction

7.3.1 Sexual Dimorphism

The color of the gonads of sexually mature individuals frequently can be used to distinguish sexes, e.g., cream ventral surface of female *Amphiporus angulatus* vs. red of male; pink to bright red intestinal region of male *Cerebratulus lacteus* (Leidy, 1851) vs. deep red to brownish red females. Immature *C. lacteus* is white or pink, but these two color phases have not been reared to sexual maturity and it is not known whether these are sexual characters. Female *Lineus viridis* and *L. ruber* are not only larger than males from the same area, but the large eggs show through the pigmented body wall and identify the sex. These eggs range from cream to yellowish orange in color.

No generalization can be made about size differences in regard to sexes in the phylum. *L. ruber* and *L. viridis* are apparently among the exceptions in this respect among strictly dioecious species. The contractility of the body would make any statistical study suspect especially since the size range in both sexes is usually approximately the same. However, Coe (1904) reported small males in addition to the large hermaphrodites of *Neonemertes* (=*Geonemertes*) *agricola* (Willemoes-Suhm, 1874).

Ehrenberg (1831) mistook the mouth for the genital pore, and Huschke (1830) considered the proboscis to be a penis. Oersted in 1844 beautifully confused the picture with a remarkable scientific analysis in which he demonstrated the presence of the organ in both sexes but concluded that it was an organ of stimulation. At the time of breeding, the gonopores become obvious features of the morphology in some species. The pores are pitlike, but depending on the state of contraction or relaxation, appear to lie at the bottom of pockets or on mounds. In pelagic species, the males (at least in preserved specimens) have external papillae associated with each testis. Such papillae appear as structures which could be called penes in *Pelagonemertes brinkmanni* Coe, 1926. However, most observations on pelagic nemertineans are from preserved material and in spite of the sporadic occurrence of these animals, adequate information on living organisms is lacking. Males of *Nectonemertes* have two lateral appendages which distinguish them from females and which have been considered to be clasping organs.

7.3.2 Hermaphroditism

The majority of the species are dioecious. *Geonemertes palaensis* Semper, 1863, *Neonemertes agricola, Tetrastemma hermaphroditicum* Keferstein, 1868, *T. kefersteinii* (Marion, 1873), *T. marioni* (Joubin, 1890), *T. caecum* Coe, 1901, the genera *Prostoma, Prosactenoporus, Poikilonemertes, Dichonemertes, Sacconemertella* and *Coenonemertes* however are monoecious. *Prosorhochmus claparedei* Keferstein, 1862, *Prostoma rubrum, Sacconemertella lutulenta* Iwata, 1970 have ovotestes and are protandrous. In her type description of *Coenonemertes caravela*, Corrêa (1966) reported that no hermaphroditic gonads occurred but that the organs appeared as pairs, distinctly separated. Anteriorly, the pairs on each side consisted of testes. In the middle and posterior regions the dorsal gonads were testes and the ventral ones ovaries. Sometimes the terminal ovaries were not accompanied by a testis and in this situation only a single gonad (ovary) occurred on each side. The species was collected in Brazil in September and showed typical protandry with mature sperm in the testes and immature ova in the ovaries. In *Dichonemertes hartmanae* Coe, 1938, a single row of gonads extends up each side of the body to just behind the brain. The anterior five to eight pairs are testes, the posterior are ovaries. The ovotestes of *Prostoma rubrum* (Leidy, 1850) may produce both eggs and sperm at the same time according to Coe (1943) but in some individuals protandry occurs. Both oogonia and spermatogonia line the gonad wall and a single oocyte normally matures at any one time. Selffertilization occurs, but it is assumed that crossfertilization is associated with protandric individuals. Mature sperm occupied the center of the gonad of the type specimen of *Sacconemertella lutulenta* and Iwata (1970) described the periphery of the gonad to be female. Apparently, oogenesis and spermatogenesis do not go on simultaneously in this species as in *P. rubrum*, and following spermatogenesis, the wall of the gonad becomes dominated by developing oocytes.

7.3.3 Anatomy of the Reproductive System

The gonads in most species are simple structures lying in the parenchyma behind the nephridial region. Basic features of the reproductive system can be found in Hyman (1951). Gonads in most free-living species lie between the diverticula of the midgut (Fig. 1). In general, a single gonad occurs in each interdiverticular space on each side. A dorsal and a ventral ovary occur in each interdiverticular space of *Oerstedia dorsalis* Abildgard, 1806. Each testis of *Arctonemertes thori* Friedrich, 1957 was described as expanded over and under the diverticula and not restricted to the interdiverticular space. In *Arenonemertes microps* Friedrich, 1933, a species lacking diverticula of the midgut, the gonads occurred in the posterior half of the body lying dorsal to the lateral nerve, but alternating in position from side to side rather than lying opposite one another. In the genus *Carinella* several gonads each with its own gonoduct and gonopore occur dorsal to the lateral nerve cords and extend above the gut in each interdiverticular space according to Bürger (1897-1907). In *Nemertopsis bivittata* (Chiaje, 1841) and *Ototyphlonemertes evelinae* Corrêa, 1948 dorsal and ventral gonads occur on each side with only dorsal ducts in *N. bivittata* according to Bürger (1895) and with dorsal and ventral ducts described by Corrêa (1948) in *O. evelinae*. The ovaries of *N. bivittata* are in the interdiverticular spaces, but in the males, the testes are outside the gut area according to Corrêa (1961). Coe (1940) described numerous gonads around the intestine of *Emplectonema gracile* (Johnston, 1837), a close relative of *Nemertopsis*, and that the gonoducts opened dorsally or ventrally according to the "position of the gonads." In the genus *Carcinonemertes*, the testes are small and scattered to form a layer between the gut and dermo-muscular sac but the ovaries are paired. In this genus the gonads extend almost from one end of the animal to the other. The gonads are scattered through the parenchyma in the genera *Gononemertes* and *Malacobdella*.

In general, the gonopores open dorsally to the lateral nerve cords but exceptions exist. In *Lineus viridis* there can be as many as three rows of gonopores on each side in females. The two most dorsal rows are above the lateral nerve cords. In males of *L. ruber* and *L. viridis*,

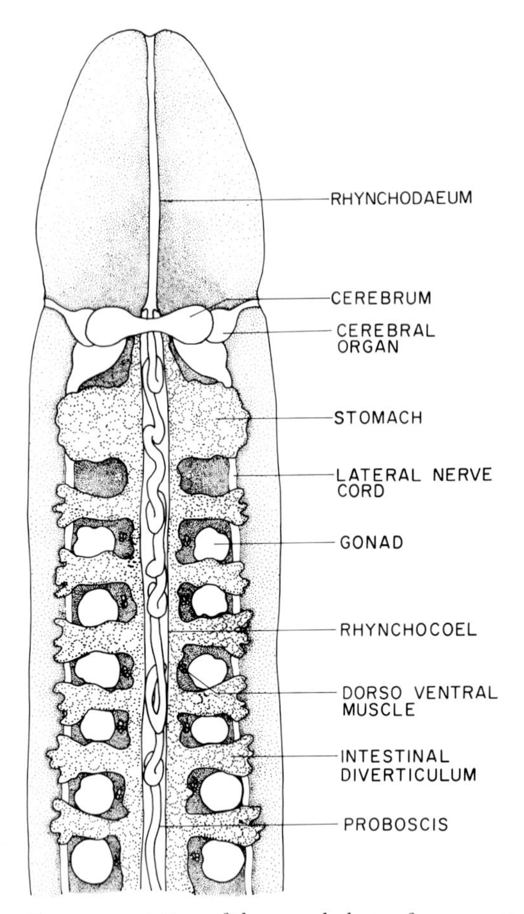

FIG. 1. Schematic representation of the morphology of a nemertinean.

the pores are ventral to the nerve cords (Fig. 2), a commonly reported situation in nemertinean males. *Dananemertes saemundssoni* Friedrich, 1957 was described from a male specimen collected in May. The testes alternated irregularly with the intestinal diverticula. In the middle and posterior part of the body, a number of gonads were present in cross sections, some with dorsal and others with ventral gonoducts. A complicated arrangement occurs in *Carcinonemertes* wherein Humes (1941) confirmed the work of Takakura (1910) in which vasa efferentia from each testis unite to form a dorsal vas deferens which expands as a seminal vesicle behind the testis zone and then continues posteriorly ending in the gut just anterior to the anus. Thus a cloaca is present in the males of this genus. The vas deferens is a permanent structure, but the vasa efferentia appear to

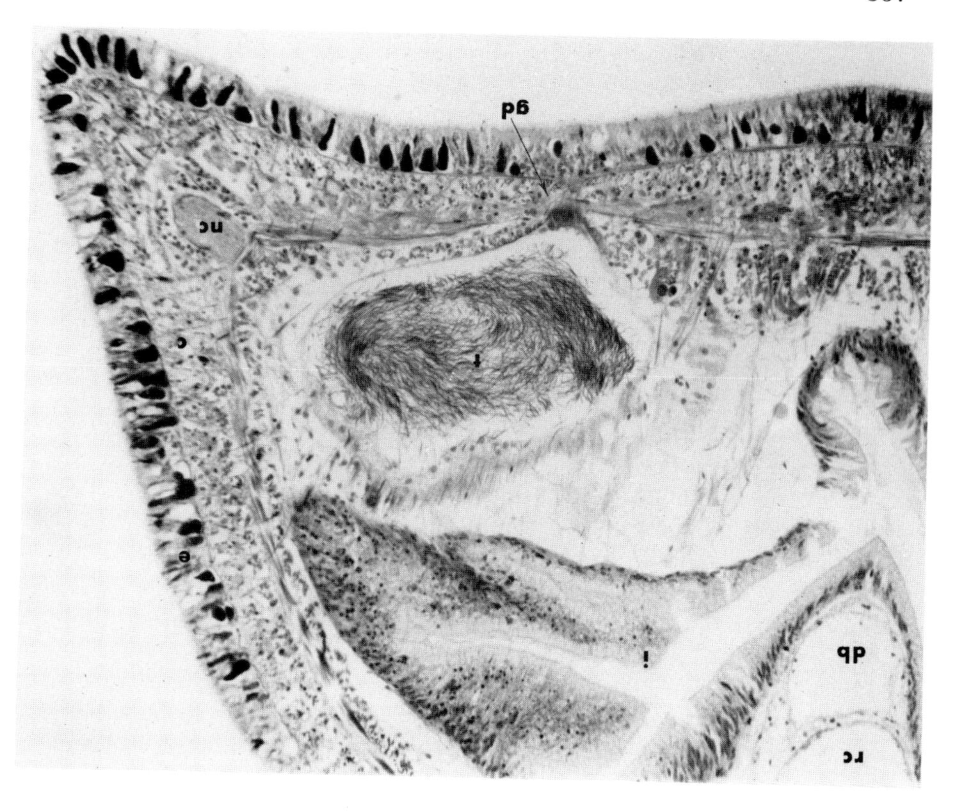

FIG. 2. Portion of *Lineus ruber* in cross section showing relationship of testis (t) and gonoduct (gd) to the lateral nerve cord (nc), gut (i), rhynchocoel (rc), dorsal blood vessel (db), epidermis (e), and cutis (c).

be transitory and arise at the time when the sperm are mature in the same way that male gonoducts arise in most species. In pelagic nemertineans, a seminal vesicle is associated with each testis.

7.3.4 Origin of the Germ Cells and Gonads

According to Olivier (1966) the ovaries of *Lineus ruber* first appear as primordial sex cells in the parenchyma as postulated by Bürger in 1895. These aggregates are reported to assemble in the interdiverticular spaces but the interaction of gut and parenchyma as organizers or inducers is still not determined. As gonial cells enlarge a membrane forms around the aggregate. He described the wall as an unstratified flattened epithelium which is the normal appearance in this species in paraffin sections.

In *Cerebratulus lacteus*, the formation of the gonad was followed by Wilson (1900) in regenerating worms. He reported that the dorsoventral muscles established the germinal pouches, and contributed the fibers of the pouch wall. He noted that after the walls of the germinal sacs formed, the wall of the gut grew outward between the sacs, and then the epithelium of the germinal sac formed. Bierne (1962a) also utilized regenerating worms but described the formation of gut diverticula preceding gonad formation in *L. ruber*.

The ovaries of *Lineus longifissus* (Hubrecht, 1887) are unique. Each consists of a stroma of spindle-shaped cells extending from the gonoduct (which is preformed) and ramifying into the connective tissue of the interdiverticular space and extending above, below, and outside the diverticula. At the beginning of vitellogenesis, pseudopods push out from the oocytes into the stroma. At the completion of vitellogenesis, each oocyte rounds up and lies in a connective tissue pouch lined by remaining stroma cells. Thus, the ovary is arborescent utilizing existing connective tissue for its wall and lacking a connective tissue wall of its own. The muscle fibers associated with the connective tissue may be incorporated for a portion of their length with the "gonad" wall.

Riepen (1933) described a middle layer of muscle fibers in the gonad of *Malacobdella grossa* (Müller, 1776) and Friedrich (1936) pointed out that a muscle layer probably existed in all cases, but had been overlooked. The fine muscles fibers associated with the wall of the gonad of nemertineans are derived from the circular muscle layer of the body wall and are primarily dorso-ventral fibers attaching the gut to the body wall. Muscle fibers extending into the area of the gonad from any part of the dermo-muscular sac are incorporated into the connective tissue sheath of the gonad. Intrinsic muscles appear to be restricted to fibers that extend inward from the body wall with the gonoduct (Fig. 2, 3). The development of the testes has been described by Lee (1887) and Gontcharoff (1951) and is similar to the formation of the ovary. Thus as Wilson (1900) noted, the sex cells develop from the epithelium of the gonad in *Cerebratulus*, while they produce the epithelium in *Lineus*.

In most species the epithelium of the gonad is masked by the gonial cells and developing gametes. It is usually visible in the mature testis, and in some species in the mature ovary. The ovary of *Oerstedia dorsalis* is characteristic for most monostyliferous hoplonemertineans, i.e., forms with a single central stylet in the proboscis. The immature ovary of these forms is hollow with a single layer of gonial cells lining the cavity and over growing the epithelial cells. The most mature oocytes are on the wall nearest to the intestine.

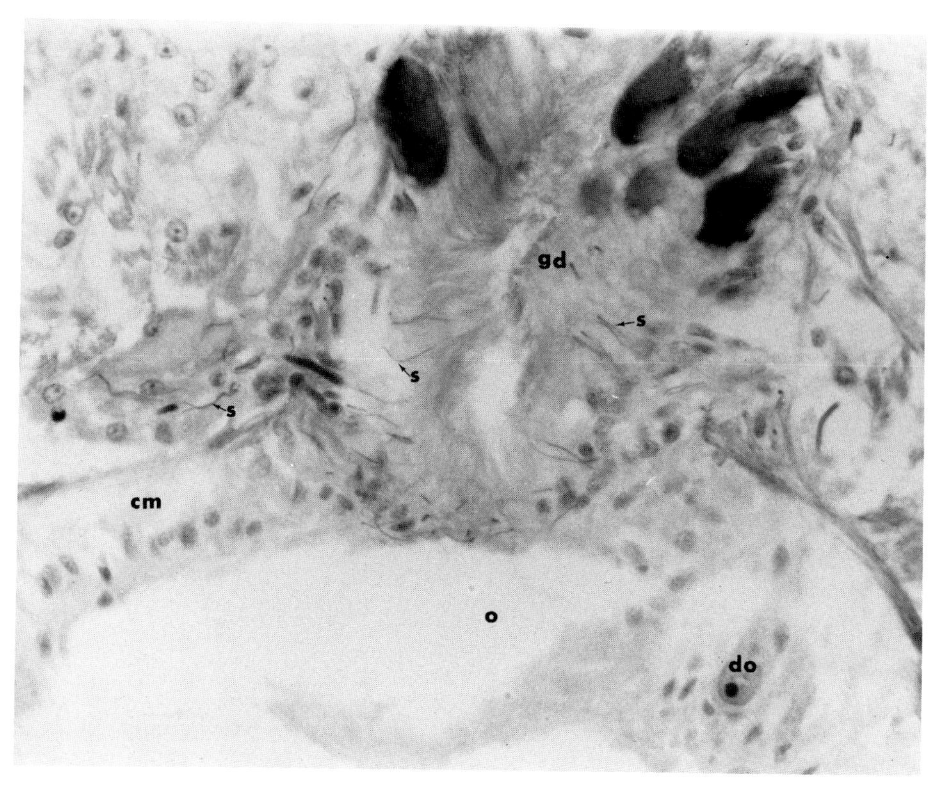

FIG. 3. Section through gonoduct (gd) of a female *Lineus viridis* immediately after spawning showing sperm (s) migrating through epidermis and in the cutis: (cm) circular muscle, (do) degenerating oogonium, (o) ovary.

Those on the peripheral side of the ovary are of rather uniform size except in the dorsal-most portion which is occupied by small gonial cells. Primitive gonial cells project from the wall of the mature ovary of *Parborlasia corrugatus* (McIntosh, 1876) but previtellogenic oocytes of large size occur in the ovarian outpocketings which lie lateral to the intestinal diverticula. The occurrence of this second generation of oocytes in prespawning worms accounts for the presence of well developed gonads 5 months later as reported by Friedrich (1970).

7.3.5 Gametogenesis

7.3.5.1 SPERMATOGENESIS

The general features of spermatogenesis have been described by McIntosh (1893), Lee (1887), Wilson (1900), and Gontcharoff (1951) as following the classical pattern. According to the last investigator the germ cells of *Lineus ruber, L. pseudo-lacteus, L. sanguineus,* and *L. viridis* begin to differentiate after the epithelial lining of the testis is formed. In *L. ruber* and *L. viridis,* the gonial cells take up a position on the testis wall, forming a hollow gonad. Usually in the other two species a lumen does not form.

Nemertinean histological and cytological investigations for many years were, and often even today are, carried out without relaxing the specimens, which are very muscular. Thick sections were hand cut and then cleared in aniline for study. Regrettably, most investigators who relaxed their specimens used magnesium chloride, a substance which while adequate for gross histology is deleterious to cytological studies because of its solating effect upon cytoplasm. The use of alcohol for dehydration and paraffin for embedding results in a major distortion of tissues and cells in all animals and is extreme in muscular organisms lacking hard parts such as the nemertineans. These technical phenomena have produced a confused literature which has discouraged workers in the field and has led to uncertain or erroneous statements. The work of Lee, Wilson, and other early investigators as well as by recent investigators must be understood in this context.

Maturing testes show a progression of development from the periphery to the center. The mature testis of most species consists of a wall enclosing a mass of spermatozoa, with spermatogonia indistinguishable from the lining cells and no spermatogenesis.

Spermiogenesis has been described in part by Retzius (1904, 1906) and in detail by Franzén (1956). Unlike the Turbellaria, the sperm of all nemertineans studied have a distinct head, midpiece and tail.

Franzén referred to the sperm of nemertinean species in which the head and midpiece were over 8 μm in length as modified. His list included *Cephalothrix rufifrons* (Johnston, 1837), *Lineus ruber*, *Emplectonema gracile*, *Amphiporus lactifloreus* (Johnston, 1828), *Carcinonemertes carcinophila* (Kolliker, 1845) and *Malacobdella grossa*. *Lineus viridis* from New England waters has sperm with the head and midpiece averaging 15 μm in length and fits this group. Franzén referred to sperm with head and midpiece less than 8 μm as unmodified; they occur in *Hubrechtella dubia* Bergendahl, 1902, *Lineus bilineatus* (Delle Chiaje, 1841), and *Micrura fasciolata* (Ehrenberg, 1831). *Amphiporus angulatus* and *Lineus torquatus* Coe, 1901 both have mature sperm with head and midpiece slightly under 3 μm in length. Iwata (1957b) reported the sperm head of *L. torquatus* from Japanese waters to be "about 0.005 mm long." Sperm morphology is indicative of reproductive behavior, and it can be assumed that the sperm of *Cerebratulus marginatus* Renier, 1804, *Oxypolella alba* Bergendahl, 1903, and *Tubulanus (=Carinella) annulatus* (Montagu, 1804) described and figured by Retzius (1906) were mature and belong to the short-headed group wherein sex products are shed directly into the water column. Gontcharoff (1951) described the sperm of *Lineus sanguineus* and *L. pseudo-lacteus* as having a shorter head and different morphology than those of *L. ruber* but neither measurements nor figures were given.

7.3.5.2 Oogenesis

Gontcharoff (1951) reported that the gonial cells of *Lineus ruber* divided once but that only four or five of the resulting cells matured as oocytes. Olivier (1966) reported only two to four of the resulting cells developing into fully mature primary oocytes. [The effete ovaries of *L. ruber* and *L. viridis* contain a few oocytelike cells which lack vitelline accumulations (Figs. 3, 8).] He divided the oocyte growth period into two stages, i.e. previtellogenic and vitellogenic. In the previtellogenic stage, the RNA and phospholipids which were concentrated in the center of the nucleolus of the gonial cell migrate to the nucleolar periphery. (A single nucleolus is present in the oocytes of *L. ruber* and *L. viridis*.) This material concentrates to form a bleb on the nucleolus and against the nuclear membrane. Following this, vitellogenesis is inaugurated. The cytoplasm contains many strands of RNA which serve as formative centers for yolk granules. These granules begin to form at the time the nucleolar bleb contacts the nuclear membrane. In addition neutral glycoprotein reserves appear in the cytoplasm at this time. The nucleolar phospholipids and RNA disappear, and the nucleolus disintegrates, both apparently into the cytoplasm of the oocyte. The loss of nucleolar basiphilia at the time of spawning of *L. viridis* has been observed in our laboratory, and the presence of the vacuolated nucleolus has been observed until the time that the nuclear membrane disappears. Vitellogenesis in *Amphiporus angulatus* and in *Amphiporus cordiceps* (Jensen, 1878) appears to be a modification of that described above (unpublished observations). In these two species, the eggs are numerous and are attached by a stalk to the wall of the ovary. A thick coating is produced by the free portion of each oocyte so that a large amount of the space in each ovary consists of capsular material and the developing cells are held widely apart. The stalk is very narrow at its base, but spreads as a thin sheet on the ovarian wall. Previtellogenic stages have not been observed, but during vitellogenesis a number of nucleolarlike bodies are spaced over the entire inner side

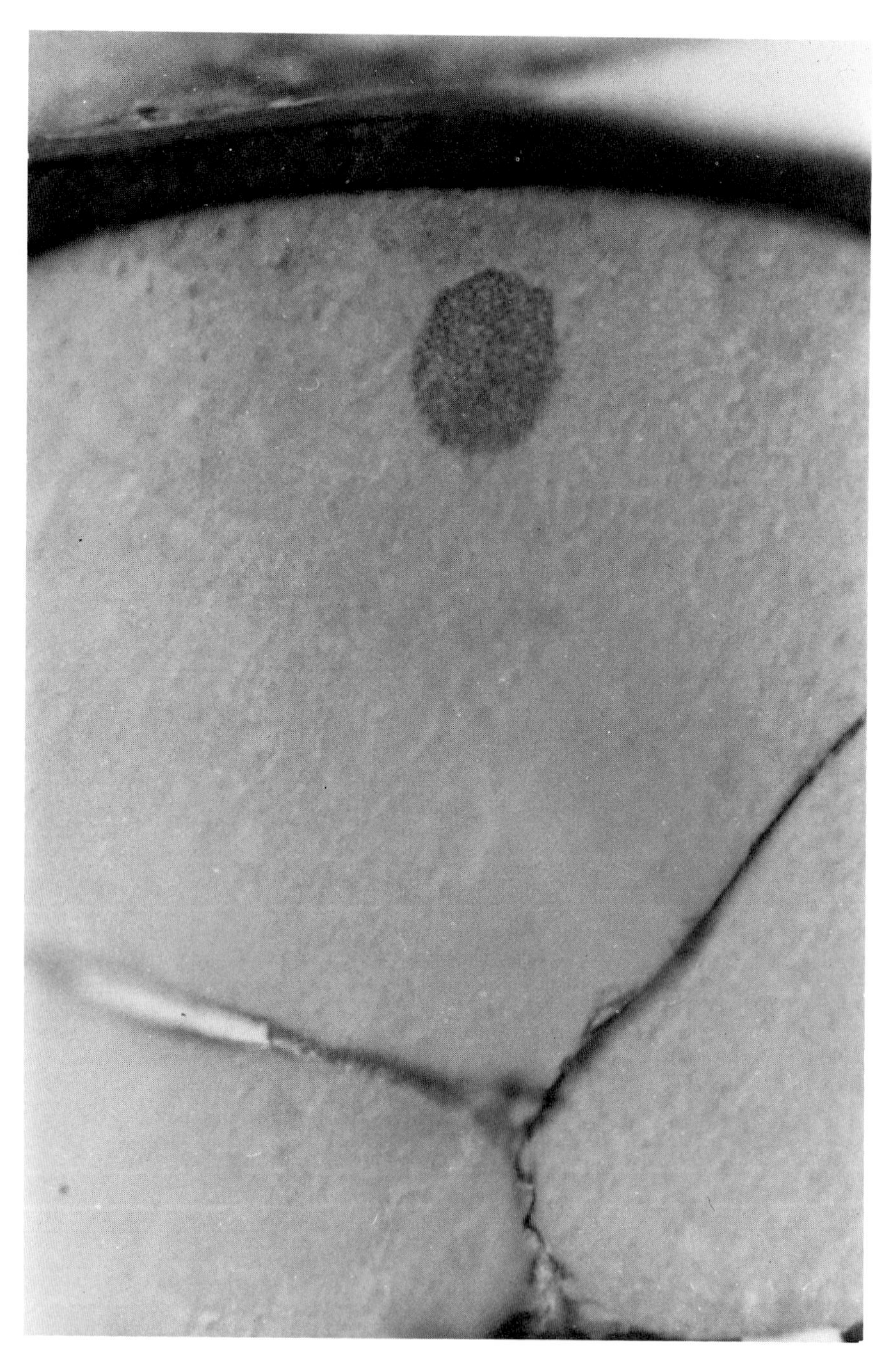

FIG. 4. Section through portions of three eggs in a cocoon of *Lineus ruber* showing chromatin of nucleus in diffuse state in one oocyte. (Hollande's fixative.)

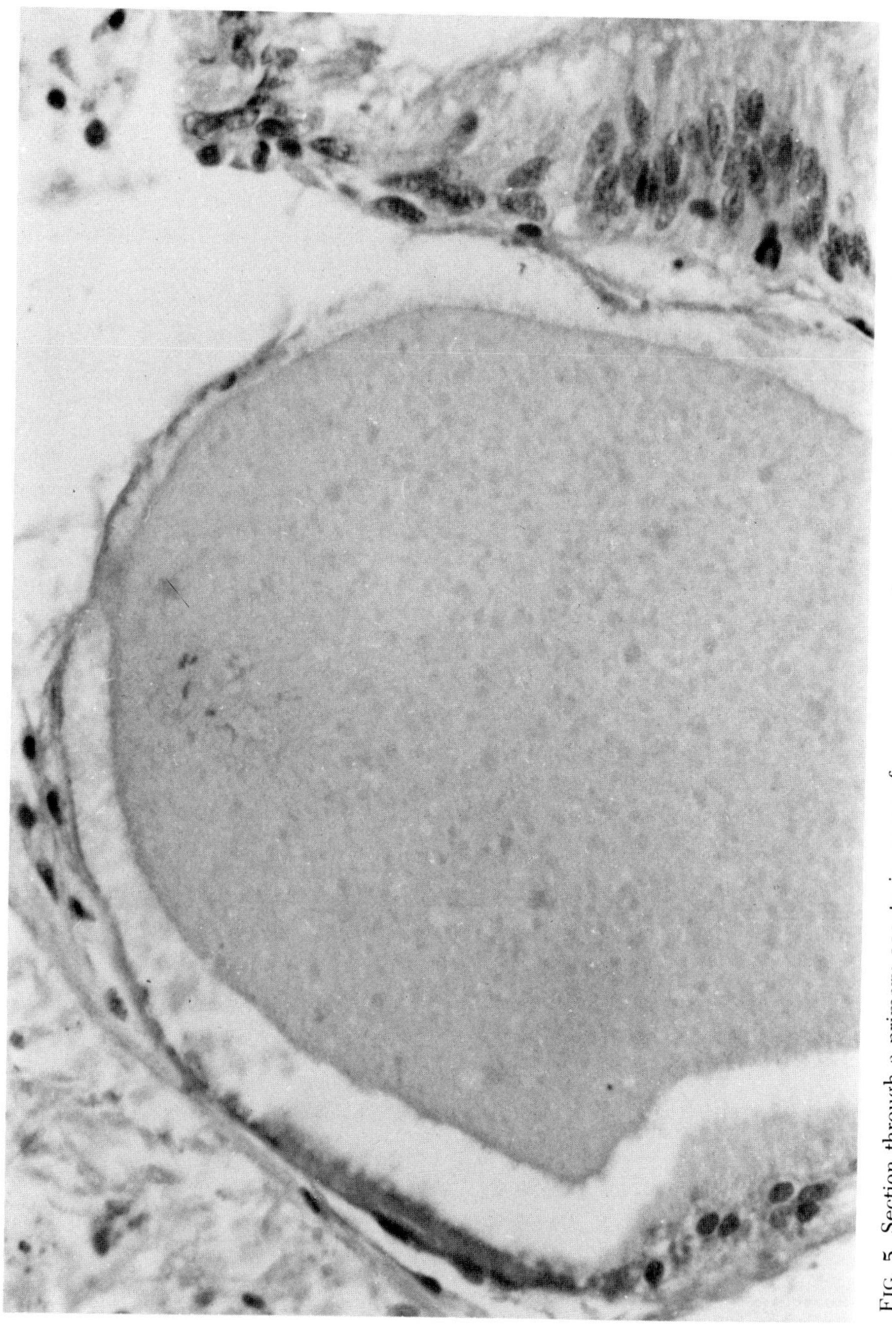

FIG. 5. Section through a primary oocyte in ovary of a spawning *L. ruber*, nucleus in late prophase, four of the chromosomes visible. (Hollande's fixative.)

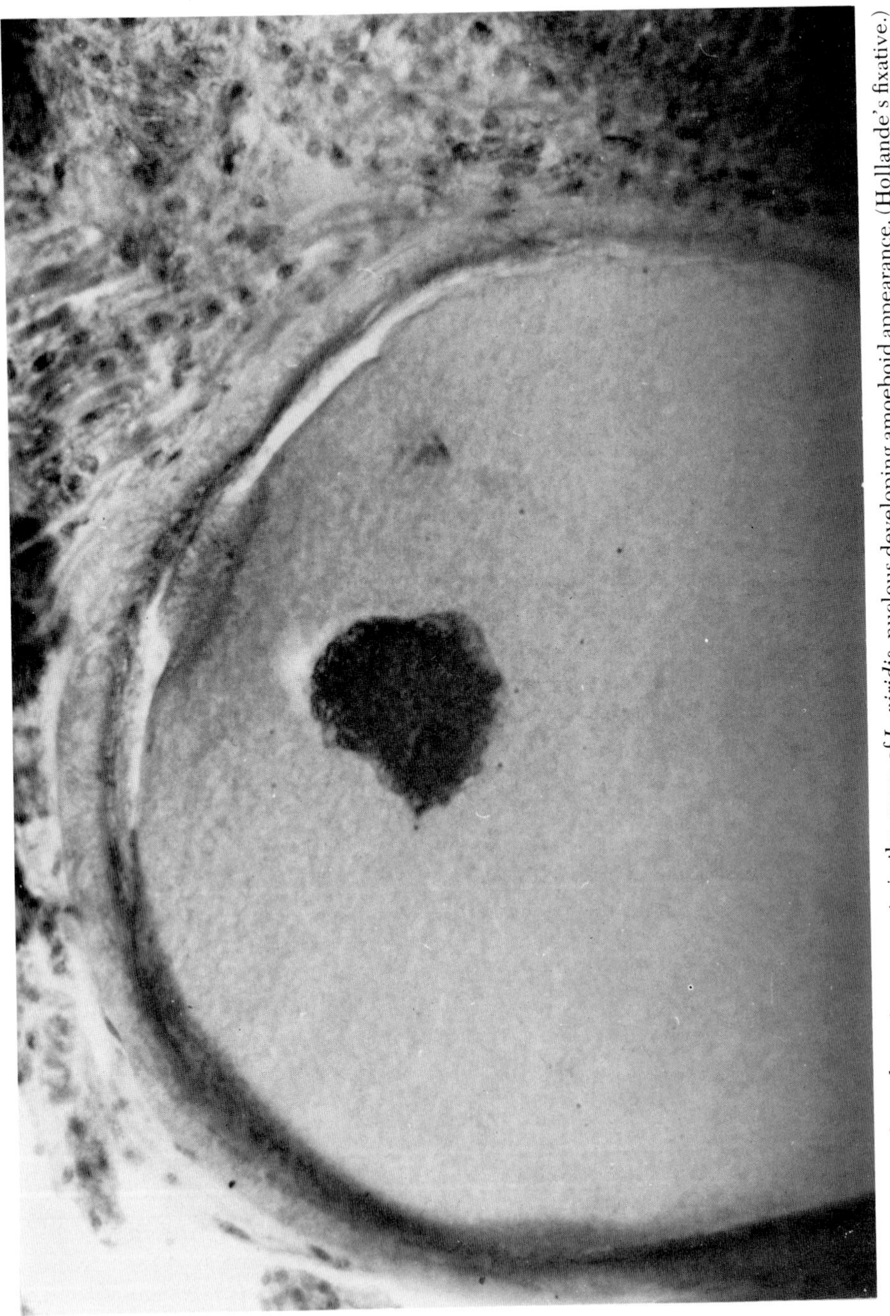

FIG. 6. Section through a primary oocyte in the ovary of *L. viridis*, nucleus developing amoeboid appearance. (Hollande's fixative.)

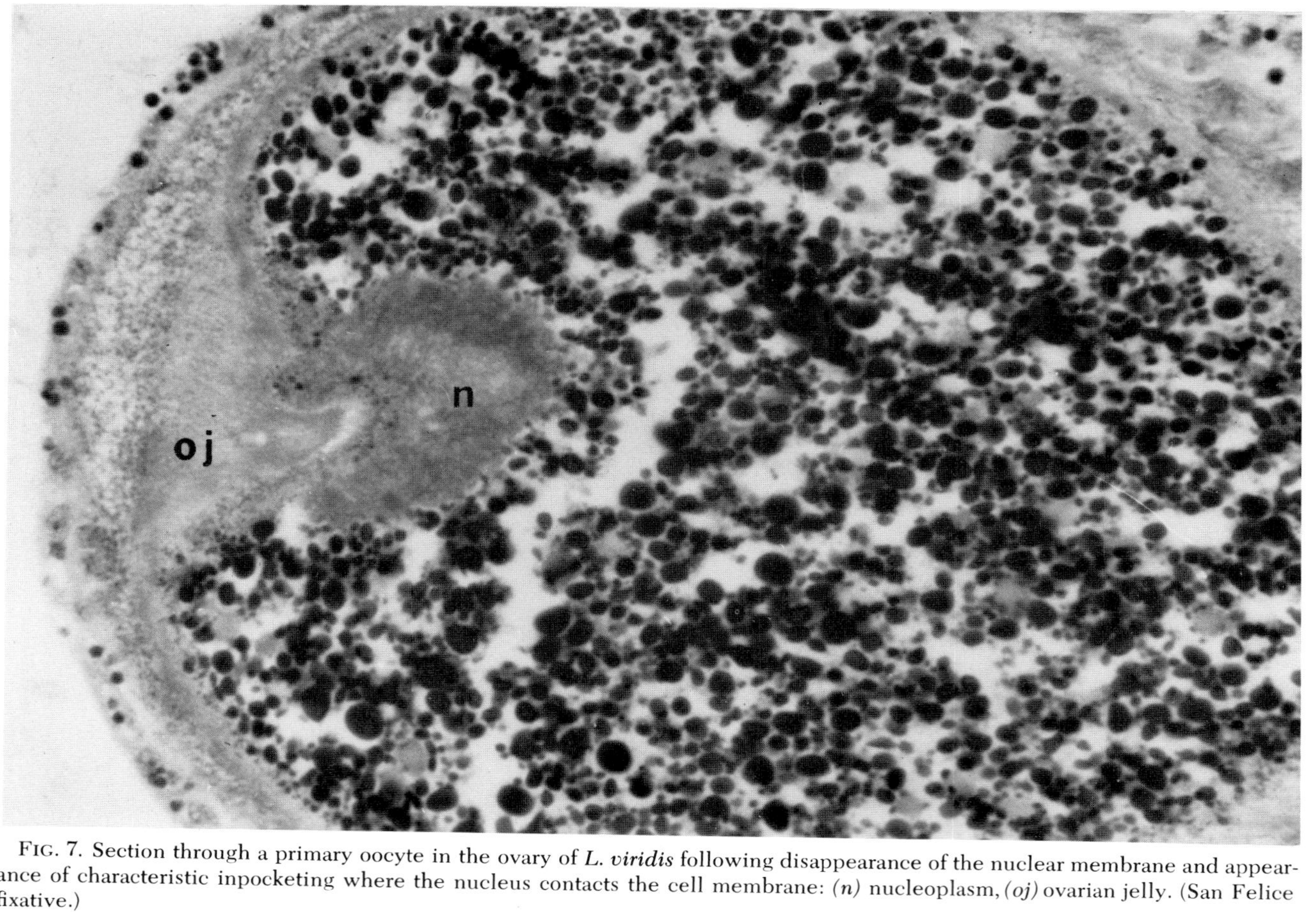

FIG. 7. Section through a primary oocyte in the ovary of *L. viridis* following disappearance of the nuclear membrane and appearance of characteristic inpocketing where the nucleus contacts the cell membrane: (*n*) nucleoplasm, (*oj*) ovarian jelly. (San Felice fixative.)

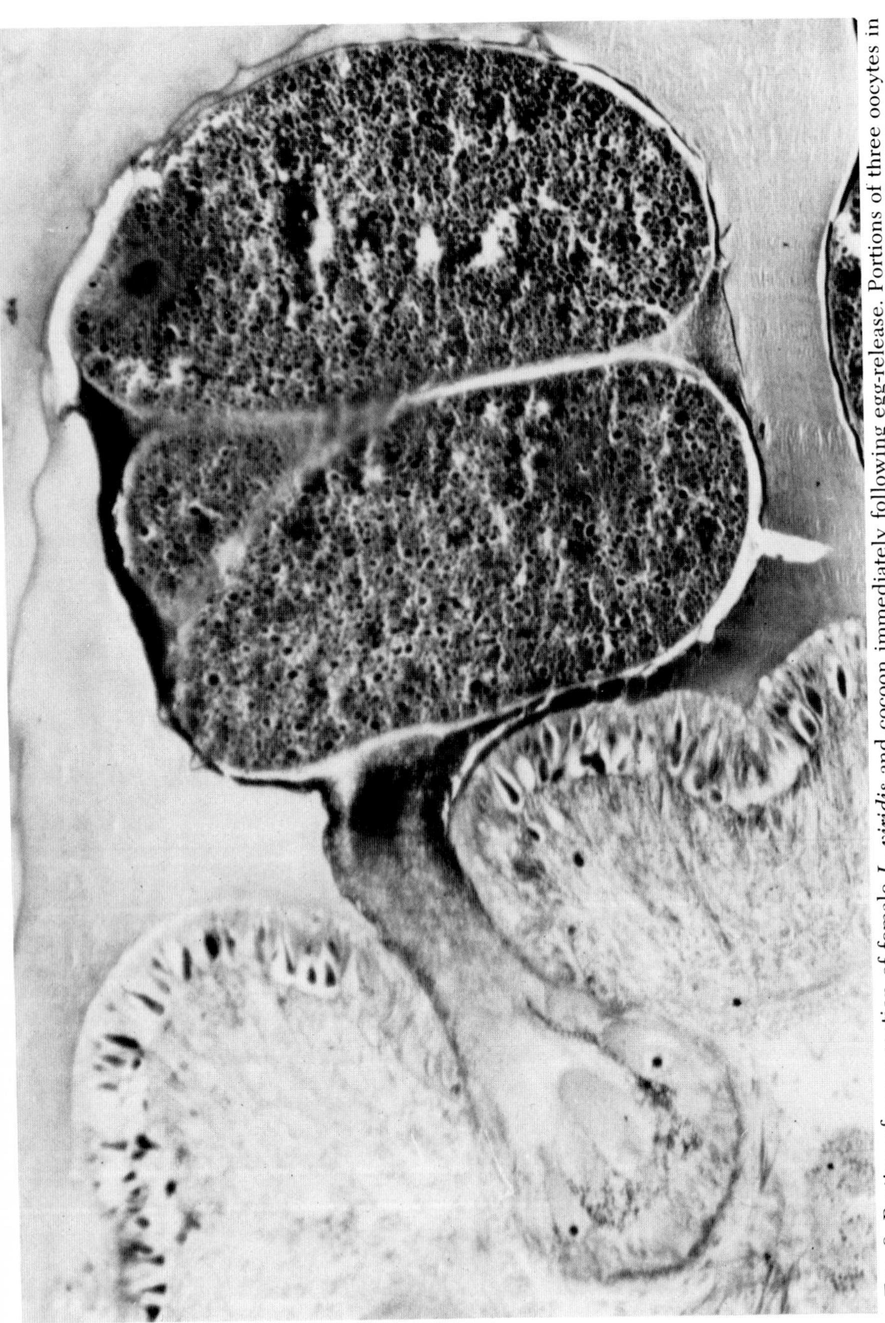

FIG. 8. Portion of cross section of female *L. viridis* and cocoon immediately following egg-release. Portions of three oocytes in cocoon. Late prophase showing in outermost oocyte. Three undifferentiated oocytes projecting from wall of ovary. Dark staining material streaming from ovary into capsule is the ovarian jelly. (San Felice fixative, Rhodanile-blue stain.)

of the nuclear membrane. These bodies range in size from 1.5-6.5 μm with the majority 4 μm in diameter. Multiple small nucleoli characterize the oocytes of most hoplonemertineans which I have investigated. In *Oerstedia dorsalis*, the nucleoli vary considerably in shape and tend to be disclike when in contact with the nuclear membrane. Previtellogenic oocytes of *Parborlasia corrugatus* have a single nucleolus 10 μm in diameter but oocytes undergoing vitellogenesis have one large nucleolus slightly under 10 μm in diameter and several small nucleoli ranging in size from 2 to 6 μm. The fully developed nucleolus of oocytes of *L. ruber* and *L. viridis* averages 6 μm in diameter. The single nucleolus of *Cerebratulus bicornis* (Joubin, 1892), a tropical Pacific species, is 31 μm in diameter and *Lineus longifissus* from antarctic waters has oocytes each containing a fully developed nucleolus 20 μm in diameter. *L. ruber* and *L. viridis* have much larger eggs than the other species mentioned above and thus the nature of the nucleolus and of stored substances in nemertinean eggs requires comparative studies patterned on the pioneering work of Olivier. It is essential to note, however, that the vitelline granules are dissolved by certain fixatives [Fig. 4,5,6, and Olivier (1966) Fig. 3.].

The nuclei of the oocytes of *L. ruber* and *L. viridis* become amoeboid (Fig. 6) at the time of spawning and migrate toward the margin of the cell. The nuclear membrane disappears (Fig. 7, 8) and the oocytes are discharged in late prophase. The chromosomes are distinct at this stage (Fig. 5) in *L. ruber* material which I have examined, but appear to be represented by a strongly basiphilic mass in *L. viridis* material (Fig. 8). Thus, the oocytes of both species are probably in metaphase at the time of sperm penetration. The chromatin material of some oocytes of *L. ruber* is in a diffuse state in the nucleus prior to spawning (Fig. 4) and remains in this condition after being extruded into the coccoon. These are probably the eggs in which development is later arrested, referred to by Schmidt (1932, 1934) and Gontcharoff (1960). Gontcharoff (1951) inadvertently reversed Schmidt's statements on egg types of *L. viridis* and *L. ruber* in her reference to his observations, but corrected this in her very important developmental study in 1960. The oocytes of *Lineus torquatus* are shed in the germinal vesicle stage according to Iwata (1957a). Gontcharoff (1951) described and figured sperm in the gonoducts and effete ovaries of *L. ruber.* In her figure she shows sperm in the epithelium of the gonoduct. I have found large numbers of sperm throughout the epidermis and dermis in the region of the gonoduct of postspawning female *L. viridis* (Fig. 3). Wilson (1900) reported that the first polar body appeared "about an hour and a quarter after fertilization" of the eggs of *Cerebratulus lacteus* and the second appeared shortly afterwards. Kostanecki (1902) described oogenesis in *Cerebratulus marginatus*. He had many problems because of polyspermy, but recognized them. Of most importance, however, he pronounced that a sperm carried its own centriole into the egg and this was the centriole of the first cleavage spindle. The controversy over the centriole of animal spermatozoa continued, even after Yatsu (1909), working with *Cerebratulus lacteus*, verified Kostanecki's observations.

Coe (1899a) and Kostanecki (1902) described a haploid chromosome number of 16 for *C. marginatus*. Yatsu (1909) reported a haploid number of 18 or 19 and a diploid number of 36 or 38 for *C. lacteus*. Gontcharoff (1961) reported a diploid number of 16 for *Lineus ruber*, *Prostoma graecense*, and *Malacobdella grossa*, and 32 for *Emplectonema gracile* and *Micrura caeca* Verrill, 1895. Lebedinsky (1897) described the reduction division in *Tetrastemma vermiculus* (Quatrefages, 1846) in which the primary oocyte had four diads, the secondary oocyte four monads, and the ootids two monads. No information exists on gametogenic cycles and almost nothing on factors influencing gametogenesis except a little on endocrine control.

7.3.6 Endocrine Control

Gontcharoff and Lechenault (1958) described spawning in December by headless specimens of *Lineus lacteus*, a species that shed sex products into the water column. They reported that they never encountered spawning except in headless forms and that spawning could be induced by decapitation. The authors posed the question of the possible role of the cerebral organs on reproduction. These observations on *L. lacteus* followed the earlier reference by McIntosh (1874) to the spawning by headless fragments of *L. marinus*, and the report of Benham (1896) on the budding off of the ripe posterior region of a species of *Tubulanus* (=*Carinella*). The latter author concluded that the "ripening of the gonads" was responsible for the fragmentation.

Precocious development of gonads and sex products following decapitation of *L. ruber* was reported by Bierne (1964) who postulated that a neurosecretion from the cerebral region passed via the nerve cords and inhibited sexual development. Fat and glycogen are stored in the gastrodermis and the parenchyma of nemertineans according to Gibson and Jennings (1969) and Jennings and Gibson (1969). The gonads of most nemertineans press against the intestine, and as the sex products mature, there is a large accumulation of fat in the maturing oocytes. The endocrine control could be acting upon the glycogen and fat reserves. [The failure of Pearse and Giese (1966) to demonstrate any "biochemical differences" between sexes or between ripe and spawned-out specimens of *Parborlasia corrugatus* can be attributed to their use of homogenates of whole animals. In this species almost 40% of the wet weight of the animal is in the nonreproductive and nonstorage preintestinal region.]

Bierne (1967b) utilized heterosexual transplants of *Lineus ruber* to demonstrate the lack of control of the head over sex determination. In addition transplants through the center of the reproductive region indicated that a hormone was elaborated by the testes and was carried throughout the body converting ovaries into ovotestes or markedly influencing ovarian differentiation. In 1970, Bierne reported the results of regeneration of heterosexual grafts of *L. ruber*. In these experiments, complete parabionts were produced by cutting sexual animals in half longitudinally and then grafting specimens together so that one side was entirely female and the other side male. Following healing, the parabionts were cut into fragments and allowed to regenerate. A preoesophageal fragment or head fragment regenerated as a gynandromorph. An oesophageal section underwent wound healing of both cut surfaces, but no regeneration. Posterior fragments regenerated with a deterioration of the ovaries followed by complete masculinization producing a "free-martin". Thus, it would appear that an androgenic factor which has a masculinizing effect is produced by the testis.

7.3.7 Spawning and Reproductive Behavior

Internal fertilization occurs in viviparous species and is assumed to occur in pelagic species. Species that shed their sex products into the surrounding water usually have primitive sperm. However, except for the sperm of *Cerebratulus marginatus* figured by Retzius (1906) all figures and references to the sperm of *C. lacteus* and *C. marginatus* which spawn into the surrounding water indicate a modified long-headed sperm. Some nemertineans lay their eggs in plaques or cocoons. Pseudocopulation is known for some of these species, and where this occurs, sperm are of the modified type. In addition to sperm type as an indicator of mating behavior and type of egg laying, the mature primary oocyte is useful in that a very thin enclosing membrane tends to be associated with an egg that is to be attached to the substrate while a thick coating on the egg is associated with an egg to be discharged into the water column.

Pseudocopulation occurs in *Lineus ruber* and *L. viridis* but remains to be proved for other species that aggregate. Knowledge of the cocoon of *Lineus ruber* dates back at least as far as Desor (1850) whose embryological studies on the species were the first for a nemertinean. Schmidt in 1932 distinguished between *L. ruber* and *L. viridis* partially on the basis of the cocoon, the type of larva, and the history of the eggs. Gontcharoff (1951) has given more detail on the process of egg-laying in the two species. In studies on the formation of the cocoon, I have observed the phenomena described by Gontcharoff on numerous occasions. Sometimes only the female is in the cocoon, at others both a male and female, and on several occasions *Lineus viridis*, as a result of crowding, has been encountered with one female and two males in the cocoon. Usually the males leave the cocoon before the female but remain in the vicinity, sometimes on the outer surface of the cocoon. Both sexes appear almost lifeless after the shedding of the sex products, and steadily deteriorate in the laboratory, dying within 2 or 3 weeks. In the females of these forms, there is also a marked hypertrophy of the gland cells of the cutis immediately before spawning. The serous glands are primarily responsible for the production of the cocoon. Gontcharoff and Lechenault (1966) investigated the ultrastructure and histochemistry of these glands and concluded that the unique staining reaction of the hypertrophied gland cells resulted from their physiological state and that they were not a special type of serous cell.

Under laboratory conditions, the author has observed *T. candidum* (Müller, 1774) females laying individual eggs in a random fashion, but also in a cocoon. In the latter case, a saddle-shaped cocoon is formed around the female and she remains within the cocoon for several days (the maximum number of days so far observed has been eleven at which time the vermiform larvae were active in their capsules). This cocoon consists of a thin outer lamella with the individual eggs attached to it by their capsules so that they are suspended from the cocoon wall (Fig. 9), quite unlike the cocoon of *L. ruber* and *L. viridis* in which the capsules are in a matrix secreted by the epidermal cells of the cutis and more than one egg normally occurs in each capsule. Eggs laid at random have seldom developed in our laboratory, but the majority of eggs in a cocoon have developed and released normal young. Humes (1942) described the males and females of *Carcinonemertes carcinophila* together in a slimy envelope at the time of reproduction; sometimes more than one pair were present in the mass. Sperm were extruded from the males into the slime. It should be noted that *Carcinonemertes* has modified sperm. Joubin (1914) postulated that the matrix containing the eggs of a new species described by him, *Amphiporus incubator,* consisted of the sloughed-off epidermis of the female enclosed in a secreted cuticle. The laminated cocoon of this species is closed at both ends, imprisoning the female, unlike *A. michaelseni* Bürger, 1896 also from the Antarctic, in which both ends are open and the female has freedom of movement.

7.3.8 Breeding Periods

Both *Lineus ruber* and *L. viridis* reproduce in the Gulf of Maine between February and May with the greatest number of cocoons appearing in March and April. From late May until mid-July, few specimens of either species can be found in nature and the majority of these are very small. The sexual condition of the large specimens of *L. viridis* which are sometimes encountered at this time of year requires further study. Schmidt and Jankovskaia (1938) reported that

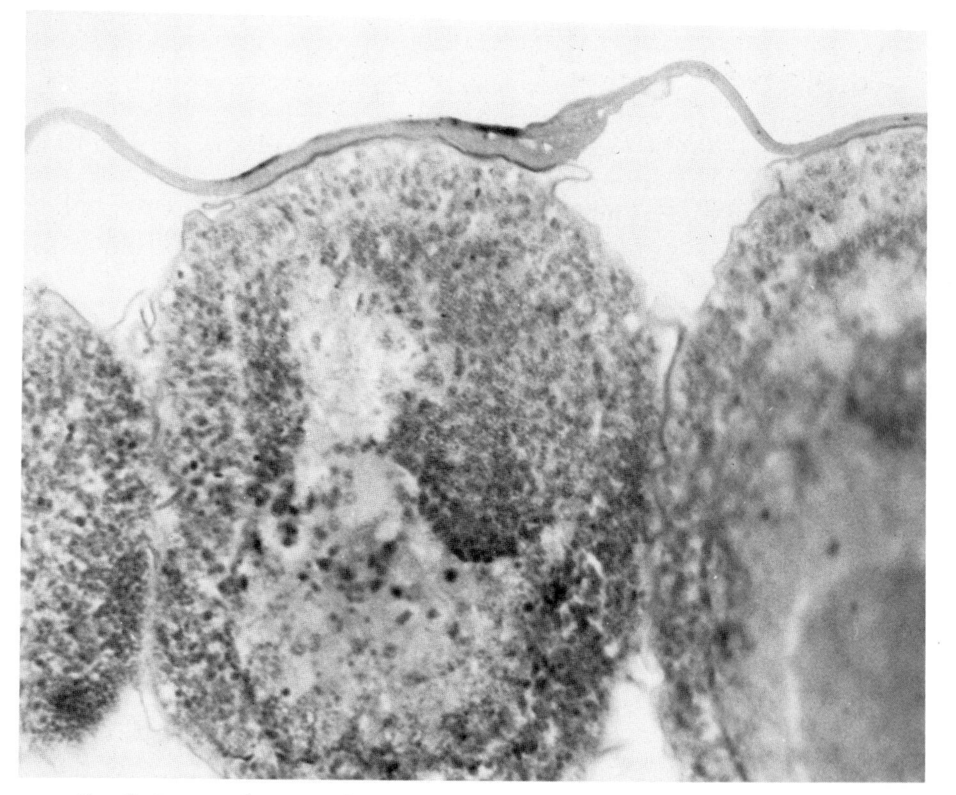

FIG. 9. Section of cocoon of *Tetrastemma candidum* showing attachment of eggs to the cocoon wall.

L. ruber reproduced in the Barents Sea in July and August, and at Roscoff from January to March. In an earlier study, (1934), Schmidt reported 8-12% development of the eggs of *L. ruber*. In the 1938 paper with Jankovskaia, the comprehensive study showed 12-13 eggs per capsule in each cocoon from the Barents Sea with 40% of the capsules containing eggs which did not develop. Cocoons from Roscoff average about five eggs per capsule but 60% of the capsules contained eggs which did not develop. Comparable observations have not been published for American material nor has there been any explanation for the sterile eggs. Gontcharoff (1960) carefully analyzed the phenomena of arrested development of eggs in the cocoons and their role in growth and survival. The breeding seasons of New England nemertineans were recorded by Coe (1899b). Bürger (1897-1907) included a table of breeding seasons for European and one for North American species. (The latter was taken primarily from the paper by Coe.) The reference to Verrill (1892) in which cocoons of "*Lineus viridis*" were found in Maine "in mid-summer" was interpreted as indicating breeding season. The cocoons in fact deteriorate slowly and can be found long after the larvae have departed. The majority of species in Kiel Bay were reported to be reproductive in the winter months by Friedrich (1935) but he gave no specific details of spawning. (Nemertineans may contain mature sex products for a long period of time before spawning, and thus literature citations of "ripe" are not too meaningful.) Iwata (1954) reported *Lineus fulvus* Iwata, 1954, *Oerstedia venusta* Iwata, 1954, and *O. dorsalis* to mature in August in Japanese waters, and the breeding season of *Lineus torquatus* to occur during July and August in the waters of Hokkaido. Müller (1962) reported *Emplectonema gracile* to reproduce in October in the Black Sea and Coe (1940) reported this species to reproduce in March on the coast of Southern California and in "early" summer farther north. *Vallencinia longirostris* Quatrefages, 1856 is sexually mature in summer according to Corrêa (1956). The type specimen of *Elcania patagonica* Moretto, 1970 collected in late October in San Julian Bay, Patagonia, was a male with formed gonoducts, with testes still containing spermatocytes. Spawned-out specimens of *Parborlasia corrugatus* were encountered by Pearse and Giese (1966) in McMurdo Sound, Antarctica, in November and December, indicating an early austral summer spawner.

Sexual reproduction has never been observed in *Lineus sanguineus*. The species reproduces asexually by fragmention. However, Gontcharoff (1950, 1951) reported a period of sexual development in this species from October to February followed by a degeneration of the sex products. Sex could not be determined in specimens collected between the months of May and October. Spermatogenesis results in modified sperm. She observed formation of a gonoduct and sperm discharge on one occasion. However she reported that the testes were frequently invaded by Orthonectids, and the oocytes were few in number and developed abnormally. Female specimens examined at intervals showed a steady disappearance of sex products. Some oocytes were found to have broken through into the gut while many ovaries contained gregarine parasites. Castration by the parasites was postulated. *Lineus socialis* is reported by Coe (1899b) to lay eggs in mucous masses in "mid-winter." In 1943 Coe stated that *L. socialis* reproduced sexually in the autumn and early spring with no additional details. This species also reproduces asexually by fission as does *Lineus vegetus*. Asexual reproduction in these forms alternates with sexual reproduction.

The note by Lassig (1964) on the reproduction of *Prostoma obscurum* Schultze, 1851 in the Baltic added some information to our knowledge ot reproduction by a viviparous species. His material was collected in June at depths of 20-80m where the salinity was 3.7-4.0 ‰. The water temperature was below 2°C in the collecting area and he reported reproduction occurring at temperatures below 1°C. The specimens described by Schultze had been collected intertidally at Greifswald and contained embryos. Friedrich (1935) described mature specimens collected at 8 m from the type locality in late April. He also reported that only young specimens of the species occurred in shallow coastal areas of Kiel Bay from June to September after which they disappeared. He postulated that the reproductive period occurred in the deep waters of the bay, and in the light of Karling's (1934) statements that sexually mature specimens were not found in the upper Baltic waters, he associated this with brackish waters. Lassig's (1964) report indicates that temperature rather than salinity is responsible for the migration to deeper water.

7.4 Development

7.4.1 Embryonic Development

Internal fertilization occurs in viviparous forms such as *Prosorhochmus claparedei* Keferstein, 1862 and *Prostoma obscurum* and possibly in some pelagic species. The eggs of *Cephalothrix galathea* (Dieck, 1874), *C. rufifrons* (Johnston, 1837) and *Carcinonemertes carcinophila* are sometimes fertilized in the ovary prior to being shed. However, external fertilization characterizes most nemertinean species.

Very little information on the embryology of this group has appeared since the reviews by Hyman (1951) and Iwata (1957a). Actually almost all of the descriptive embryology was published prior to 1920 and only a few of the conclusions have been confirmed. Cleavage is spiral and, after the four-cell stage, is determinant. The initial four blastomeres are capable of producing miniature normal larvae, but larvae obtained from isolated blastomeres at the eight-cell stage are abnormal. The term micromere is utilized topographically in nemertinean embryology and does not refer to cell size since a marked size distinction between the first eight blastomeres does not occur in most species although the D cell of *Lineus ruber* was described by Nusbaum and Oxner (1913) as larger than the other three cells of the first quartet. A coeloblastula is formed and gastrulation is by emboly or unipolar ingression or a combination of the two processes. The mesoblasts are in close proximity to the entoblasts and in *Malacobdella* according to Hammarsten (1918) are derived from cells $2a^{111}$, $2b^{111}$, $2c^{111}$, $2d^{111}$, and according to Nusbaum and Oxner (1913) from 4d in *L. ruber*. There is a question as to whether mesodermal bands form. The entire problem of the mesoderm requires serious investigation. Organogenesis in those forms which undergo direct development is normally completed except for anus formation prior to hatching.

7.4.2 Larvae

Anarchic or necrobiotic larvae are produced by some heteronemerteans. These are of two types, i.e., pilidia larvae and Desor's larvae. The latter type has only been described for *Lineus viridis*. In both larval types ectodermal invaginations separate from the larval ectoderm and spread to enclose the larval gut and thus form the definitive epidermis. The larval ectoderm is shed at metamorphosis. Cantell (1966) agreed with earlier reports that the thin, brittle, inflated envelope of "*pilidium auriculatum*" type larvae was shed by the emerging worm and described the ingestion of the larval pilidial envelope by six "species" of metamorphosing worms. Pilidia are known for species of *Cerebratulus* and *Micrura*, and for *Lineus lacteus* and *L. torquatus*. Coe (1899c) described the pilidia of some New England heteronemerteans. Thorson (1946) reported that pilidia constitute a very small portion of the larvae of bottom forms encountered in his study. He described and figured a number of types, but knowledge is too incomplete at present to relate pilidia to adult species.

Pelagic lecithotrophic larvae with direct development are known for *Cephalothrix linearis* (Rathke, 1799), *C. rufifrons* and *Malacobdella grossa*. There is no information on the settling of pelagic nemertinean larvae at the present time. The nonpelagic larvae of *L. ruber* show no phototaxis while those of *L. viridis* exhibit a positive phototaxis. This reaction to light could be a result of the "well-fed" condition of *L. ruber* larvae resulting from their feeding on the "nurse-eggs" and thus allowing them to begin a bottom dwelling existence upon hatching. Gontcharoff (1959, 1960) noted that *L. ruber* larvae began to feed upon leaving the cocoon, but she was unable to induce feeding by *L. viridis* larvae until they lost their positive phototaxis.

After spawning, the walls of the intestine and ovary of *Amphiporus incubator* break down, and cells pass through the gonoducts (which are ventral) into the matrix of the brood chamber. These cells as well as the mucous matrix and sloughed-off epidermal cells were absent from cocoons containing fully formed young worms and Joubin (1914) postulated that this material had been ingested by the young worms.

7.5 References

Benham, W. B. (1896). Fission in nemerteans. *Quart. J. Microsc. Sci.* **39,** 19-31.

Bierne, J. (1962a). La régénération des gonades chez la Némerte *Lineus ruber* Müller. *C. R. Acad. Sci. Paris* **255,** 185-187.

Bierne, J. (1962b). La régénération de la trompe chez les Némertes. *Bull. Biol. Fr. Belg.* **96,** 481-504.

Bierne, J. (1964). Maturation sexuelle anticipée par décapitation de la femelle chez l'Hétéronémerte *Lineus ruber* Müller. *C. R. Acad. Sci. Paris* **259,** 4841-4843

Bierne, J. (1967a). Viabilité, stabilité phénotypique, croissance et régénération postérieure de chimères interspécifiques obtenues par la greffe chez des Némertiens Lineidae adultes. *C. R. Acad. Sci. Paris* **264,** 1080-1083.

Bierne, J. (1967b). Sur le contrôle endocrinien de la différenciation du sexe chez la Némerte *Lineus ruber* Müller. La masculinisation des ovaires des chimères Hétérosexuées. *C. R. Acad. Sci. Paris* **265,** 447-450.

Bierne, J. (1967c). Régénération postérieure des chimères interspécifiques chez les Lineidae (Hétéronémertes). *Bull. Soc. Zool. Fr.* **92**(2), 351-359.

Bierne, J. (1970). Influence des facteurs hormonaux gonado-inhibiteur et androgène sur la différenciation sexuelle des parabionts hétérosexués chez un némertien. *Ann. Biol.,* **9,** 395-400.

Bürger, O. (1895). Nemertinen. *Fauna Flora des Golfs von Neapel, Monogr.* **22,** 1-343.

Bürger, O. (1897-1907). Nemertini. *Bronn's Klas. Ord. d. Tierreichs. Suppl.* **4**, 1-542.

Cantell, C.-E. (1966). The devouring of the larval tissues during the metamorphosis of pilidium larvae (Nemertini). *Ark. Zool.* **18**, 489-493.

Coe, W. R. (1899a). The maturation and fertilization of the egg of *Cerebratulus. Zool. Jahrb. Anat. Ont.* **12**, 425-476.

Coe, W. R. (1899b). Notes on the times of breeding of some common New England nemerteans. *Science* **9**, 167-169.

Coe, W. R. (1899c). On the development of the pilidium of certain nemerteans. *Trans. Conn. Acad. Arts Sci.* **10**, 235-262.

Coe, W. R. (1904). The anatomy and development of the terrestrial nemertean *(Geonemertes agricola)* of Bermuda. *Proc. Boston Soc. Natur. Hist.* **31**, 531-570.

Coe, W. R. (1929a). Regeneration in nemerteans. *J. Exp. Zool.* **54**, 411-459.

Coe, W. R. (1929b). The regeneration of minute sectors cut from the bodies of nemertean worms. *Science* **69**, 502.

Coe, W. R. (1930a). Asexual reproduction in nemerteans. *Physiol. Zool.* **3**, 297-308.

Coe, W. R. (1930b). Regeneration in nemerteans. II. Regeneration of small sections of the body split or partially split longitudinally. *J. Exp. Zool.* **57**, 109-144.

Coe, W. R. (1931). A new species of nemerteans *(Lineus vegetus)* with asexual reproduction. *Zool. Anz.* **94**, 54-60.

Coe, W. R. (1932). Regeneration in nemerteans. III. Regeneration in *Lineus pictifrons. J. Exp. Zool.* **61**, 29-41.

Coe, W. R. (1934a). Analysis of the regenerative processes in nemerteans. *Biol. Bull.* **66**, 304-315.

Coe, W. R. (1934b). Regeneration in nemerteans. IV. Cellular changes involved in restitution and reorganization. *J. Exp. Zool.* **67**, 283-314.

Coe, W. R. (1940). Revision of the Nemertean fauna of the Pacific Coasts of North, Central, and Northern South America. *Allan Hancock Pac. Expedit.* **2**(13), 247-322.

Coe, W. R. (1943). Biology of the nemerteans of the Atlantic Coast of North America. *Trans. Conn. Acad. Arts Sci.* **35**, 129-328.

Corrêa, D. D. (1948). *Ototyphlonemertes* from the Brazilian Coast. *Com. Zool. Mus. Montevideo* **2** (49), 1-12.

Corrêa, D. D. (1956). Estudo de Nemertinos Mediterrâneos (Palaeo e Heteronemertini). *An. Acad. Brasil* Ciênc. **28**, 195-214.

Corrêa, D. D. (1961). Nemerteans from Florida and Virgin Islands. *Bull. Mar. Sci. Gulf Carib.* **11**, 1-44.

Corrêa, D. D. (1966). A new hermaphroditic nemertean. *An. Acad. Brasil Ciênc.* **38**(2), 365-370.

Dawydoff, C. (1909). Sur la Régénération de l'extremite Postérieure chez les Némertiens. *Bull. Acad. Imp. Sci. St. Petersbourg* **3**, 301-311.

Dawydoff, C. (1910). Restitution von Kopfstücken, die vor der Mundöffnung abgeschnitten waren, bei den Nemertinen *(Lineus lacteus). Zool. Anz.* **36**, 1-6.

Dawydoff, C. (1942). Régénération créatrice chez les némertes. *Bull. Biol. Fr. Belg.* **76**, 58-141.

Desor, A. (1850). On the embryology of *Nemertes* with an appendix on the embryonic development of *Polynoë*; and remarks upon the embryology of marine worms in general. *Boston J. Natur. Hist.* **6**, 1-18.

Ehrenberg, C. G. (1831). Symbolae physicae seu icones et descriptiones corporum naturalium novorum aut minus cognitorum. Phytozoa turbellaria. Berolini.

Franzén, Å. (1956). On spermiogenesis, morphology of the spermatozoon, and biology of fertilization among invertebrates. *Zool. Bidr. Uppsala* **31**, 355-482.

Friedrich, H. (1933). Morphologische Studien an Nemertinen der Kieler Bucht, I und II. Z. *Wiss. Zool.* **144,** 496-509.

Friedrich, H. (1935). Studien zur Morphologie, Systematik und Ökologie der Nemertinen der Kieler Bucht. *Arch. Naturgesch.* **4,** 293-375.

Friedrich, H. (1936). Nemertini. *In* "Die Tierwelt der Nord- und Ostsee" (G. Grimpe and E. Wagler, eds), Vol. IVd, pp. 1-69. Leipzig.

Friedrich, H. (1957). Beiträge zur Kenntnis der arktischen Hoplonemertinen. *Viden. Medd. Dansk. natur. Foren.* **119,** 129-154.

Friedrich, H. (1970). Nemertinen aus Chile. *Sarsia* **40,** 1-80.

Gibson, R., and Jennings, J. B. (1969). Observations on the diet, feeding mechanisms, digestion and food reserves of the entocommensal rhynchocoelan *Malacobdella grossa. J. Mar. Biol. Ass. U. K.* **49,** 17-32.

Gontcharoff, M. (1950). Sur la ReproductionSexuée chez *Lineus sanguineus (Lineus ruber β).C. R. Acad. Sci. Paris* **230,** 233-234.

Gontcharoff, M. (1951). Biologie de la Régénération et de la Reproduction chez quelques Lineidae de France. *Ann. Sci. Natur. (Zool.)* **13,** 149-235.

Gontcharoff, M. (1959). Rearing of certain Nemerteans (Genus *Lineus*). *Ann. N. Y. Acad. Sci.* **77,** 93-95.

Gontcharoff, M. (1960). Le Développement Post-Embryonnaire et la Croissance chez *Lineus ruber* et *Lineus viridis* (Nemertes, Lineidae). *Ann. Sci. Natur. (Zool.) suppl. 12* **2,** 225-279.

Gontcharoff, M. (1961). Embranchement des Némertiens. *In* "Traité de Zoologie" (P. Grassé, ed.), Vol. 4(1), pp. 785-886. Masson, Paris.

Gontcharoff, M. and Lechenault, H. (1958). Sur le Déterminisme de la Ponte chez *Lineus lacteus. C. R. Acad. Sci. Paris* **246,** 1630-1632.

Gontcharoff, M., and Lechenault, H. (1966). Ultrastructure et histochimie des glandes sous épidermiques chez *Lineus ruber* et *Lineus viridis. Histochemie* **6,** 320-335.

Hammarsten, O. D. (1918). Beitrag zur Embryonalentwicklung der *Malacobdella grossa* (Müll). *Inaug. Diss.* 1-96. 10 pl. Uppsala.

Hubrecht, A. W. (1887). The relation of the Nemertea to the Vertebrata. *Quart. J. Microsc. Sci.* **27,** 605-644.

Humes, A. G. (1941). The male reproductive system in the nemertean genus *Carcinonemertes. J. Morphol.* **69,** 443-454.

Humes, A. G. (1942). The morphology, taxonomy and bionomics of the nemertean genus *Carcinonemertes. Ill. Biol. Monogr.* **18**(4), 1-105.

Huschke, E. (1830). Beschreibung und Anatomie eines neuen an Sicilien gefundenen Meerwurms,*Notospermus drepanensis. Isis* **23,** 681.

Hyman, L. H. (1951). The Acoelomate Bilateria - Phylum Rhynchocoela. *In* "The Invertebrates" Vol. 2, Chapter 11, pp. 459-531. McGraw-Hill, New York.

Iwata, F. (1954). The fauna of Akkeshi Bay. XX. Nemertini in Hokkaido. *J. Fac. Sci. Hokkaido Univ. suppl. 6,* **12,** 1-39.

Iwata, F. (1957a). Nemertini. *In* "Invertebrate Embryology" (M. Kumé and K. Dan, eds.), pp. 144-158. Clearinghouse Fed. Sci. Tech. Inf., TT67-58050. 1968.

Iwata, F. (1957b). On the early development of the nemertine *Lineus torquatus* Coe. *J. Fac. Sci. Hokkaido Univ. suppl. 6* **13,** 54-58.

Iwata, F. (1970). On the brackish water nemerteans from Japan, provided with special circulatory and nephridial organs useful for osmoregulation. *Zool. Anz.* **184,** 133-154.

Jennings, J. B., and Gibson, R. (1969). Observations on the nutrition of seven species of rhynchocoelan worms. *Biol. Bull.* **136,** 405-433.

Joubin, L. (1914). Némertiens. *In* "Deuxième Expédition Antarctic Francaise 1908-

1910," pp. 1-33. Masson, Paris.

Karling, T. G. (1934). Ein Beitrag zur Kenntnis der Nemertien des Finnischen Meerbusens. *Mem. Soc. Fauna Flora Fennica* **10**, 76-90.

Kipke, S. (1932). Studien über Regenerationserscheinungen bei Nemertinen. (*Prostoma graecense* Böhmig). *Zool. Jahrb. Allgem.* **51**, 1-66.

Kostanecki, M. C. (1902). Über die Reifung und Befruchtung des Eies von *Cerebratulus marginatus*. *Bull. Akad. Sci. Cracovie, Math.-Natur. Kl.* pp. 270-277.

Lassig, J. (1964). Notes on the occurrence and reproduction of *Prostoma obscurum* Schultze (Nemertini) in the inner Baltic. *Ann. Zool. Fennici* **1**, 146.

Lebedinsky, J. (1897). Entwicklungsgeschichte der Nemertinen. *Arch. Mikrosk. Anat. Entwicklungsmech.* **49**, 503-556.

Lee, A. B. (1887). La spermatogenèse chez les némertiens. *Rec. Zool. Suisse* **4**, 409-430.

McIntosh, W. C. (1873). A monograph of the British annelids. Part 1. The Nemerteans, pp. 1-96. Ray Soc., London.

McIntosh, W. C. (1874). A monograph of the British annelids. Part 1. cont'd. The Nemerteans, pp. 97-213. Ray Soc., London.

Monastero, S. (1928). Esperienze sulla rigenerazione dei Nemertini (*Lineus nigricans* Bürger 1892 e *Prostoma melanocephalum melan.* Johnst. 1837). *Boll. Ist. Zool. Univ. Palermo* **2**, 1-16.

Moretto, H. J. A. (1970). Sobre un Hoplonemertino Monostilifero de la Bahia de San Julian (Patagonia). *Neotropica* **16**, 17-34.

Müller, G. I. (1962). Contributii La Studiul Nemertienilor Din Marea Neagra (Litoralul Romînesc). *Stu. Cerc. Biol. Ser. Biol. Anim.* **14**, 371-384.

Nusbaum, J., and Oxner, M. (1910a). Beiträge zur Kenntnis der Regenerationserscheinungen bei den Nemertinen. *Bull. Int. Acad. Sci. Cracovie Cl. Sci. Math. Nat. suppl. B*, 1-11.

Nusbaum, J., and Oxner, M. (1910b). Über die Ungleichartigkeit des Regenerationsrhythmus in verschiedenen Körperregionen desselben Tieres. (Beobachtungen an der Nemertine *Lineus ruber* Müll.). *Bull. Int. Acad. Sci. Cracovie, Cl. Sci. Math. Natur. suppl. B*, 439-447.

Nusbaum, J., and Oxner, M. (1910c). Studien über die Regeneration der Nemertinen. I. Regeneration bei *Lineus ruber* (Müll.), Teil I-III. *Arch. Entw. Mech.* **30**, 74-132.

Nusbaum, J., and Oxner, M. (1911a). Bildung des ganzen neuen Darmkanals durch Wanderzellen mesodermalen Ursprungs bei der Kopfrestitution des *Lineus lacteus* (Grube) (Nemertine). *Zool. Anz.* **37**, 302-315.

Nusbaum, J., and Oxner, M. (1911b). Weitere Studien über die Regeneration der Nemertinen. I. Regeneration bei *Lineus ruber* Müll. Teil IV. u. V. *Arch. Entw. Mech.* **32**, 349-396.

Nusbaum, J., and Oxner, M. (1912a). Fortgesetzte Studien über die Regeneration der Nemertinen. II. Regeneration des *Lineus lacteus*. *Arch. Entw. Mech.* **35**, 236-308.

Nusbaum, J., and Oxner, M. (1912b). Zur Regeneration der Nemertinen. Verh. VIII. Internat. Zool. Kongr. 1910, pp. 631-635. Graz.

Nusbaum, J., and Oxner, M. (1913). Embryonalentwicklung des *Lineus ruber* Müll. Ein Beitrag zur Entwicklungsgeschichte der Nemertinen. *Z. Wiss. Zool.* **107**, 78-197.

Oersted, A. S. (1844). Entwurf einer systematischen Eintheilung und speciellen Beschreibung der Plattwürmer auf mikroskopische Untersuchung gegründet, 76 pp. Kopenhagen.

Olivier, J. (1966). Cytochimie de l'Ovocyte au Cours de la Vitellogenese chez *Lineus ruber* (Némerte). *Ann. Univ. Reims L'Arers* **4**, 158-165.

Oxner, M. (1909). Sur deux modes différents de régénération chez *Lineus ruber. C. R. Acad. Sci. Paris* **148**, 1424-1426.

Oxner, M. (1910a). Étude sur la régénération chez les némertes. I. La régénération chez "*Lineus ruber*" (Müll.). *Ann. Inst. Ocean. Monaco,* **1**(8), 1-32.

Oxner, M. (1910b). Analyse biologique du phénomène de la régénération chez *Lineus ruber* (Müll.) et *L. lacteus* (Rathke). *C. R. Acad. Sci. Paris* **150**, 1618-1620.

Oxner, M. (1911). Analyse biologique d'une serie d'expériences concernant l'avenement de la maturité sexuelle, la régénération et l'inanition chez les némertiens, *Lineus ruber* (Müll). et *L. lacteus* (Rathke). *C. R. Acad. Sci. Paris* **153**, 1168-1171.

Oxner, M. (1912). Contribution à l'analyse biologique du phénomène de la régénération chez les némertiens. *Bull. Inst. Ocean. Monaco* No. 236.

Pearse, J. S., and Giese, A. C. (1966). The organic constitution of several benthonic invertebrates from McMurdo Sound, Antarctica. *Comp. Biochem. Physiol.* **18**, 47-57.

Retzius, G. (1904). Zur Kenntnis der Spermien Evertebraten. 1. *Retzius Biol. Untersuch. N. F.* **XI**, 1-33.

Retzius, G. (1906). Die Spermien der Enteropneusten und der Nemertien. *Retzius Biol. Untersuch. N. F.* **XIII**, 37-40.

Riepen, O. (1933). Anatomie und Histologie von *Malacobdella grossa* Müll. *Z. Wiss. Zool.* **143**, 323-496.

Sabatier, A. (1883). Spermatogenese chez les némertiens. *Rev. Sci. Nat. Montpelier* **2**, 165-181.

Schmidt, G. A. (1932). Dimorphisme embryonnaire de *Lineus ruber* de la côte Mourmane et de Roscoff. *Bull. Inst. Ocean. Monaco* No. 595.

Schmidt, G. A. (1934). Ein zweiter Entwicklungstypus von *Lineus gesserensis-ruber* O. F. Müller. *Zool. Jahrb. Anat. Ontog.* **58**, 607-660.

Schmidt, G. A., and Jankovskaia, L. A. (1938). Biologie de la réproduction de *Lineus gesserensis-ruber* de Roscoff et du Golfe de Kola. *Arch. Zool. Exp. Gener.* **79**, 487-513.

Takakura, U. (1910). Kisei himomushi no ichi shinshu. (On a new species of parasitic nemertean.) *Dobuts. Zasshi Tokyo* **22**, 111-116.

Thorson, G. (1946). Reproduction and larval development of Danish marine bottom invertebrates with special reference to the planktonic larvae in the sound (Oeresund). *Medd. Komm. Havund. Kjobenh.(Plankton)* **4**, 1-523.

Tucker, M. (1959). Inhibitory control of regeneration in nemertean worms. *J. Morphol.* **105**, 569-599.

Verrill, A. E. (1892). The marine nemerteans of New England and adjacent waters. *Trans. Conn. Acad. Arts. Sci.* **8**, 382-456.

Wilson, C. B. (1900). The habit and early development of *Cerebratulus lacteus. Quart. J. Microsc. Sci.* **43**, 97-198.

Yatsu, N. (1909). Observations on ookinesis in *Cerebratulus lacteus* Verrill. *J. Morphol.* **20**, 353-401.

CHAPTER 8

NEMATODA

W. D. Hope

8.1 Introduction

Nematodes are small, spindle-shaped metazoans possessing digestive, reproductive, and nervous systems within a pseudocoelomic body cavity. The body wall consists of numerous elongate, longitudinally oriented muscles, hypodermal tissue, additional elements of the nervous system, and an external nonchitinous cuticle bearing various sensory and secretory structures, i.e., pores, papillae, and setae.

There are an estimated 10,000 nominal species of nematodes which dwell as free-living inhabitants of soil and marine and freshwater sediments, or as parasites in nearly every major group of multicellular plants and animals. Naturally, the economically important parasites, most of which are included in the class Secernentea, have been the subject of much more research than the free-living forms. Thus, while numerous studies have been made on the reproduction of nematodes parasitizing man and domestic plants and animals, some of which led to the founding of important general principles underlying sexual reproduction (Bütschli, 1875; van Beneden, 1883; Boveri, 1892, 1899, 1909), the knowledge of reproduction among free-living nematodes, and especially marine species, is sparse and largely limited to taxonomic descriptions and studies of comparative morphology. While this is an attempt to review in detail the relatively limited information concerning the reproduction of marine species, the literature of the soil and parasitic species has also been liberally drawn upon to document that which appears to be of fundamental importance in, or unique to, the reproduction of nematodes as a group. This is not intended to be an exhaustive treatment of nematode reproduction since, of necessity, much detail has been deleted, but it is hoped that it will serve as a useful introduction for future research on the reproduction of marine species.

Each of the fourteen orders of nematodes is represented by species mentioned in the text. As an aid to understanding the taxonomic relationships of these species, an abbreviated classification of the Nematoda is given in Table I, and each species is listed under its appropriate superfamily name. The principal habitat and mode of existence (free-living or parasitic) typical of the representatives of each order are also indicated.

8.2 Asexual Reproduction

There are no known cases in which nematodes reproduce by fission or polyembryony, but asexual reproduction by parthenogenesis and hermaphroditism has been documented clearly for some soil and parasitic species through cytological studies. Parthenogenesis in these forms is accomplished through one of three variations in oogenesis which are reviewed here because similar variations may occur among the marine forms. Two of these variations represent meiotic and the third mitotic parthenogenesis. Arranged with increasing deviation from the usual amphimictic oogenesis, they are as follows: (A) Oogenesis proceeds as in normal amphimictic species, with synapsis and two maturation divisions, the first a disjunction of homologous (paired) chromosomes, and the second equational separation of split chromosomes. However, the pronucleus of what would be the

TABLE I

ABBREVIATED CLASSIFICATION OF THE NEMATODA WITH LIST OF GENERA AND SPECIES MENTIONED IN TEXT

ADENOPHOREA VON LINSTOW, 1905

Enoplida (marine and freshwater sediments)

Enoplina

Enoploidea

Anticoma limalis Bastian, 1865; *Anticoma pellucida* Bastian, 1865; *Anticoma typica* Cobb, 1891; *Cylicolaimus magnus* (Villot, 1875) de Man, 1889; *Deontostoma californicum* Steiner and Albin, 1933; *Enoploides amphioxi* Filipjev, 1918; *Enoplus communis* Bastian, 1865; *Enoplus groenlandicus* Ditlevsen, 1926; *Enoplus michaelseni* von Linstow, 1896; *Lauratonema* Gerlach, 1953; *Lauratonemoides* de Coninck, 1965; *Leptosomatum acephalatum* Chitwood, 1936; *Mesacanthion* Filipjev, 1926; *Oxystomina cylindraticauda* (de Man, 1922) Filipjev, 1921; *Phanoderma* Bastian, 1865; *Pseudocella triaulolaimus* Hope, 1967; *Rhabdodemania scandinavica* Schuurmans-Stekhoven, 1946; *Rhaptothyreus typicus* Hope and Murphy, 1969; *Synonchus* Cobb, 1893; *Thoracostoma coronatum* (Eberth, 1863) Bütschli, 1874 (*Thoracostoma figuratum* Bastian, 1865 =); *Tuerkiana comes* (Türk, 1903) Platanova, 1970 (*Thoracostoma comes* Türk, 1903 =); *Tuerkiana strasseni* (Türk, 1903) Platanova, 1970 (*Thoracostoma strasseni* Türk, 1903 =).

Triploidina

Tripyloidea

Prismatolaimus microstomus Daday, 1905; *Tobrilus gracilis* (Bastian, 1865) Andrassy, 1959 (*Trilobus gracilis* Bastian, 1865 =).

Oncholaimina

Oncholaimoidea

Adoncholaimus fuscus (Bastian, 1865) Filipjev, 1918 (*Oncholaimus fuscus* Bastian, 1865 =); *Anoplostoma* Bütschli, 1874; *Eurystomina* Filipjev, 1921 (*Eurystoma* Marion, 1870 =); *Kreisoncholaimus* Rachor, 1969; *Metoncholaimus pristiurus* (Strassen, 1894) Cobb, 1932; *Meyersia* Hopper, 1967; *Oncholaimium appendiculatum* Cobb, 1930; *Oncholaimellus* de Man, 1890; *Oncholaimus* Dujardin, 1845; *Pelagonema* Cobb, 1893; *Pontonema zernovi* (Filipjev, 1916) Cobb and Steiner, 1934 (*Paroncholaimus zernovi* Filipjev, 1916 =); *Symplocostoma* Bastian, 1865; *Viscosia carnleyensis* (Ditlevsen, 1921) Kreis, 1932 (*Oncholaimus carnleyensis* Ditlevsen, 1921 =); *Viscosia macramphida* Chitwood, 1951.

Dorylaimida (soil and freshwater sediments; parasites of plants and invertebrates)

Dorylaimina

Actinolaimoidea

Actinolaimus tripapillatus (Daday, 1905) Steiner, 1916.

Dorylaimoidea

Longidorus africanus Merny, 1966; *Longidorus laevicapitatus* Williams, 1959; *Longidorus macrosoma* Hooper, 1961; *Xiphinema index* Thorne and Allen 1950.

Leptonchoidea

Tyleptus striatus Thorne, 1939.

Continued

TABLE I (*Continued*)

Mermithoidea
Agamermis decaudata Cobb et al., 1923; *Mermis subnigrescens* Cobb, 1929.

Monochoidea
Anatonchus tridentatus (de Man, 1876) de Coninck, 1939; *Iotonchus* Cobb, 1916.

Trichinellida (parasites of animals)
Trichinelloidea
Trichinella spiralis (Owen, 1835) Raillet, 1895.

Chromadorida (marine and freshwater sediments)
Chromadorina
Chromadoroidea
Chromadora axi Gerlach, 1951; *Chromadora quadrilinea* Filipjev, 1918; *Chromadorina epidemos* Hopper and Myers, 1967; *Euchromadora* de Man, 1886; *Hypodontolaimus* de Man, 1886; *Neochromadora poecilosoma* (de Man, 1893) Micoletzky, 1924 (*Chromadora poecilosoma* de Man, 1893 =); *Sabatieria hilarula* de Man, 1922; *Spilophorella paradoxa* (de Man, 1888) Filipjev, 1918.

Cyatholaimina
Choanolaimoidea
Halichoanolaimus robustus (Bastian, 1865) de Man, 1888; *Halichoanolaimus microspiculum* Allgen, 1929; *Latronema* Wieser, 1954.

Cyatholaimoidea
Biarmifer Wieser, 1954; *Cyatholaimus* Bastian, 1865; *Paracanthonchus* Micoletzky, 1924; *Paracyatholaimus* Micoletzky, 1922.

Desmodorida (marine sediment)
Desmodorina
Desmodoroidea
Desmodora greenpatchi Allgén, 1953.
Monoposthioidea
Monoposthia costata (Bastian, 1865) de Man, 1889; *Nudora lineata* Cobb, 1920.

Spirinoidea
Onyx perfectus Cobb, 1891; *Spirinia parasitifera* (Bastian, 1865) Gerlach, 1963 (*Spirinia parasitifera* (Bastian, 1865) Filipjev, 1918 =).

Draconematina
Draconematoidea
Draconema cephalatum Cobb, 1913; *Epsilonema* Steiner, 1927.

Monhysterida (marine and freshwater sediments)
Monhysterina
Linhomoeoidea
Desmolaimus zeelandicus de Man, 1880.
Monhysteroidea
Monhystera disjuncta Bastian, 1865; *Monhystera filicaudata* Allgén, 1929; *Monhystera parelegantula* de Coninck, 1943; *Monhystera stagnalis* Bastian, 1865; *Monhystera stenosoma* de Man, 1907; *Pseudosteineria horridus* (Steiner, 1916)

Wieser, 1956 (*Theristus horridus* Steiner, 1916 =); *Sphaerolaimus asetosus* Allgén, 1952; *Sphaerolaimus duplex* Allgén, 1952; *Theristus pertenuis* Bresslau and Schuurmans-Stekhoven, 1935; *Theristus setosus* (Bütschli, 1874) de Man, 1922; *Tripylium carcinicolum* (Baylis, 1915) Cobb, 1920 (*Monhystera carcinicola* Baylis, 1915 =).

Araeolaimida (marine and freshwater sediments and soil)
Araeolaimina
Araeolaimoidea
Diplopeltis Cobb, 1905.
Axonolaimoidea
Axonolaimus setosus Filipjev, 1918; *Bastiania* de Man, 1880.
Leptolaimoidea
Aphanolaimus de Man, 1880.
Plectoidea
Plectus granulosis Bastian, 1865 (*Anaplectus granulosis* (Bastian, 1865) Schuurmans-Stekhoven, 1933 =).
Tripyloidoidea
Halanonchus Cobb, 1920; *Tripyloides gracilis* (Ditlevsen, 1919) Allgén, 1935.

Desmoscolecida (marine)
Desmoscolecina
Desmoscolecoidea
Desmoscolex Claparède, 1863 (*Eudesmoscolex* Steiner, 1916 =) (*Eutricoma* Allgén, 1939 =) (*Prodesmoscolex* Staufer, 1924 =).

Greeffielloidea
Greeffiella Cobb, 1922.

SECERNENTEA VON LINSTOW, 1905

Rhabditida (freshwater sediments and soil)
Rhabditina
Diplogasteroidea
Cylindrocorpus curzii (T. Goodey, 1935) T. Goodey, 1939; *Cylindrocorpus longistoma* (Stefanski, 1922) T. Goodey, 1939; *Mesodiplogaster lheritieri* (Maupas, 1919) Goodey, 1963 (*Diplogaster lheritieri* Maupas, 1919 =).

Rhabdiasoidea
Rhabdias bufonis (Schrank, 1788) Stiles & Hassal, 1905 (*Ascaris bufonis* Schrank, 1788 =); (*Ascaris nigrovenosa* Goeze, 1800 =) [*Rhabditis nigrovenosa* (Goeze, 1800) Goette, 1882 =].

Rhabditoidea
Acrobeles ciliatus (Linstow, 1877) Thorne, 1925 (*Cephalobus ciliatus* Linstow, 1877 =); *Acrobeles complexus* Thorne, 1925; *Bunonema* Jägerskiöld, 1905; *Caenorhabditis briggsae* (Dougherty & Nigon, 1949) Dougherty, 1953; *Caenorhabditis dolichura* (Schneider, 1866) Dougherty, 1955 (*Leptodera dolichura* Schneider, 1866 =); *Diploscapter coronata* (Cobb, 1893) Cobb, 1913; *Mesorhabditis belari* (Nigon, 1949) Dougherty, 1953 (*Rhabditis monohystera* Bělăr, 1923 =) (*Rhabditis belari* Nigon, 1949 =); *Panagrolaimus rigidus* (Schneider, 1866)

Continued

TABLE I (*Continued*)

Thorne, 1937; *Pelodera strongyloides* (Schneider, 1860) Schneider, 1866; *Pelodera teres* Schneider, 1866; *Rhabditis aberrans* Kruger, 1913; *Rhabditis marina* Bastian, 1865; *Rhabditis terrestria* Stephenson, 1942; *Turbatrix aceti* (Müller, 1783) Peters, 1927 (*Anguillula aceti* (Müller, 1783) Ehrenberg, 1838 =).

Tylenchida (soil and parasites of plants)
Aphelenchina
Aphelenchoidea
Aphelenchus avenae Bastian, 1865; *Seinura oxura* (Paeslor, 1957) J. B. Goodey, 1960.

Criconematoidea
Hemicycliophora arenaria Raski, 1958; *Paratylenchus nanus* Cobb, 1923; *Tylenchulus* Cobb, 1913.

Heteroderoidea
Heterodera rostochiensis Wollenweber, 1923; *Meloidogyne arenaria* (Neal, 1889) Chitwood, 1949; *Meloidogyne hapla* Chitwood, 1949; *Meloidogyne incognita* (Kofoid and White, 1919) Chitwood, 1949; *Meloidogyne javanica* (Treub, 1885) Chitwood, 1949.

Tylenchoidea
Anguina tritici (Steinbuch, 1799) Filipjev, 1936; *Bradynema rigidum* (Siebold, 1836) Strassen, 1892; *Ditylenchus destructor* Thorne, 1945; *Ditylenchus dipsaci* (Kühn, 1857) Filipjev, 1936; *Helicotylenchus* Steiner, 1945; *Hirschmaniella gracilis* (de Man, 1880) Luc and Goodey, 1963; *Pratylenchus scribneri* Steiner, 1943; *Tylenchorhynchus claytoni* Steiner, 1937.

Strongylida (parasites of animals)
Metastrongylina
Protostrongyloidea
Angiostrongylus cantonensis (Chen, 1935) Dougherty, 1946.
Strongylina
Ancylostomatoidea
Ancylostoma caninum (Ercolani, 1859) Hall, 1913; *Bunostomum trigonocephalum* (Rudolphi, 1808) Raillet, 1902; *Necator* Stiles, 1903.

Strongyloidea
Metastrongylus apri (Gmelin, 1790) Molin, 1861 (*Strongylus paradoxus* Mehlis, 1831 =).
Trichostrongylina
Trichostrongyloidea
Cooperia curticei (Giles, 1892) Ransom, 1907; *Haemonchus contortus* (Rudolphi, 1803) Cobb, 1898; *Trichostrongylus* Looss, 1905.
Ascarida (parasites of animals)
Ascaridina
Ascaridoidea
Ascaridia lineata (Schneider, 1866) Raillet and Henry, 1912; *Ascaris lumbricoides* Linnaeus, 1758; *Contracaecum incurvum* (Rudolphi, 1819) Baylis &

Daubney, 1922 (*Ascaris incurva* Rudolph, 1819 =); *Parascaris equorum* (Goeze, 1782) York and Maplestone, 1926 (*Ascaris megalocephala* Cloquet, 1824 =); *Phocanema decipiens* (Krabbe, 1878) Meyers, 1959; *Toxocara canis* (Werner, 1782) Johnston, 1916.

Spirurida (parasites of animals)
Camallanina
Camallanoidea
Camallanus lacustris (Zoega, 1776) Raillet & Henry, 1915 (*Cucullanus elegans* Zeder, 1800 =); *Paracamallanus sweeti* (Moorthy, 1938) Campana-Rouget, 1961 (*Camallanus sweeti* Moorthy, 1938 =).

---

second polar body apparently fuses with the pronucleus of the ovum to restore the somatic number of chromosomes. (B) Normal meiotic oogenesis proceeds as above until telophase II, which stops short of the formation of the second polar body pronucleus, and the two groups of telophase chromosomes become enclosed in a single pronucleus, thus re-establishing the 2N condition. (C) No synapsis of homologous chromosomes takes place and the somatic (2N) condition is maintained throughout the first division. A second maturation division does not occur.

The majority of the species that are known to reproduce by parthenogenesis appear to reproduce solely by obligatory mitotic parthenogenesis (Triantaphyllou, 1971). In other instances, different populations within the same nominal species may reproduce by different methods. Most populations of *Mesorhabditis belari* reproduce by pseudogamy, although they reproduce rarely by amphimixis (Bělăr, 1923; Nigon, 1949). Triantaphyllou (1966) has shown that one race of *Meloidogyne hapla* reproduces by amphimixis and meiotic parthenogenesis while another race reproduces by mitotic parthenogenesis. Roman and Triantaphyllou (1969) have also determined that one population of *Pratylenchus scribneri* reproduced by meiotic parthenogenesis while in another population it was by mitotic parthenogenesis. Finally, at least one instance is known in which a hermaphroditic species reproduces by mitotic parthenogenesis and pseudogamy (Krüger, 1913). Pseudogamy, the necessity for sperm to penetrate the egg to initiate cleavage divisions, but without karyogamy, occurs in several species of the family Rhabditidae.

It is suspected that the majority of parthenogenetic species are polyploid, although some appear not to be (Hechler, 1968; Roman and Triantaphyllou, 1969) and at least one strictly amphimictic species is thought to be tetraploid (Hechler and Taylor, 1966a).

Owing to the absence of detailed studies on the reproduction of marine nematodes, there is no direct evidence that any reproduce asexually. However, the ratio of males to females may at least be an indirect indication of the mode of reproduction for a given species, as amphimictic nematodes usually have a fairly equal proportion of males and females. Parthenogenetically and hermaphroditically reproducing species on the other hand, would be expected to have few if any males at all. While a dearth of males does not exclude the possibility of protandric reproduction, it is in such populations that a search for parthenogenesis would begin.

Probably the earliest indication of the possibility of parthenogenesis in marine nematodes is to be found in a statement by Marion (1870) that ". . . chez certaines espèces je n'ai jamais pu rencontrer un seul individu mâle; bien que j'eusse souvent les femelles à profusion." While Marion failed to attach significance to this observation, Maupas (1900) regarded Marion's statement as an indication that the species referred to are either hermaphroditic or parthenogenetic. Throughout the literature concerning systematics of marine nematodes there is additional evidence of disproportionate sex ratios (Cobb, 1918; Schuurmans-Stekhoven, 1931), but in the vast majority of these instances too few specimens of the same species were obtained to eliminate sampling error.

More convincing evidence of disproportionate sex ratios and the possibility of parthenogenesis has been obtained by Hopper and Myers (1966) who studied populations of marine nematodes cultured on fungus-infested cellulose matrix. They found that while *Chromadorina epidemos* was represented in the field by both sexes (sex ratio not given), males were not present in the cultures. They regarded this as an indication that this species might reproduce both amphimictically and parthenogenetically. Likewise, while *Viscosia macramphida* are found in their natural habitat with relatively equal occurrence of both sexes, after 10 days in culture all sexually mature individuals were females. Because several generations of only females were produced, they concluded that this species is probably capable of both amphimictic and parthenogenetic reproduction. Finally, *Monhystera parelegantula* also appeared to reproduce in culture without males and was regarded as having parthenogenetic capabilities.

Further indication that parthenogenetic reproduction may occur in the genus *Monhystera* was found by Tietjen (1967); in his cultures of the marine species *M. filicaudata*, only 5% of the population consisted of males, which corresponds to the ratio that was found in their natural habitat. Wieser (1953, 1956) also comments that males of certain species of *Monhystera* and *Phanoderma* are uncommon.

8.3 Sexual Reproduction

8.3.1 Sexual Dimorphism

Sexual dimorphism occurs among the vast majority of bisexual nematodes. Secondary sexual characters, described in detail in Section 8.3.5, vary widely among males. Typically, however, males possess paired copulatory structures (spicula), and a guiding mechanism for the spicula (gubernaculum). In addition they frequently possess various sensory receptors and sometimes glands that presumably aid in copulation. While males of the Strongylida typically have a copulatory bursa, and while it is common in the Rhabditida, it is rare among marine forms. Females may also have slightly longer setae (sensory receptors) in the region of the vulva and sometimes particularly large, glandlike cells in the lateral hypodermal chord near the level of the vulva, especially in the Enoploidea.

A notable exception to the generalization above is to be found in the family Eurystominidae, comprised of the subfamilies Eurystomininae and Symplocostomatinae (=Enchelidiinae, Filipjev, 1934). Members of the former subfamily are normal with regard to sexual dimorphism and males and females each have a large, well developed stoma composed of two consecutive chambers separated by one or several transverse annules containing or formed of denticles. In addition, the stoma in both sexes contains three large teeth. By contrast, while the stoma of the females and juveniles of the Symplocostomatinae is well developed in a manner rather similar to the Eurystomininae, the males are provided with but a narrow tubelike vestige of a stoma. Further, while the amphids (paired, lateral chemoreceptive organs in the head region) are rather small in females and juveniles, they may occupy nearly the entire width of the male head.

A second exception may prove to exist in the deep-sea nematode *Rhaptothyreus typicus*. Males are characterized by such unusual features as the absence of a well developed digestive system and the amphids are exceptionally large and shield-shaped. Thus far, females have not been identified in the same samples, and while it may be that they are parasites, or are to be found in some other specialized but as yet unsampled habitat, it may also be that the females are present with the males but have not been recognized due to extensive sexual dimorphism.

8.3.2 Intersexes

Intersexes are known to occur in species of all major groups of nematodes. Those that have been reported for marine and closely related freshwater species are listed in Table II. Additional intersexes of soil and freshwater species are given by Nigon (1965) and still others have been recently reported upon for *Tyleptus striatus* by Jairajpuri and Siddiqi (1964), *Longidorus macrosoma* by Aboul-Eid and Coomans (1966), *L. laevicapitatus* by Merny (1966), and *L. africanus* by Cohn and Mordechai (1968).

Intersexes are most commonly functional females with male secondary characters and/or occasionally with rudimentary testes. That they are functional as females is frequently evidenced by the presence of eggs in the uterus. The male secondary characters that may appear on intersexes are the spicula, gubernaculum, ventromedian supplements, subventral supplements, or copulatory muscles (see Section 8.3.5). As may be seen from Table II, some of these male characters may appear alone on female intersexes, or they may occur in any one of several combinations. Various degrees of maleness with different combinations of male features may occur between individual intersexes of the same species, as Steiner (1923) has shown for *Agamermis decaudata.*

Male intersexes with female secondary characters have also been reported, but are less common. At least in the case of *Meloidogyne javanica*, it appears that the male intersexes were derived from second stage female larvae that had undergone sex reversal. For this species sex reversal seems to be induced by such unfavorable environmental conditions as overcrowding. Christie (1929) also proposed that maleness could be produced in females by certain environmental forces, and when the force was insufficient to completely transform the females into a functional male, they became female intersexes with male secondary characters. Steiner (1923) has suggested the possibility that mermithid intersexes resulted from hybridization of closely related genotypes, and he later (1937) considered that nematode intersexes might result from incomplete transformation of male protandric hermaphrodites to females. Hirschmann and Sasser (1955) have also reviewed some genetic factors that are believed to be operative in the production of intersexes in other animals. But at present, the cause or causes of intersexes in nematodes is in need of further study.

8.3.3 Sex Determination

Again, in the absence of any detailed studies of their reproduction, nothing definitive can be stated regarding the mechanism of sex determination in marine nematodes. However, sex determination has been studied in free-living freshwater and soil forms (Maupas, 1900;

TABLE II

INTERSEXES OF MARINE AND RELATED FRESHWATER SPECIES

Species	Reference	Functional sex and secondary characters of opposite sex
Anticoma limalis	Schuurmans Stekhoven and Adam (1931), p. 11	♀ with eggs + spicula
Axonolaimus setosus	Filipjev (1918), p. 232[a]	
Deontostoma californicum	Hope (unpublished)	♀ with normal female reproductive system and eggs + ventromedian supplements, subventral supplements, spicula, and gubernaculum; testes absent
Desmodora greenpatchi	Allgén (1953), p. 97	♀ with spicula, gubernaculum, preanal papillae and a short testis[b]
Enoplus communis	Schneider (1866)[c]	
Enoplus michaelseni	de Man (1904), p. 24	♀, normal and well developed + spicula and gubernaculum.
Enoplus sp.	Mendes (1942), p. 255	♀ + spicula and gubernaculum
Halichoanolaimus microspiculum	Allgén (1929), p. 142	♂ with ovary, developing eggs and precursor of vulva.
Monhystera (?) *carcinicola*	Baylis (1915), p. 420	♂ functional testis, spicula and gubernaculum + ovary but no eggs or vulva (protandric hermaphrodite?)
Monhystera stagnalis	Micoletzky (1914), p. 413	♀ with eggs + rudimentary gubernaculum
Neochromadora poecilosoma	de Man (1893), p. 99	♀ with ovaries + spicula and gubernaculum
Prismatolaimus microstomus	Daday (1905), p. 57	♀ + preanal papillae
Rhabdodemania scandinavica	Allgén (1958), p. 318	♀ with eggs + spicula and gubernaculum
Sphaerolaimus asetosus	Allgén (1952), p. 190	♀ + spicula and gubernaculum
Sphaerolaimus duplex	Allgén (1952), p. 190	♀ (?) + spicula and gubernaculum
Thoracostoma coronatum	de Man (1893), p. 111	♀ with eggs + spicula and gubernaculum
Tobrilus gracilis	Daday (1905), p. 55	♀ with eggs + spicula, gubernaculum and preanal papillae
	Ditlevsen (1912), p. 233	♀ with eggs + preanal supplements only

[a] Filipjev (1921) lists several reports of intersexes including one he allegedly described earlier (1918, p. 232), but reference to that page reveals no such description.

[b] Juvenile female.

[c] Publication not seen.

Krüger, 1913; Bělăr, 1923, 1924; Honda, 1925; Kröning, 1923; Nigon, 1949; Hechler, 1970a, b), plant parasitic forms (Triantaphyllou, 1966; Triantaphyllou and Hirschmann, 1966, 1967) and animal parasitic nematodes (Edwards, 1910; Schleip, 1911; Goodrich, 1914, 1916; Walton, 1918, 1940; Gouilliart, 1932; Nigon and Roman, 1952). This subject has been reviewed recently by Triantaphyllou (1971). Certain generalizations are evident from these works which, in part at least, will doubtless prove to be applicable to marine nematodes.

In the majority of nematodes that have been studied, separate and distinct sex chromosomes have been demonstrated. In these cases there is an XO-XX mechanism for sex determination and the male is consistently heterogametic. Thus the males have at least one less chromosome than the females and produce two kinds of gametes. This system may prove to be the more usual mechanism for nematodes in general. However, in the case of a soil inhabiting nematode, *Mesorhabditis belari*, Nigon (1949) found one pair of chromosomes at diakinesis of primary spermatocytes which were comprised of one chromosome comparable in size to the autosomes and a considerably larger second chromosome. Nigon proposed that these are X and Y chromosomes, respectively. He was unable to demonstrate heterosomes in the free-living soil nematode *Panagrolaimus rigidus*, but concluded that a genetic system of sex determination exists, as there is very nearly a 1:1 sex ratio. He speculated that X and Y chromosomes may be present, but indistinguishable from the autosomes. Some plant parasitic nematodes may also have an XY-XX mechanism for Triantaphyllou (1966) and Triantaphyllou and Hirschmann (1966) were unsuccessful in demonstrating sex chromosomes in *Meloidogyne hapla* and *Anguina tritici*, respectively. They also concluded that, while probably present, the sex chromosomes could not be distinguished from autosomes on morphological or behavioral bases.

Goodrich (1916), in his observations on the late prophase of *Contracaecum incurvum* spermatocytes, found a Y chromosome paired with a long X component. In addition, there were seven univalent X elements. A multiplicity of X chromosomes in the absence of a Y chromosome is fairly prevalent among the ascarids.

Epigenetic factors control sex in some nematodes, especially members of the insect parasitic family Mermithidae and the plant parasitic genus *Meloidogyne*. Christie (1929) found that when 1-3 specimens of *Mermis subnigriscens* occur in a single grasshopper they were always females; when 4-14 were present they were males and females, and when infestation exceeded fourteen per host all were males. Triantaphyllou (1960) also found that when relatively low populations of *Meloidogyne incognita* infested root tips of tomato plants most were females, with few or no males. Conversely, when overcrowding resulted from high population densities, nearly all specimens were males. Even juvenile females that had begun to develop ovaries underwent sex reversal to produce phenotypic males.

8.3.4 Hermaphroditism

Although hermaphroditism has not been demonstrated among marine nematodes, like parthenogenesis, it may occur in those populations of marine nematodes that are comprised predominantly of individuals displaying female secondary sexual characters. Marine species so characterized are mentioned in the section on parthenogenesis (8.2). However, whether these species are parthenogenetic, hermaphroditic, or have an apparent unequal sex ratio for other reasons must await further study.

Protandric hermaphroditism has been well documented among soil nematodes (Maupas, 1900; Potts, 1910; Krüger, 1913; Honda, 1925; Nigon, 1949; Nigon and Dougherty, 1949; Hirschmann, 1951; Osche, 1952, 1954). It is generally regarded that among nematodes hermaphroditism has evolved from gonochorism. This belief is based largely on the numerous degrees of hermaphroditism extant among the rhabditids. Further evidence lies in the observations of Hirschmann (1951), who obtained hermaphrodites through inbreeding experiments with the soil nematode *Mesodiplogaster lheritieri*. Nigon (1949) having studied the hermaphrodites of *Caenorhabditis elegans* postulated that they are derived from normal bisexual species through mutation or hybridization. Additional information regarding hermaphroditism may be obtained from Triantaphyllou and Hirschmann (1964) and Triantaphyllou (1971).

8.3.5 Anatomy of the Reproductive System

Present knowledge of the reproductive systems of marine nematodes has largely resulted from gross anatomical studies by de Man (1886), Jägerskiöld (1901), Türk (1903), Filipjev (1921), Kreis (1934) and Rachor (1969) supplemented by numerous taxonomic studies. Detailed histological studies are generally lacking, but significant contributions have been made by Jägerskiöld (1901) and Türk (1903) in their studies on some of the larger marine nematodes of the family Leptosomatidae. Chitwood and Chitwood (1950) have reviewed the anatomy of nematodes, including the reproductive system, with original observations of their own. More recent reviews, pertaining at least in part to the reproductive system, are those of Hirschmann (1960), de Coninck (1965) and Bird (1971).

The following account is primarily based on the earlier studies but is supplemented by my own observations on *Deontostoma californicum*. Specimens used in this study were either wholemounts in anhydrous glycerine or embedded in polyethylene glycol by a method described elsewhere (Hope, 1969) and sectioned at 5 μm. Sections were either stained with hematoxylin by the method of Craig and Wilson (1937), Mallory's Triple stain, PAS for glycogen, or Fuelgen for DNA.

8.3.5.1 Female Reproductive System

This system typically consists of two tubular gonads, each continuous with a separate gonoduct opening to the exterior by way of a common vagina and vulva (Fig. 1). When gonads and associated ducts are paired in this manner the system is designated as didelphic. Commonly one gonad and gonoduct is directed anteriorly while the other branch extends in the opposite direction and this condition is termed amphidelphic (Fig. 1). At or near the juncture of the ovary and gonoduct the latter is flexed, i.e., turned back upon itself (Fig. 1).

The vulva is a ventral cleft usually oriented transversely to the longitudinal body axis (but longitudinally oriented in *Desmoscolex*), and it is usually located slightly posterior to the midbody. The position of the vulva is not fixed, however, as in a few species of *Oxystomina* it is shifted much farther anteriorly in which case the reproductive system is monodelphic and opisthodelphic, i.e., the anterior branch of the reproductive system has been lost leaving but a single, posteriorly directed branch. When monodelphic, the flexure is usually absent and the ovary is said to be outstretched. More frequently, as in the orders Oncholaimina and Monhysterina, the displaced vulva is posterior to its usual midposition with the reproductive system again being monodelphic, but in this case it is the anterior branch that is retained, a condition designated prodelphic.

Extremes in the posterior displacement of the vulva are found in the marine nematode family Lauratonematidae. In the genus *Lauratonemoides* the vulva is located just anterior to the anus, while in the genus *Lauratonema* the vulva actually opens into the proctodaeum, the latter being a true cloaca.

8.3.5.1.1 *Ovary.* Ovaries of nematodes in general may be identified as one of two types depending upon the site of the production of

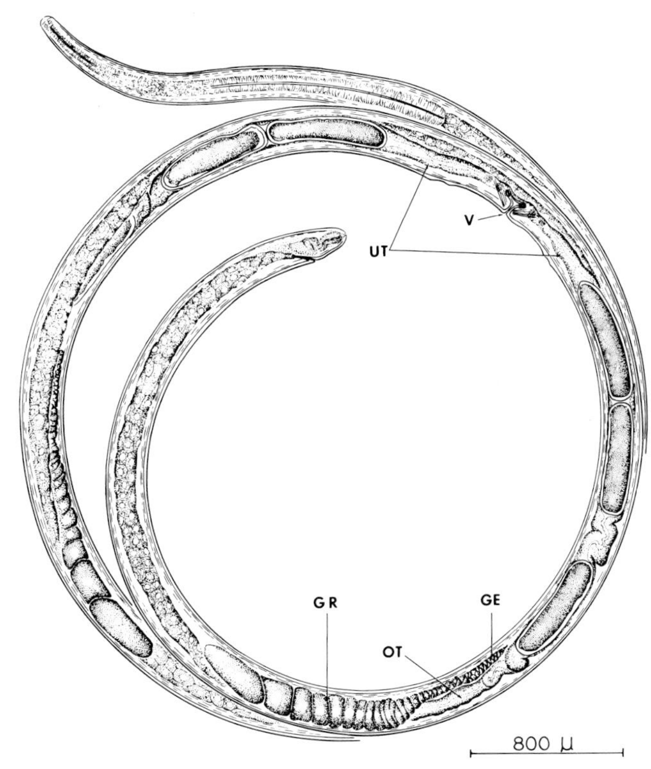

FIG. 1. Female of *Pseudocella triaulolaimus* showing the vulva (V), uterus (UT), oviduct (OT) and the ovary which consists of germinal (GE) and growth (GR) zones [from Hope (1967b)].

oogonia. In those regarded as hologonic, the oogonia are produced throughout the length of the ovary; only members of the parasitic nematode superfamilies Trichuroidea and Dioctophymatoidea are known to be of this type. All other nematodes have telogonic ovaries, in which the production of oogonia is limited to a short region at the distal end of the ovary; this end is nearest the vulva when the gonoduct is flexed and farthest from it when outstretched. A narrow cylindrical structure termed the rachis commonly exists along the median axis of the ovaries in ascarids, oxyurids, strongylids and spirurids, and the developing oogonia are radially arranged around it. Chitwood and Chitwood (1950) have questioned the importance of the rachis since it occurs in some species but may be absent in other closely related forms. Foor (1967, 1968) has shown that the rachis consists of non-nucleated tissue containing such organelles as lamellar bodies, lipid droplets, dense granules and what may be ribo-

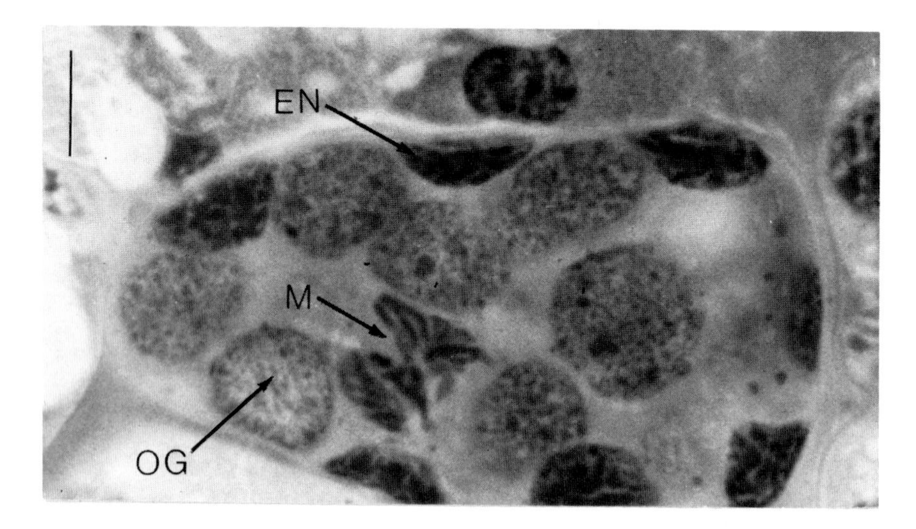

FIG. 2. Transverse section through the germinal zone of the ovary of *Deontostoma californicum*. Oogonia (OG) are proliferated in the germinal chord by mitotic divisions (M). The entire ovary is enclosed by epithelial cells (EN). Scale line indicates 10 μm.

somes. He further showed that the oocytes are attached to the rachis until a short distance before the beginning of the oviduct, and here the detached oocytes lie free in the lumen of the gonoduct. Foor believes the function of the rachis is to maintain the spatial arrangement of the oocytes. Ovaries of marine nematodes, although telogonic, are not known to possess a rachis.

Histologically the telogonic ovary of marine nematodes consists of a chord of germinal tissue enveloped in a single layer of epithelial cells, as exemplified in the case of *Deontostoma californicum* (Fig. 2). In this species the nuclei of the epithelial cells may be readily distinguished from nuclei of the oogonia by their smaller size, flattened form, and more basophilic, and coarser nucleoplasm. Filipjev (1921) regarded the distal end of the ovary as being syncytial, although this is likely a misimpression owing to the indistinctness of cell boundaries seen with light microscopy.

Musso (1930), in studying the ovary of *Ascaris lumbricoides* and *Parascaris equorum*, has suggested that an epithelial cell at the end of the ovary functions as an undifferentiated germinal cell giving rise to the oogonia. Most investigators, according to Chitwood and Chitwood (1950), have failed to find evidence of nuclear division in this cell. Such evidence in the case of *D. californicum* also is absent.

The ovary consists of germinal and growth zones, the former occupying but a small distal portion (Fig. 1 and 3) where small spherical oogonia are proliferated. The beginning of the growth zone is

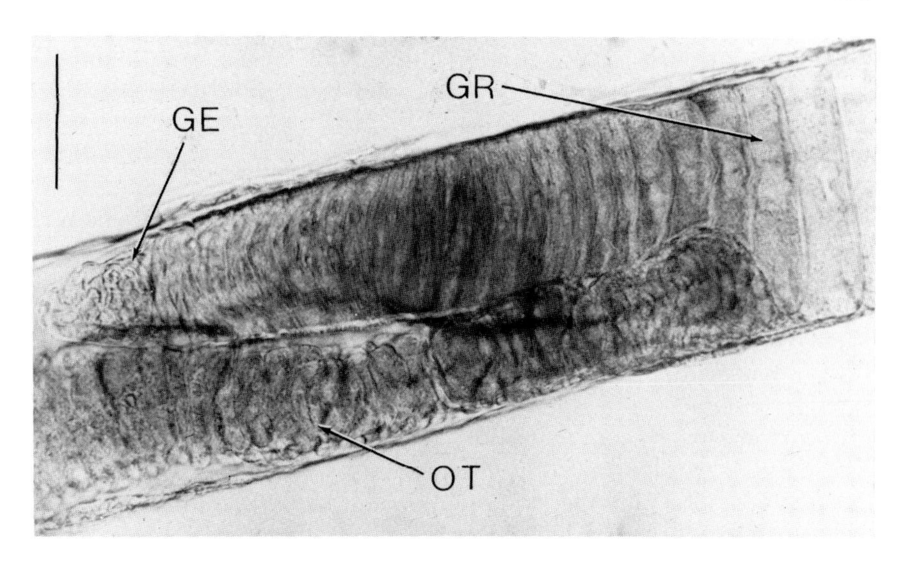

FIG. 3. Germinal zone (GE) and a portion of the growth zone (GR) of an ovary dissected from *Deontostoma californicum*. The oviduct (OT) is connected to the dorsal side of the ovary not far from the germinal zone. Scale line indicates 50 μm.

evidenced by the abrupt transition of the oocytes from spherical to disc-shaped, each with their flat surfaces facing adjacent oocytes (Fig. 3). As the oocytes move proximally they greatly increase in size, and the full grown oocytes are at the proximal end of the ovary. In forms with outstretched gonoducts the oocytes continue to progress in the same direction, moving into the gonoduct, whose walls are continuous with the epithelial envelope of the ovary. On the other hand, when the gonoduct is flexed, the oogonia most likely make a 180 degree turn as is assumed to be the case in *Chromadora* sp. whose female reproductive system has been illustrated by Chitwood and Chitwood (1950). However, in the case of *Deontostoma californicum* and the other leptosomatids, the ovary is in reality an elongated epithelial sac closed at both ends, but with an opening into the oviduct located on one side of the ovary fairly close to the juncture between germinal and growth zones (Fig. 3). In such cases, as Türk (1903) has observed, the largest oocyte enters the gonoduct by reversing its direction of movement and slipping back past the less mature oocytes and into the opening of the gonoduct. Attempts to separate oocytes from one another in the growth zone of a freeze-dried ovary dissected from *D. californicum* resulted in fracture of the oocytes themselves rather than separation at intercellular junctions. This suggests that each developing oocyte adheres to its neighbor until it reaches the distal end of the ovary, thus preventing the immature oocytes from slipping out of line and prematurely entering the oviduct. The means by which the full grown oocytes are directed into the orifice of the oviduct is not known.

8.3.5.1.2 *Gonoduct.* The primary function of the gonoduct is to transport full grown oocytes from the ovary to the vagina. But even in the earliest studies of nematodes it has been evident that the gonoduct is divisible into structurally, and by implication, functionally distinct regions.

De Man (1886) recognized in *Enoplus communis, Adoncholaimus fuscus,* and *Anticoma pellucida,* a region of the gonoduct extending from its juncture with the ovary to near the level of the tip of the flexed ovary where the epithelial cells were spindle-shaped, their long axis perpendicular to the longitudinal axis of the gonoduct. He designated this region the oviduct and called the remaining proximal portion the uterus. The latter is comprised of polyhedral epithelial cells, and, in at least *Enoplus* and *Adoncholaimus,* there is an outer muscle layer which usually extends onto part of the oviduct. In *Anticoma* and *Euchromadora,* de Man found the oviduct to be set off from the uterus by a cone or funnel-shaped extension of the wall of the oviduct projecting into the lumen of the uterus. A similar structure has been reported for *Enoploides amphioxi* and *Pontonema zernovi.*

Jägerskiöld (1901), in his study of *Cylicolaimus magnus,* found that the distal end of the gonoduct lying adjacent to the flexed ovary was convoluted, that its walls were constructed of a flattened but irregular epithelium, and that its lumen contained many sperm. This region he surmised to function as a seminal receptacle and simply referred to it as a "tube." The remainder of the gonoduct, which he termed the uterus, was without convolution, harbored few sperm, and the cells of the wall projected irregularly into the lumen when the latter was empty. Contrary to de Man's findings, no muscle fibers were found in the wall of the gonoduct, except a ringlike muscle at the juncture of the gonoduct and ovary, nor does Jägerskiöld mention the cone-shaped extension of the oviduct into the uterus.

The gonoduct of *Tuerkiana strasseni,* according to Türk (1903), is very similar to those described by Jägerskiöld. However, sperm occur in large quantities not only in the convoluted oviduct, but also in the ovarial sac where Türk believed fertilization occurs. Türk also found longitudinal muscles in the gonoduct wall, but made no mention of the cone-shaped structure.

The gonoduct of *Leptosomatum acephalatum* was described by

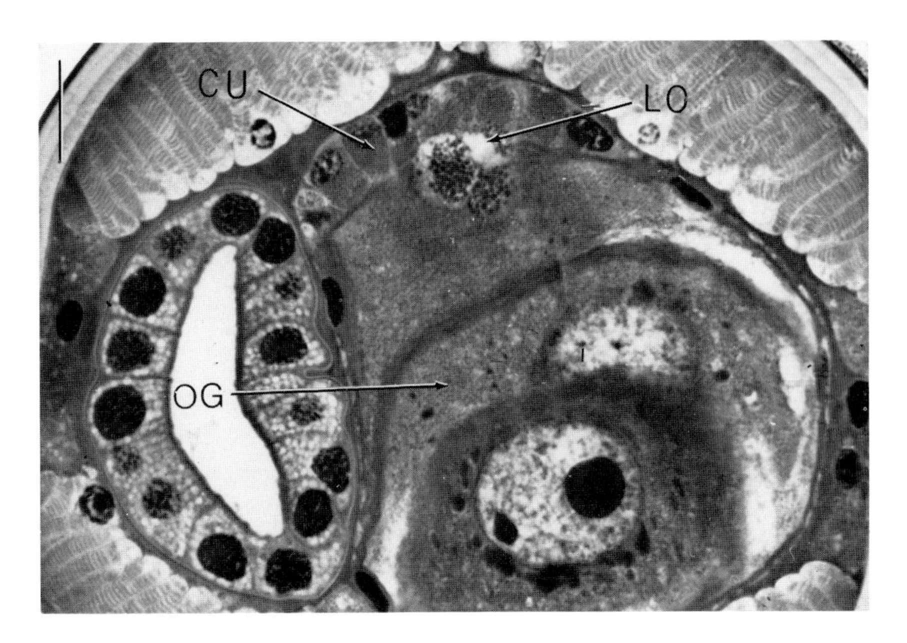

FIG. 4. Transverse section of the ovary of *Deontostoma californicum*. The cuboidal epithelium (CU) on the dorsal side of the ovary marks the beginning of the oviduct. The lumen of the oviduct (LO), partially filled with secretory granules, lies between disc-shaped oogonia (OG) and the cuboidal epithelium. Scale line indicates 25 μm.

Timm (1953) as being comprised of "collapsed spongy tissue" without evidence of a lumen. The uterus is formed of flat epithelial cells, without muscle fibers "except at the proximal end." Presumably, this is the sphincter muscle of the vagina to be described later.

The gonoduct of *Deontostoma californicum* consists of two morphologically distinct regions which, as a matter of convention, are here termed the oviduct and uterus. Each oviduct arises from the dorsal side of its respective ovary not far from the level of the transition between germinative and growth zones (Fig. 3). The point of origin is indicated in cross sections by an abrupt transition of the epithelium of the ovary from squamous to cuboidal epithelium (Fig. 4), and the cytoplasm of the latter is filled with granules. Only in occasional sections is there evidence of a lumen, and then it is often filled with the same granules. The diameter of the oviduct continues to increase slightly as it proceeds toward the uterus and the epithelial cells become columnar with the long axis of the cell approximately transverse to the long axis of the duct (Fig. 5), just as described by de Man (1886). In the proximal region of the oviduct the cells are upright, and here cell membranes are only evident in longitudinal sec-

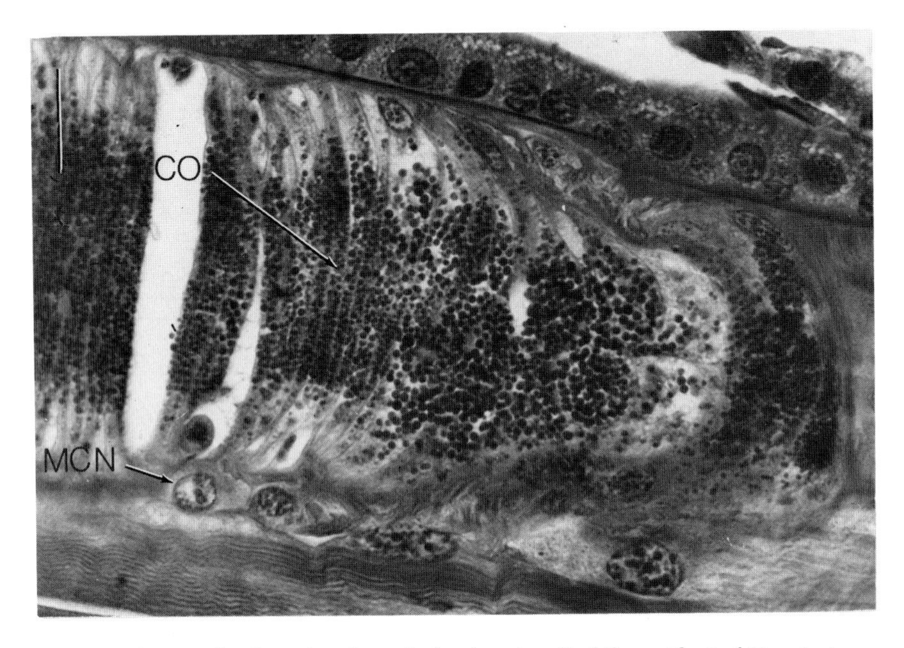

FIG. 5. Longitudinal section through the dorsal wall of the oviduct of *Deontostoma californicum*, near its attachment to the ovary. The inner cells (CO) of the duct wall are narrow, columnar and filled with secretory granules. The nucleus of each cell in the peripheral layer is within a bag-like process (MCN). Scale line indicates 25 μm.

tions (Fig. 5) since cross sections tend to be tangential to the cell surface. The lumen is not consistently evident, especially in transverse sections, except where occupied by sperm. In most sections it is presumably occluded by the epithelial cells. This interpretation was confirmed through observing with the scanning electron microscope the fractured surfaces of a freeze-dried oviduct (Fig. 6). Here the folds of the cell surfaces clearly converge on the remnant of the lumen.

Not far from where the oviduct merges with the ovary, a second layer of cells appears in the wall of the oviduct, peripheral to the epithelium. The cells of this layer are striking in that the nucleus of each cell lies within a baglike process extending outward from the wall of the oviduct (Fig. 5, 7, and 8); these may be the same cells referred to by Timm (1953) as having "bulging nuclei," or those described by Filipjev (1921) as "short cells with large nuclei." Further, each cell has deep infoldings in its surface membranes and numerous fibers which seem to span the cell in all directions (Fig. 7). In the absence of fine structure studies, it cannot be stated for certain whether these are contractile filaments or tonofilaments. However, if

FIG. 6. Scanning electron micrograph of the fractured surface of a freeze-dried oviduct dissected from *Deontostoma californicum*. The folds in the cell walls converge on the narrow lumen. Scale line indicates 25 μm.

they are tonofilaments, presumably they would give structural support to the cells, thus preventing them from tearing as oocytes passed through the oviduct. If this were the case, one would also expect to find tonofilaments in the epithelium of the ovary, the medial layer of columnar cells in the oviduct, and especially at the proximal end of the uterus where passing oocytes have attained their greatest diameter and exert the greatest stress on the gonoduct; however, except in the proximal region of the oviduct and the distal region of the uterus, no filaments were observed. On the other hand, oocytes in fixed specimens are either observed to be within the ovary or as far toward the proximal end of the uterus as is possible. This would suggest that oocytes move quickly through the oviduct and distal end of the uterus, and would lend support to the supposition that they are moved by muscular contractions. However, if these filaments are contractile, their radial arrangement within the fiber makes them unlike other fibers associated with the gonoducts of either sex wherein the filaments are parallel.

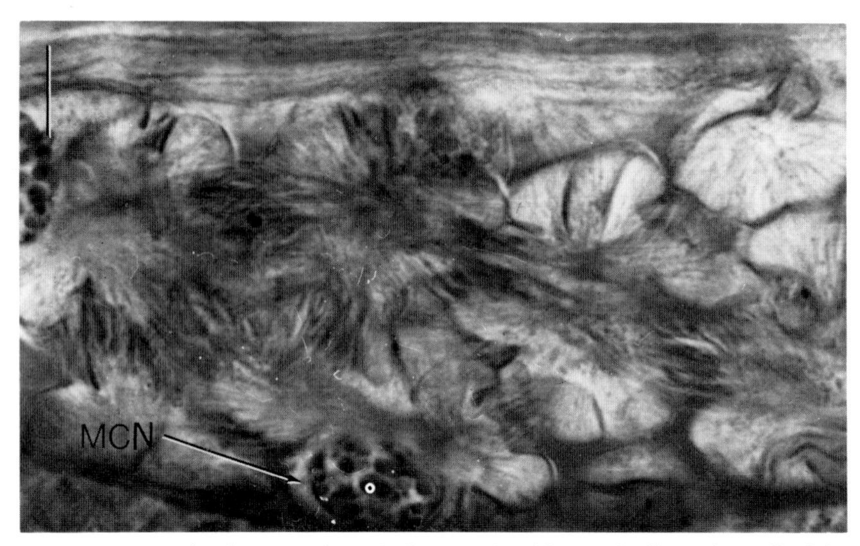

FIG. 7. Longitudinal section through the peripheral layer of cells in the wall of the oviduct of *Deontostoma californicum.* Numerous fibers span these irregularly shaped cells and the nucleus (MCN) is within a baglike protrusion. Scale line indicates 10 μm.

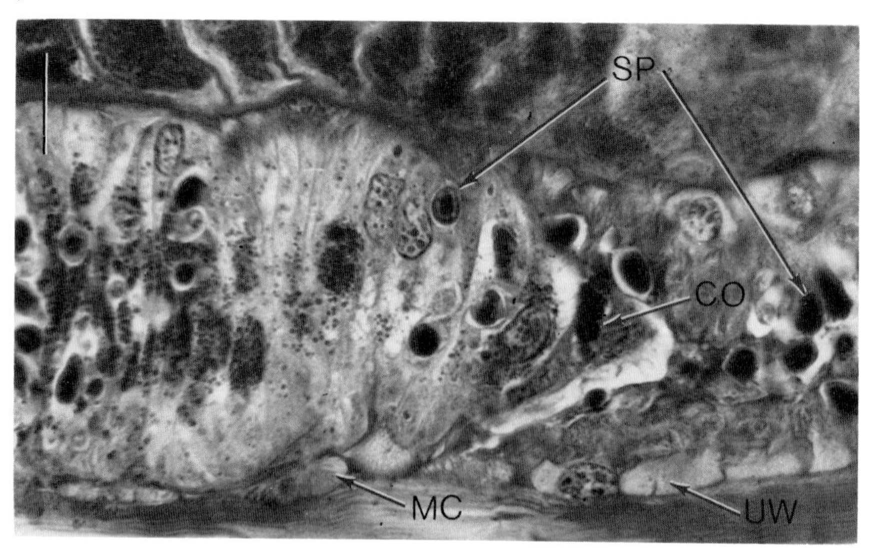

FIG. 8. Longitudinal section of the juncture between the oviduct and uterus of *Deontostoma californicum.* The inner layer of cells (CO) at the proximal end of the oviduct extends conelike into the distal end of the uterus. The peripheral cells (MC) of the oviduct wall extend proximally as the uterine wall (UW). Spermatozoa (SP) are evident in both the oviduct and uterus. Scale line indicates 25 μm.

At the juncture of the oviduct and uterus, the columnar epithelial tissue separates from the peripheral layer of cells and projects as a short conical mass of tissue into the lumen of the distal end of the uterus (Fig. 8). This is doubtless the same structure described by de Man (1886) and possibly that mentioned by Filipjev (1921). A function cannot readily be ascribed to this structure. Contrary to what one might suspect from its valvelike appearance, it does not prevent entry of sperm into the oviduct, as sperm were equally abundant in the oviduct and distal region of the uterus.

The filament-bearing cells that had formed the peripheral layer of cells in the oviduct continue proximally to form the wall of the distal portion of the uterus. Proximally, the cells of the uterine wall are cuboidal (Fig. 9) and then squamous near the vagina (Fig. 10).

8.3.5.1.3 *Seminal Receptacle.* A seminal receptacle in the form of a structural modification of the gonoduct does not exist in the representatives of Enoplida thus far studied. In *Leptosomatum acephalatum* sperm were found among oocytes in the oviduct, uterus, and shell gland (Timm, 1953). Sperm were abundant throughout the gonoduct in *Deontostoma californicum* as well. Jägerskiöld (1901) found most sperm limited to the convoluted region of the gonoduct (oviduct?) and regarded this region as a seminal receptacle. Türk (1903) found sperm in the proximal ends of the ovaries where he believed fertilization takes place.

In many other species the seminal receptacle is simply a widened region at the proximal end of the oviduct or distal end of the uterus. But in at least certain members of the Comesomatidae and Axonolaimidae the seminal receptacle exists as blind out-pocketings of the uterus (Chitwood and Chitwood, 1950).

In monodelphic forms the lost branch of the reproductive system may persist as a blind sac that serves as a seminal receptacle, as in the case of *Oxystomina cylindraticauda* and *Pseudosteineria horridus*.

If the interpretations of Rachor (1969) are correct, the most complex seminal receptacle is the demanian system which occurs in the genera *Adoncholaimus, Kreisoncholaimus, Meyersia, Oncholaimellus, Oncholaimium, Oncholaimus,* and *Viscosia*, all members of the family Oncholaimidae. This system was first described by de Man (1886) and has since been described in more detail by Cobb (1930), Strassen (1894) and Rachor (1969) among others.

The morphology of the demanian system varies considerably from one genus to another (Fig. 11), but basically it consists of two ducts designated by Cobb as the uterine and enteric efferents, although the enteric efferent is absent in *Viscosia* and *Oncholaimellus.* According

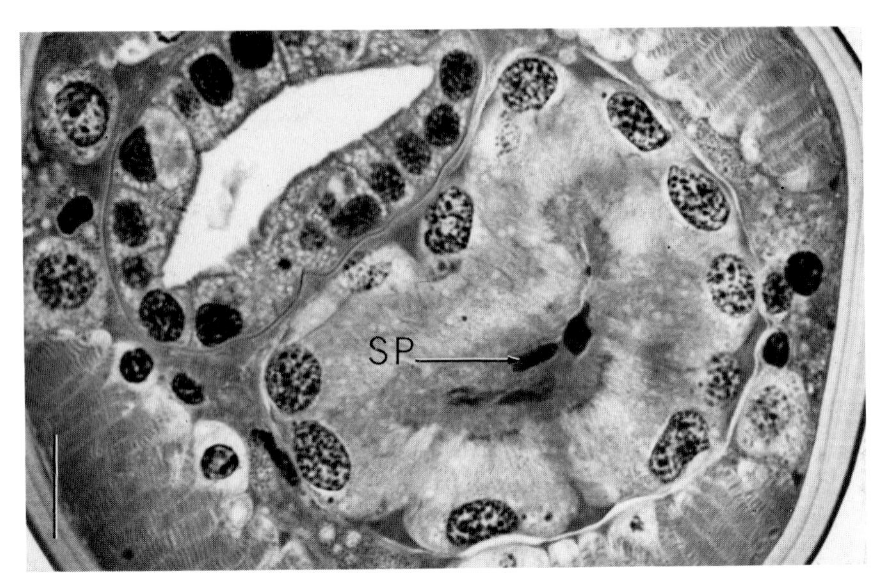

FIG. 9. Transverse section through the midregion of the uterus of *Deontostoma californicum*. The lumen contains spermatozoa (SP). Scale line indicates 25 μm.

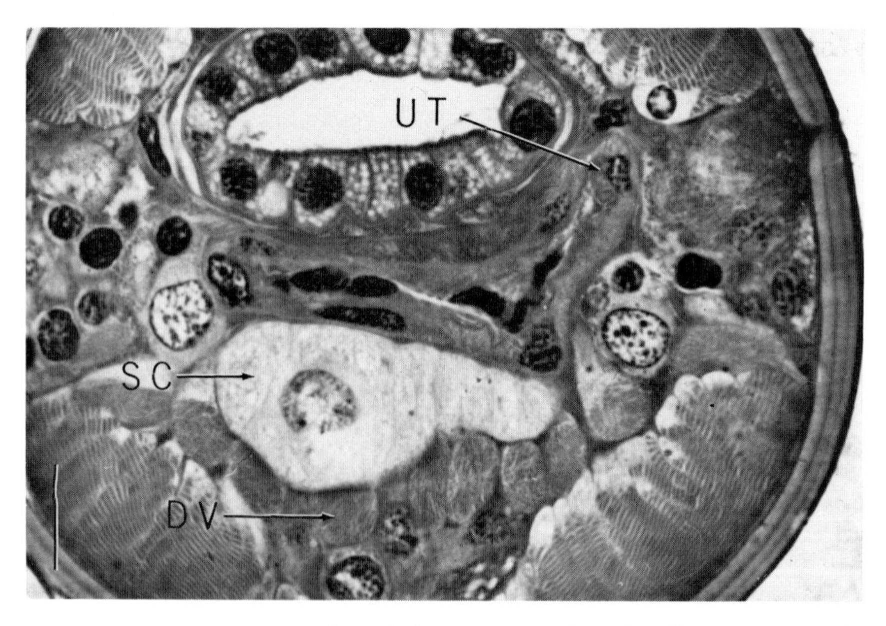

FIG. 10. Transverse section through the uterus near the vulva of *Deontostoma californicum*. Here the uterus (UT) is comprised of squamous epithelium, the sarcoplasmic portion of the constrictor muscle of the vagina (SC) and the dilator muscles of the vulva (DV). Scale line indicates 25 μm.

to Cobb, the uterine efferent (Du) has a closed end attached to the wall of the uterus. Special glandular cells in the uterine wall secrete a substance into the lumen of the uterine efferent. The opposite end of the uterine efferent is attached to the side of the enteric efferent (De) and the lumina of these two ducts are continuous by way of a small pore in the enteric efferent wall. A group of cells, collectively designated the uvette (Uv), surrounds the pore. These cells may be arranged radially *(Adoncholaimus fuscus)* or form an ampulla *(Oncholaimium appendiculatum).*

The enteric efferent also has a closed end, the osmosium (Os), which is attached to the intestinal wall. According to Cobb, the cells in the wall of the intestine underlying the osmosium are structurally unique, in comparison to other cells of the intestine. There is no open communication between the enteric efferent and the intestine, but Cobb suggests that the specialized cells of the intestine extract a substance from the intestine which is transported by way of the osmosium into the enteric efferent. The opposite end of the enteric efferent opens through the body wall to the exterior by way of one or two openings in the posterior body region (Pt). (A single pore opens into the vagina in the case of *Kreisoncholaimus* and *Meyersia.)* The enteric efferent is accompanied by a group of cells, the moniliform gland (not illustrated), anterior to the external openings.

Strassen and Cobb reasoned that fluid in the demanian system flows from the uterus and intestine, through the moniliform glands to the exterior where the secretion serves as an adhesive to aid copulation. Filipjev (1918) and Cobb also proposed that the secretion might be an embedding matrix to protect zygotes after deposition, while Kreis (1934) added to the above the possibilities that it might attract males, or offer protection to the female.

Filipjev (1918), in reviewing de Man's description of the system, implied that sperm might be present in the uterine efferent. Rachor (1969), having studied histological sections of several oncholaims, has repeatedly demonstrated large quantities of sperm in the enteric efferent as well as in the uterine efferent, and he concluded that this system is a seminal receptacle. Among other factors supporting this conclusion are that the demanian system, although variable in other respects, always has an osmosium which must transport nutrients from the intestine to nourish the sperm, and nearly always communicates with the distal end of the uterus where it is presumed that fertilization takes place. The narrow duct between the main tube and uterine efferent permits passage of but one spermatozoan at a time and, therefore, increases the chances of each egg being fertilized but once. Rachor further states that an important function can not be assumed for the terminal ducts because of their heterogeneity and variability.

8.3.5.1.4 *Vagina and Vulva.* The vagina, in usual circumstances, is readily distinguished from the uterus by a thin layer of cuticle on the medial surface of the vaginal wall (Fig. 12). This cuticle is continuous with the somatic cuticle, but differs in that it is comprised of but a single layer. The vagina may be an undivided tube in which case the two uteri are continuous with one another as, according to Filipjev (1921), is exemplified by *Pelagonema, Viscosia, Symplocostoma, Enoplus, Diplopeltis,* and all species of Chromadoridae. Or the vagina may be divided at its distal end, each branch continuous with one of the two separate uteri, as in *Deontostoma, Eurystomina, Pontonema,* and *Axonolaimus.*

The vagina is typically enclosed in a sheath of constrictor muscle which, according to Chitwood and Chitwood (1950), is continuous with and of the same general type as the uterine muscle. The vaginal muscles of *Deontostoma californicum,* and perhaps of other leptosomatids, are not similar to any cells of the uterine wall which may have contractile properties. Rather, the contractile filaments of the vaginal muscles are parallel with one another as is usual in the case of nematode muscle in general. In *D. californicum* there is a prominent constrictor muscle comprised of two muscle fibers that surround the main trunk of the vagina (Fig. 12). The sarcoplasmic portion of these two muscle cells, each with a nucleus, extends baglike from the vagina for a short distance along the ventral surface of the uterus. A loop of muscle tissue, one each from the anterior dorsal and posterior dorsal regions of the main constrictor muscle, ensheathes respectively the anterior and posterior branches of the vagina.

In addition to the constrictor muscles of the vagina, nematodes commonly have a complex of dilator muscles that extend in a radial fashion from the vagina to the body wall (de Man, 1886; Jägerskiöld, 1901; Türk, 1903; Filipjev, 1921; Chitwood and Chitwood, 1950; Timm, 1953), which presumably facilitates copulation and oviposition. Thirty-six dilator muscles are attached to the vagina of *Deontostoma californicum,* 24 of which have their origin on the subventral body wall, 12 anterior and as many posterior to the vulva (Fig. 12). At least the most dorsal of these subventral fibers has its origin at the ventral margin of the lateral hypodermal chord. The remaining subventral fibers are divided into several small branches at their origin, each branch passing independently between somatic muscle fibers and finally terminating at the hypodermis.

Each dilator muscle has its own sarcoplasmic region and nucleus near its midlength. The insertion of these muscles is only partly on

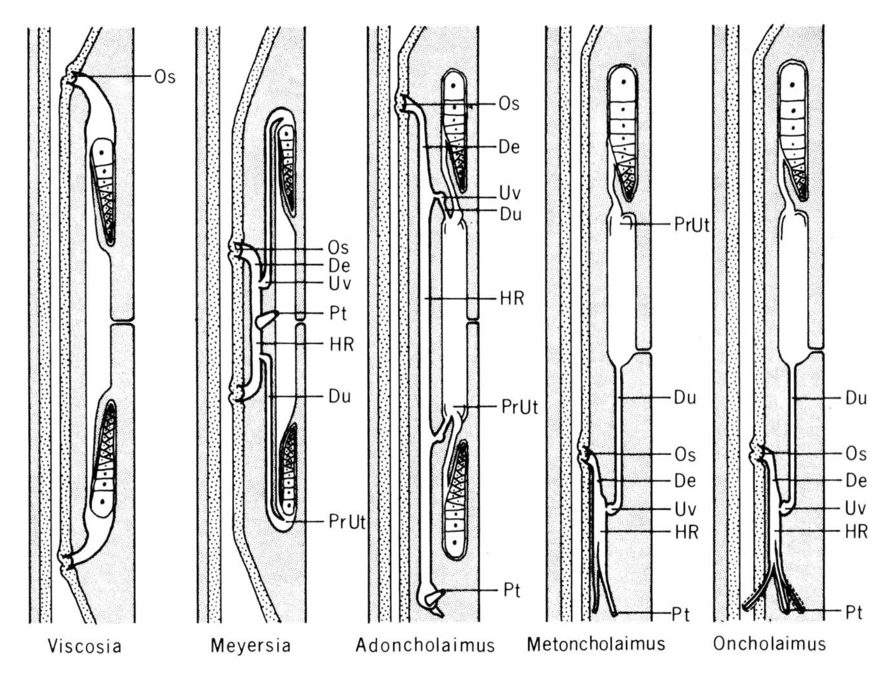

FIG. 11. Diagrams of the basic types of demanian systems [redrawn from Rachor (1969)]: osmosium (Os); enteric efferent (De); uvette (Uv); external pore (Pt); main tube (HR); uterine efferent (Du); pre-uterus (Pr Ut).

the constrictor muscle, for they also pass between the constrictor muscle and the wall of the vagina and are apparently attached to the latter as well. The subventral dilators also branch to the cuticle in the region of the vulva where they attach to cuticular apodemes that project into the body cavity from the lip region of the vulva (Fig. 12).

Multicellular and unicellular glands opening into the vagina have been described for most species studied in detail. While numerous cells occur in the region of the vagina of *D. californicum,* none appears to be glandular. Rather, it is likely that these are either nerve cells or the sarcoplasmic portions of muscle cells. Aside from the cells of the oviduct wall, the only ones appearing to have an excretory function are the several large cells in the lateral field near the vulva that discharge their products through the body wall to the exterior. These cells, which also occur less abundantly elsewhere along the entire length of both lateral fields, may discharge metabolic wastes and/or sex attractants.

Evidence that gland cells are associated with the female reproductive system of at least certain members of the Enoplida lies in Hopper's (1961) observations that eggs of *Enoplus communis* are depos-

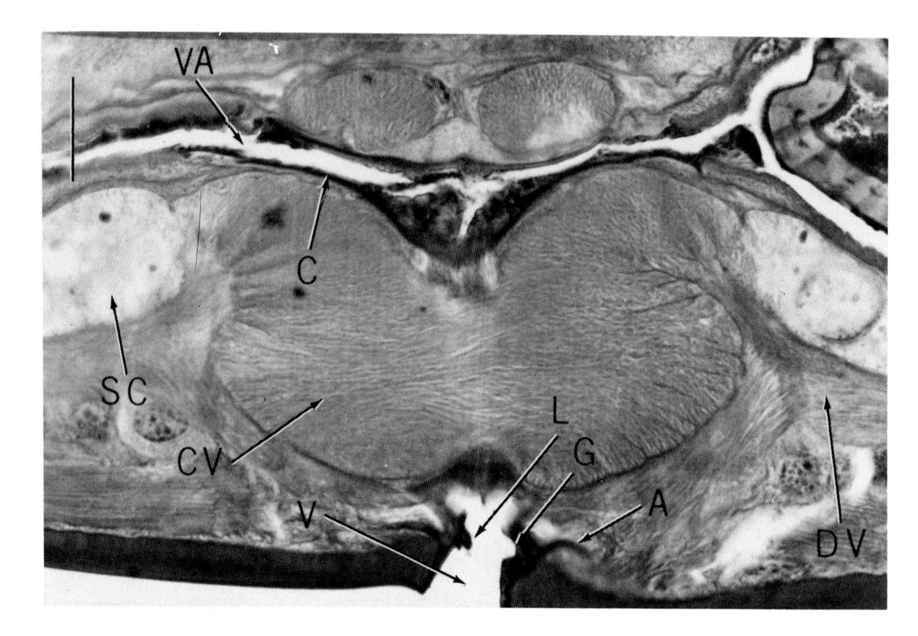

FIG. 12. Parasaggital section of a female *Deontostoma californicum* at the level of the vagina (VA) and vulva (V). A constrictor muscle (CV) comprised of two cells surrounds the neck of the vagina. One sarcoplasmic extension (SC) each on the anterior and posterior sides of the neck contains the muscle cell nuclei. Muscle cells also pass over the dorsal side of the vagina. Dilator muscles of the vagina (DV) are inserted on the constrictor muscle and cuticular apodemes (A) at the vulva. The vagina is lined with cuticle (C). A lip (L) and groove (G) exist on opposite sides of the vulva. Scale line indicates 25 μm.

ited in a stringlike mass. The entire mass is enclosed by an adhesive sheath which presumably anchors the eggs to the substrate. Hopper was unable to detect vaginal or vulval glands and so the source of this adhesive material is unknown. Some plant parasitic species deposit eggs in a gelatinous matrix. Maggenti and Allen (1960) have shown that this matrix is produced by rectal glands and is discharged through the anus in species of *Meloidogyne*. But in species of *Tylenchulus* the matrix is produced by the excretory cell (Maggenti, 1962).

The vulva, aside from its position on the body, seems to be very much the same in many nematodes and generally lacks specialization. However, longitudinal sections of *Deontostoma californicum* reveal a small cuticular ridge that protrudes slightly from each lip of the vulva into the lumen. Both ridges extend the full breadth of the vulva, and the edge of one ridge is turned toward the gonopore while the edge of the opposing ridge is turned inward, thus giving the impression that one ridge might interlock with the other so as to seal the gonopore when it is closed (Fig. 12). These ridges stain bright red in Mallory's Triple and are therefore dissimilar to the surrounding cuticle which under the same circumstances stains blue. The cuticle of the ridges extends into the body cavity on each side of the vulva to form the previously mentioned apodemes to which the dilator muscles are attached.

8.3.5.1.5 *Setae.* Many nematodes have sensory and/or secretory setae distributed over the general body surface. At least in some members of the Leptosomatidae sensory setae may be slightly more abundant and longer near the vulva where they may aid in coordinating copulatory behavior. Setae on or near the vulva in some members of the genus *Desmoscolex* are especially long and curved posteriorly, apparently providing a means by which ova may be carried against the venter of the female (Timm, 1970).

8.3.5.2 Male Reproductive System

This system lies free in the body cavity between the gut and ventral body wall and it consists of one (monorchic) or, more frequently, of two (diorchic) tubular testes (Fig. 13), each of which is continuous with a separate seminal vesicle. At the point where the two seminal vesicles converge there begins a common vas deferens and/or ejaculatory duct which opens into the hind gut (cloaca) and then to the exterior by way of a slitlike, transverse vent.

8.3.5.2.1 *Testes.* Males of the class Secernentea are normally monorchic, according to Chitwood and Chitwood (1950), but within the class Adenophorea monorchic and diorchic species may be found within the same family. The number of testes is not necessarily correlated with the number of ovaries in females of the same species.

When paired, one testis and its seminal vesicle is directed anteriorly from its point of transition with the vas deferens while the other testis and seminal vesicle proceeds posteriorly (Fig. 13B, D-G, I, and J). Reportedly, *Anticoma typica* is exceptional in that both testes are directed anteriorly and their positions are tandem to one another (Fig. 13A). Rarely, the testis is folded back upon itself as in the case of the monorchic species *Onyx perfectus* and *Monhystera stenosoma.* The diorchic condition, which according to Chitwood and Chitwood (1950) is exemplified by *Draconema cephalatum, Theristus setosus, Desmolaimus zeelandicus* (Fig. 13D), *Spilophorella paradoxa,* species of *Aphanolaimus* and *Bastiania,* as well as some axonolaims, comesomatids, and cyatholaims, is more prevelant among the orders of marine nematodes and is regarded by Filipjev (1921) to be more primitive than the monorchism. The existence of a single testis, and

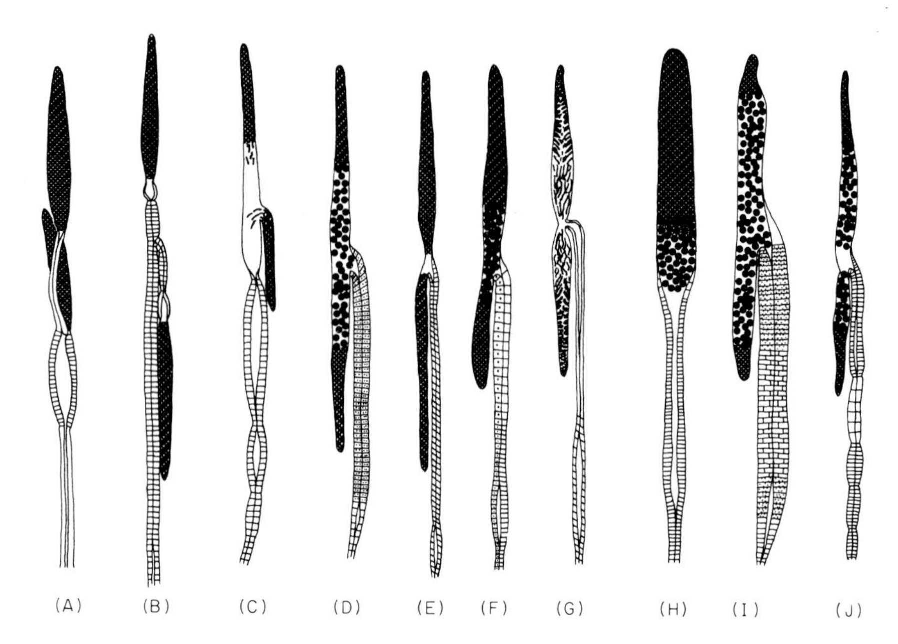

FIG. 13. Male reproductive systems of *Anticoma typica* (A), *Agamermis decaudata* (B), *Enoplus communis* (C), *Desmolaimus zeelandicus* (D), *Metoncholaimus pristiuris* (E), *Anaplectus granulosus* (F), *Tobrilus gracilis* (G), *Chromadora quadrilinea* (H), *Halichoanolaimus robustus* (I), and *Sabatieria hilarula* (J). The reproductive system of *C. quadrilinea* is monorchic and the rest are diorchic. The testes of *A. typica* are tandem and the testes of the other diorchic species are opposed [redrawn from Chitwood and Chitwood (1950)].

it is almost always the anterior one that prevails, is to be found in the genera *Chromadora* (Fig. 13H), *Euchromadora, Monoposthia, Spirinia, Epsilonema, Tripyloides, Desmoscolex, Greefiella, Monhystera,* and *Cyatholaimus*. According to Filipjev (1921), a single posterior testis exists in *Monhystera stenosoma*.

Whereas the seminal vesicles, in addition to the testes, are typically paired and opposed in a symmetrical fashion, the male system of *Enoplus communis* and most members of Leptosomatidae thus far studied are in this regard asymmetrical. That is, instead of paired seminal vesicles, there is but one vesicle which is, at its distal end, continuous with the anterior testis and at the proximal end with the ejaculatory duct. A vas deferens is absent. The posterior testis is attached to the midregion of the seminal vesicle (Fig. 13C).

The above asymmetry is sometimes accentuated by unequal size between testes. In *Cylicolaimus magnus, Tuerkiana comes, Enoplus communis, Halichoanolaimus robustus,* and *Sabatieria hilarula* the anterior testis is the larger one, while the posterior one is larger in *Metoncholaimus pristiuris*. In *Deontostoma californicum* the posterior testis is 2.46 mm long, the anterior testis 1.90 mm.

Histological sections reveal that the testis is enveloped in an extremely thin glycocalyx (also termed the tunica propria or lamina propria in earlier literature). Underlying this is a single layer of squamous epithelium, each cell containing a single nucleus. This epithelium has been regarded as syncytial (Türk, 1903; Filipjev, 1921), but, as was mentioned previously with regard to the ovary, the cell membranes and intercellular spaces are so tenuous as to be imperceptible with the light microscope.

The epithelium of the testes encloses a germinal chord; the latter is telogonic but without a rachis in marine nematodes. The distal end of the germinal chord is comprised of spermatogonia (Fig. 14).

The spermatogonia undergo mitotic divisions near the distal end of the testes. Here the interphase chromatin threads and metaphase chromosomes are readily evident (Fig. 14). The resultant primary spermatocytes have little cytoplasm and contain a nucleus that stains darkly with hematoxylin or methylene blue. Nearly 37% of the posterior testis is occupied by these cells.

The region in which meiotic divisions occur begins about 1.20 mm from the distal end of the testis and extends proximally for a distance of about 150 μm. Within a single section (Fig. 15) taken near the middle of this region, several stages of meiosis can be observed. The paired (bivalent) chromosomes appear in the larger nuclei. As metaphase ensues the chromosomes contract and become much thicker. The second division follows producing relatively small spermatids. The chromosomes uncoil and the nucleus of the spermatids becomes densely packed with chromatin material.

Spermiogenesis occurs in the remainder of the testis. The resultant mature sperm are spindle-shaped, measuring approximately 8 × 18 μm with a nucleus 3 × 15 μm (Fig. 16). The nucleus, which seems to have a spiral structure, is Fuelgen positive.

The lumen at the proximal end of each testis in *D. californicum* contains not only mature sperm but also a substance of uncertain identity. In other sections it also appears to envelope degenerating sperm. While it is possible that the substance in question might be amoeboid processes of epithelial cells that phagocytize excess sperm, it was never possible to demonstrate clearly continuity between the epithelium and the processes. Nor is it likely that this is the cytoplasm of cells free in the lumen since they do not seem to have their own nuclei. The increased thickness of the epithelium and presence of blue intracellular granules suggests what is, perhaps, a more likely possibility, namely that the substance in the lumen is a secretion of the epithelium.

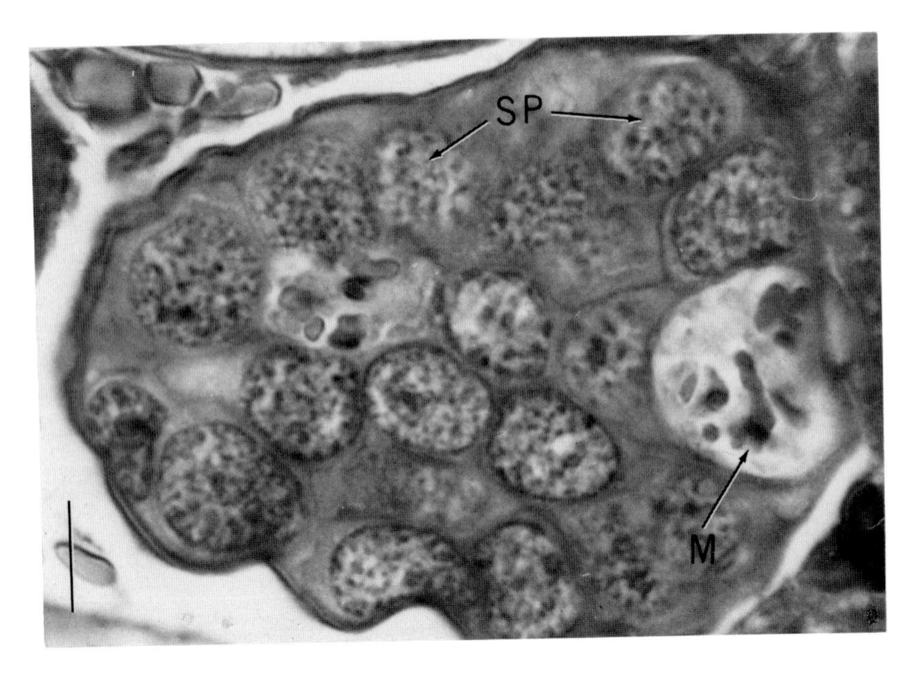

FIG. 14. Transverse section of the distal end of a testis of *Deontostoma californicum*. The germinal chord is comprised of spermatogonia (SP) which are proliferated by mitotic divisions (M). Scale line indicates 10 μm.

At its proximal end the posterior testis narrows gradually, as it does in other leptosomatids, to form a duct 30 μm wide and 200 μm long (Fig. 17). It loops posteriorly before merging with the seminal vesicle. The duct has no counterpart in the anterior testis.

Except for the size difference and the duct connecting the posterior testis to the seminal vesicle, the anterior and posterior testes are entirely comparable in all respects. The transition from anterior testis to seminal vesicle occurs without change in the diameter of the duct or other outward evidence, and in this respect they again resemble other leptosomatids. Histologically, it is evident by the gradual displacement of the squamous epithelium of the testis by the cuboid cells of the seminal vesicle (Fig. 17).

8.3.5.2.2 *Gonoduct.* The male gonoduct is an epithelial tube continuous with the testes at their anterior ends and opening into the cloaca posteriorly. It serves both to store sperm and convey them to the exterior at the time of copulation. Like the female gonoduct, it is comprised of regions, each structurally and, by implication, functionally unique. According to Chitwood and Chitwood (1950), the typical

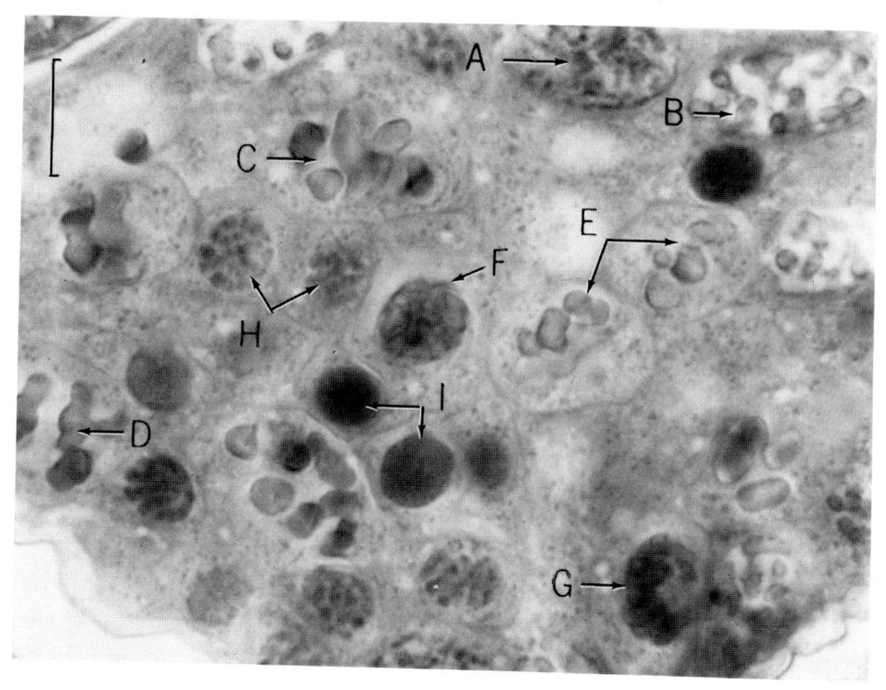

FIG. 15. Transverse section through the region of meiosis in the posterior testis of *Deontostoma californicum*. Stages of meiosis represented are leptonema or zygonema of prophase I (A), pachynema or diplonema of prophase I (B), prometaphase or metaphase I (C), anaphase I (D), telophase I (E), metaphase II (F), early anaphase II (G), telophase II (H), and young spermatids (I). As spermiogenesis proceeds the nucleus becomes densely packed with chromatin material. Scale line indicates 10 μm.

gonoduct in Adenophorea consists of a pair of opposed seminal vesicles, each filled with sperm received from its respective testis, and proximally each seminal vesicle attaches to the distal extremity of a single, common vas deferens, typically glandular. A short, nonglandular, posterior portion ensheathed with muscle serves as an ejaculatory duct (Fig. 13). As noted above, the enoplids deviate from the typical arrangement in that they possess but a single seminal vesicle which receives sperm from both testes, the vas deferens is absent, and the ejaculatory duct is much more muscular. Further comparison of other free-living adenophoreans is not possible as so little is known in the case of the latter group. There is a remarkable degree of similarity among the gonoducts of enoplids, and they so closely resemble the male gonoduct of *Deontostoma californicum* that a description of the latter will suffice as a review.

The seminal vesicle of *Deontostoma californicum* is continuous at its distal end with the anterior testis and receives the posterior testis

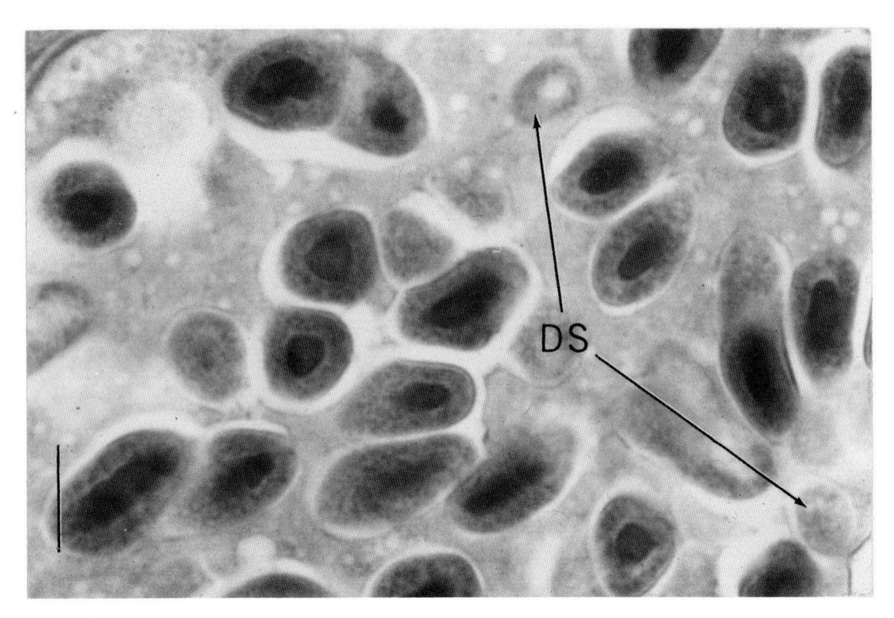

FIG. 16. Transverse section of the proximal region of the testis of *Deontostoma californicum*. The spermatozoa are sectioned transversely and obliquely. The remainder of the lumen is filled with an amorphic mass and envelops what may be degenerating sperm (DS). Scale line indicates 10 μm.

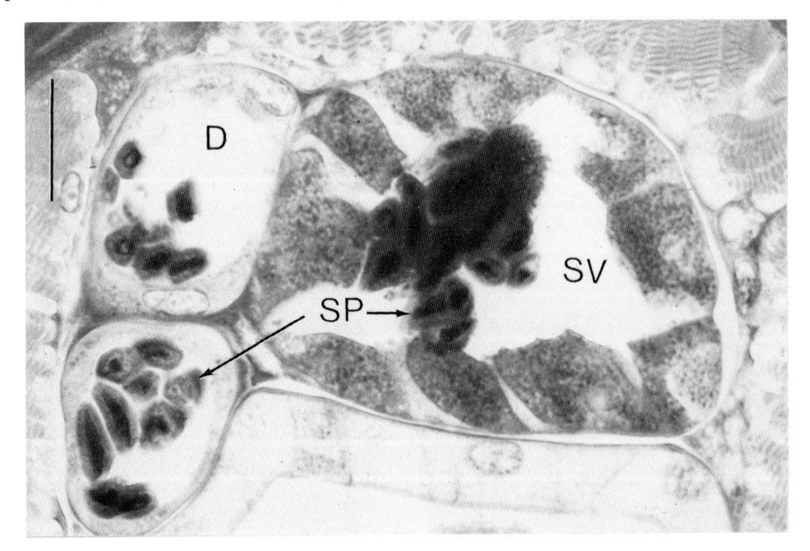

FIG. 17. Transverse section through the seminal vesicle (SV) and looped duct (D) of the posterior testis of *Deontostoma californicum*. Both contain sperm (SP). Scale line indicates 25 μm.

near its midlength. The margin between the distal extremity of the seminal vesicle and the proximal end of the anterior testis is not evident by gross examination. Rather, the squamous epithelium of the anterior testis is gradually replaced by cuboidal epithelium. The cubodial cells are packed with granules (Fig. 17), but none appear in the lumen of the seminal vesicle. The seminal vesicle gradually tapers toward its proximal end, finally attaining a minimum diameter of about 45 μm. As it tapers, the cells of the epithelium continue to decrease in size.

The ejaculatory duct is the second major region of the gonoduct and it delivers sperm from the seminal vesicle to the cloaca. Throughout its length the walls are comprised of a peripheral layer of muscle and a medial layer of epithelial cells. The structure of these layers differs so that seven histological regions can be identified. Two of these regions are sphincters and both are characterized by a very narrow epithelial tube surrounded by a heavy wall of musculature (Fig. 18). The epithelium of the remaining regions appears to have a secretory function as these cells are commonly packed with granular inclusions (Fig. 19). The entire ejaculatory duct is sheathed with a layer of muscle tissue, but while the muscle fibers of the sphincters are arranged ringlike around the epithelial tube, they are V-shaped elsewhere. The base of the V, which contains the nucleus, is located at the lateral sides of the duct. From the base one arm of the fiber extends obliquely in a posterior and dorsal direction to the mid-dorsal line of the duct where it is attached to the end of a comparable muscle fiber on the opposite side of the duct. In a similar fashion the other arm of the V-shaped fiber extends posteriorly and ventrally where it also joins a fiber from the opposite side. Each fiber is succeeded by another and since the angle of the V is acute, the arms of several fibers are present in each transverse section of the duct (Fig. 19). The nucleus is in a sarcoplasmic region at the base of the V and here each fiber is overlaid externally by the posterior tip of the fiber anterior to it.

Posteriorly the ejaculatory duct merges with the proctodaeum, which is in fact a cloaca, and sperm pass through it to the exterior.

8.3.5.2.3 *Spicula.* The spicula are the copulatory organs of male nematodes. They are not a true intromittent organ in that they do not possess a lumen through which sperm are conveyed. Rather, they are typically paired, cuticularized, and shaped somewhat like a scimitar (Fig. 20A and B). Each is located in a dorsal pouchlike evagination of the cloaca, termed the spicular pouch. Basically, the spicula consist of a knoblike proximal end (capitulum), cylindrical or bladelike shaft

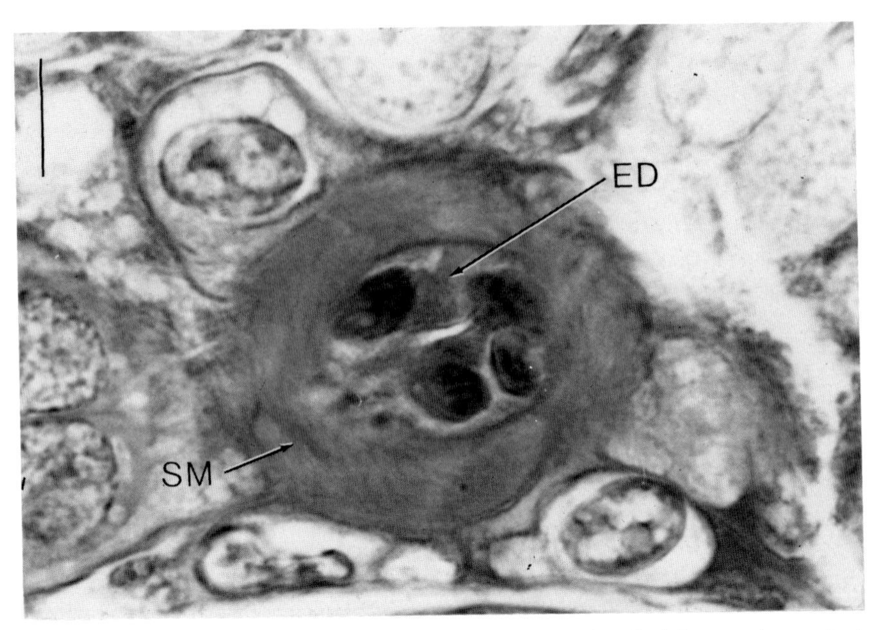

FIG. 18. Transverse section of the sphincter at the distal end of the ejaculatory duct of *Deontostoma californicum*. The diameter of the epithelial portion of the duct (ED) and its lumen are reduced. The muscle (SM) layer is well-developed and its fibers extend ringlike around the duct. Scale line indicates 10 μm.

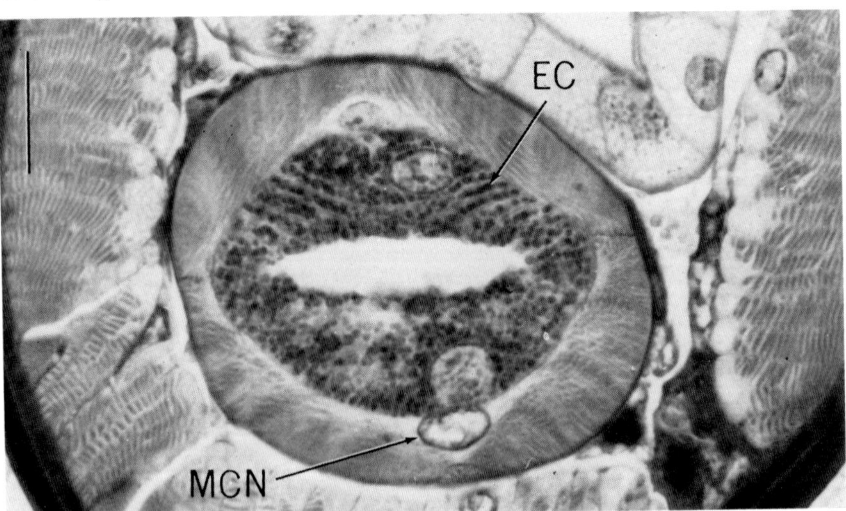

FIG. 19. Transverse section through the ejaculatory duct of *Deontostoma californicum*. The granular inclusions suggest that the epithelial cells (EC) may have a secretory function. The nucleus of the muscle cells (MCN) is at the lateral side of the duct. Scale line indicates 25 μm.

(calomus) and laterally compressed blade (lamina) (Fig. 20B). Occasionally the spicula of a pair may differ from one another as in *Enoplus groenlandicus.* Others may have but a single spiculum as in *Rhaptothyreus typicus*, or none at all as in species of *Monoposthia.* The structural variations of the spicula are too extensive to be given full consideration here. For further information on this subject, reference should be made to Chitwood and Chitwood (1950).

During copulation spicula apparently assist in dilating and in holding the male vent and the female gonopore in juxtaposition. To perform these functions they are provided with muscles by which they may be protruded through the cloacal vent or withdrawn into the spicular pouch. In general, as many as five muscles may be involved in moving each spiculum. They are as follows:

1. Laterodorsal retractor—extends anteriorly from the proximal end of the spiculum to the body wall at the dorsal margin of the lateral chord. This muscle is commonly paired, at least among enoplids.
2. Lateroventral retractor—extends anteriorly from the proximal end of the spiculum to the body wall at the ventral margin of the lateral chord.
3. Anterior ventral protractor—extends posteriorly from the proximal end of the spiculum to the subventral body wall anterior to the cloaca.
4. Posterior ventral protractor—extends posteriorly from the external area of the proximal end of the spiculum to the subventral wall of the tail.
5. Dorsal protractor—extends from the proximal end of the spiculum to the dorsal wall of the tail.

The posterior ventral protractor commonly attaches first to the apophysis of the gubernaculum, when the latter is present (see below) and then proceeds to the ventral body wall. De Coninck (1965) also mentions an internal protractor muscle extending from the internal surface of the proximal end of the spiculum to the subventral wall of the tail, but acknowledges its uncertain nature.

8.3.5.2.4 *Gubernaculum.* The gubernaculum is also a cuticularized structure commonly associated with the distal ends of the spicula. This structure is, in part at least, a cuticularization of the dorsocaudal region of the spicular pouch. In its simplest form it is little more than a small plate. In other cases the gubernaculum possesses a cuneus which is a medial process that extends into the spicular pouch between the spicula. According to Chitwood and Chitwood (1950), the posterior part of the gubernaculum is then termed the corpus; lateral pieces may also be present and are designated the crura (Fig. 20). Structural modification of the gubernaculum has been extensive among marine nematodes and it is often difficult to identify homologous parts. A point of interest in common for the crura of many enoplids, is its cytoplasmic core and the papillalike protrusion at the distal end of each. The tissue in the core may in part be nervous and the protrusion may be a sensory receptor.

Other variations in the structure of the gubernaculum are given by Chitwood and Chitwood (1950) and de Coninck (1965).

Where the gubernaculum occurs as a platelike thickening its function may be little more than to prevent the spicula from penetrating the wall of the spicular pouch. But complex gubernacula have attached muscles and are moveable. Three pairs of muscles are commonly involved and they are as follows:

1. Retractor—extends from the distal extremity of the gubernaculum to the dorsal side of the lateral chord.
2. Protractor—paired muscles extending posteriorly from the proximal extremity of the gubernaculum to the ventral body wall.
3. Seductor—extends from the proximal extremity of the gubernaculum to the latero-dorsal region of the body wall.

Variations exist in the number, origin and insertion of muscles attached to the gubernaculum. In *Deontostoma californicum*, for example, the seductor muscle is absent and the retractor muscle is inserted on the lateral surface of the anterior arm of the crus. Also, the protractor muscle of the spiculum is inserted on the apophysis of the corpus. Such variations not only assist in the protraction of the spicula, but provide for the protrusion and retraction of the gubernaculum as well. Protrusion of the distal ends of the gubernaculum through the cloacal vent has been observed in specimens of *Deontostoma californicum*, *Synonchus* sp., and *Enoplus groenlandicus*. In such cases the gubernaculum may assist in clasping the female and serve in sensory reception as well. In the case of the last species, the right and left halves of the gubernaculum differ structurally from one another.

8.3.5.2.5 *Copulatory Muscles*. A series of modified somatic muscles occur anterior to the cloaca on both sides of the body of male nematodes. Each fiber extends from the body wall at the dorsal margin of the lateral chord to the subventral body wall (Fig. 20). The nucleus of these fibers is located in a small sarcoplasmic belly on the medial surface at the dorsal end of the fiber. Contraction of these muscles enables the male to coil around the female at the time of copulation.

8.3.5.2.6 *Bursa*. The bursa is a flaplike, sublateral extension of the

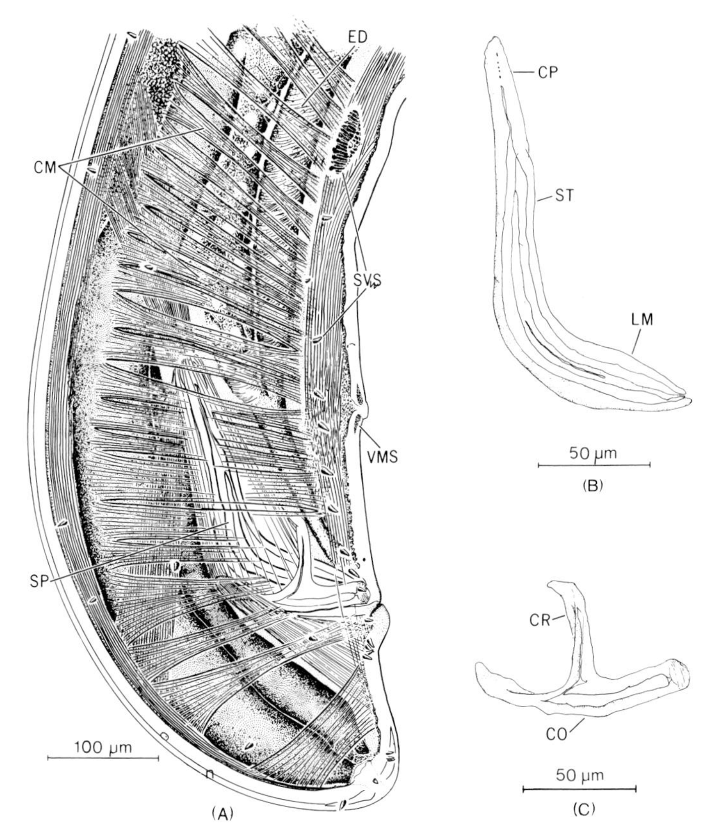

FIG. 20. (A) Male tail of *Deontostoma californicum.* (B) Right spiculum. (C) Right lateral view of gubernaculum. Spiculum (SP). Capitulum (CP). Shaft (ST). Blade (LM). Corpus (CO). Crus (CR). Ventromedian supplement (VMS). Subventral supplements (SVS). Copulatory muscles (CM). Ejaculatory duct (ED). Scale to the right applies to both B and C [from Hope (1967a)].

somatic cuticle on each side of the body at the level of the cloaca. While they are relatively common among parasitic species, they are uncommon among those that are marine. Examples of a bursa in marine forms are to be found in the genera *Anoplostoma, Oncholaimellus,* and in *Rhabditis marina.* The bursa is generally regarded as functioning to assist in holding the copulating pair together and in sealing the dilated gonopore from the external environment.

8.3.5.2.7 *Copulatory Supplements.* This term is used here to designate the ventromedian and subventral papillae and setae, and depressions and other intracuticular structures anterior and posterior to the cloacal vent (Fig. 20). The ventro-median supplement may be a small cuplike, intracuticular structure, as in *Deontostoma;* a narrow cuticularized tube protruding from the somatic cuticle into the body cavity, as in *Anticoma, Phanoderma,* and *Mesacanthion;* or a large intracorporeal funnel, as in some species of *Enoplus.* Several ventromedian supplements of a given form may also be present on the same animal, i.e., cuplike intracuticular depressions as in *Biarmifer,* setae as in *Paracyatholaimus,* intracorporeal tubes as in *Paracanthonchus,* large, cuplike depressions with moveable anterior and posterior intracorporeal extensions as in *Eurystomina,* or external, suckerlike discs as in *Latronema.* They appear to be absent in members of the Desmoscolecida and many Oncholaimina.

The subventral supplements commonly occur in a longitudinal series on each side of the body and they are nearly always in the form of papillae or setae. In some cases they are located upon hemispherical swellings of the cuticle *(Deontostoma).* The supplements occur in a ring enclosing the cloacal vent in most oncholaims, while they are absent in Desmoscolecida.

The function of copulatory supplements has not been conclusively demonstrated, but it is suspected that the setae and papillae are tangoreceptors, while the depressions and tubes may be either chemoreceptors or secretory pores. The ventromedian supplements of *Eurystomina* may be adhesive organs.

8.3.6 Origin of Germ Cells and Gonads

The embryogeny of the reproductive system has not been studied in the case of free-living marine nematodes. However, considerable information regarding the development of this system has resulted from the early classic studies in the cytology and germ cell lineage of certain nematodes parasitic in animals (Boveri, 1892, 1899, 1910; Spemann, 1895; Zoja, 1896; Strassen, 1896; Neuhaus, 1903; Müller, 1903; Martini, 1903, 1906) and of the free-living nematode *Turbatrix aceti* by Pai (1928). These studies have been reviewed by Hyman (1951), de Coninck (1965), and Tadano (1968). The relatively recent interest in plant parasitic nematodes has resulted in additional information, especially on postembryonic gonad development (Van Gundy, 1958; Weerdt, 1960; Triantaphyllou, 1960; Yuksel, 1960; Hirschmann, 1962; Dickerson, 1962; Yuen, 1966; Hirschmann and Triantaphyllou, 1967; Roman and Hirschmann, 1969). Similar studies have also been made on free-living soil nematodes by Hechler (1968, 1970b). The early success in following the germ cell lineage is substantially attributable to the facts that the germ cell line is in some ascarids segregated from the somatic cells early in development, and that these lines can be distinguished readily from one another. The distinction lies in the occurrence of chromatin diminution in the somatic line, a phenomenon absent in the germ cell line.

By the above distinction it has been determined that the stem line cell classically designated P_3 divides to produce stem line cell P_4 and the somatic cell S_4 (see Section 8.4.1). P_4 then divides to produce the paired primordial germ cells G_1 and G_2 which, after gastrulation, come to lie in the blastocoel (pseudocoelom) at or slightly posterior to the midbody length. A pair of somatic cells, one each on opposite ends of the paired primordial cells, is invariably present and these give rise to the epithelium of the gonad and gonoduct (Fig. 21A). This pair of somatic cells may be derived from somatic cell S_4, although this is not well established (Hyman, 1951; Tadano, 1968) for the parasitic species. In the case of *Turbatrix aceti*, however, Pai (1928) found that S_4 is ectodermal, giving rise to the hind gut, while P_4 divides to produce P_5 and S_5, the former being the definitive primordial germ cell and the latter giving rise to the epithelium of the gonad and gonoduct. As in the case of P_4 and S_4 of ascarids, P_5 and S_5 of *T. aceti* each divide one more time during early embryogeny so as to become paired. Further development of the gonads and gonoducts is arrested until the first stage juvenile is fully formed.

With the completed development of the first-stage juvenile and, usually, its hatching from the egg, the genital primordia resume development and continue through the subsequent juvenile stages and molts, although such development is restricted to periods of molting in certain species (Hirschmann, 1971).

Paired primordial germ and epithelial cells may develop into adults with one or two gonads, depending upon the species. In the case of females with two ovaries, the germ cell nuclei are gradually separated by developing gonoducts and each gonad receives a single primordial germ cell (Fig. 21C and H). Where but a single ovary develops, the primordial epithelial cells are on the posterior side of the germ cells. The germ cells remain together and so the single developing gonad is endowed with two primoridal germ cells (Fig. 21B, D, G). The female becomes monodelphic and prodelphic. The gonad development of opisthodelphic species has not been studied.

Males normally with two testes are restricted to the class Adenophorea and development of such individuals has not been studied. Males of the class Secernentea typically have a single testis which commonly develops with two primordial germ cells. In this case both cells remain together at the posterior end of the developing gonad (Fig. 21D, E, and F). The gonoduct at first elongates in an anterior direction and then turns end for end so that the mature testis is anteriorly directed (Hirschmann, 1971).

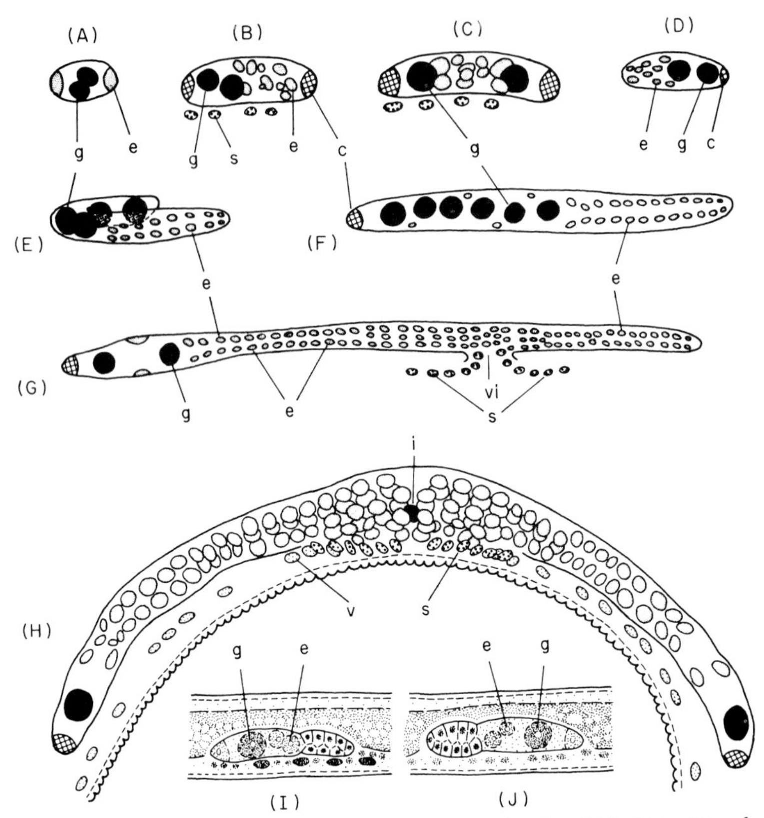

FIG. 21. Development of the reproductive system of Tylenchida [after Hirschmann (1971)]. (A-G) *Pratylenchus* sp.; (H) *Helicotylenchus* sp.; (I) and (J) *Ditylenchus* sp. (A) Genital primordium of second-stage juvenile. (B) Development of monodelphic gonad, third-stage female. (C) Development of amphidelic gonad, third-stage female. (D) Genital primordium, third-stage male. (E) Male gonad reversing direction of growth during third molt. (F) Reversed male gonad of fourth-stage male. (G) Monodelphic gonad of fourth-stage female. (H) Amphidelphic gonad of fourth-stage female. (I) Genital primordium with one germinal nucleus, third-stage female. (J) Genital primordium with one germinal nucleus, third-stage male. Cap cell nuclei (c); epithelial nuclei (e); germinal nuclei (g); ventral chord nuclei (v); specialized ventral chord nuclei (s); vaginal invagination (vi); and "I" nucleus [vaginal organizer?] (i).

In contrast, the genital primordium may consist of but a single germ cell as in the case of *Ditylenchus destructor* (Fig. 21I and J). In this instance the juvenile will become a male if the germinal cell is at the posterior end of the gonad in its stages of development, and a female if at the anterior end. As above, the developing testis and its gonoduct eventually undergo reversion (Anderson and Darling, 1964).

The sex of the developing juveniles can usually be determined in the late second or early third-stage juveniles by the presence of the primordia of the vulva and vagina or primordia of the male copulatory apparatus. The former arises from specialized cells of the ventral hypodermal chord and the latter from the rectal epithelium (Hirschmann, 1971).

8.3.7 Gametogenesis

Previous observations on the gametogenesis of free-living marine nematodes is limited to a description of spermatogenesis in *Spirinia parasitifera* by Cobb (1928). However, according to Chitwood and Chitwood (1950), Cobb's interpretations are not accurate.

A brief description of spermatogenesis and a partial description of oogenesis of *Deontostoma californicum* has been given in the section of this chapter pertaining to the anatomy of the gonads. While an accurate count of chromosomes was not feasible, it was observed that spermatocytes undergo two nuclear divisions in rapid succession without an intermediate interphase as would occur in normal meiosis. This, in conjunction with a nearly equal sex ratio, suggests that developing gametes of *D. californicum* undergo reduction division and that this species reproduces by amphimixis.

Gametogenesis of nematode parasites of plants and animals has been studied extensively and recently reviewed by Nigon (1965), Bird (1971) and Triantaphyllou (1971). Gametogenesis of these forms is usually by normal meiosis, and amphimixis is the common mode of reproduction, although, as mentioned in Section 8.2, parthenogenesis is not uncommon in some genera of plant parasitic species. Spermatogonia and oogonia undergo mitotic divisions in the germinal zone, and the beginning of the growth zone occurs at the level where the mitotic divisions cease, and the chromatin material of the gametocytes becomes compacted into a densely staining network or mass which indicates synapsis (Triantaphyllou, 1971). The oocytes, whose volume increases substantially in the growth zone, are in prometaphase I as they leave the ovary and enter the spermatheca where fertilization normally occurs. With penetration of the sperm into the oocyte, the nucleus advances to metaphase I and the vitelline membrane is transformed into a shell by "impregnation with and deposition of materials condensed from the cytoplasm" (Triantaphyllou, 1971). After shell formation, a substance within cytoplasmic vacuoles of the oocyte is exuded through membranes enveloping the oocyte. As this occurs, the volume of the oocyte diminishes creating a space between it and its chorion which is simultaneously filled with the exuded substance—the perivitelline fluid (Tadano, 1968). A thin lipoid membrane is then separated from the cytoplasm and becomes apparent in the perivitellous space (Triantaphyllou, 1971). The first maturation division is then completed and the second follows, without an intermediate interphase, and two polar bodies result. The pronuclei of the ovum and sperm migrate toward one another and upon contact move together toward the center of the zygote.

Returning to the development of sperm, the spermatocytes of plant parasitic species reach metaphase of the first maturation division upon leaving the testes and entering the gonoduct. Here normal meiosis proceeds resulting in four haploid sperm (Triantaphyllou, 1971). Mature nematode sperm are generally amoeboid, spherical, conical or spindle-shaped. Further, they are thought to be nonflagellated (Bird, 1971) although there are unconfirmed reports of flagellate sperm in species of *Halanonchus*, *Tobrilus*, and *Leptosomatum acephalatum* (Chitwood and Chitwood, 1950; Timm, 1953). The only information to date on the fine structure of sperm is limited to that of nematode parasites of animals, and these have been reviewed by Bird (1971). A description of the sperm of *Deontostoma californicum* has been published elsewhere (Wright *et al.*, 1973).

The fecundity of plant and animal parasitic nematodes is in general very high, as would be expected in overcoming the vicissitudes of parasitism, and is, thus, meaningless in terms of their free-living marine relatives. Throughout the taxonomic literature there exist isolated observations on the number of eggs in adult female marine nematodes, typically ranging from one in the case of desmoscolicids to several in enoplids and oncholaims, but total productivity during the life span is known only in a few cases. Tietjen *et al.* (1970) has shown that a female *Rhabditis marina* will deposit 70-100 eggs during a life span, while Gerlach and Schrage (1971) determined that *Theristus pertenius* will produce at least 30 eggs over its total productive period of 120 days.

8.3.8 Gametogenic Cycles

Nematodes, as individuals, are not known to pass through successive cycles characterized by periods of gametogenesis interspersed with periods of gonad recovery. Rather, after once attaining sexual maturity, their gametogenic potential is expended during a single reproductive period after which death usually ensues. The period of egg laying is uninterrupted in most species, but in *Agamermis decaudata* egg laying in the fall may be stopped by the onset of cold weather and resumed the following spring (Christie, 1936). Tietjen *et al.* (1970) observed that about 60% of the females of *Rhabditis marina* died after apparently completing egg deposition. Those that did not die resumed oviposition after copulating a second time. The temporary cessation of oviposition may have been due to a depletion of sperm before the female had attained the end of her egg laying potential. This suggestion seems reasonable in view of the statement by Triantaphyllou (1971) that gametogenesis may be halted if the first oocytes produced reach the uterus without being fertilized.

Although reproductive activity of individuals is not known to be cyclic, it may be seasonal for the species as evidenced by the studies of Wieser and Kanwisher (1961), Hopper and Meyers (1967) and Tietjen (1969). Tietjen, for example, found *Nudora lineata, Tripyloides gracilis, Spirinia parasitifera, Monoposthia costata, Chromadora axi, Chromadora quadrilinea,* and species of *Hypodontolaimus* abundant as sexually mature adults during the spring and summer on the coast of Rhode Island. During fall and winter the same species were either absent or significantly less abundant. While it is possible that populations of such species largely die out during unfavorable periods and are then reestablished through recruitment from other areas, it seems more plausible that such species are present throughout the year but pass adverse seasons in a cryptic but resistant stage, such as a young juvenile or egg.

8.3.9 Factors Influencing Gametogenesis

Very few data exist with regard to the effect of physiological and environmental factors on the production of gametes of marine nematodes, but in the case of *Rhabditis marina**, which will reproduce in 0-75% salinity, Tietjen *et al.* (1970) found that the life span was increased and reproductive potential decreased at extremes of salinity in the laboratory. Also, population growth was fastest at 25°-30°C. However Tietjen *et al.* surmise that biotic factors are more important limiting factors in reproduction than physical ones.

8.3.10 Reproductive Behavior

8.3.10.1 Sex Attractants

Since most nematodes are dioecious, fertilization is internal, both sexes are commonly mobile, and all appear to have sensory recep-

*The genus *Rhabditis* is largely comprised of soil and freshwater species and this species is believed to have secondarily returned to the marine environment. Its wide range of salinity tolerance is a major factor in the success of this transition, but it should not be regarded as a typical marine nematode.

tors, it is not unreasonable to assume that for a given species, one or both individuals would be attracted to the other for purposes of reproduction. Recent investigations on the animal parasitic nematodes *Ancylostoma caninum* by Beaver *et al.* (1964) and *Trichinella spiralis* by Bonner and Etges (1967), on plant parasitic nematodes of the genus *Heterodera* by Green (1966) and Green and Plumb (1970), and on the free-living soil nematodes *Panagrolaimus rigidus* by Greet (1964), on *Pelodera teres* by Jones (1966) and on *Cylindrocorpus longistoma* and *C. curzii* by Chin and Taylor (1969) have given evidence that this attraction is chemically mediated, and that members of one or both sexes will migrate to the general locality of the other. In all of the above cases the males are attracted to females of the same species and females are attracted to males in *Ancylostoma caninum, Trichinella spiralis, Panagrolaimus rigidus,* and in species of *Heterodera.* Attraction of females to males was greater than males to females in the case of *T. spiralis.*

Greet (1964) has shown that the chemical attractant involved is water soluble, its molecular size is sufficiently small to pass through a cellophane barrier and it will diffuse 2 cm in 1.5% water-agar over a period of 17 hours. Chin and Taylor (1969) have also shown this attractant to be species specific in the case of the two species of *Cylindrocorpus.* Green and Plumb (1970), however, found that the ten species of *Heterodera* with which they were concerned could be divided into three subgeneric groups on the basis of interspecific attraction of males to females. They concluded that at least six distinct substances were present in the attractants of these species.

Santos (1969), according to Green (1971) was unable to demonstrate that females of *Meloidogyne arenaria* produced an attractant for the males. Notwithstanding, it seems likely that chemical attraction between opposite sexes of the same species will prove to be the rule rather than the exception, even among marine species.

8.3.10.2 Mating

For nematodes, in general, it is known that copulation takes place with the tail and posterior region of the body of the male coiled about the vulval region of the female. Exceptions might be the stouter bodied species such as *Desmoscolex.* The insertion of the spicula into the vagina insures alignment and close adherence of the female gonopore with the male cloacal vent. Where present, the male bursa may also aid in holding and in sealing the open gonopore from the external environment. Threlkeld (1941), in his observation of copulating pairs of nematodes parasitic in sheep, noted that males of *Haemonchus contortus* have a large bursa with muscular rays which enable the males to firmly grasp the females. But in *Bunostomum trigonocephalum* the bursa is too small to function as a holdfast organ; union between the mating pairs is assured by copius secretions from the male which cement the bursa over the vulva. Chitwood (1929) has also reported the secretion of a cement by copulating males of *Pelodera strongyloides* which he believes serves to hold the mating pair in copula. This secretion is retained by the fertilized females as a large, nearly transparent mass at the vulva which Chitwood calls the copulatory saccus. Eggs are forced into this mass as they are laid, and the mass stretches as the eggs accumulate. Eggs may escape from the mass through the hole left by the spicula.

Details are especially sparse in the literature on marine nematodes. While the opening of the vulva is another function commonly attributed to the spicula, it would seem that this function could be attributed to the muscles of the vagina and vulva in *D. californicum* and related species.

Chin and Taylor (1969) have observed that females of *Cylindrocorpus longistoma* and *C. curzii* became receptive to males upon completion of the final molt. Contact between the posterior halves of both bodies was necessary to elicit a copulatory posture in the male. Once the male had coiled around the female, she continued to feed moving backward and forward until the spicula came into contact with the vulva. At this time the spicula were inserted with rapid action and ejaculation ensued.

After sperm have entered the female gonoduct they migrate, presumably by amoeboid movement, to the spermatheca or distal end of the gonoduct where fertilization occurs.

8.4 Development

8.4.1 Embryonic Development

Studies on the embryogeny of nematodes began during the latter part of the nineteenth century. In addition to contributing to the knowledge of nematodes, these studies provided new information of fundamental importance to understanding reproduction and embryonic development of animals in general. According to Triantaphyllou (1971), for example, it was Bütschli (1873) who first observed that two nuclei are present in fertilized eggs of *Caenorhabditis dolichura* and, according to Chitwood and Chitwood (1950), Bütschli (1875) was also the first to witness formation of polar bodies. But it was van Beneden (1883) who, in his study on *Parascaris equorum* first discovered meiosis, found that the somatic number of chromosomes was reduced by half in the egg nucleus, and that the other half was eliminated in the polar body. He also discovered that the diploid number was reestablished with union of the egg and sperm pronuclei (Wilson, 1898). Also, according to Triantaphyllou (1971), of major significance was the concept developed by van Beneden (1883) and Boveri (1888) that basic chromosomal organization does not change during interphase when chromosomes are not visible. Further, Boveri (1887) discovered chromosome fragmentation and chromatin diminution and, finally, Krüger (1913) first described pseudogamy (see Section 8.2) as a normal and necessary aspect of asexual reproduction in *Rhabditis aberrans*.

This period in the history of nematology is also noted for the detailed embryology and cell lineage studies of *Parascaris equorum* made by Strassen (1896), Boveri (1899), and Müller (1903), of *Camallanus lacustris* by Martini (1903, 1906) and *Turbatrix aceti* by Pai (1928). Nematodes proved to be particularly suitable material for such investigations since their development is highly determinate, although cleavage is not typically spiral, and the number of somatic cells is relatively small. Further, the number of somatic cells, excluding those that comprise the epithelial gonoducts, remains very nearly constant after embryonic development.

During the first half of the twentieth century research on nematode embryogeny declined as attention was turned to animals whose eggs were more favorable for experimental manipulation. However, during this period Pai (1928) published the first detailed cell lineage study of a free-living nematode, *Turbatrix aceti*, and this work was followed shortly with a study by Kreis (1930) on the reproduction and embryogeny of the soil dwelling nematode *Actinolaimus tripapillatus*. With an awareness of their economic importance, considerable attention is being given now to the embryonic and postembryonic development of plant parasitic species of nematodes as well as some of their free-living relatives (Van Gundy, 1958; Triantaphyllou, 1960; Weerdt, 1960; Yuksel, 1960; Fassuliotis, 1962; Hirschmann, 1962; Anderson and Darling, 1964; Seshadri, 1964; Thomas, 1965; Yuen, 1966; Clark, 1967; Hirschmann and Triantaphyllou, 1967; Roman and Hirschmann, 1969; Siddiqui and Taylor, 1970; Wang, 1971). Because these nematodes and their eggs are very small and development is rapid, detailed cell lineage studies have not yet been attempted. Rather, the above studies are concerned primarily with very early embryonic and postembryonic development.

Marine nematodes have not been the subject of embryogeny studies. However, from such investigations of various nonmarine species as mentioned above, information of fundamental and possibly general importance has been obtained. This is reviewed below and is supplemented where possible with unpublished, preliminary observations on the embryogeny of the marine form *Deontostoma californicum.*

The fertilized eggs of nematodes may begin development *in utero* in some species, such as *Rhabditis marina*, according to Tietjen *et al.* (1970) and in *Monhystera disjuncta*, according to Chitwood and Murphy (1964). In older females development and hatching may occur within the female (ovoviviparous), often resulting in her death, a phenomenon designated *endotokia matricida.* Often, however, development does not begin until the eggs have been oviposited and environmental conditions are favorable (oviparous). When conditions are unfavorable, *Ascaris* eggs, for example, may remain dormant for long periods of time. Some marine nematodes, such as *Viscosia carnleyensis* and *Monoposthia costata* are known to occur in temperate latitudes as adults and juveniles from February to October while being absent in these stages of life through the remaining winter months (Tietjen, 1969). These species may pass this period as resistant eggs. Adult (including gravid females) and juvenile *Deontostoma californicum* are present during all months of the year on the northern coast of California where temperatures vary little from 15°C throughout the year.

When eggs are deposited during conditions favorable for development, the time interval preceding the first cleavage appears to vary considerably among species. Six to eleven hours are necessary in species of *Pratylenchus* (Roman and Hirschmann, 1969) and 4-6 hours in the case of *Ditylenchus dipsaci* according to Yuksel (1960). However, in *Deontostoma californicum* the first cleavage occurs 26-35 hours after entering seawater, although, part of this delay may be attributable to the fact that the observed eggs were removed prematurely from the female.

After the fertilized egg has completed its maturation divisions, the female pronucleus moves away from the edge of the egg where the second polar body had been formed, and the pronuclei of the male and female gametes then move toward each other. Where the pronuclei contact one another, their surfaces disappear and the two pronuclei are joined. The chromosomes form and then unite on a common equatorial plate. As soon as the nuclei join they begin to migrate toward the center of the zygote and, as they migrate, they commonly rotate from 90° to as much as 360° in the case of *Ditylenchus destructor,* according to Anderson and Darling (1964). Finally, the spindle elongates and the cleavage furrow forms.

Where the zygote is oval, and this seems to be the usual case, the first division is typically transverse to the long axis. This results in two blastomeres, one commonly a little larger than the other, but they are equal in *Deontostoma californicum*. When subequal in size, it is generally because the mitotic spindle is more in the vegetal region of the zygote (Tadano, 1968). The animal pole becomes the anterior S_1 blastomere and the vegetal pole the posterior P_1. In *Toxocara canis* and *Parascaris equorum* the stem cell (P_1) is usually the smaller one (Walton, 1918). The letters used in designating these and the blastomeres to follow, are after the system of nomenclature used by Boveri (1899), Martini (1903), and Pai (1928) and are still used widely today in preference to other systems. The letters *S* (or sometimes *AB*) and *P* stand for "somazellen" and "propagation zelle" to which their respective descendants give rise, and they reflect the highly determinate nature of nematode development. Thus, the *S* cells and other offshoots of the stem line ultimately produce the somatic tissues, while the final division in the stem line itself (*P* cells) gives rise to the primordial germ cells.

The second cleavage in various species of *Ascaris* and in *Rhabdias bufonis* is brought about by a mitotic division which, in one respect, is fundamentally different in the S_1 cell as compared to P_1. The former undergoes chromosome fragmentation and chromatin diminution. These phenomena occur at metaphase whereupon the middle portion of each chromosome fragments into several small pieces, called karyosomes, and two larger and thicker end fragments, the latter constituting ½–⅔ of the total chromatin material. According to Walton (1918), 60–72 of the smaller particles are produced in *T. canis*. As anaphase ensues each karyosome divides and their halves are distributed proportionately to the daughter blastomeres, thus behaving as normal chromosomes. As the new cell membranes begin to form, the larger end pieces divide in a plane coinciding with that of the karyosomes and half goes to each daughter cell where they lie in the cytoplasm close to the newly formed nuclear membrane. Here they undergo dissolution without further apparent activity (Walton, 1918). It has been thought that the end pieces do not contain genes, although Tadano (1968) finds that they give a very strong Fuelgen reaction.

While most investigators have found that chromatin diminution takes place during the division of the S_1 blastomere into its daughter cells *A* and *B*, it may be delayed until the next cleavage when *A* and *B* give rise to a, α, b, and β (Zoja, 1896; Walton, 1918). Unfortunately, little information regarding chromosome fragmentation and chromatin diminution has been obtained in recent studies of the plant parasitic and soil nematodes. Without this information, and not having followed the fate of each blastomere through organogenesis, recent investigators have resorted to the identification of S_1 on the basis that, at least in the animal parasitic species thus far studied, this cell and its progeny usually pass through the next few divisions at a faster rate than those of P_1.Ten hours, for example, occur between division of what is believed to be S_1 and P_1 in *Pratylenchus* (Roman and Hirschmann, 1969). The time lapse, however, between the division of the blastomeres in the two-cell stage of *D. californicum* varied from as little as 4 minutes to no more than 15 minutes. Such a short interval provides little confidence in using this method to distinguish between S_1 and P_1. Further, according to Strassen (1892), the order in which these two cells divide is reversed in *Bradynema rigidum.*

Typically, at least for ascarids, the division of P_1 follows that of S_1 and, as mentioned previously, is without chromosome fragmentation and diminution. One of the daughter blastomeres of P_1 is designated P_2 and, as the letter designation implies, is the line that gives rise to the propagation cells. The other daughter blastomere is S_2, or *EMSt*, which through subsequent divisions will yield the endoderm, primary mesoderm and stomodaeum. The daughter blastomeres of S_1, as mentioned previously, are *A* and *B* and, as they give rise to primary ectoderm, they are termed ectoblasts.

Whereas the first cleavage is always transverse to what will be the long axis of the embryo (Fig. 22A), the plane of cleavage for S_1 may or may not be the same as P_1, depending upon the species, and for either blastomere the plane may not be the same in the corresponding blastomere of another species. For example, the cleavage plane of S_1 in ascarids is parallel to the long axis of the embryo while the cleavage plane of P_1 is transverse to it, resulting in a T-shaped arrangement (Fig. 22B and C). P_2, which is at the base of the "T", then moves into the angle formed by *B* and S_2 *(EMSt)* to form the rhomboid shape (Figs. 22D and E). This transitory T-shaped arrangement is not known among nematodes other than ascarids. In several other species cleavage patterns lead to the rhomboid arrangement without the T-shaped configuration. As described for *Rhabdias bufonis* by Goette (1882) and Ziegler (1895), *Metastrongylus apri* by Wandolleck (1892), and *Turbatrix aceti* by Pai (1928), S_1 and P_1 change shape to have a diagonal common surface and then cleave in parallel to form a rhomboid configuration. Work on *Ditylenchus* sp. by Yuksel (1960), Hirschmann (1962), and Anderson and Darling

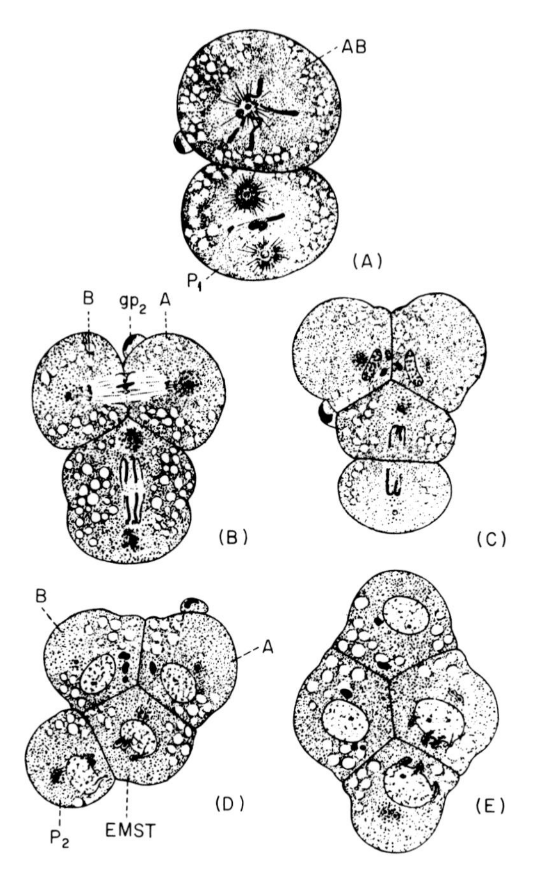

FIG. 22. Development of *Parascaris equorum* [from Nigon (1965)]. (A) two-cell stage; (B) and (C) T-stage; (D) and (E) formation of rhomboid stage.

(1964), on *Pratylenchus* sp. by Roman and Hirschmann (1969), and on *Acrobeles complexus* by Thomas (1965) has shown that in these species S_1 and its descendants divide in advance of P_1 to form 6-celled embryos, and the resulting somatoblasts shift positions to form a rhomboid.

Strassen (1892) found that blastomeres in the 4-cell stage of *Bradynema rigidum* may be in either of two possible arrangements. More commonly P_1 first divides transverse to the long axis of the egg; this is followed by a comparable division of S_1 without an intervening shift in the position of the blastomeres. Thus, the blastomeres of the 4-cell stage are in tandem. Less commonly the transverse division of P_1 is followed by an oblique division of S_1 with S_2 shifting ventrad and *B* dorsad to form the familiar rhomboid configuration. However, the blastomeres in tandem shift position so that by the 8-cell stage the arrangement of blastomeres is essentially identical in both cases.

Variation in the arrangement of the blastomeres of the 4-cell stage was also observed in *Deontostoma californicum*. Basically two cleavage patterns are evident. In the first type, which occurred in 20 of 26 developing embryos, the first cleavage was transverse and the cleavage planes of the daughter blastomeres, S_1 and P_1, were longitudinal and at right angles to each other. In the six remaining cases the first cleavage was as before and the second cleavages are also at right angles to the first, but parallel to each other. With two exceptions the positions of the blastomeres remained unchanged through subsequent cleavages. The exceptions occurred among those of the former pattern and consisted of a shift to a distinct rhomboid arrangement.

Both of these arrangements of blastomeres in the 4-cell stage (excluding those that shifted to a rhomboid) are unknown in other forms of nematodes. Possibly *D. californicum* is like *Bradynema* in having at least two possible configurations of the 4-cell stage whose differences are overcome during subsequent development. However, as it was not possible to clearly identify a common configuration of blastomeres at the 8-cell stage, either their differences are overcome in still later stages of development, or normal development was interferred with in the course of obtaining eggs and preparing them for study. Unfortunately, none of the eggs developed much beyond about the 64-cell stage when held in chambers for observation with the microscope. Eggs expressed from bisected females will complete development in about 30 days when incubated in seawater in sealed BPI dishes at 15°C (Viglierchio and Johnson, 1971).

While much additional information is needed with regard to even the early development of nematodes, it is clear from the above that early ascarid development is not typical of all nematodes and that the kinds of early cleavage patterns seem more numerous than is generally recognized.

Embryonic development beyond the 8-cell stage is best known for *Camallanus lacustris*, *Turbatrix aceti*, and, especially, *Parascaris equorum*, and, accordingly, the following account of further nematode development is based on studies of these species. The third cleavage in these species begins at what will become the anterior and dorsal regions of the animal. In *Parascaris*, ectoblasts *A* and *B* divide longitudinally to produce blastomeres each on the right and left sides of the embryo's saggital plane. With these divisions the bilateral symmetry of the embryo is established. In accord with the same system of nomenclature, Greek letters are assigned to daughter blastomeres on the left and Roman letters to those on the right. Thus the ectoblasts *A* and *B* give rise to α and *a* and β and *b* respectively. The position of each of these daughter cells is not entirely symmetrical to its sister cell on the opposite side. The α and β blastomeres are slightly anterior and below *a* and *b* in ascarids and *C. lacustris.* Further, all four of these cells are shifted slightly toward the left of the embryo (Fig. 23A, B, and D), giving the embryo a slight secondary asymmetry. By the late 8-cell stage of *Parascaris equorum* and *Metastrongylus apri* blastomere *a* has moved dorso-medially and now separates *b* from β. At this time *b* is ventral to *a* while β is posterior to α and posterior and ventral to *a* (Fig. 23D).

Blastomere S_2 usually divides next with its cleavage plane at right angles to the longitudinal axis of the embryo giving rise to a posterior cell *E* which ultimately produces endoderm (Fig. 23C and E). Its anterior sister cell will produce mesoderm and the stomodaeum (esophagus) and is termed *MSt.* This cleavage, being the first to occur after S_2 was produced from the division of a cell in the stem (propagation) line, is characterized by chromatin diminution (Fig. 23E). The 8-cell stage is finally achieved with the division of P_2 (Fig. 23C), whose cleavage plane is transverse but also somewhat oblique. Of the two daughter cells, P_3 is adjacent to *E* while *C* (=S_3) is dorsal to its sister cell (Fig. 23E). P_3 continues the stem line while *C* gives rise to secondary ectoderm and tertiary mesoderm. At this stage the primordial cells are extant for each of the germ layers.

The fourth cleavage in *Parascaris equorum* and *Camallanus lacustris*, like the third cleavage, begins with the ectoblasts (Fig. 23E and F). The cleavage planes of α, β, and *b* approximately parallel the frontal plane of the embryo, but are slightly oblique, with the posterior margin of each plane more dorsal than the anterior margin. The cleavage plane of *a*, however, is tilted in the opposite direction with its anterior margin more dorsal. As a result of these cleavages the four ectoblasts on the left side of the embryo, αI, αII, βI, and βII, form a rhomboid configuration, while on the right side their counterparts, *a*I, *a*II, *b*I, and *b*II are in the form of a "T" (Fig. 23H and I). Blastomeres αII and *a*II, along with *C*, P_3, *E*, and *MSt* form a band that circumscribes the embryo along its saggital plane in *P. equorum* (Fig. 23I). In *C. lacustris* the arrangement differs slightly; αII is shifted to the left and its area of contact with *a*I is located in the saggital plane. Thus, while αII is in contact with *b*I, and *a*II and *a*I are separated from one another in *P. equorum*, αII and *b*I are not in contact with each other, and *a*II and *a*I are adjacent to each other in *C. lacustris.*

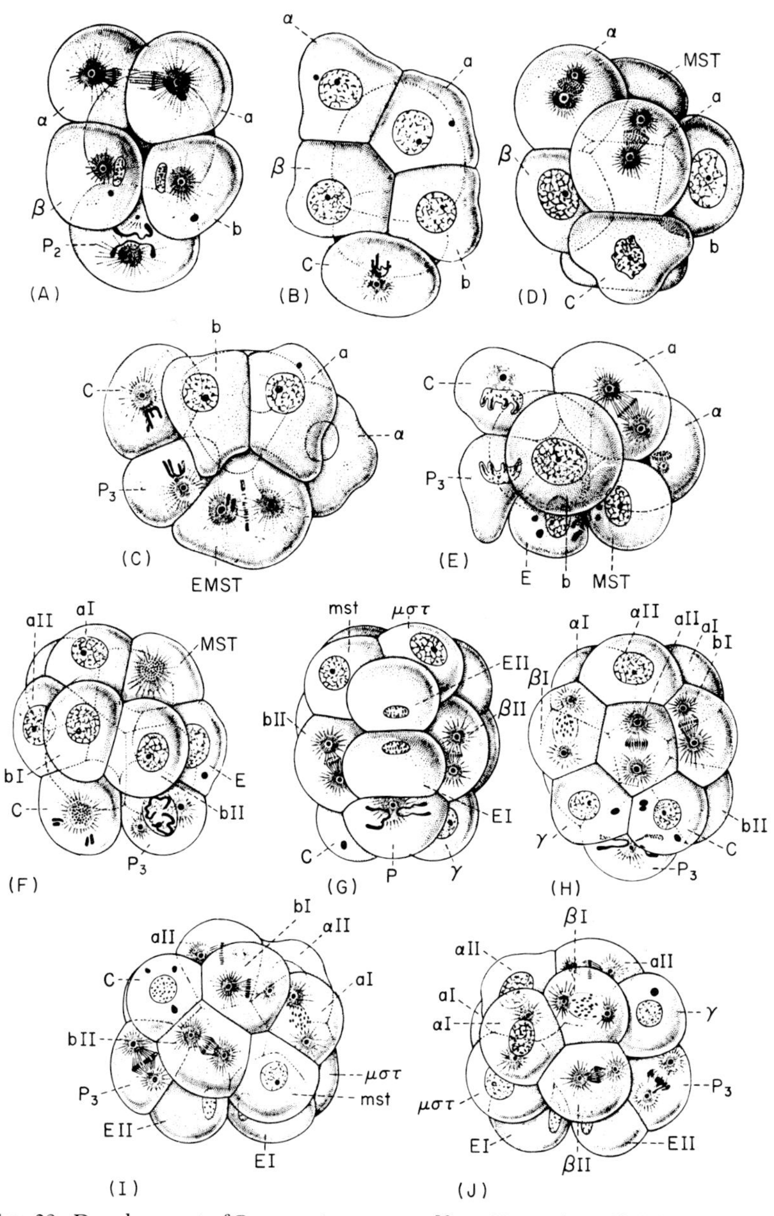

FIG. 23. Development of *Parascaris equorum* [from Nigon (1965)]. (A) six-cell stage, dorsal surface; (B) a stage less advanced than (A); (C) a more advanced stage, lateral view; (D) 8-cell stage, dorsal surface; (E) 8-cell stage, lateral view; (F) 12-cell stage, lateral view; (G) 15-cell stage, ventral surface; (H) 15-cell stage, dorsal surface; (I) 15-cell stage, view of right side; (J) 15-cell stage, view of left side.

However, the above mentioned rhomboid and T-shaped arrangements of ectoblasts exist in the embryos of both species, these cells forming the right and left sides of the embryo. In both species, the adjusted position of the ectoblasts tends to restore bilateral symmetry and equalizes the number of cells in right and left halves.

Although the cell lineage has not been described and illustrated in the same detail in *Rhabdias bufonis* and *Metastrongylus apri*, it appears that the relative positions of the ectoblasts vary somewhat from *P. equorum* and *C. lacustris*.

The fourth cleavage continues with the divisions of *MSt*, whose cleavage plane lies within and parallels the saggital plane, to form *mst* and $\mu\sigma\tau$ which lie respectively on the right and left side of the saggital plane (Fig. 23G). Endoblast *E* divides to produce *E*I and *E*II, the former anterior to the latter, and both lie in the saggital plane (Fig. 23I). Blastomere P_3 divides in a similar fashion to produce P_4 and $D(=S_4)$ the latter dorsal and slightly posterior to the former (Fig. 24A). *D* is a quaternary somatic cell and will give rise to secondary mesoblasts. P_4 continues the stem line. The cleavage of ectomesoblast *C* is similar to *MSt*, its daughter blastomeres *c* and γ on the right and left respectively, although here chromosome diminution occurs (Fig. 24A). The blastomeres of the fully formed 16-cell stage retain their respective positions, except that *mst* and $\mu\sigma\tau$ begin to elongate and shift posteriorly, with *E*I and *E*II coming between them.

The fifth cleavage begins with the ectoblasts (Fig. 24B), the cleavage plane being transverse to the anterior-posterior axis of the embryo in all but two cases, which thus elongates the embryo. Endoblasts *E*I and *E*II now divide transversely, their daughter cells *e*I and ϵI lying anterior to *e*II and ϵII, respectively (Fig. 24B). Mesoblasts *mst* and $\mu\sigma\tau$ continue to shift posteriorly until they lie on either side of the endoblasts and then divide to produce *st* and $\sigma\tau$ which are anterior to their respective sister cells *m* and μ (Fig. 24C). Strassen (1896) regarded all four cells as being mesodermal in *P. equorum*, while Boveri (1899) believed *st* and $\sigma\tau$ to be ectodermal constituents of the stomodaeum. The quartet of endoblasts then rotate clockwise and shift into a rhomboid configuration with *e*I anterior (Fig. 24D), ϵII posterior and *e*II and ϵI between them (Fig. 24C). Mesoblast *D* divides transversely (Fig. 24C), while undergoing chromosome diminution to form *d* and δ, and ectomesoblasts *c* and γ divide with their cleavage planes nearly parallel to the frontal plane so that *c*I and γI are dorsal to their respective sister cells *c*II and γII (Fig. 24D). The fifth division is not completed until the division of P_4 which is delayed (Fig. 24E). In the meantime, the ectoblasts have undergone

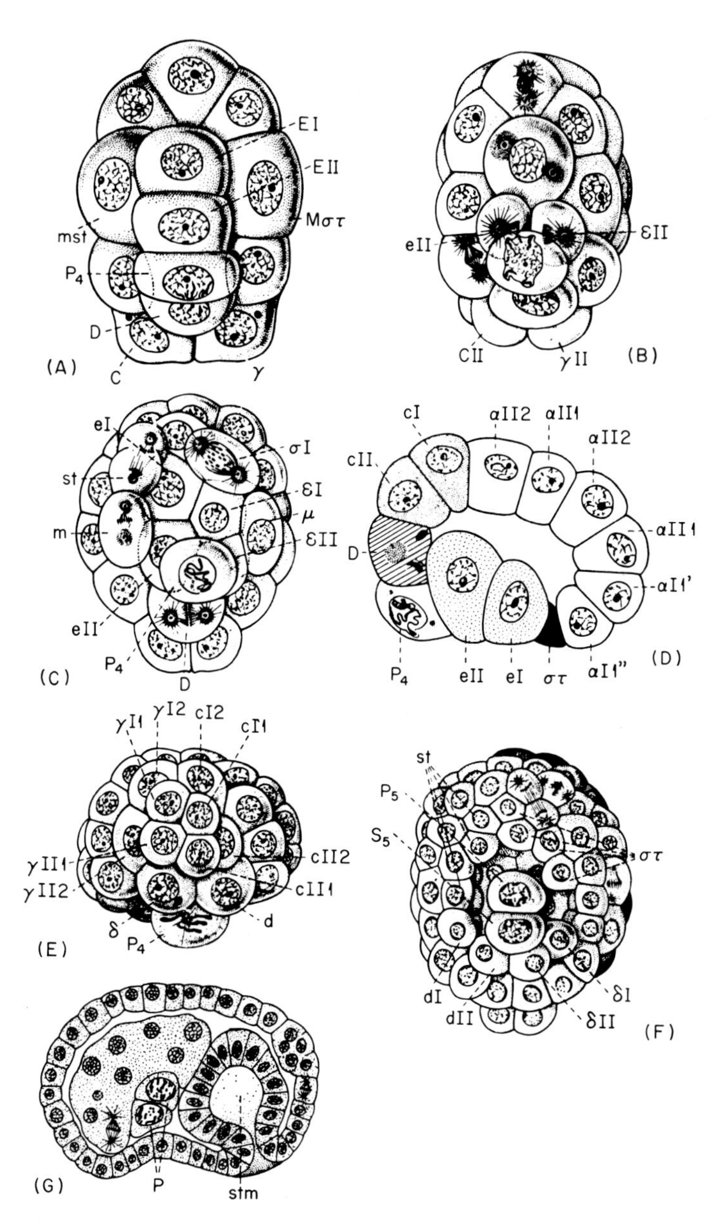

FIG. 24. Development of *Parascaris equorum* [from Nigon (1965)]. (A, B, and C) successive stages of cleavage, ventral view, beginning with 24-cell stage; (D) longitudinal vertical section of a stage corresponding to (C); (E) a stage later than the preceding, posterior surface; (F) still later stage, ventral surface; (G) longitudinal section showing the formation of the stomodaeum (stm) after gastrulation.

their sixth cleavage and the embryo has by this time become a blastula (Fig. 24D).

Gastrulation is generally believed to be by the process of epiboly since an archenteron is absent, although Strassen (1896) was of the opinion that it was by invagination. The endoblasts begin to sink into the blastocoel (Fig. 24C, D, and F) as the primary mesoblast cells and the stomatoblasts, especially, move toward the midventral line, gradually causing the endoblasts to slip into the blastocoel. Here they will continue to divide to give rise to the midgut (Fig. 24G). The lip of the blastopore consists of *m*I, *m*II, *st*I, *st*II, μI, μII, $\sigma\tau$I, $\sigma\tau$II, *d*, δ, and P_4. The latter cell moves to a position ventral to the endoderm cells, and the ectomesoblasts (descendants of *C*) increase in number and occupy the posterior region of the embryo from the dorsal to the ventral surfaces (Fig. 25A-D, and F).

Gastrulation continues as the primary mesoblasts enter the blastocoel behind the endoblasts. The secondary mesoblasts then divide and their daughter cells *d*I, *d*II, δI, and δII occupy the posterior and posterior-lateral lip of the blastopore, thus embracing P_4 (Fig. 25A). Blastomere P_4 then divides to produce the primordial germ cells *G*I and *G*II (or S_5 and P_5) which, along with the secondary mesoderm cells are subsequently carried into the blastopore as invagination progresses (Fig. 25A-C). The stomatoblasts increase to eight ($st\text{I}_1$, $st\text{I}_2$, $st\text{II}_1$, $st\text{II}_2$, $\sigma\tau\text{I}_1$, $\sigma\tau\text{I}_2$, $\sigma\tau\text{II}_1$, and $\sigma\tau\text{II}_2$) and the posterior stomatoblasts ($st\text{II}_2$ and $\sigma\tau\text{II}_2$) come into contact with the most anterior of the secondary ectoblasts (γII_2 and $c\text{II}_2$) so that these two groups of cells now form the lip of the blastopore (Fig. 25C). Meanwhile, the endoblasts and primary mesoblasts continue their divisions, the mesoblasts contributing to the lateral wall of the embryo (Fig. 25F).

Next, the stomatoblasts begin to form a pocketlike invagination, the stomodaeal pocket, in the anterior portion of the blastopore (Fig. 24G and 25F). Concurrently, the proliferating ectoblasts begin closing the posterior and lateral regions of the blastopore, spreading over *G*I and *G*II as they do so. As the blastopore closes and the ventral surface of the embryo becomes smooth, the stomodaeal invagination is reduced to a small opening (Fig. 25E) that later shifts to the anterior end of the embryo and becomes the stoma. Posteriorly the stomodaeum makes contact with the endodermal tissue, progenitor of the midgut. As development continues the embryo elongates and begins to assume the form of the juvenile. In summary, the principal organs of the juvenile, and the blastomeres that contribute to their formation are as follows:

Epithelium. Blastomeres *A* and *B* gives rise to $\frac{3}{4}$-$\frac{4}{5}$ of the epithelium, the remainder coming from blastomere *C*.

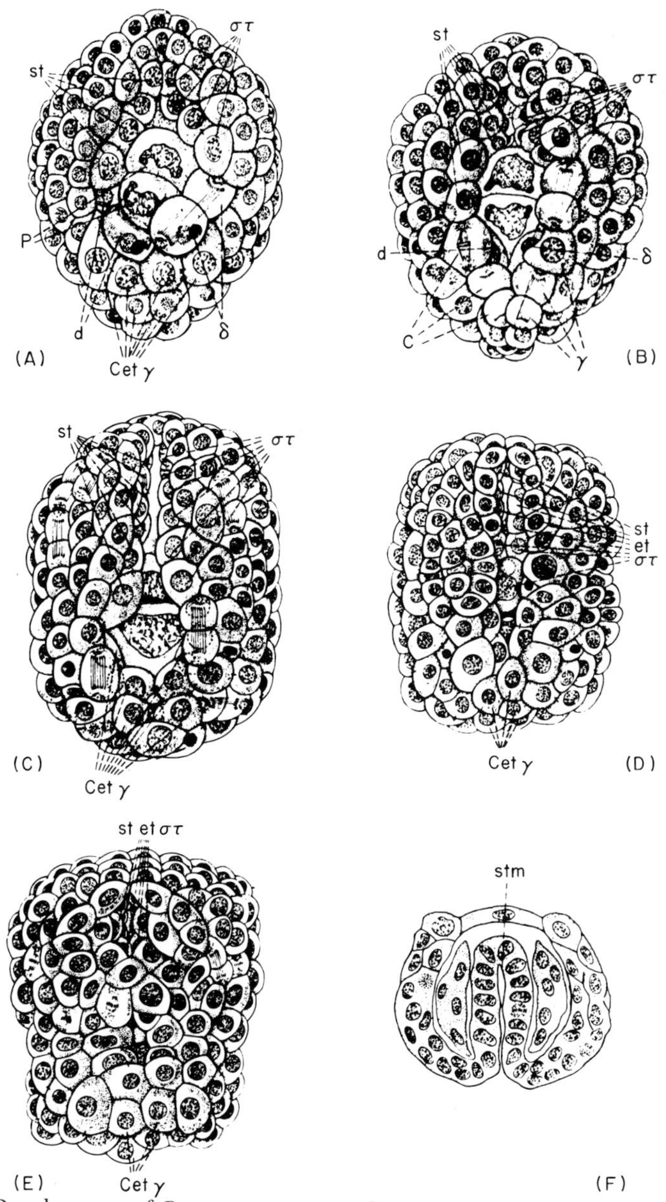

FIG. 25. Development of *Parascaris equorum* [from Nigon (1965)]. (A-E) ventral views showing the development of the stomodaeum and the proctodaeum; (F) transverse section at the level of the stomodaeum (stm).

Body wall musculature. Blastomeres m and μ develop as narrow bands, one in each side of the embryo, between the primordial gut and epithelium (Fig. 25F). The cells of these bands soon disperse and become attached to the epithelium where they form the somatic musculature.

Central nervous system. This system develops from primary ectoblasts proliferated into the interior from the anterior rim of the blastopore when the stomodaeum is formed. These cells produce the circumesophageal commissure.

Excretory system. According to Pai, this system is derived from primary ectoblasts (*A*,*B*) while Strassen and others believe it to be derived from mesoderm.

Esophagus. Strassen regarded the stomodaeum as being formed by the stomatoblasts (*st* and $\sigma\tau$). Boveri and Martini held that ectoblasts (*A* and *B*) contribute the nonmuscular portion of the esophagus.

Midgut. This organ is derived entirely from the endoderm (*e* and ϵ).

Proctodaeum. The hind gut is believed to be formed by an infolding of the secondary mesoderm (*d* and δ) at the posterior end of the embryo, although derivation of this region needs further study.

Body cavity. The body cavity is a pseudocoelom, i.e., a remnant of the blastocoel, and the inner body wall and the internal organs are not lined with epithelium as they are in animals with a true coelom.

Reproductive organs. The system is discussed in Section 8.3.5. The cell lineage of *Parascaris equorum* is summarized in Fig. 26.

8.4.2 Hatching and Postembryonic Development

Hatching, in the case of nematodes, is the emergence of a fully developed but sexually immature juvenile from the confines of the egg membranes in which embryonic development occurred. This process involves mechanical force and/or chemical breakdown of at least a portion of the chorion. Frequently, mechanical force is a function of body movements of the first or second-stage juveniles. Stylet-bearing nematodes, as exemplified by *Heterodera rostochiensis* may rend the chorion by repeatedly puncturing it with the stylet (Doncaster and Shepherd, 1967). Chemical weakening and eventual rupture of the egg membranes has been demonstrated in the hatching of *Ascaris lumbricoides* (Rogers, 1958, 1960; Fairbairn, 1961). At the time of hatching, juveniles of this species secrete lipase, chitinase and protease (Rogers, 1960).

Even though an embryo has achieved the development necessary for hatching, it may not do so until triggered by the appropriate environmental stimuli. This delay may be prolonged and the stimuli may be complex. Juveniles of *Heterodera* species may not hatch until stimulated by exudates from the roots of a host plant. In the case of

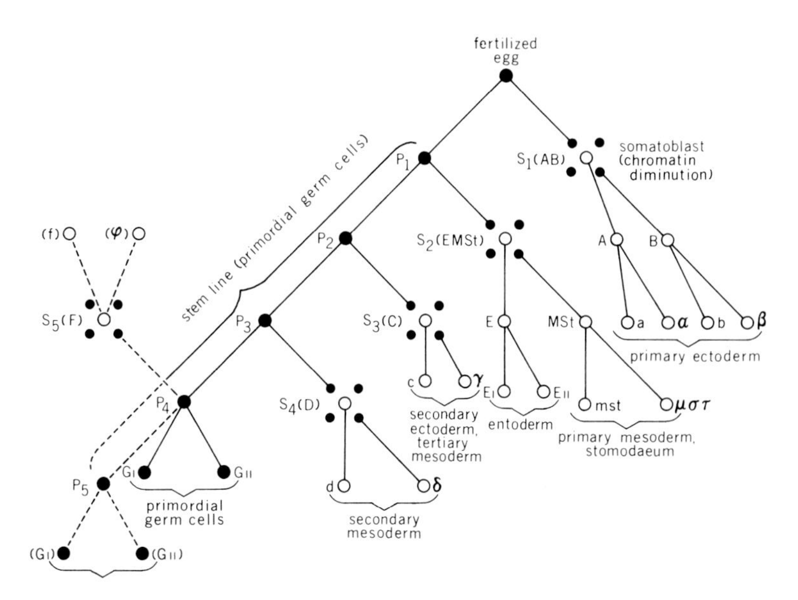

FIG. 26. Summary of cell lineage of *Parascaris equorum* (after Tadano, 1968).

Ascaris lumbricoides, hatching does not occur until stimulated by gaseous carbon dioxide and/or undissociated carbonic acid. Hydrogen ion concentration, redox potential, type and concentration of electrolytes, and presence of surface active agents influence the action of the stimulus (Rogers, 1958, 1960, 1962).

To what degree stimuli are important to the hatching of free-living species, and what their nature might be, are not known. Stimuli from the environment may be unimportant to marine nematodes inhabiting abyssal sediments where most environmental factors are stable and there is little or no seasonal change. In shallower sediments throughout temperate latitudes seasonal factors may affect hatching, in which case it may be the direct effect of temperature or chemical changes in the water brought about by metabolic activities of other seasonal organisms.

After the juvenile has hatched and before it reaches sexual maturity, it typically undergoes a phase of growth. As stated previously, growth in nematodes is not generally accomplished by cellular division. Exceptions are the growth and development of the gonad in all species, and growth of the midgut in at least some cases. In *Paracamallanus sweeti*, Moorthy (1938) found that intestinal cells increase from 35 in young juveniles to about 200 in the adult. In other organs the number of cells remains constant during postembryonic morphogenesis, and growth results from an increase in cell size (Lee, 1965;

Nigon, 1965; Crofton, 1966). As a result certain cells in the larger nematodes may reach considerable sizes, up to 10 mm in the case of muscle cells, for example (Hyman, 1951).

Growth of nematodes is not uncommonly allometric. Disproportionate growth of the various body parts may result in subtle or sometimes striking and important changes in the form of the body. An example common to many nematodes is the change in the length of the esophagus relative to the total body length. The esophagus of newly hatched *Caenorhabditis elegans* is ⅓ the total body length whereas in the adult it is only 1/6 of the body length (Nigon, 1965). In the plant parasitic species of Tylenchulidae, Heteroderidae, Naccobidae, and some Tylenchorhynchidae and Paratylenchidae, the insect parasitic species of Sphaerulariidae and Entaphelenchidae, and the bird parasites of Tetrameridae, the gravid females become swollen to the extent that their vermiform shape as juveniles is transformed to a rotund or saclike shape.

Temperature and nutrition may affect the degree of postembryonic growth. The mean total body length of adult female *Pelodera teres* may vary from 1000 to 1650 μm, and in *Rhabditis terrestris* from 1050 to 2150 μm depending upon the nutritional quality of the culture medium (Nigon, 1965).

Nematodes usually undergo four molts which divide the postembryonic period of the life cycle into four juvenile stages and the adult stage. The onset of molting is typically indicated by the cessation of body movements which lasts until molting has been completed. Johnson *et al.* (1970) have demonstrated that during this period the somatic muscles are broken down and then reformed, which they believe accounts for the period of quiesence.

The first molt usually occurs after the first stage juvenile hatches from the egg, although one or more molts may occur in the egg. Sexual maturity is attained with the last molt at which time the copulatory organs are formed, and at approximately this time the gonads reach maturity. Molting involves the loosening of the old cuticle from the body wall, accompanied by the formation of a new cuticle, and escape from the loose cuticle.

The cuticle is a condensation at the outer edge of the hypodermis and is comprised principally of a secreted collagen (Bird and Bird, 1969). The fine structure of the cuticle consists of an external cortical layer which is commonly subdivided into internal and external cortical layers, a median layer, and an internal basal layer. Detailed features of the layers differ from species to species and between different stages of development of the same species (Shepherd and

Clarke, 1971). Bird (1957) has postulated that the cuticle is metabolically active. Reports that the cuticle of *Ascaris* contains esterase (Lee, 1961, 1962, 1965; Anya, 1966), adenosine triphosphate and acid phosphates (Anya, 1966), hemoglobin (Lee, 1965), ribonucleic acid and ascorbic acid (Anya, 1966) lend support to this hypothesis. Kan and Davey (1968a) were unable to demonstrate enzymes in the cuticle of *Phocanema decipiens* and, because the cuticle does not contain cellular organelles, it is regarded by Shepherd and Clarke (1971) to be a nonliving structure.

Cuticle not only covers the external surface of the body but also lines the stoma, esophagus, vulva and vagina, terminal portion of the excretory pore in some species, external duct of certain sensory receptors (amphids), and the proctodaeum. In *Cylindrocorpus* species, the cuticular lining of the anterior portion of the stoma is shed as part of the cuticle covering the body. The remainder of the cuticle lining the esophagus is passed through the intestine and out through the anus (Chin and Taylor, 1969). The same is true of *Acrobeles complexus* according to Thomas (1965), and of *Diploscapter coronata* according to Hechler (1967). But in *Anatonchus tridentatus* and *Iotonchus* species the lining of the esophagus is carried out intact through the stoma with the external cuticle at the time of molting (Clark, 1960). The stylet-bearing tylenchids shed only the anterior portion of the stylet which is then replaced *in situ*. The anterior portion of the stylet is also shed in the Dorylaimida, but its replacement migrates from a special cell farther posterior in the wall of the esophagus.

The contention that the hypodermis is involved in both the formation of the new cuticle and breakdown of the old one is substantiated by cyclical changes in the fine structure and histochemical properties of the hypodermis during molting. With regard to the formation of the new cuticle, the hypodermis of *Meloidogyne* becomes granular and increases in thickness at the start of the molt. The granules resemble ribosomes and histochemical tests at this stage reveal the presence of RNA (Bird and Bird, 1969). Johnson *et al.* (1970) have demonstrated the presence of discrete globular structures, termed molting bodies, only during the production of the new cuticle. They were unable to demonstrate molting bodies in *Aphelenchus avenae* and *Hirschmanniella gracilis* although the hypodermis stained more intensely at that time. Davey (1965) and Kan and Davey (1968a, b) have shown that, at the onset of molting in juvenile *Phocanema decipiens*, the nuclei and nucleoli of the hypodermis enlarge, changes occur in nuclear morphology, mitochondria become more numerous, and cytoplasmic RNA increases. Johnson *et al.*(1970) have also demonstrated the appearance of globular structures 0.2–1.2 μm in diameter in the hypodermis of *Hemicycliophora arenaria* at the outset of molting and their disappearance thereafter.

Concerning the breakdown of the old cuticle, Roggen *et al.* (1967) claim to have demonstrated leucine aminopeptidase (LAP) in the hypodermis of juvenile *Xiphinema index*, where it undergoes quantitative fluctuations correlated with molting cycles. During the intermolt period it increases and reaches a maximum immediately prior to molting. But, during the actual process of molting, it disappears and the old cuticle is loosened from the body wall while the new cuticle is formed. While LAP is also believed to be present in adults of this species its quantity remains stable. These authors concluded that this enzyme is involved with the breakdown of the inner layers of the old cuticle and its separation from the body. They point out that their observations are in agreement with those of Rogers (1965) who also claims to have demonstrated the same enzyme in the molting fluid of *Haemonchus* and *Trichostrongylus* species. As will become evident in the discussion of exsheathing, however, the presence of LAP in molting fluid and, if present, its function still remains questionable.

Before molting is complete, juveniles must escape from the loosened old cuticle or sheath, as it is also termed. In many species, especially those that are free-living, escape (exsheathing) occurs very soon, if not immediately, after the old cuticle is loosened. But at least in the case of several species parasitic in animals, notably members of *Trichostrongylus* and *Haemonchus*, the exsheathing process may be delayed and the sheath retained presumably as a protective cover, until environmental conditions are favorable for continued development. As in the case of *Ascaris* eggs, the stimuli for exsheathing of infective *Trichostrongylus* larvae seem to be undissociated carbonic acid or dissolved carbon dioxide, but temperature, eH, pH and the presence of salts may also be important (Shepherd and Clarke, 1971).

There appears to be variation between major taxa in the means by which the sheath is opened to permit escape of the juvenile. In species of *Seinura* the head withdraws slightly from the lip region of the old cuticle, whereupon the head begins periodic movements from side to side. Subsequently, the old cuticle becomes pliable and loosens around the body, and the head frequently presses against the old cuticle just behind the lips where the cuticle finally gives way and provides the juvenile with an avenue of escape. The shed cuticle remains in one piece. A secretion, possibly originating in the esophageal gland, may first soften the cuticle before it is torn by mechanical force. The stylet of the juvenile is not known to function in breaking through the old cuticle (Hechler and Taylor, 1966b). In the case of *Aphelenchus avenae,* irregular breaking of the cuticle suggests it is removed by tension. In his study of this species, Fisher (1970) found that this was accomplished in more than one way. In approximately half of the observed cases, the removal was effected by increases in the body length of the juvenile. In other cases, pressure against the culture medium (agar) helped remove the cuticle and this was much more rapid when food was available due to differences in movement or use of the stylet. The cuticle in this species is also cast in one piece. Juveniles of *Tylenchorhynchus claytoni* escape as a result of complete transverse or circular separation in the loosened cuticle either in the neck or tail region. Thus, the shed cuticle is in two parts. The point of separation may be foretold by the localized accumulation of an unidentified fluid between the old and new cuticles, and this material may be responsible for localized breakdown of the sheath (Wang, 1971). The sheaths in the case of *Caenorhabditis briggsae* (Jantunen, 1964) and in species of *Haemonchus* and *Trichostrongylus* also separate along a circular line, but near the head region only. In the case of *Trichostrongylus* juveniles, ligation and selective irradiation studies have provided evidence that an exsheathing fluid is secreted by cells at the base of the esophagus and released through the excretory pore (Rogers, 1963). Excretion of this substance and its localized action on the sheath results first in the appearance of a light refractive ring about 19 μm from the anterior end, and then separation of the sheath at this level.

Several studies have been conducted recently attempting to identify the substance that attacks and opens the sheath of *H. contortus* juveniles. Rogers (1963, 1965, 1970) has provided evidence that the substance in question is LAP. He claimed that the enzyme is secreted by juveniles at the time of, but not before or after, molting; its action on sheaths and other substrates required Mg^{++} or Mn^{++} as a cofactor; it hydrolysed L-leucinamide; its action was strongly inhibited by diaminoethanetetra-acetic acid (DAEA), Hg^{++} and Cu^{++}; its optimum pH was 9.5 and it was unstable below pH 7.0; it lost much of its activity when lyophilized or when subjected to ultrafiltration at 5°C; its activity was destroyed at 80°C for 15 minutes; and fractionation of the exsheathing (metabolic) fluid on acrylamide columns revealed that only those fractions which had LAP activity were active against sheaths. However, Ozerol and Silverman (1969, 1970, 1972) have obtained quite different results in their studies of the exsheathing (metabolic) fluid of the same species. They found LAP in somatic tissue of both ensheathed and exsheathed juveniles, but not in the exsheathing fluids; the exsheathing fluid was active without Mg^{++} and Mn^{++} cofactors and, in fact, these ions were slightly inhibitory; DAEA and Hg^{++} were inhibitory at $10^{-2}M$, but not at $10^{-3}M$; Cu^{++} had no inhibitory effect; its effective pH range was from 4 to 10; it was not affected by lyophilization; it was thermostable from 38° to 70°C and recovered from exposure to even higher temperature; and no LAP could be demonstrated in some metabolic fluids that produced ring formation (exsheathing). Their conclusions were that LAP is not the active component of exsheathing fluid. Rather, the exsheathing factor is probably a small protein comprised principally of glycine, proline and alanine and without tertiary structure, lipid or carbohydrate. By further contrast, Slocombe and Whitlock (1971a,b), having used Kjeldahl nitrogen and Lowry protein assays as estimates of the maximum amount of protein possible in exsheathing fluid from *H. contortus,* concluded that only traces of protein are found in the exsheathing factor, and that products described by other workers are more likely to be metabolites than the exsheathing factor. Whitlock (1971) has suggested that the exsheathing phenomenon may be an example of a Szent-Györgyi (1969) zipper which unfastens with the reception of electrons, and it may be that the material found in supernatant fluids immediately after ecdysis could provide the necessary vehicle in these types of experiments for a charge transfer.

While the entire process of molting may be regulated by an endocrine system, much as in insects, current evidence for such a system stems mainly from studies of the exsheathing process. Rogers (1962) and Rogers and Sommerville (1963) proposed that an endocrine system links reception of the stimulus and release of the exsheathing fluid. Davey (1966) and Davey and Kan (1967) have shown that in fourth-stage infective juveniles, when exposed to favorable conditions *in vitro,* nerve cells in the dorsal and ventral ganglia associated with the circumesophageal commissure actively produce a neurosecretion. The cycle of secretion is closely correlated with molting. Davey and Kan (1968) have further shown that if the juveniles are cultured in saline only, a new cuticle is formed, but the neurosecretory cells remain inactive and exsheathing fails. When held in a complete culture medium simulating the digestive system of the definitive host, the neurosecretory cells are activated, and exsheathing ensues. Further, when partially isolated excretory glands were incubated with extracts of heads of worms cultured for 1-5 days in saline only, LAP activity was weak. But when incubated in extracts of heads of worms from the complete culture medium (with favorable environmental stimuli), neurosecretion occurred, and the excretory glands gave a strong positive test for the enzyme.

While molting in the Arthropoda is primarily a growth accommodating phenomenon, this does not necessarily appear to be the case in nematodes. This is not to imply that growth does not occur in nematodes, for adult plant parasitic nematodes are commonly 3-10 times longer than the juveniles and *Ascaris* up to 400 times longer (Shepherd and Clarke, 1971). Rather, most of the growth of nematodes occurs between each molt or after the molting phase of development has passed. For example, growth occurs between molts in *Turbatrix aceti, Ascaridia lineata*, in species of *Cooperia* and *Ancylostoma, Ditylenchus dipsaci, Caenorhabditis briggsae* and *Aphelenchus avenae* (Pai, 1928; Ackert, 1931; Sommerville, 1960; Blake, 1962; Hansen *et al.*, 1964; Shepherd and Clarke, 1971; Fisher, 1970). On the other hand, at least three molts are known to occur in *Ascaridia lineata* by the end of the 22nd day of postembryonic development, during which time it grows from less than 1 mm to about 28 mm; but after these molts the juvenile continues to grow, nearly tripling its length by the end of the 50th day. The life cycle of *Angiostrongylus cantonensis* involves two hosts. Juveniles grow from about 0.3 mm to just over 0.45 mm in the first host, molt twice, and then remain inactive until ingested by the second host. At this time the third and fourth molts occur, and at the end of the last molt the juveniles are about 2 mm long, but during the next 50 days the young adult males increase to about 20 mm and the females to about 30 mm (Mackerras and Sanders, 1955). Juveniles of *Cooperia curticei* grow from just under 1 mm in length at the time of hatching to just over 2 mm by the end of the last molt. The adults continue to grow attaining a final length of from nearly 5 mm to more than 6 mm (Sommerville, 1960).

One known example in which growth seems to be limited to each molt is that of *Acrobeles ciliatus*, according to Maupas (1899).

The structure of a given nematode is fundamentally similar in all of its stages of development, and the juveniles basically resemble the adults. Metamorphosis does not occur, and so, as Hyman (1951) has pointed out, sexually immature nematodes should be referred to as juveniles and not larvae, as is commonly the case. However, this is not to imply that, aside from the development of primary and secondary sexual characters, no changes in morphology occur between the first juvenile and adult stages. Growth of various organs and body parts may be allometric, as previously mentioned, and at the time of molting certain anatomical features may become modified, disappear or appear *de novo*, and such changes may affect any of the body organs. An example of such changes in the body wall is the progressive development of modified regions of cuticle, termed cordons, near the head in members of the subfamily Acuariinae. The cordons of the younger juveniles are often less complex than they are in the adults, and their development in some cases suggests evolutionary relationships within the subfamily (Chabaud, 1954). Prominent swellings or tubercles, arranged in one or two lateral series in adults of the genus *Bunonema*, are absent in the young juveniles, but become permanent structures during the last two molts (Cobb, 1915). The body wall in adults of the genus *Desmoscolex* is comprised of a series of large, raised granular rings usually bearing concretion particles, and these rings are separated from one another by annulated interzones. The granular rings may possess subdorsal or subventral setae. Although juveniles had never been identified or described, it had been assumed that they were very similar to the adults. Lorenzen (1971), however, has shown that the juveniles do not have granular rings or annulated interzones but rather have small transverse striations; subdorsal setae are never present. In one molting fourth-stage juvenile, Lorenzen found the adult cuticle with concretion rings forming under the last juvenile cuticle. Lorenzen further concluded that because of the striking difference between juveniles and adults of this genus, all previously described juveniles of *Desmoscolex* had been erroneously placed in the genera *Eudesmoscolex, Eutricoma,* and *Prodesmoscolex,* and juveniles only were known for these genera.

With regard to the digestive system, juveniles of both sexes of species within the Enchelidiinae, a taxon of free-living marine nematodes, have a large open stoma with numerous teeth. During the final molt of those juveniles destined to become males the stoma is reduced to a narrow unarmed vestige. However, transformations of the digestive system are more common and more prevalent in parasitic species, especially those occurring in other animals. The juveniles and females of the plant parasitic genus *Paratylenchus* have hollow stylets used in feeding on root tissue. During the final molt of juvenile males the stylet is lost and not replaced by a new one. First- and second-stage free-living juveniles of the Strongylida and Rhabdiasoidea typically have a relatively small stoma of simple structure and a relatively complex esophagus provided with median and posterior bulbs resembling the esophagus typical of those found in members of the order Rhabditida. Because of the nature of the esophagus, these are termed rhabditiform juveniles. The second molt commonly results in the development of a more complex stoma, sometimes provided with teeth or cutting plates (*Ancylostoma* and *Necator* species), and the esophagus becomes cylindrical with at most a slight basal swelling. These third-stage larvae, which commonly retain the sheath or cast cuticle of the second-stage, are known as strongyliform or filariform larvae and are the infective stage prepared for entry into their host where development to sexual maturity is completed. Likewise, the infective juveniles of many members of the superfamily Mermithoidea, parasites of Arthropoda, and especially insects, possess a dorylaimoid stylet which appears to be used in the penetration by the parasite through the body wall of the host. The stylet is shed during a subsequent molt and not replaced.

Adult males of *Rhaptothyreus typicus*, a free-living marine nematode that occurs at least as adult males in abyssal sediments, have an extremely large amphid, a vestigial stoma and esophagus, and what may be a trophosome in lieu of a typical gut (Hope and Murphy, 1969). The specialized condition of these organs suggests that the young males and possibly the juveniles, for at least a portion of their development, are parasitic in other marine organisms. Here, then, is a marine species in which it seems likely that rather profound changes occur during one or more molts.

Much more research is needed to understand the mechanism and significance of molting. Present knowledge, however, suggests that molting is not necessarily a growth accommodating function. The changes that occur in nematodes at the time of molting, and especially the more profound changes peculiar to parasitic species, suggest that this phenomenon enables nematodes to occupy different ecological niches at different stages of development. Thus, species that remain in the same ecological niche during all juvenile and adult stages would be expected to undergo little change at each molt. By contrast, those that must change habitats and niches to complete their life cycles may do so by way of structural and, presumably, physiological changes that to a substantial degree take place during the molting process.

Acknowledgments

I am grateful to Dr. A. C. Triantaphyllou, North Carolina State University, Raleigh, Dr. K. A. Wright, University of Toronto, and Dr. M. L. Jones, Smithsonian Institution, for their reading of this manuscript, and to Mrs. Teresa Smith and Mrs. Vernetta Williams, Smithsonian Institution, for their assistance and patience. I am further indebted to the following for permission to reproduce certain of the figures herein: Academic Press, New York (Fig. 21); The American Microscopical Society (Fig. 20); The Helminthological Society of Washington (Fig. 1); Masson Co., Paris (Fig. 22, 23, 24, and 25); Nolit, Publishing House, Belgrad (Fig. 26); and Springer Verlag, New York (Fig. 11).

8.5 References

Aboul-Eid, H. Z., and Coomans, A. (1966). Intersexuality in *Longidorus macrosoma. Nematologica* **12,** 343-344.

Ackert, J. E. (1931). The morphology and life history of the fowl nematode *Ascaridia lineata* (Schneider). *Parasitology* **23,** 360-379.

Allgén, C. A. (1929). Über einen merkwürdigen Fall von Hermaphroditismus bein *Halichoanolaimus microspiculum* Allgén (Nematodes: Chromadoridae, Choanolaiminae). *Zool. Anz.* **81,** 139-143.

Allgén, C. A. (1952). Über das Vorkommen von Hermaphroditismus bei zwei südlichen Arten der Gattung *Sphaerolaimus* Bastian. *Zool. Anz.* **149,** 189-191.

Allgén, C. A. (1953). Über einen Fall von Hermaphroditismus in der Gattung *Desmodora* de Man (Chromadoroidea: Nematodes). *Zool. Anz.* **151,** 95-98.

Allgén, C. A. (1958). Zwei weitere Falle von Bisexualität bei schwedischen freilebenden marinen Nematoden. *Zool. Anz.* **161,** 317-319.

Anderson, R. V., and Darling, H. M. (1964). Embryology and Reproduction of *Ditylenchus destructor* Thorne, with emphasis on gonad development. *Proc. Helminthol. Soc. Wash.* **31,** 240-256.

Anya, A. O. (1966). The structure and chemical composition of the nematode cuticle. Observations on some oxyurids and *Ascaris. Parasitology* **56,** 179-198.

Baylis, H. A. (1915). Two new species of *Monhystera* (Nematodes) inhabiting the gill-chambers of land-crabs. *Ann. Mag. Natur. Hist. Ser. 8* **16,** 414-421.

Beaver, P. C., Yoshida, Y., and Ash, L. R. (1964). Mating of *Ancylostoma caninum* in relation to blood loss in the host. *J. Parasitol.* **50,** 286-293.

Bĕlar, K. (1923). Über den Chromosomenzyklus von parthenogenetischen Erdnematoden. *Biol. Zentralbl.* **43,** 513-518.

Bĕlar, K. (1924). Die Cytologie der Merospermie bei freilebenden *Rhabditis*- Arten. *Z. Zellen-U. Gewebelehre* **1** (1), 1-21.

Beneden, E. van (1883). L'appareil sexuel femelle de l'*Ascaride megalocephale. Arch. Biol. Gand.* **4**(1), 95-142.

Bird, A. F. (1957). Chemical composition of the nematode cuticle. Observations on individual layers and extracts of these layers in *Ascaris lumbricoides* cuticle. *Exp. Parasitol.* **6,** 383-403.

Bird, A. F. (1971). "The Structure of Nematodes." Academic Press, New York and London.

Bird, A. F., and Bird, J. (1969). Skeletal structures and integument of Acanthocephala and Nematoda. *In* "Chemical Zoology" (M. Florkin and B. T. Scheer, eds.), Vol. 3, pp. 253-288. Academic Press, New York and London.

Blake, C. D. (1962). The etiology of tulip-root disease in susceptible and resistant varieties of oats infested by stem nematode, *Ditylenchus dipsaci* (Kühn) Filipjev. II. Histopathology of tulip root and the development of the nematode. *Ann. Appl. Biol.* **50,** 713-722.

Bonner, T. P., and Etges, F. J. (1967). Chemically mediated sexual attraction in *Trichinella spiralis. Exp. Parasitol.* **21,** 53-60.

Boveri, T. (1887). Die Bildung der Richtungskörper bei *Ascaris megalocephala* and *Ascaris lumbricoides. Zellen-stud. Jena.* **1,** 1-93.

Boveri, T. (1888). Die Befruchtung und Teilung des Eies von *Ascaris megalocephala. Zellen-stud. Jena.* **2,** 1-198.

Boveri, T. (1892). Über die Entstehung des Gegensatzes zwischen den Geschlect-zellen und den somatischen Zellen bei *Ascaris megalocephala*, nebst Bemerkungen über die Entwicklungsgeschichte der Nematoden. *Sitzber. Ges. Morph. Phy. München* **8**, 114-125.

Boveri, T. (1899). Die Entwicklung von *Ascaris megalocephala* mit besonderer Rücksicht auf die Kernverhältnisse. Festschr. C. v. Kupffer. Jena. pp. 383-430.

Boveri, T. (1909). Die Blastomerenkerne von *Ascaris megalocephala* und die Theorie der Chromosomenindividualität. *Arch. Zellforsch.* **3**, 181-268.

Boveri, T. (1910). Die Potenzen der *Ascaris-Blastomeren* bei abgeanderter Furchung. Zugleich ein Beiträg zur Frage qualitativungleicher Chromosomen-Teilung. *Festschr. Sechzigst. Geburtst. Richard Hertwigs* (München) **3**, 133-214.

Bütschli, O. (1873). Beiträge zur Kenntnis der freilebenden Nematoden. *Nova Acta* **36**, 1-144.

Bütschli, O. (1875). Verläufige Mitteilung über Untersuchungen betreffend die ersten Entwickelungsvorgange im befruchteten Ei von Nematoden und Schnecken. *Z. Wiss. Zool.* **25**, 201-213.

Chabaud, A. G. (1954). Sur le cycle évolutif des Spirurides et de Nématodes ayant une biologie comparable. Valeur systématique des caractères biologiques. *Ann. Parasitol.* **29**, 40-48, 206-249, 358-426.

Chin, D. A., and Taylor, D. P. (1969). Sexual attraction and mating patterns in *Cylindrocorpus longistoma* and *C. curzii* (Nematoda: Cylindrocorporidae). *J. Nematol.* **1**, 313-317.

Chitwood, B. G. (1929). Notes on the copulatory sac of *Rhabditis strongyloides* Schneider. *J. Parasitol.* **15**, 282.

Chitwood, B. G., and Chitwood, M. B. (1950). "An Introduction to Nematology." Monumental Printing, Baltimore, Maryland.

Chitwood, B. G., and Murphy, D. G. (1964). Observations on two marine monhysterids - their classification, cultivation and behavior. *Trans. Amer. Microl. Soc.* **83**, 311-329.

Christie, J. R. (1929). Some observations on sex in the Mermithidae. *J. Exp. Zool.* **53**, 59-76.

Christie, J. R. (1936). Life history of *Agamermis decaudata*, a nematode parasite of grasshoppers and other insects. *J. Agr. Res.* **52**, 161-198.

Clark, S. A. (1967). The development and life history of the false root-knot nematode, *Nacobbus serendipiticus*. *Nematologica* **13**, 91-101.

Clark, W. C. (1960). The oesophago-intestinal junction in the Mononchidae (Enoplida: Nematoda). *Nematologica* **5**, 178-183.

Cobb, N. A. (1915). The asymmetry of the nematode *Bunonema inequale*, n. sp. "Contributions to a Science of Nematology," Vol. 3, pp. 101-112. Waverly Press, Baltimore, Maryland.

Cobb, N. A. (1918). Filter-bed nemas: Nematodes of the slow and filter-beds of American cities with notes on hermaphroditism and parthenogenesis. "Contributions to a Science of Nematology," Vol. 7, pp. 189-212. Waverly Press, Baltimore, Maryland.

Cobb, N. A. (1928). Nemic spermatogenesis. *J. Wash. Acad. Sci.* **18**, 37-50.

Cobb, N. A. (1930). The demanian vessels in nemas of the genus *Oncholaimus;* with notes on four new oncholaims. *J. Wash. Acad. Sci.* **20**, 225-241.

Cohn, E., and Mordechai, M. (1968). A case of intersexuality and occurrence of males in *Longidorus africanus*. *Nematologica* **14**, 591-593.

Coninck, L. A. de (1965). Systématique des nématodes, *In* "Traité de Zoologie" P. P. Grassé, ed.), Vol. 4, pp. 586-681. Masson, Paris.

Craig, B., and Wilson, C. (1937). The use of buffered solutions in staining: Theory and practice. *Stain Tech.* **12,** 99-109.

Crofton, H. D. (1966). "Nematodes." Hutchinson, London.

Daday, E. von (1905). Untersuchungen über die Süsswasser-mikrofauna Paraguays. *Zoologica* **18** (44), 48-87, 327.

Davey, K. G. (1965). Molting in a parasitic nematode, *Phocanema decipiens.* I. Cytological events. *Can. J. Zool.* **43,** 997-1003.

Davey, K. G. (1966). Neurosecretion and molting in some parasitic nematodes. *Amer. Zool.* **6,** 243-249.

Davey, K. G., and Kan, S. P. (1967). Endocrine basis for ecdysis in a parasitic nematode. *Nature (London)* **214,** 737-738.

Davey, K. G., and Kan, S. P. (1968). Molting in a parasitic nematode, *Phocanema decipiens.* IV. Ecdysis and its control. *Can. J. Zool.* **46,** 893-898.

Dickerson, O. J. (1962). Gonad development in *Pratylenchus crenatus* Loof and observations on the female genital structure of *P. penetrans. Proc. Helminthol. Soc. Wash.* **29,** 173-176.

Ditlevsen, H. (1912). Danish free-living nematodes. *Vidensk. Medd. Naturhist. Foren. Kjobenhavn* **63,** 213-256.

Doncaster, C. C., and Shepherd, A. M. (1967). The behavior of second stage *Heterodera rostochiensis* larvae leading to their emergence from the egg. *Nematologica* **13,** 476-478.

Edwards, C. L. (1910). The idiochromosomes in *Ascaris megalocephala* and *Ascaris lumbricoides. Arch. Zellforsch.* **5,** 422-429.

Fairbairn, D. (1961). The in vitro hatching of *Ascaris lumbricoides* eggs. *Can. J. Zool.* **39,** 153-162.

Fassuliotis, G. (1962). Life history of *Hemicriconemoides chitwoodi* Esser. *Nematologica* **8,** 110-116.

Filipjev, I. N. (1918). Svobodnozhivushchiya morskiya Nematody okrestnostei Sevastopolya. I [Free-living marine nematodes of the Sevastopol area. I.]. *Tr. Osoboi Zool. Lab. Sevastop. Biol. Stantsii Rossiiskoi Akad. Nauk* **2**(4), 1-350.(Transl. to English. Israel Program for Scientific Transl., Jerusalem, 1968. TT 67-51338.)

Filipjev, I. N. (1921). Svobodnozhivushchiya morskiya Nematody okrestnostei Sevastopolya. II. [Free-living marine nematodes of the Sevastopol area. II.]*Tr. Osoboi Zool. Lab. Sevastop. Biol. Stantsii Rossiiskoi Akad. Nauk* **2**(4), 351-610.(Transl. to English. Israel Program for Scientific Transl., Jerusalem, 1968. TT 69-55054.)

Fisher, J. M. (1966). Observations on moulting of fourth-stage larvae of *Paratylenchus nanus. Aust. J. Biol. Sci.* **19,** 1074-1079.

Fisher, J. M. (1970). Growth and development of *Aphelenchus avenae* Bastian. *Aust. J. Biol. Sci.* **23,** 411-419.

Foor, W. E. (1967). Ultrastructural aspects of oocyte development and shell formation in *Ascaris lumbricoides. J. Parasitol.* **53,** 1245-1261.

Foor, W. E. (1968), Cytoplasmic bridges in ovary of *Ascaris lumbricoides. Bull. Tulane Univ. Med. Fac.* **27**(1), 23-30.

Gerlach, S. A., and Schrage, M. (1971). Life cycles in marine meiobenthos. Experiments at various temperatures with *Monhystera disjuncta* and *Theristus pertenuis* (Nematoda). *Mar. Biol.* **9,** 274-280.

Goette, A. (1882). Abhandlungen zur Entwickelungsgeschichte der Tiere. Erstes Heft. Untersuchungen zur Entwickelungsgeschichte der Würmer, *Rhabditis nigrovenosa.* Beschreibender Teil. Hamburg u. Leipzig. 104 p.

Goodrich, H. B. (1914). The maturation divisions in *Ascaris incurva. Biol. Bull.* **27,** 147-150.

Goodrich, H. B. (1916). The germ cells in *Ascaris incurva. J. Exp. Zool.* **21,** 61-99.

Gouilliart, M. (1932). Le comportement de l'Hétérochromosome dans la spermatogenèse et l'ovagenèse chez un *Ascaris megalocephala* hermaphrodite. *C. R. Soc. Biol.* **110,** 1176-1179.

Green, C. D. (1966). Orientation of male *Heterodera rostochiensis* Woll. and *H. schachtii* Schm. to their females. *Ann. Appl. Biol.* **58,** 327-339.

Green, C. D. (1971). Mating and host finding behavior of plant nematodes. *In* "Plant Parasitic Nematodes" (B. M. Zuckerman, W. F. Mai, and R. A. Rhode, eds.), Vol. 2, pp. 247-266. Academic Press, New York and London.

Green, C. D., and Plumb, S. C. (1970). The interrelationships of some *Heterodera* spp. indicated by the specificity of the male attractants emitted by their females. *Nematologica* **16,** 39-46.

Greet, D. N. (1964). Observations on sexual attraction and copulation in the nematode *Panagrolaimus rigidus* (Schneider). *Nature (London)* **204,** 96-97.

Hansen, E., Buecher, E. J., Jr., and Yarwood, E. A. (1964). Development and maturation of *Caenorhabditis briggsae* in response to growth factor. *Nematologica* **10,** 623-630.

Hechler, H. C. (1967). Morphological changes during the molt of *Diploscapter coronata* (Nematoda: Rhabditidae). *Proc. Helminthol. Soc. Wash.* **34,** 151-155.

Hechler, H. C. (1968). Postembryonic development and reproduction in *Diplogaster coronata* (Nematoda: Rhabditidae). *Proc. Helminthol. Soc. Wash.* **35,** 24-30.

Hechler, H. C. (1970a). Chromosome number and reproduction in *Mononchoides changi* (Nematoda: Diplogasterinae). *J. Nematol.* **2,** 125-130.

Hechler, H. C. (1970b). Reproduction, chromosome number, and postembryonic development of *Panagrellus redivivus* (Nematoda: Cephalobidae). *J. Nematol.* **2,** 355-361.

Hechler, H. C., and Taylor, D. P. (1966a). The life histories of *Seinura celeris, S. oliveirae, S. oxura,* and *S. steineri* (Nematoda: Aphelenchoididae). *Proc. Helminthol. Soc. Wash.* **33,** 71-83.

Hechler, H. C., and Taylor, D. P. (1966b). The molting process in species of *Seinura* (Nematoda: Aphelenchoididae), *Proc. Helminthol. Soc. Wash.* **33,** 90-96.

Hirschmann, H. (1951). Uber das Vorkommen zweier Mundhöhlentyp bei *Diplogaster lheritieri* Maupas und *Diplogaster biformis* n. sp. und die Entstehung dieser hermaphroditisch art aus *Diplogaster lheritieri. Zool Jahrb. Abt. Syst. Okol. Geogr. Tiere* **80**(1-2), 132-170.

Hirschmann, H. (1960). Reproduction of nematodes. *In* "Nematology." (J. N. Sasser, and W. R. Jenkins, eds.), pp. 140-167. Univ. North Carolina Press, Chapel Hill, North Carolina.

Hirschmann, H. (1962). The life cycle of *Ditylenchus triformis* (Nematoda: Tylenchida) with emphasis on post-embryonic development. *Proc. Helminth. Soc. Wash.* **29,** 30-43.

Hirschmann, H. (1971). Comparative morphology and anatomy. *In* "Plant Parasitic Nematodes" (B. M. Zuckerman, W. F. Mai, and R. A. Rhode, eds.), Vol. 1., pp. 11-63. Academic Press, New York and London.

Hirschmann, H., and Sasser, J. N. (1955). On the occurrence of an intersexual form in *Ditylenchus triformis,* n. sp. (Nematoda, Tylenchida). *Proc. Helminthol. Soc. Wash.* **22,** 115-123.

Hirschmann, H., and Triantaphyllou, A. C. (1967). Mode of reproduction and development of the reproductive system of *Helicotylenchus dihystera. Nematologica* **13,**

558-574.

Honda, H. (1925). Experimental and cytological studies on bisexual and hermaphrodite free-living nematodes, with special reference to problems of sex. *J. Morphol. Physiol.* **40**, 191-225.

Hope, W. D. (1967a). Free-living marine nematodes of the genera *Pseudocella* Filipjev, 1927, *Thoracostoma* Marion, 1870, and *Deontostoma* Filipjev, 1916 (Nematoda: Leptosomatidae) from the west coast of North America. *Trans. Amer. Microscop. Soc.* **86**, 307-334.

Hope, W. D. (1967b). A review of the genus *Pseudocella* Filipjev, 1927 (Nematoda: Leptosomatidae) with a description of *Pseudocella triaulolaimus* n. sp. *Proc. Helminth. Soc. Wash.* **34**, 6-12.

Hope, W. D. (1969). Fine structure of the somatic muscles of the free-living marine nematode *Deontostoma californicum* Steiner and Albin, 1933 (Leptosomatidae). *Proc. Helminthol. Soc. Wash.* **36**, 10-29.

Hope, W. D., and Murphy, D. G. (1969). *Rhaptothyreus typicus* n. g., n. sp., an abyssal marine nematode representing a new family of uncertain taxonomic position. *Proc. Biol. Soc. Wash.* **82**, 81-92.

Hopper, B. E. (1961). Occurrence of an egg-string in *Enoplus communis* Bastian, (Nematoda: Enoplidae). *Nature (London)* **189** (4761), 331-332.

Hopper, B. E., and Meyers, S. P. (1966). Aspects of the life cycle of marine nematodes. *Sond. Tir. Helog. Meer. Wiss.* **13**, 444-449.

Hopper, B. E., and Meyers, S. P. (1967). Population studies on benthic nematodes within a subtropical seagrass community. *Mar. Biol.* **1**, 85-96.

Hyman, L. H. (1951). "The Invertebrates: Acanthocephala, Aschelminthes, and Entoprocta." McGraw-Hill, New York.

Jägerskiöld, L. (1901). Weitere Beiträge zur Kenntnis der Nematoden. *Kgl. Svenska Vetenskapsakad. Handl.* **35**(2), 1-80.

Jairajpuri, M. S., and Siddiqi, A. H. (1964). Intersexuality in *Tyleptus striatus*. *Nematologica* **10**, 182-183.

Jantunen, R., (1964). Molting of *Caenorhabditis briggsae* (Rhabditidae). *Nematologica* **10**, 419-424.

Johnson, P. W., Van Gundy, S. D., and Thomson, W. W. (1970). Cuticle formation in *Hemicycliophora arenaria*, *Aphelenchus avenae* and *Hirschmanniella gracilis*. *J. Nematol.* **2**, 59-79.

Jones, T. P. (1966). Sex attraction and copulation in *Pelodera teres*. *Nematologica* **12**, 518-522.

Kan, S. P., and Davey, K. G. (1968a). Molting in a parasitic nematode, *Phocanema decipiens*. II. Histochemical study of the larval and adult cuticle. *Can. J. Zool.* **46** (1), 235-241.

Kan, S. P., and Davey, K. G. (1968b). Molting in a parasitic nematode, *Phocanema decipiens*. III. The histochemistry of cuticle deposition and protein synthesis. *Can. J. Zool.* **46**, 723-727.

Kreis, H. A. (1930). Die Entwicklung von *Actinolaimus tripapillatus* (v. Daday). Ein Beiträg zur postembryonalen Entwicklung der freilebenden Nematoden. *Z. Morphol. Okol. Tiere Berlin* **18**, 322-346.

Kreis, H. A. (1934). Oncholaiminae Filipjev, 1916 eine Monographische Studie. *Capita Zool.* **4**(5), 1-271.

Kröning, F. (1923). Studien zur Chromatinreifund der Keimzellen. Die Tetrapenbildung und die Reifeteilungen bei einigen Nematoden. *Arch. Zellforsch.* **17** (1), 63-85.

Krüger, E. (1913). Fortpflanzung und Keimzellenbildung von *Rhabditis aberrans*. *Z. Wiss. Zool.* **105**, 87-124.

Lee, D. L. (1961). Localization of esterase in the cuticle of the nematode *Ascaris lumbricoides*. *Nature (London)* **192**, 282-283.

Lee, D. L. (1962). The histochemical localization of leucine aminopeptidase in *Ascaris lumbricoides*. *Parasitology* **52**, 533-538.

Lee, D. L. (1965). "The Physiology of Nematodes." Oliver and Boyd, Edinburgh and London.

Lorenzen, S. (1971). Jugendstadien von *Desmoscolex*-arten (Nematoden, Desmoscolecidae) und deren Bedeutung für die Taxonomie. *Mar. Biol.* **10**, 343-345.

Mackerras, M., and Sandars, D. F. (1955). The life history of the rat lung-worm, *Angiostrongylus cantonensis* (Chen) (Nematoda: Metastrongylidae). *Aust. J. Zool.* **3**, 1-21.

Maggenti, A. R. (1962). The production of the gelatinous matrix and its taxonomic significance in *Tylenchulus* (Nematoda: Tylenchulinae). *Proc. Helminthol. Soc. Wash.* **29**, 139-144.

Maggenti, A. R., and Allen, M. W. (1960). The origin of the gelatinous matrix in *Meloidogyne*. *Proc. Helminthol. Soc. Wash.* **27**, 4-10.

Man, J. G. de (1886). "Anatomische Untersuchungen über freilebende Nordsee Nematoden." Leipzig.

Man, J. G. de (1893). Cinquième note sur les nématodes libres de la Mer du Nord et de la Manche. *Mem. Soc. Zool. France* **6**, 81-125.

Man, J. G. de (1904). Nematodes Libres. Résultats du voyage du S. Y. Belgica. Rapports Scientifiques. Zoologie. Buschmann, Anvers. 51 pp.

Marion, M. A. F. (1870). Recherches zoologiques et anatomiques sur des Nématoides non parasites marines. *Ann. Sc. Natur. Zool.* **13**, 1-102.

Martini, E. (1903). Über Furchung und Gastrulation bei *Cucullanus elegans* Zed. *Z. Wiss. Zool.* **74**, 501-556.

Martini, E. (1906). Über Subcuticula und Seitenfelder einiger Nematoden. *Z. Wiss. Zool.* **81**, 699-766.

Maupas, E. (1899). La mue et l'enkystement ches les nématodes. *Arch. Zool. Exp. Gen. Ser. 3*, **7**, 563-629.

Maupas, E. (1900). Modes et formes de reproduction des nématodes. *Arch. Zool. Exp. Gen. Ser. 3*, **8**, 463-624.

Mendes, M. V. (1942). Anomalia sexual num Nemátode marinho. *Bol. Fac. Fil. Cien. Letr. Univ. Sao Paolo Zool.* **25** (6), 255-265.

Merny, G. (1966). Nématodes d'Afrique tropicale: un nouveau *Paratylenchus* (Criconematidae), deux nouveaux *Longidorus* et observations sur *Longidorus laevicapitatus* Williams, 1959 (Dorylaimidae). *Nematologica* **12**, 385-395.

Micoletzky, H. (1914). Freilebende Süsswasser-Nematoden der Ost-Alpen mit besonderer Berücksichtung des Lunzer Seengebietes. *Zool. Jahrb. Abt. Syst. Oekol. Geogr. Tiere* **36**, 331-546.

Moorthy, V. N. (1938). Observations on the life history of *Camallanus sweeti*. *J. Parasitol.* **24**, 323-342.

Müller, H. (1903). Beitrag zur Embryonalentwickelung der *Ascaris megalocephala*. *Zool. Stuttgart* **41** (17), 1-30.

Musso, R. (1930). Die genitalröhren von *Ascaris lumbricoides* und *megalocephala*. *Z. Wiss. Zool.* **137** (2), 274-363.

Neuhaus, C. (1903). Die postembryonale Entwickelung der *Rhabditis nigrovenosa*. *Jen. Z. Naturwiss.* **37** (4), 653-690.

Nigon, V. (1949). Les modalités de la reproduction et le déterminisme du sexe chez quelques nématodes libres. *Ann. Sci. Natur.* **11**, 1-132.
Nigon, V. (1965). Développement et reproduction des Nématodes. *In* "Traité de Zoologie" (P. P. Grassé, ed.), Vol. 4, p. 218-386. Masson, Paris.
Nigon, V., and Dougherty, E. C. (1949). Reproductive patterns and attempts at reciprocal crossing of *Rhabditis elegans* Maupas, 1900, and *Rhabditis briggsae* Dougherty and Nigon, 1949, (Nematoda: Rhabditidae). *J. Exp. Zool.* **112**, 485-503.
Nigon, V., and Roman, E. (1952). Le déterminisme du sexe et le développement cyclique de *Strongyloides ratti. Bull. Biol. Fr. Belg.* **86**, 405-448.
Osche, G. (1952). Systematik und Phylogenie der Gattung *Rhabditis* (Nematoda). *Zool. Jahrb. Abt. Syst. Oekol. Geogr. Tiere* **81**, 190-280.
Osche, G. (1954). Ein Beitrag zur Kenntnis mariner *Rhabditis*-Arten. *Zool. Anz.* **152** (9-10), 242-251.
Ozerol, N. H., and Silverman, P. H. (1969). Partial characterization of *Haemonchus contortus* exsheathing fluid. *J. Parasit.* **55**, 79-87.
Ozerol, N. H., and Silverman, P. H. (1970). Further characterization of active metabolites from histotropic larvae of *Haemonchus contortus* cultured in vitro. *J. Parasit.* **56**,, 1199-1205.
Ozerol, N. H., and Silverman, P. H. (1972). Exsheathment phenomenon in the infective-stage larvae of *Haemonchus contortus. J. Parasit.* **58**, 34-44.
Pai, S. (1928). Die Phasen des Lebenscyclus der *Anguillula aceti* Ehrbg. und ihre experimentell-morphologische Beeinflussung. *Z. Wiss Zool.* **131** (2), 293-344.
Potts, F. A. (1910). Notes on the free-living nematodes. I. The hermaphrodite species. *Quart. J. Microsc. Sci.* **55**, 433-484.
Rachor, E. (1969). Das de Mansche Organ der Oncholaimidae, eine genito-intestinale Verbindung bei Nematoden. *Z. Morphol. Tiere* **66**, 87-166.
Rhoades, H. L., and Linford, M. B. (1959). Molting of preadult nematodes of the genus *Paratylenchus* stimulated by root diffusates. *Science* **130** (3387), 1476-1477.
Rogers, W. P. (1958). Physiology of the hatching of eggs of *Ascaris lumbricoides. Nature (London)* **181**, 1410-1411.
Rogers, W. P. (1960). The physiology of infective stages: The stimulus from the host. *Proc. Roy. Soc. London* **152**, 367-386.
Rogers, W. P. (1962). "The Nature of Parasitism." Academic Press, New York and London.
Rogers, W. P. (1963). Physiology of infection: Some effects of the host stimulus on infective stages. *Ann. N. Y. Acad. Sci.* **113**, 208-216.
Rogers, W. P. (1965). The role of leucine aminopeptidase in the moulting of nematode parasites. *Comp. Biochem. Physiol.* **14**, 311-321.
Rogers, W. P. (1970). The function of leucine aminopeptidase in exsheathing fluid. *J. Parasit.* **56**, 138-143.
Rogers, W. P., and Sommerville, R. I. (1963). The infective stages of nematode parasites and its significance in parasitism. *In* "Advances in Parasitology" (B. Dawes, ed.), pp. 109-177. Academic Press, New York and London.
Roggen, D. R., Raski, D. J., and Jones, N. O. (1967). Further electron microscopic observations of *Xiphinema index. Nematologica* **13**, 1-16.
Roman, J., and Hirschmann, H. (1969). Embryogenesis and postembryogenesis in species of *Pratylenchus* (Nematoda: Tylenchidae). *Proc. Helminthol. Soc. Wash.* **36**, 164-174.
Roman, J., and Triantaphyllou, A. C. (1969). Gametogenesis and reproduction of seven species of *Pratylenchus. J. Nematol.* **1**, 357-362.
Santos, M. S. N. de A. (1969). Rothamsted Exp. Sta. Rep. 1968, Pt. 1, pp. 155-156.

Schleip, W. (1911). Das Verhalten des Chromatins bei *Angiostomum (Rhabdonema) nigrovenosum*. *Arch. Zellforsch.* **7,** 87-138.

Schneider, A. (1866). "Monographie der Nematoden." Reimer, Berlin.

Schuurmans-Stekhoven, J. H., Jr. (1931). Okologische und morphologisch Notizen über Zuiderseenematoden. I. Die westliche Hälfte der Zuidersee. *Z. Morphol. Okol. Tiere* **20,** 613-678.

Schuurmans-Stekhoven, J. H., Jr., and Adam, W. (1931). The free-living marine nemas of the Belgian Coast. *Mém. Musée Roy. Hist. Natur. Belg.* No. 49.

Seshadri, A. R. (1964). Investigations on the biology and life cycle of *Criconemoides xenoplax* Raski, 1952. (Nematoda: Criconematidae). *Nematologica* **10,** 540-562.

Shepherd, A., and Clarke, A. J. (1971). Molting and hatching stimuli. *In* "Plant Parasitic Nematodes" (B. M. Zuckerman, W. F. Mai, and R. A. Rohde, eds.), Vol. 2, pp. 267-287. Academic Press, New York and London.

Siddiqui, I. A., and Taylor, D. P. (1970). The biology of *Meloidogyne naasi*. *Nematologica* **16,** 133-143.

Slocombe, J. O. D., and Whitlock, J. H. (1971a). Analyses of supernatant fluids from exsheathing infective *Haemonchus contortus* larvae. J. Parasit. **57,** 794-800.

Slocombe, J. O. D., and Whitlock, J. H. (1971b). Further analyses of supernatant fluids from exsheathing infective *Haemonchus contortus cayugensis* larvae. *J. Parasit.* **57,** 801-807.

Sommerville, R. I. (1957). The exsheathing mechanism of nematode infective larvae. *Exp. Parasitol.* **6,** 18-30.

Sommerville, R. I. (1960). The growth of *Cooperi curticei* (Giles, 1892), a nematode parasite of sheep. *Parasitology* **50,** 261-267.

Spemann, H. (1895). Zur entwicklung des *Strongylus paradoxus*. *Zool. Jahrb. Anat. Ontog.* **8,** 301-317.

Steiner, G. (1923). Intersexes in nematodes. *J. Hered.* **14,** 147-158.

Steiner, G. (1937). Intersexuality in two new parasitic nematodes, *Pseudomermis vanderlindei* n. sp. (Mermithidae) and *Tetanonema strongylurus* n. g. n. sp. (Filariidae). Papers on Helminthology, Skrjabin Memorial Volume, Moscow.

Strassen, O. zur (1892). *Bradynema rigidum* v. Sieb. *Z. Wiss. Zool.* **54,** 655-747.

Strassen, O. zur (1894). Über das röhrenförmige Organ von *Oncholaimus*. *Z. Wiss. Zool.* **58,** 460-474.

Strassen, O. zur (1896). Embryonalentwickelung des *Ascaris megalocephala*. *Arch. Entwickelungsmech. Organ.* **3,** 27-105.

Szent-Györgyi, A. (1969). Molecules, electrons and biology. *Trans. N. Y. Acad. Sci. Ser. II* **31,** 334-340.

Tadano, M. (1968). Nemathelminthes. *In* "Invertebrate Embryology" (M. Kumé and K. Dan, eds.; J. C. Dan, translator), pp. 159-191. Nolit, Belgrade.

Thomas, P. R. (1965). Biology of *Acrobeles complexus* Thorne, cultivated on agar. *Nematologica* **2,** 395-408.

Threlkeld, W. L. (1941). Notes on copulation of certain nematodes. *Virginia J. Sci.* **2,** 31-34.

Tietjen, J. H. (1967). Observations on the ecology of the marine nematode *Monhystera filicaudata* Allgén, 1929. *Trans. Amer. Microsc. Soc.* **86,** 304-306.

Tietjen, J. H. (1969). The ecology of shallow water meiofauna in two New England estuaries. *Oecologia* **2,** 251-291.

Tietjen, J. H., Lee, J. J., Rullman, J., Greengart, A., and Trompeter, J. (1970). Gnotobiotic culture and physiological ecology of the marine nematode *Rhabditis marina* Bastian. *Limnol. Oceanog.* **15,** 535-543.

Timm, R. W. (1953). Observations on the morphology and histological anatomy of a marine nematode *Leptosomatum acephalatum* Chitwood, 1936, new combination (Enoplidae: Leptosomatinae). *Amer. Midl. Natur.* **49**, 229-248.

Timm, R. W. (1970). A revision of the nematode order Desmoscolecida Filipjev, 1929. *Univ. Calif. Pub. Zool.* **93**, 1-115.

Triantaphyllou, A. C. (1960). Sex determination in *Meloidogyne incognita* Chitwood, 1949 and intersexuality in *M. javanica* (Treub, 1885) Chitwood, 1949. *Ann. Inst. Phytopath. Benaki, N. S.* **3**, 12-31.

Triantaphyllou, A. C. (1966). Polyploidy and reproductive patterns in the rootknot nematode *Meloidogyne hapla*. *J. Morphol.* **118**, 403-414.

Triantaphyllou, A. C. (1969). Gametogenesis and the chromosomes of two rootknot nematodes, *Meloidogyne graminicola* and *M. naasi*. *J. Nematol.* **1**, 62-71.

Triantaphyllou, A. C. (1971). Oogenesis and the chromosomes of the cystoid nematode, *Meloidodera floridensis*. *J. Nematol.* **3**, 183-188.

Triantaphyllou, A. C., and Hirschmann, H. (1964). Reproduction in plant and soil nematodes. *Ann. Rev. Phytopathol.* **2**, 57-80.

Triantaphyllou, A. C., and Hirschmann, H. (1966). Gametogenesis and reproduction in the wheat nematode *Anguina tritici*. *Nematologica* **12**, 437-443.

Triantaphyllou, A. C., and Hirschmann, H. (1967). Cytology and reproduction of *Helicotylenchus dihystera* and *H. erythrinae*. *Nematologica* **13**, 575-580.

Türk, F. (1903). Über einige im Golfe von Neapel frei lebende Nematoden. *Mittheil. Zool. Sta. Neapel.* **16**, 281-348.

Van Gundy, S. D. (1958). The life history of the citrus nematode *Tylenchulus semipenetrans* Cobb. *Nematologica* **3**, 283-294.

Viglierchio, D. R., and Johnson, R. N. (1971). On the maintenance of *Deontostoma californicum*. *J. Nematol.* **3**, 86-88.

Walton, A. C. (1918). The oogenesis and early embryology of *Ascaris canis* Werner. *J. Morphol.* **30**, 527-603.

Walton, A. C. (1940). Gametogenesis. *In* "An Introduction to Nematology" B. G. Chitwood and M. B. Chitwood, (eds.), Sect. II, Pt. I, pp. 205-215. Monumental Printing, Baltimore, Maryland.

Wandolleck, B. (1892). Zur embryonalentwicklung des *Strongylus paradoxus*. *Arch. Naturgesch.* **1** (2), 123-148.

Wang, L. H. (1971). Embryology and life cycle of *Tylenchorhynchus claytoni* Steiner, 1937 (Nematoda: Tylenchoidea). *J. Nematol.* **3**, 101-107.

Weerdt, L. G. van (1960). Studies on the biology of *Radopholus* (Cobb, 1893) Thorne, 1949. *Nematologica* **5**, 43-51.

Whitlock, J. H. (1970). Ecdysis of *Haemonchus* hypotheses. *Cornell Vet.* **61**, 349-361.

Wieser, W. (1953). Free-living marine nematodes. I. Enoploidea. *Lunds Univ. Arsskrift (Avdelning 2)* **49** (6), 1-155.

Wieser, W. (1956). Free-living marine nematodes III. Axonolaimoidea and Monhysteroidea. *Lunds Univ. Arsskrift (Avdelning 2)* **52** (13), 1-115.

Wieser, W., and Kanwisher, J. (1961). Ecological and physiological studies on marine nematodes from a small salt marsh near Woods Hole, Mass. *Limnol. Oceanogr.* **6**, 262-270.

Wilson, E. B. (1898). "The Cell in Development and Inheritance." Macmillan, New York.

Wright, K. A., Hope, W. D., and Jones, N. O. (1973). "The ultrastructure of the sperm of *Deontostoma californicum*, a free-living marine nematode. *Proc. Helminthol. Soc. Wash.* **40**, 30-36.

Yuen, P. H. (1966). Further observations on *Helicotylenchus vulgaris* Yuen. *Nematologica* **11**, 623-637.

Yuksel, H. N. S. (1960). Observations on the life cycle of *Ditylenchus dipsaci* on onion seedlings. *Nematologica* **5**, 289-296.

Ziegler, H. E. (1895). Untersuchungen über ersten Entwicklungsvorgänge der nematoden. *Z. Wiss. Zool.* **60**, 351-410.

Zoja, R. (1896). Untersuchungen über Entwicklung der *Ascaris megalocephala*. *Arch. Mikrosk. Anat.* **47**, 218-261.

CHAPTER 9

ROTIFERA

Anne Thane

9.1 Introduction

The rotifers are microscopic and among the smallest of all metazoa. They are usually considered as a class within the phylum Aschelminthes, or as a distinct phylum, Rotifera. The rotifers are thought to be of limnetic origin and by far the largest number of species are found in fresh water. Representatives of a number of genera are, however, found in brackish water and a few species live in the sea.

Only the order Seisonidea with one genus, *Seison*, is strictly marine. Among the remaining orders, Bdelloidea (Digononta) and Monogononta, only the bdelloid genus *Zelinkiella* is found exclusively in sea or brackish water. All other genera represented in the sea belong to families which have their main distribution in fresh water, and only the genera *Encentrum* and *Synchaeta* have more species in brackish and seawater than they do in fresh water. In brackish water the number of species of rotifers decreases strongly with increasing salinity (Thane-Fenchel, 1968).

Rotifers are found as a constituent of both the plankton and the microbenthos. They have adapted to a number of habitats, and creeping and swimming as well as sessile forms are known. A few forms are epizoic or parasitic (Remane, 1929; Thane-Fenchel, 1966).

The rotifers show some histological peculiarities. Almost all cell structures are syncytial. This condition arises from the congruence of originally separate cells during development. Another feature is the constant number of cells or rather nuclei in the adult animal. In any given species each organ contains a characteristic number of cells. Consequently cells do not divide after development is completed. Unfortunately the literature on the reproduction and development of marine rotifers is very restricted and much of the information given in the present chapter is based on studies of freshwater forms. Probably most aspects of reproduction are in common with marine forms but many details concerning the biology of marine rotifers still need to be clarified.

9.2 Asexual Reproduction

With the exception of the order Seisonidea, asexual reproduction is the most common method of reproduction and is found in all species. Rotifers are exclusively dioecious but within the order Bdelloidea males are unknown and reproduction is always asexual, by ameiosis and parthenogenesis.

Within the order Monogononta males have been found in about 10% of the species (Wesenberg-Lund, 1923) but it has been proposed that males occur in all species (Hyman, 1951). Whatever the case may be, males do not occur constantly throughout the year but show a cyclical or seasonal occurrence, and asexual reproduction only is found through the larger part of the year. Details on the oogenesis and reproductive organs are given in Section 9.3.3.

Within the seisonids, males are constantly present although the sex ratio is not known (Remane, 1932). Here asexual reproduction apparently does not occur since Remane (1932) found that females are always fertilized.

9.3 Sexual Reproduction

9.3.1 Sexual Dimorphism

Sexual dimorphism is found in the two orders in which males are known although it is not very strongly developed within the seisonids. Thus the seisonid males are fully developed with the same degree of organization as the females but can be recognized from the latter by being shorter and by the position of the cloaca.

Among the monogononts sexual dimorphism appears as various degrees of reduction in some structures of the male. The males are always much smaller than the females (from ½ to less than ¹⁄₁₀ of the female body size). The digestive system can be rather reduced; anus and cloaca are absent. Furthermore, the bladder and excretory system may be lacking, and the general body shape may differ from that of the female. The structure of the corona is usually simplified. In general males are comparable in structure to early developmental stages of females. The most reduced males seem to be found among planktonic and sessile forms. A detailed account of the morphology of the males in marine forms is given by Remane (1932). Different types of males are shown in Fig. 1.

9.3.2 Sex Determination

In the order Monogononta two types of females are found, mictic and amictic females. Both types are diploid and morphologically similar, but they produce different types of eggs. Amictic females produce eggs which only form one polar body during development; the maturation cleavage is ameiotic and thus gives rise to diploid eggs. These develop into new females parthenogenetically. Mictic females produce haploid eggs through meiotic divisions and the eggs have two polar bodies. If these eggs are not fertilized, they develop into males. If they are fertilized, so-called resting eggs are formed which, after diapause periods of varied lengths, develop into amictic females (Gilbert and Thompson, 1968). The eggs of both types of females are, in their early stages, identical and bipotent in development. Thus the mictic oocytes may develop parthenogenetically or after fertilization and the oocytes of amictic females may develop into either mictic or amictic females (Buchner *et al.*, 1969). Whether a given female will produce mictic or amictic eggs is determined at a certain moment during the development of the oocytes since in one ovary all eggs develop either through meiosis or mitosis.

It has been shown that only mictic eggs from young animals (a few hours old) of the freshwater form of *Asplanchna sieboldi* can be fertilized (Buchner and Kiechle, 1967). Eggs from older females can not be fertilized and always develop into males. This mechanism prevents a newborn male from fertilizing its mother and probably also secures a more constant supply of males during the period of sexual reproduction.

9.3.3 Anatomy of Reproductive Systems and Gametogenesis

In general the gonads of rotifers may be described as one or two sacs with openings into the cloaca. In forms without an intestine (e.g., *Asplanchna* and in most males) a genital pore is found which, however, is considered to be homologous to a cloaca (Remane, 1932).

In the bdelloids or digononts the reproductive system consists of two syncytial germovitellaria, each consisting of a small ovary with minute nuclei, and a larger vitellarium with large nuclei. The germovitellaria are situated lateral to the intestine on each side and a common oviduct opens into the cloaca anterior to the nephridial entrance.

The female monogononts have a single ventrally situated ovary and vitellarium held together by a common membrane (Fig. 2). The membrane continues as a simple tubular oviduct entering the cloaca. According to Remane (1932) this single gonad is homologous to both ovaries of the bdelloids since the gonad of the monogononts often contains 8 eggs while each ovary of the bdelloids contains 4 eggs. The paired ovary of *Asplanchna priodonta* is considered as an anomaly (Lehmensick, 1926).

Both the ovary and the vitellarium are syncytial. After hatching or, in the case of viviparous forms, after birth no more cell divisions take place in these organs. For further details on the anatomy of the gonads see Remane (1932).

In the seisonids a paired ovary is found while vitellaria are absent. The oocytes of the syncytial ovary are covered by a mantle which functions as a seminal receptacle (Fig. 2). The sperms, however, are not found in the cavity but are embedded in the cytoplasm. Remane (1932) suggests that the mantle is homologous with the vitellarium of the two other orders of rotifers.

Oogenesis of bdelloids has been studied in a number of cases. Zelinka (1891) described the origin of germ cells and gonads of *Mniobia russeola*, a freshwater form. Hsu (1956a) has described oogenesis in *Philodina roseola*, which also occurs in brackish water. Oogenesis is initiated by two equational divisions. No synapsis has been observed between any 2 of the 13 chromosomes which are found to be in a condensed stage even in the youngest oocytes. The anaphase chromosomes of the oogonial division do not despiralize when the nuclei of the syncytial ovary are formed. After the last oogonial division the chromosomes still remain condensed. By progressively packing together, they first form a ring and end up by forming a homogeneous and spherical mass of chromatin in the center of the nucleus. The chromosomes remain in this state while the nucleus is

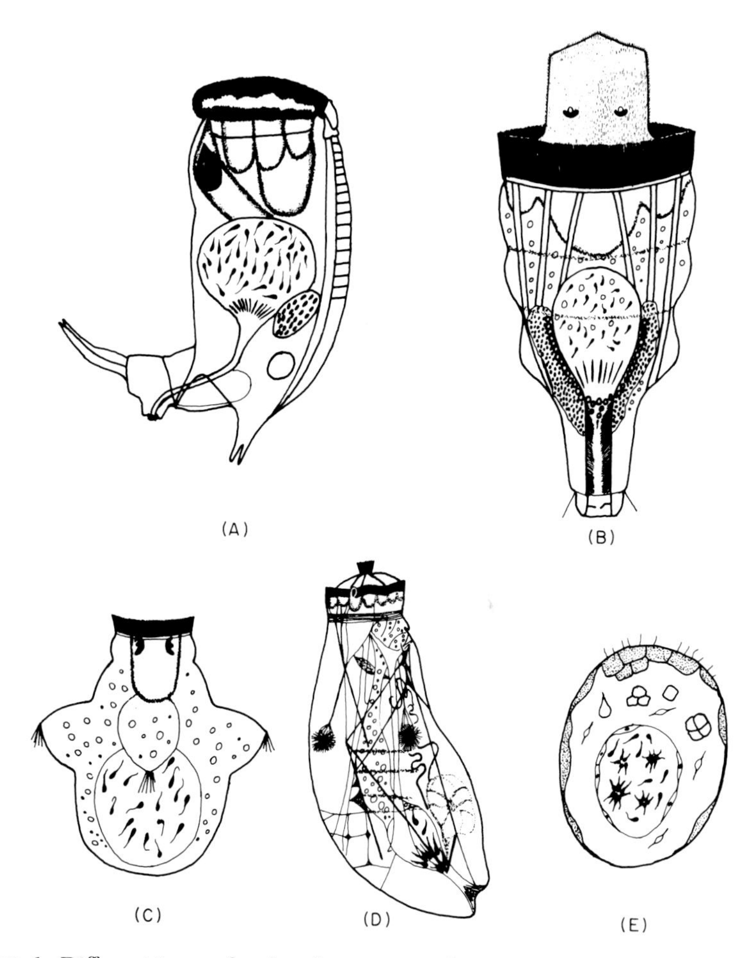

FIG. 1. Different types of males showing varied degrees of reduction. (A) *Mytilina mucronata*, (B) *Filinia longiseta*, (C) *Pedalia mira*, (D) *Asplanchna priodonta*, (E) *Asplanchna sieboldi* [after Remane (1929)].

isolated from the ovary to form an oocyte. With increasing size of the germinal vesicle the chromosomes separate and eventually the 13 condensed chromosomes can be counted. Three of these can easily be distinguished from the rest, two being dot-shaped and the third being very long. Hsu (1956a) suggests that the chromosomes in this group of obligatory parthenogenetical animals have lost their homology.

It cannot at present be stated whether the above described is typical for all bdelloids but Hsu (1956b) found this to be the case for two other species belonging to two different families. The bdelloids may

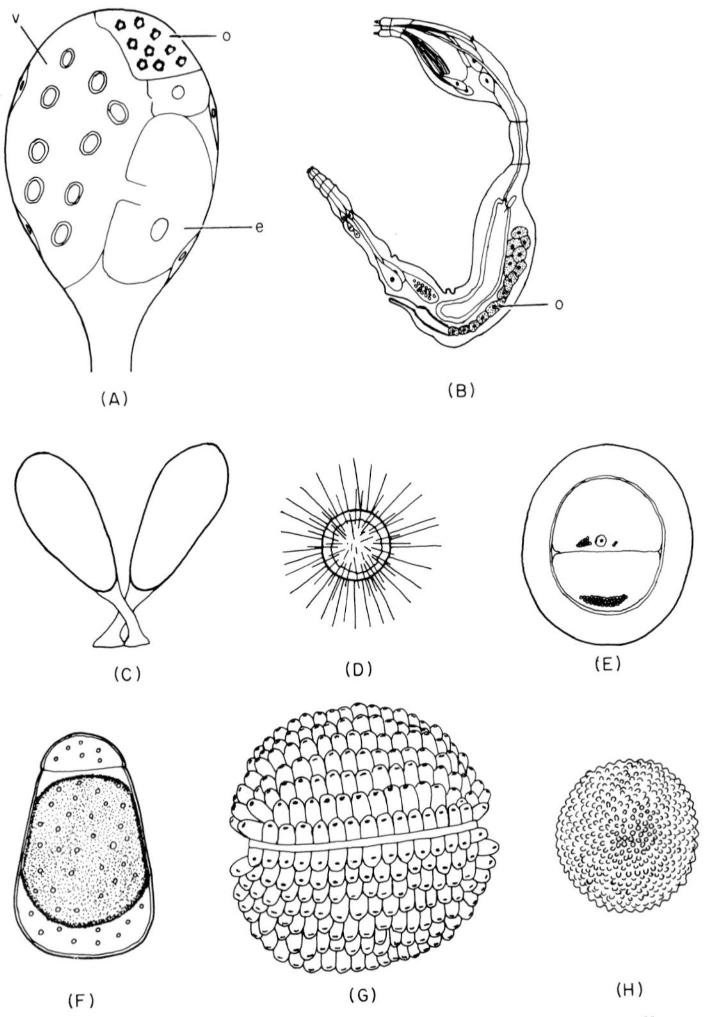

FIG. 2. (A) Female reproductive system of a monogonont, o, ovary: v, vitellarium: e, developing egg showing the feeding tube from the vitellarium [after Remane (1929)]. (B) Seisonid female showing reproductive system, o, ovary [after Hyman (1951)]. (C-H) Different types of eggs, (not to scale). (C) Stalked eggs of *Seison*. (D, E) Amictic eggs, developing parthenogenetically. (F-H) Mictic fertilized resting eggs [after Remane (1929, 1932)].

prove to be amenable to the study of the evolutionary significance of aneuploidy and possibly polyploidy in animals.

During oogenesis in the monogononts, the oocytes separate one by one from the ovary and join the vitellarium during maturation. After disintegration of the membranes, the vitellarium penetrates the eggs with a tubelike protuberance in order to provide the eggs with nutritional substance (Lehmensick, 1926; Peters, 1931) (Fig. 2). The fertilized mictic eggs particularly are provided with much material from the vitellarium, which in this case is entirely emptied of cytoplasm and at last only nuclei remain in this organ. When eggs are produced, the vitellarium is often filled with a fatty, red-brownish substance.

Mictic eggs undergo normal maturation divisions with a prolonged prophase and the formation of two polar bodies. Amictic eggs undergo only one maturation division and have a short prophase with the formation of only one polar body (Remane, 1929).

Oogenesis in the seisonids resembles in principle that of the mictic eggs of the monogononts. However, there is a long time interval between the termination of the first maturation division and the initiation of the second division, which takes place when growth has stopped (Remane, 1932).

The three types of eggs (amictic, and fertilized and unfertilized mictic eggs) found within the monogononts often differ from each other morphologically. Amictic eggs have a thin primary egg-membrane. In *Ploesoma* and *Euchlanis* and in some *Synchaeta* species they are often also provided with a gelatinous membrane and in other species of *Synchaeta* they are provided with spines apparently serving as floatation devices. Amictic eggs of planktonic forms may also be provided with oil drops which probably serve the same purpose.

Unfertilized mictic eggs developing into males can generally be recognized due to their small size. Fertilized mictic eggs evolving into resting eggs are characterized by the presence of a thick outer membrane which may be sculptured or ornamented. This membrane as well as the thinner inner membrane both originate from the egg.

Within the bdelloids only one type of thin-shelled egg occurs (Remane, 1932). The eggs of the seisonids, which always require fertilization, are also of only one type and are characterized by a stalk. Different types of rotifer eggs are shown in Fig. 2.

The male reproductive system of the monogononts consists of a single sacciform testis with a ciliated sperm duct leading to the male gonopore. One or more prostatic glands are present. The terminal part of the sperm tract may be protrusive as a cirrus or it can be lined with a hard cuticle forming a protrusive penis (Fig. 3). Within the genera *Polyarthra*, *Filinia*, and *Pedalia* (all comprising marine forms) the males are strongly reduced and devoid of a penis; during copulation sperm is transferred directly from the distal part of the vas deferens to the female.

According to Beauchamp (1965) very little is known about spermatogenesis. Since spermatogenesis takes place during embryonic development, the testis contains only spermatids and mature sperm.

Two kinds of sperms are found among monogononts. Typically they are flagellated and with a long or rounded head and the tail is provided with an undulating membrane. Atypical sperms are rod-shaped and are considered to help penetrate the cuticle of the female (Fig. 3). Whitney (1917) suggested that the atypical sperms arise from an additional division giving one normal and one atypical sperm. Insemination is mostly hypodermic so that the sperm enter the pseudocoel of the female; more rarely sperms enter the cloaca.

The seisonid males have paired testes and are devoid of copulatory organs. Dorsal to the intestine the sperm duct enters a syncytial mass. Within this mass the duct is coiled and provided with ciliated enlargements and here the sperms, which are of the flagellate type, are cemented together forming spermatophores (Illgen, 1914). The genital pore enters the cloaca in accordance with the fact that seisonid males possess an intestine (Fig. 3).

9.3.4 Gametogenic Cycles and Factors Influencing Them

As previously discussed the monogononts show heterogony, i.e. there is an alternation between parthenogenetic reproduction and reproduction involving fertilization. Knowledge of the factors controlling the occurrence of the two types of reproduction is almost completely based on observations of freshwater forms but many of the results can probably be generalized to include marine forms.

In all forms, populations consisting only of amictic females reproducing parthenogenetically by producing diploid eggs are dominant through the larger part of the year, and only in certain periods do populations contain mictic females producing haploid eggs and males. This cycle may be controlled either by endogenous or exogenous factors. The evidence for inherited, endogenous factors controlling the cycle consists of observations showing that within certain species some strains will produce a higher percentage of mictic females than will other strains under similar environmental conditions (Buchner *et al.*, 1969; Pourriot, 1965).

External factors such as temperature, dessication, light, amount or quality of food and the presence of various chemicals have all been shown to influence the occurrence of mictic females (for references see Hyman, 1951; Gilbert, 1968). Generalizing from the numerous observations on freshwater forms, it seems that factors connected with a degrading of the environmental quality for the rotifer popula-

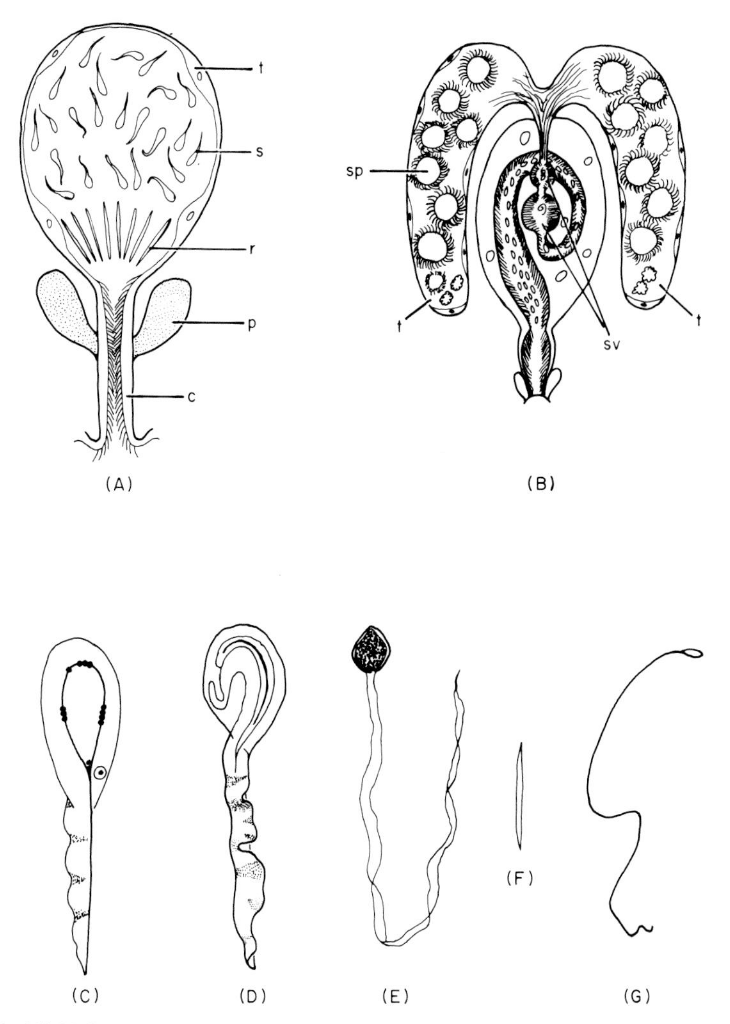

FIG. 3. (A) Male system of monogonont: t, testis; s, sperm; r, rod sperm; p, prostatic gland; c, cirrus. (B) Male system of seisonid: sp, spermatophore; t, testis; sv, spermatophoral vesicles [after Remane (1929)]. (C-E) Typical sperms of monogononts. (F) Rod sperm of monogonont. (G) Sperm of seisonid [after Remane (1932)].

tion in question may induce the presence of mictic females. Thus high population densities and unfavorable climatic changes through the year are conditions under which mictic females can be expected. In some species sexual reproduction occurs only once a year, others have two or more cycles per year, while still others are quite acyclic.

Within the marine monogononts males are known for a number of forms (Remane, 1929; Rousselet, 1902) but there exists very little evidence indicating which mechanisms are responsible for the occurrence of mictic females and males although Remane (1929) observed that males of species of the genera *Synchaeta*, *Brachionus*, and *Colurella* occur after maxima in population sizes. Considering that fluctuations in environmental factors in general must be less severe in the sea than is the case in fresh water, experimental data on the mechanisms inducing sexual reproduction in the marine forms would be of interest.

9.3.5 Fertilization and Depositing of Eggs

Mictic females may be fertilized by males either hypodermically or through the cloaca (Fig. 4). Within the genus *Brachionus* both types of fertilization may take place (Remane, 1929).

Eggs are dropped on the substratum or are attached to it either individually or in groups. Planktonic forms (e.g., *Brachionus*) often carry their eggs attached to the body or they attach the eggs to other plankton organisms or floating objects. In certain species within some genera amictic eggs develop inside the mother; examples of such viviparous forms can be found within *Asplanchna* and *Lindia*.

Dormant (fertilized mictic) eggs are either carried outside or inside the body of the mother (Fig. 4). In the latter case they are freed after the death of the mother. In most cases mictic eggs sink to the bottom but they may later form air bubbles and float.

Within the bdelloids eggs are normally deposited on the substratum although some forms are viviparous. The parasitic seisonids attach their eggs to the gills of their hosts by means of a stalk (Fig. 2).

9.4 Development

9.4.1 Embryonic Development

Until recently knowledge of the embryology of rotifers was based mainly on the works of Nachtwey (1925) and Tannreuter (1920) on the development of amictic eggs of two species of *Asplanchna*. These studies gave the impression that the embryological development of rotifers resembles that of acoelous turbellarians rather than that of other pseudocoelomates (Hyman, 1951). However, *Asplanchna* is in several respects an aberrant genus.

Beauchamp (1956) and Pray (1965) described the embryology of *Ploesoma hudsoni* and *Monostyla cornuta*, more typical rotifers.

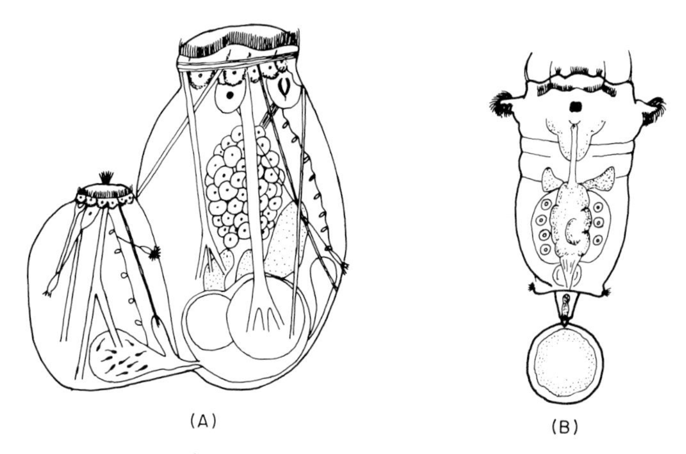

FIG. 4. (A) Copulation in *Asplanchna* [after Hyman (1951)]. (B) *Synchaeta*, a planktonic form with egg attached to the body [after Remane (1929)].

These works indicate that embryological development is more closely related to that of other Aschelminthes, i.e. a modified spiral cleavage of determinate type.

Cleavage is holoblastic. In the four cell stage (Fig. 5) there are three small and one large blastomere. The D quadrant divides unequally and the smaller of the daughter cells d_1 comes to lie at the same level as cells A, B, and C. These three cells also divide unequally and give rise to three smaller, granular blastomeres which together with d_1 form a germ ring, and to three larger, clear blastomeres above them. The ectoderm is formed by the descendants of the clear blastomeres and may easily be recognized in the later stages as may the descendants of the granulated D cell. The blastopore forms opposite to the ectodermal cells. Through epiboly the clear ectodermal cells cover the surface while the descendants of the D cell and of the germ ring move anteriad. As the ectodermal cells meet, the blastopore is eventually closed. The germovitellarium is formed by the progeny of the D cell.

In general newly hatched individuals have the shape of the adult form. Within the sessile suborders Collothecacea and Flosculariacea the young forms are free swimming and are often considered as a special larval form since they undergo a considerable transformation when changing to the sessile way of life (Hyman, 1951).

9.4.2 Life Span and Population Growth Rate

Population parameters such as life span, birth rate, developing time and intrinsic rate of natural increase have been relatively well

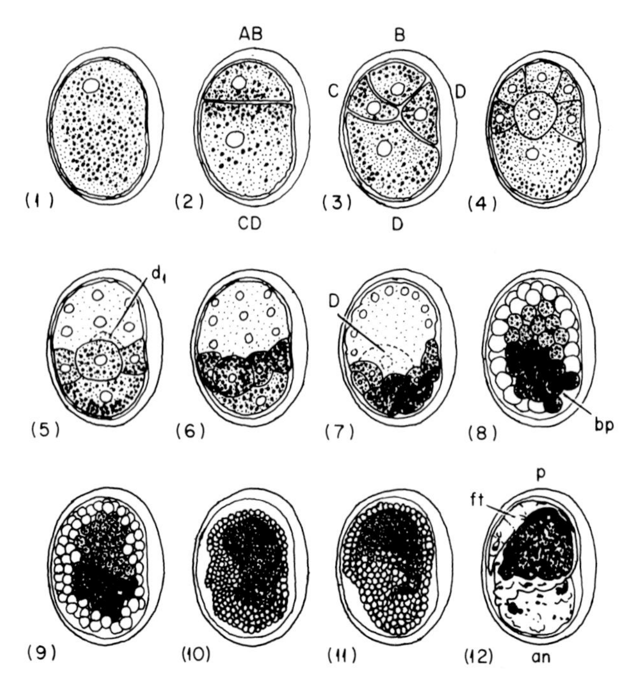

FIG. 5. (1 through 4). Early cleavage and the formation of the eight-cell stage. (5) Illustrates the position of the d_1 cell on the opposite side of the embryo from the plane of view. (6 and 7). Epiboly and beginning of involution of the micromeres. (8) The involution of the yolk-laden cells and the position of the blastopore. The D cell progeny have relatively smaller amounts of yolk granules than the involuted micromeres. (9, 10, and 11) Internal migration of the products of the D cell and the involuted micromeres. (12) Illustrates the orientation of the anterior-posterior axis of the embryo relative to the position of the blastopore.

AB, AB micromere; CD, CD macromere; A, A micromere; B, B micromere; C, C micromere; D, D macromere; d_1, d_1 micromere; bp, blastopore; ft, foot; p, posterior; an, anterior [after Pray (1965)].

studied in the rotifera compared to other groups of small metazoa. However, as is the case in other aspects of rotifer biology, information is primarily from freshwater forms.

Survivorship curves of laboratory and field populations are close to the rectangular type (Edmondson, 1945; King, 1966 and papers cited therein). For a natural population of *Floscularia conifera,* Edmondson (1945) found a maximum life span of about 11 days and a mean length of life of about 5 days. Aggregating specimens were found to have a longer life expectancy than solitary individuals. The values found by Edmondson are of the same magnitude as in other published life tables of rotifers.

Other studies are those of Hillbricht (1960) on laboratory populations of *Philodina citrina* and of Edmondson (1960, 1965, 1968) who studied natural populations of the planktonic *Keratella cochlearis* and some other forms. It was found that the ratio of eggs:females was proportional to the rate at which eggs were laid, and knowing the development time and population densities at different times, several population parameters could be calculated for different environmental conditions and mathematical models of population growth developed. Development of the eggs of *Ploesoma truncatum* was found to take about 1 day at 25°C and about 5 days at 11°C.

In general rotifers can be said to have a short life expectancy and high reproductive potential; under optimal conditions several forms may more than double their populations every 2 days.

9.5 References

Beauchamp, P. de. (1956). Le développement de *Ploesoma hudsoni* (Imhof) et l'origine des feuillets chez les Rotifères. *Bull.Soc.Zool. Fr.* **81**, 374-383.

Beauchamp, P. de (1965). Classe des Rotifères. *In* "Traité de Zoologie" (P.-P. Grassé, ed.), Vol. 4 (3), pp. 1225-1379. Masson, Paris.

Buchner, H. and Kiechle, H. (1967). Die Determination der Männchen und Dauereiproduktion bei *Asplanchna sieboldi. Biol.Zentralbl* **86**, 599-621.

Buchner, H., Kiechle, H., and Tiefenbacher, L. (1969). Untersuchungen über die Bedingungen der heterogonen Fortplanzungsarten bei den Rädertieren. I. Die miktische Reaktion ihre Beziehungen zum Populationsdynamik und ihre Abhängigkeit vom Milieu. *Zool.Jb.Physiol.* **74**, 329-426.

Edmondson, W.T. (1945). Ecological studies of sessile rotatoria.II. Dynamics of population and social structures. *Ecol. Monographs* **15**, 141-172.

Edmondson, W.T. (1960). Reproductive rates of rotifers in natural populations. *Mem.Ist.Ital.Idrobiol.* **12**, 21-177.

Edmondson, W.T. (1965). Reproductive rate of planktonic rotifers as related to food and temperature in nature. *Ecol. Monographs* **35**, 61-111.

Edmondson, W.T. (1968). A graphical model for evaluating the use of the egg ratio for measuring birth and death rates. *Oecologia* **1**, 1-37.

Gilbert, J.J. (1968). Dietary control of sexuality in the rotifer *Asplanchna brightwelli* Gosse. *Physiol.Zool.* **41**, 14-43.

Gilbert, J. J., and Thompson, G. A. (1968). Alpha tocophenol control of sexuality and polymorphism in the rotifer *Asplanchna. Science* **159**, 734-736.

Hillbricht, A. (1960). Population dynamics of *Philodina citrina* Ehr. (Rotatoria) bred in aquaria. *Ekol. Polska Ser. B* **6**, 161-170.

Hsu, W.S. (1956a). Oogenesis in the Bdelloidea rotifer, *Philodina roseola. La Cellule* **57**, 283-296.

Hsu, W. S. (1956b). Oogenesis in *Habrotrocha tridens* (Milne). *Biol. Bull.* **111**, 364-394.

Hyman, L. H. (1951). "The Invertebrates, Vol. III, Acanthocephala, Aschelminthes and

Entoprocta. The pseudocoelomate bilateria." McGraw-Hill, New York.
Illgen, H. (1914). Zur Kenntnis der Spermatogenese und Biologie bei *Seison grubei* Claus. *Zool. Anz.* **44**, 550-554.
King, C. E. (1966). Food, age and the dynamics of a laboratory population of rotifers. *Ecology* **48**, 111-128.
Lehmensick, R. (1926). Zur Biologie, Anatomie und Eireifung der Rädertiere. *Z. Wiss. Zool.* **128**, 37-113.
Nachtwey, R. (1925). Untersuchungen über die Keimbahn, Organogenese und Anatomie von *Asplanchna priodonta* Gosse. *Z. Wiss. Zool.* **126**, 239-492.
Peters, F. (1931). Anatomie und Zellkonstanz von *Synchaeta*. *Z. Wiss. Zool.* **139**, 1-119.
Pray, F. A. (1965). Early development of the rotifer *Monostyla*. *Trans. Amer. Microsc. Soc.* **84**, 210-216.
Pourriot, R. (1965). Sur le déterminisme du mode reproduction chez les Rotifères. *Schweiz Z. Hydrol.* **27**, 76-87.
Remane, A. (1929). Rotatoria. *In* "Die Tierwelt der Nord- und Ostsee" (G. Grimpe and E. Wagler, eds.). Lief. 16, Teil VIIe, 156 pp. Akad. Verlagsges., Leipzig.
Remane, A. (1932). Rotatoria. *In* "Klassen und Ordnungen des Tierreichs" (H. G. Bronn, ed.), Bd. IV, Abt. II, Buch 1, Lief. 1-4, pp. 284-448. Akad. Verlagsges., Leipzig.
Rousselet, C. F. (1902). *Synchaeta*, a monographic study. *J. Roy. Microsc. Soc. Ser. 2* **22**, 148-154.
Tannreuther, G. (1920). Development of *Asplanchna ebbesborneii*. *J. Morphol.* **33**, 389-437.
Thane-Fenchel, A. (1966). *Proales paguri* sp. nov. a rotifer living on the gills of the hermit crab *Pagurus bernhardus* (L.). *Ophelia* **3**, 93-97.
Thane-Fenchel, A. (1968). Distribution and ecology of non-planktonic brackish-water rotifers from Scandinavian waters. *Ophelia* **5**, 273-297.
Wesenberg-Lund, C. (1923). Contributions to the biology of the Rotifera. I. The males of the Rotifera. *Kgl.Dan.Vidensk.Selsk.Skr.,Nat.Math.,Afd.8,Raekke* **4**(3), 191-345.
Whitney, D.D. (1917). The production of functional and rudimentary spermatozoa in Rotifers. *Biol.Bull.* **33**, 305-315.
Zelinka, C. (1891). Studien über Rädertiere III. *Z. Wiss. Zool.* **53**, 1-159.

CHAPTER 10

GASTROTRICHA

William D. Hummon

10.1 Introduction

Gastrotricha are small free-living pseudocoelomates, which typically move about by means of ventral cilia. They vary from strap- to bottle-shaped animals and, as adults, range from 60 to 600 μm in total length, though adults of some species may reach several millimeters in length. Gastrotrichs are common members of both benthic and epiphytic meiofaunal communities in fresh water, estuarine, and marine habitats, often forming one of the numerically abundant components of these communities. Though bacteria probably form the largest single item in their diet, gastrotrichs appear to be more or less generalized omnivores.

There are two orders, the Macrodasyida and the Chaetonotida. The former is currently comprised of about 130 valid species in 25 genera which are grouped into six families (Dactylopodolidae, Lepidodasyidae, Macrodasyidae, Planodasyidae, Thaumastodermatidae, and Turbanellidae); the latter has some 260 valid species in 22 genera, grouped into seven families (Chaetonotidae, Dasydytidae, Dichaeturidae, Neodasyidae, Neogossiidae, Proichthydidae, and Xenotrichulidae). All macrodasyids plus members of two chaetonotid families, Neodasyidae and Xenotrichulidae, are marine. In addition, at least one member of each genus in the largest of the families, the Chaetonotidae, is marine, bringing the tally of marine species to nearly one-half of the total number of species known. Taxonomic work is proceeding, with major emphasis on the marine forms. This trend, if continued, will quickly increase the proportion of known marine to freshwater species above its current level.

10.2 Asexual Reproduction

Although asexual reproduction by fission or fragmentation is not known among gastrotrichs, most freshwater and some marine species appear to be obligate thelytokous parthenogens, that is, females are invariably produced from unfertilized eggs. However, in only one instance (Sacks, 1964) are data on this type of asexual reproduction supported by rigorous experimental evidence. In a few cases data are available by inference from clonal cultures (Packard, 1936; Brunson, 1949; Goldberg, 1949; Robbins, 1963) or from anatomical studies (Goldberg, 1949). In most cases, parthenogenesis is simply inferred from the absence of visible testes (all sexually reproducing forms reported to date are hermaphrodites) or from the taxonomic position of the animal in question (similarity of reproductive method is usually presumed among closely related taxonomic entities). The derivation of parthenogenesis from hermaphroditism rather than from separate sexes is attested to by the occasional discovery of a residual bursa in what otherwise would appear to be a parthenogenically reproducing female. The homolog of this organ is present in most hermaphroditic chaetonotids, along with rudimentary or complete male reproductive systems.

Hennig's (1966) contention that uniparental reproduction is a derived condition, referrable back to a larger biparental phylogenetic unit, is supported by evidence from various animal groups as summarized in Suomalainen (1950) and White (1954). This sort of assumption led Pennak (1963) to postulate a marine origin for the phylum Gastrotricha and led Hummon (1966) to postulate a marine origin for the largest chaetonotid family Chaetonotidae.

According to information presented by Sacks (1964), the meiotic cycle associated with parthenogenic reproduction in *Lepidodermella squammata* appears to consist of a first meiotic division, resulting in a first polar body, followed by a mitotic division. This is not an uncommon type of parthenogenic cycle (Suomalainen, 1950; White,

1954) and represents Route 1c of Beatty (1957): diploid parthenogenesis after suppression of the second polar body of the unfertilized egg. Parthenogenic gastrotrichs probably have the added complexity of polyploidy. Remane (1936) determined the diploid chromosome number to be eight in each of two hermaphroditic marine genera, *Dactylopodola* (family Dactylopodolidae) and *Platydasys* (family Thaumastodermatidae). On the other hand, Sacks (personal communication), studying chromosomes in Feulgen stained material from the freshwater parthenogenic *Lepidodermella squammata* (family Chaetonotidae), states that they are too small and numerous for an accurate count, but that there must be at least 20-30.

The genetic implications of parthenogenesis result from its trend toward increasing homozygosity, which in turn decreases phenotypic variability with successive generations (Suomalainen, 1950; White, 1954). Polyploidy when combined with obligate parthenogenesis functions in protecting more or less homozygous but successful strains from the phenotypic effects of heterozygosity reintroduced by mutation.

Of the various strains which develop, those which are incompatible with the selective environment are eliminated. Successful uniparental strains, possessing a unitary evolutionary role in a given adaptive zone, have qualities of a species, in the sense of Simpson (1961). They have continuity of inheritance, a pool of genes and a differential flow of genes from one environment to another. Based on criteria of morphological discontinuities clustered about adaptive peaks, Sonneborn (1957) and Mayr (1957, 1963, 1969) both agree to the taxonomic subdivision of obligate parthenogenic entities into species. Robbins (1967) in reviewing the uniparental species concept in the Gastrotricha adds his concurring opinion. In using the term evolutionary "blind alleys" with respect to these taxonomic entities, Darlington (1958) and White (1961) unfortunately fail to add that specialized adaptations can be expected to be shorter lived as a group than generalized adaptations. It is true that obligate parthenogenesis is adaptively far less flexible than most other reproductive systems, but it is also true that such species may persist as long as the environmental zone to which they are adapted persists. I believe, in the case of the chaetonotid gastrotrichs at least, that the presence and variety of obligate parthenogens testifies to the evolutionary stability of this adaptive zone.

Other aspects of the reproductive and developmental biology of parthenogenic gastrotrichs will be taken up in the appropriate subsections of Section 10.3.

10.3 Sexual Reproduction

10.3.1. Sex Determination and Hermaphroditism

While many species of Gastrotricha have been inadequately studied, available data indicate that sexual reproduction in the group is basically hermaphroditic. All of the families which are strictly marine apparently are also entirely hermaphroditic. Many members of the family Chaetonotidae are marine, but in only a few of these has hermaphroditism been demonstrated (Remane, 1936; Wilke, 1954; Hummon, 1966, 1969a; Renaud-Mornant, 1968). Interestingly, one freshwater member of the family Chaetonotidae, *Polymerurus oligotrichus*, has also been reported to be hermaphroditic (Remane, 1927, 1936).

Several questions have been raised about whether some species, such as in *Dactylopodola*, have sequential hermaphroditism, separate sexes, or both. Remane (1936) suggested that cases of males, hermaphrodites, and females, whose presence correlates with increasing body size, probably represent instances of sequential protandric hermaphroditism. It is likely that the case presented by Hummon (1966) for *Chaetonotus testiculophorus*, but thought by Renaud-Mornant (1967) to be a seasonal proliferation of testes, belongs in this category (Renaud-Mornant, 1968). Recently, Teuchert (1968) has shown that the situation for the group as a whole is quite complex. Some species are now known to be simultaneous hermaphrodites, others simple sequential hermaphrodites and still others alternating sequential hermaphrodites. Both protandry and protogyny have been found, as have cases of overlapping and more or less nonoverlapping sequences of alternation.

In explaining the evolution of hermaphroditism, Ghiselin (1969) has touched upon possible selective pressures leading from simultaneous hermaphroditism to sequential hermaphroditism and parthenogenesis. Of the three models he postulated, two—size advantage and low density—are relevant to this issue. Simple sequential hermaphroditism might result if, under low density conditions, the reproductive potential could be increased by shortening the life cycle. The life cycle could be compressed by: (1) reduction of body size of the ovigerous female without altering the growth rate, (2) speeding of the growth rate leading to the former body size, or (3) a combination thereof. Both the first and third of these alternatives seem to have been successfully adopted by at least some Gastrotricha. While body size of the ovigerous female can be reduced somewhat without necessitating a complete alteration in the reproductive processes, sizable reductions are often accompanied by changes resulting from lack of internal body space. One adjustment would be precocious development of the testes, retaining the hermaphroditic reproductive capabilities while allowing the full spatial resources of the adult to be used in ovaproduction. Another adjustment, which might follow directly from the previous one, would be the elimination of male parts altogether, resulting in parthenogenic reproduction. If this last step actually occurred in the gastrotrichs, it would be difficult to say where and when it happened. There is little evidence to indicate whether the presumed parthenogenic marine chaetonotids developed parthenogenesis in the marine habitat prior to invasion of fresh water, or whether they are the result of reinvasion of marine waters after the parthenogenic mode of reproduction had developed in the freshwater habitat.

Alternating sequential hermaphroditism, though not considered by Ghiselin (1969), would best fit his size advantage model. I think it can be shown that in a tiny but elongate animal, producing large, moderately yolky eggs which develop directly into free-swimming juveniles without the benefit of a larval stage, economy of reproductive effort can be achieved by pulsating reproductive activity. If, as was argued earlier, for a certain reproductive effort a life cycle can be shortened by shifting from simultaneous to simple sequential hermaphroditism, the converse is also true, that for a constant life cycle, reproductive effort can be increased by shifting from simultaneous to alternating sequential hermaphroditism. Given a specific amount of time, space and energy, total gamete output can plausibly be maximized by utilizing wavelike pulses wherein the combination of vitellogenesis and spermatogenic divisions alternate with the combination of ovogenic divisions and spermiogenesis.

10.3.2. Anatomy of the Reproductive System

Reproductive organs are rather simple. The female reproductive system (Fig. 1) is made up at most of single or paired ovaries, single oviduct (not uterus), seminal receptacle, copulatory bursa, and feminine antrum, paired glands and gonopore, though the entire sequence is seldom present in a single species. The ovary consists of a cluster of more or less distinct oogonia and small oocytes, and may be sheathed by the same delicate oviductal membrane in which the developing oocytes lie. Ovaries are posterior to the oviduct, except in the Lepidodasyidae, Neodasyidae, and Planodasyidae, and are

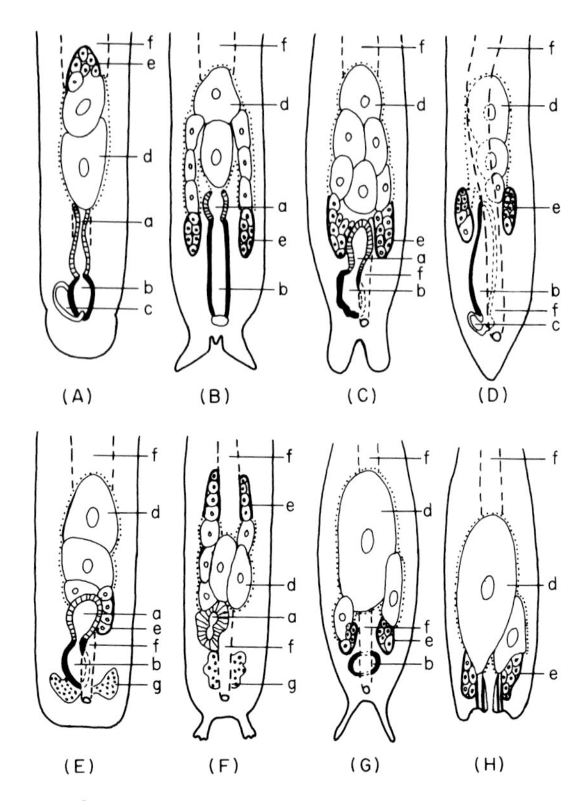

FIG. 1. Schematic diagram of female genital apparatus (dorsal view): (A) *Cephalodasys*, (B) *Turbanella*, (C) *Dactylopodola*, (D) *Macrodasys*, (E) Thaumastodermatidae, (F) *Neodasys*, (G) *Chaetonotus*, (H) *Neogossea*. a, Seminal receptacle; b, copulatory bursa; c, feminine antrum; d, large primary oocyte in oviduct; e, oocyte proliferation site of ovary; f, intestine; g, gland. [Redrawn from Remane (1936).]

paired, except in the Lepidodasyidae and Thaumastodermatidae. The relationship between oviduct, seminal receptacle, and copulatory bursa varies, assuming all are present. They can lie in a linear series with sperm penetrating from gonopore through bursa and receptacle to the oviduct and the eggs being spawned in reverse from oviduct through receptacle, bursa, and gonopore, or as in most cases simply released by rupture of the body wall; they can lie in a cyclic arrangement, with sperm penetrating from gonopore through bursa and receptacle to the oviduct, but with the eggs being spawned directly from the oviduct through the gonopore; or, finally, they can lie in a V-shaped arrangement, with sperm being stored in a blind receptacle after penetrating from gonopore through the bursa and the eggs being spawned directly from the oviduct through the gonopore.

In most cases the oviduct lies dorsal to the intestine, though in some cases the two are reversed in position. A feminine antrum, if present, is generally located between the base of the copulatory bursa and gonopore; glandular tissue, if present, lies near the gonopore, but is of unknown function. A medial female gonopore is located posterioventrally, often in the vicinity of the anus.

The male reproductive system (Fig. 2) is usually made up of elongate testes, vasa deferentia and gonopore. Testes are located in the trunk region, adjacent to the fore part of the intestine, and, except in most Thaumastodermatidae, where only the right one occurs, they are paired. Testes are more or less rudimentary when present in the Chaetonotidae and are not known from members of the Dasydytidae, Dichaeturidae, and Proichthydidae. The vas deferens is a simple tube leading from testis to gonopore, though in some cases a seminal vesicle is present along its length. A medioventral male gonopore is generally present, its location ranging from the region of the anterior end (Turbanellidae) to that of the posterior end of the intestine (Lepidodasyidae, Thaumastodermatidae). In some Lepidodasyidae and Macrodasyidae, the vasa deferentia may terminate in a penis, which is capable of being protruded through the male gonopore. Testes of the Neodasyidae differ in opening via short vasa deferentia through separate mediolateral gonopores. A more detailed review of gastrotrich reproductive anatomy can be found in Remane (1936).

10.3.3 Gametogenesis and Gametogenic Cycles

Neither karyokinesis nor cytokinesis has been seen in the ovary during the posthatching growth period, implying that the individual oogonia are produced during embryogenesis and that their number is thereafter fixed. Oogonium numbers range from 4-5 to 15-22, with the maximum occurring in the families Dactylopodolidae and Turbanellidae (Remane, 1936).

Information is unavailable concerning vitellogenesis. Up to one-third of the body volume is comprised of ovarian yolk in some ripe chaetonotids, despite the fact that large vitellaria are not identifiable. While a lesser proportion of the total body volume is represented by yolk in macrodasyids, ths actual amount of yolk may exceed that found in chaetonotids.

Vitellogenesis initially proceeds along with growth, and as oocytes increase in size they move out into the oviduct. This is generally a sequential process, with those located furthest in the oviduct containing the most yolky material. It usually involves only a few oo-

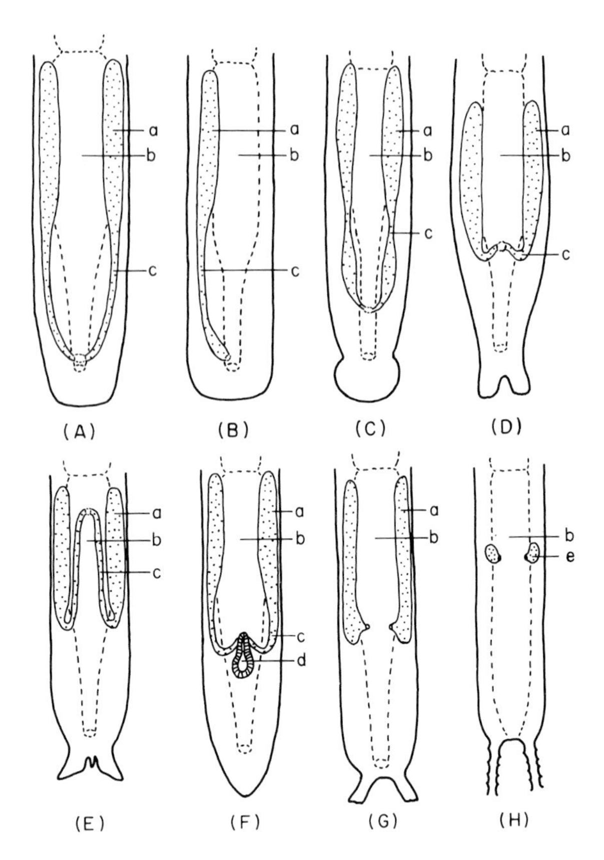

FIG. 2. Schematic diagram of male genital apparatus (dorsal view): (A) *Cephalodasys*, (B) Thaumastodermatidae, (C) *Lepidodasys*, (D) *Dactylopodola*, (E) *Turbanella*, (F) *Macrodasys*, (G) *Neodasys*, (H) *Polymerurus oligotrichus*. a, Testis; b, intestine; c, vas deferens; d, penis; e, rudimentary testis. [Redrawn from Remane (1936).]

cytes at a time, typically one to several in macrodasyids and one in chaetonotids. Occasionally chaetonotids show what is probably aberrant development, with 5-9 oocytes concurrently receiving nearly equal amounts of yolk material (Metschnikoff, 1864; Zelinka, 1889; Remane, 1926; Hummon, unpublished data). Developing oocytes are round to elipsoid and contain a sizable germinal vesicle (12-16 μm in diameter) prior to spawning. When full grown, oocytes measure approximately 60 × 40 μm in both macrodasyids and chaetonotids, though those of some genera vary sufficiently to provide a long axis ranging from 120 μm in *Mesodasys* to only 35 μm in *Dactylopodola* (Teuchert, 1968). Each full grown oocyte is surrounded by a thin, clear membrane.

Spawning takes place while the germinal vesicle is present, with meiosis following shortly thereafter. Teuchert (1968) is of the opinion that meiosis has been completed at the time of spawning, though the description and illustrations of events which follow spawning indicate that each of her "Blasenbildung" stages probably corresponds to one of the meiotic divisions. This interpretation is consistent with data given by Sacks (1955) for incomplete meiosis in the parthenogenic chaetonotid *Lepidodermella squammata* (Section 10.2). It would be interesting to know more about the processes which are encompassed by the Blasenbildung stages in the hermaphroditic forms: (a) what happens to the sperm pronucleus after penetration, (b) when does fusion of the pronuclei occur, and (c) does the clockwise meridional rotation of the mitotic spindle prior to first cleavage occur in hermaphroditic forms in the same manner as reported by Sacks (1955) and noted subsequently by myself for parthenogenic forms?

Testicular development apparently takes place with general body maturation. During this period the testes develop from compact organs, filled with spermatocytes, into hollow tubes filled with sperm, which result from accompanying spermatogenesis. Sperm often reach 120-180 μm in length (in an animal whose total length is perhaps 160-320 μm) and lie parallel in packets or bundles which are often looped back upon themselves in the testis. A short acrosome is followed by an elongate midpiece, having a spiralled undulating membrane, and a tail piece of moderate length.

Though little is known of special environmental factors which may induce gametogenesis, several, such as salinity, temperature, and moisture, can inhibit reproduction if not at or near optimal levels. Some species appear to be strictly marine in distribution, as for instance *Turbanella ambronensis* (synonyms: *T. italica*, *T. cirrata*, and *T. digitifera*) (Remane, 1943; Gerlach, 1953; Papi, 1957; Renaud-Debyser, 1964; d'Hondt, 1965, 1968; Hummon, 1969b; Schmidt and Teuchert, 1969). Members of this species from Woods Hole, when acclimatized at approximately 30 ‰, have physiological tolerances for exposure to salinities ranging from 10 to 50 ‰ for 24 hours (Hummon, unpublished data). The restricted distribution of this species indicates that its reproductive tolerances and habitat preferences are more narrowly restricted with respect to salinity. Other species have broader, marine-estuarine distributions. *Turbanella cornuta* from Woods Hole, for example, when acclimatized at approximately 30 ‰, can withstand salinities ranging from 6 to 60 ‰ for 24 hours (Hummon, unpublished data). The Elbe River population of *T. cornuta*, however, is known to live and presumably to reproduce in salinities as low as 1 ‰ (Riemann, 1966). Regardless of the set of factors whose values make up the niche hypervolume for a species (see Hutchinson, 1957), it is clear that "reproductive hypervolume" does not exceed and in many cases is considerable smaller than "physiological hypervolume" for members of the same population. Knowledge is lacking with regard to the effects of long term acclimatization on niche hypervolume in gastrotrichs and with respect to the effect of parameter shifts on gametogenesis.

10.3.4 Reproductive Behavior, Mating, and Spawning

The usual method of sperm transfer in macrodasyid gastrotrichs is copulation, the account which follows being based on that of Teuchert (1968) for *Turbanella cornuta*. An animal in male phase raises its posterior end and waves it perpendicularly to and fro (Fig. 3A). This behavior is repeated periodically until responded to by an animal in female phase. The two reproductive partners entwine their abdomens (Fig. 3B) and, with anterior adhesive tubes attached to the substratum, draw in opposing directions until their intercoiled abdomens are stretched considerably in length (Fig. 3C). Finally, after approximately a minute, the two animals loosen their hold and swim away with abdomens still coiled (Fig. 3D), each returning to its normal shape within several minutes. Although the act of sperm transfer has not been seen, it apparently occurs during the intercoiling period when reproductive openings of the respective partners come in close proximity to one another. Shortly after copulation sperm can be seen in the seminal receptacle, though not yet in the oviduct. Animals in male phase seem capable of copulating several times, since they do not release all their mature sperm at once.

Dactylopodola baltica, on the other hand, represents a smaller group of gastrotrichs which transfer sperm encompassed in spermatophores (Teuchert, 1968). The spermatophores, constant in size (26-30 μm long) and shape, are located adjacent to the copulatory bursa, dorsolaterally on the right side, midway down the length of the intestine. An animal in male phase apparently seeks out another in female phase, extrudes the spermatophore and attaches it in a comparable position on the second animal. Less often the spermatophore is attached to a sand grain, later to be removed by an animal in female phase. Sperm then penetrate to the copulatory bursa from which they migrate to the oocyte. Some 20-25 days are necessary for sex reversal from male to female phase.

Eggs are typically released to the exterior and are attached to a sand grain or other piece of substratum by means of a clear adhesive material secreted from epidermal or ovarian gland cells. Oviposition

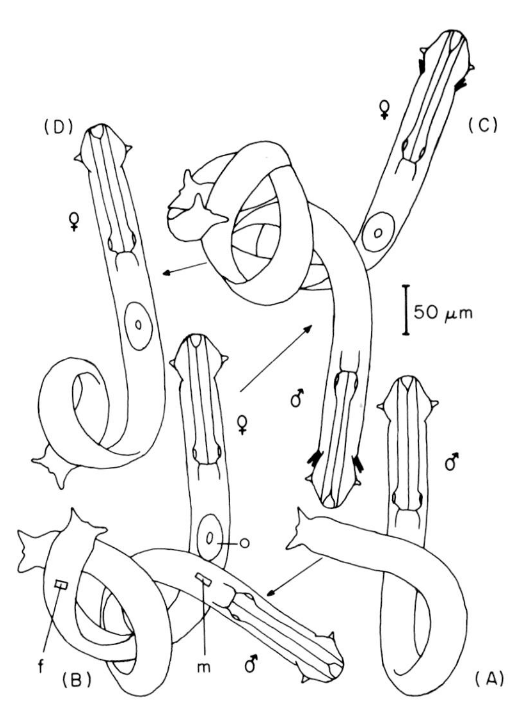

FIG. 3. Copulation in *Turbanella cornuta*. See text for explanation. ♂, animal in male phase; ♀, animal in female phase; m, male gonopore; f, female gonopore; o, oocyte. [Modified from Teuchert (1968).]

occurs by means of contracting muscle bands which force the release of one or more eggs, while the animal is attached to the substratum by means of its adhesive tubules. In most macrodasyids oviposition is accomplished by rupture of the body wall (Teuchert, 1968). It is possible that some armored forms, such as *Acanthodasys* or members of the family Thaumastodermatidae, have an oviduct opening to the exterior as in most chaetonotids, through which the egg is released. In one species, *Urodasys viviparus*, the egg is retained within the oviduct, where the embryo develops ovoviviparously, and the juvenile is live born. Occasionally, oviposition is aborted in other species and development proceeds within the oviduct (Sacks, 1955). The results of these abnormal processes are not given.

Once deposited, the egg, at least in freshwater chaetonotids, produces an outer shell which often expands to form humps, spines or other outward projections. The outer surfaces of developing marine eggs are usually smooth or at most granular. So far as is known the division into two types of eggs, termed tachyblastic and opsiblastic by Brunson (1949), is restricted to parthenogenic forms. Others have only tachyblastic eggs which begin development immediately after oviposition. It is perhaps the opsiblastic, or late developing eggs, which provide the only record of gastrotrich activity in the recent geologic past. A summary of reports from lake sediments can be found in Colinvaux (1964) and Frey (1964).

Swedmark (1958) reported reproductive activity throughout the year for most species of interstitial fauna on the Brittany coast of France, with some reduction in intensity during the winter months. Only a few species are limited to late spring and summer reproduction. Teuchert (1968) indicated that gastrotrichs on the Isle of Sylt along the North Sea are reproductive from March to October. Species such as *Mesodasys laticaudatus*, *Macrodasys caudatus*, *Dactylopodola baltica*, and *Acanthodasys aculeatus* reproduce more or less continuously during this period, whereas others such as *Turbanella cornuta*, *T. hyalina*, and *Cephalodasys maximus* appear to have peak reproductive periods in late spring and fall.

My work with marine gastrotrichs from Woods Hole, Massachusetts (1969b), indicates that many if not most species reproduce throughout the year. Based on gross population numbers, there are three basic patterns of peak density present. One species, *Tetranchyroderma papii,* is multicyclic, with maximum densities occurring whenever there is a relaxation of environmental severity. Other species are unicyclic. *Heteroxenotrichula squamosa* and *Xenotrichula beauchampi,* for instance, have late summer-early fall peaks, whereas *Cephalodasys* sp., *Pseudostomella roscovita*, *Turbanella ambronensis*, *Halichaetonotus batillifer*, and *Xenotrichula sp.* have late spring peaks. In each case density seems to be controlled more by changes in mortality rate than by changes in reproductive rate, though rate differences in both mortality and reproduction vary from species to species.

10.4 Development

10.4.1 Embryonic Development

Embryonic development of parthenogenic chaetonotids is described in the works of Beauchamp (1930), Brunson (1949), and Sacks (1955); that for macrodasyids is given by Swedmark (1955) and Teuchert (1968). Nothing comparable has as yet been published on the hermaphroditic chaetonotids.

The following discussion utilizes Sack's modification of the Kofoid

(1894) cell numbering system. The letter designates descendants of each of the blastomeres resulting from second cleavage; the unit number designates the number of cleavages which have occurred; and the decimal designates the position of the blastomere in the embryo, lower numbers referring to blastomeres lying more anterior, ventral or to the right. This system was also used by Teuchert.

First cleavage is equatorial and adequal (Fig. 4A,M), future anterior end of the animal being marked by the presence of polar bodies (Fig. 4M). Second cleavage is meridional, adequal, and slightly asynchronous, with the anterior blastomere cleaving slightly in advance of the posterior. This asynchrony is ever more pronounced with succeeding cleavages.

Between the first and fourth cleavages there occurs a 45° clockwise equatorial rotation of the anterior blastomeres relative to the posterior blastomeres (Fig. 4B,N). Teuchert (1968) reports it as occurring prior to second cleavage in *Turbanella cornuta, Macrodasys caudatus,* and *Cephalodasys maximus.* Brunson (1949) observed it during or after second cleavage but prior to third cleavage in *Lepidodermella squammata* and *Chaetonotus tachyneusticus,* while Sacks (1955) noted its occurrence after second but prior to third cleavage in *L. squammata.* My own observations indicate that, while this rotation usually occurs early, it can even occur during the fourth cleavage in *L. squammata,* confirming Beauchamp's original observation of relatively late rotation in *Neogossea antennigera.*

Some basic differences now occur between embryos of the two groups. In macrodasyids the posterior duet shifts forward, with its c blastomere joining the anterior duet dorsally to form a triplet and leaving the d blastomere posteriorly (Fig. 4C,D). No such migration occurs in parthenogenic chaetonotids. The critical part played by early rotation of the blastomeric duets in macrodasyids can now be appreciated, whereas chaetonotid development remains constant regardless of when between first and fourth cleavage this rotation occurs.

Third cleavage, at least in the a and b lineages, is equatorial in macrodasyids. After the cleavage is complete (Fig. 4E,F), the d3.1 daughter blastomere (=entoblast) shifts forward into a medioventral position, leaving only its d3.2 counterpart posteriorly (Fig. 4G,H). The results are two dorsal, five frontal, and one ventral blastomeres. No such shift occurs in parthenogenic chaetonotids, where third cleavage is meridional. Two quartets result, one anterior and one posterior, which are oriented at a 45° angle to one another (Fig. 4O,P). This and subsequent cleavages are indicative of a modified

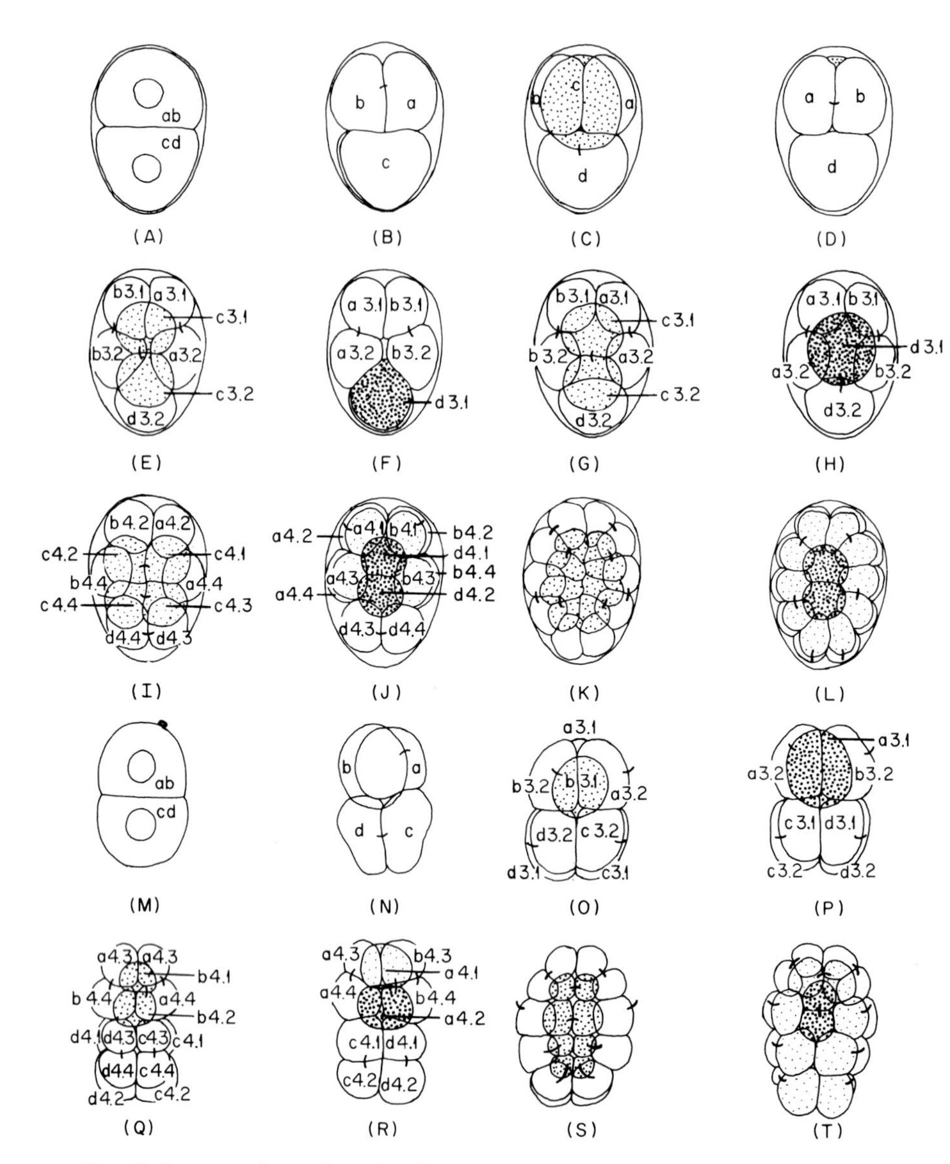

FIG. 4. Stages in the early embryology of *Turbanella cornuta* (A-L) and *Lepidodermella squammata* (M-T); (A,M) dorsal view, 2 blastomeres; (B,N) dorsal view, 4 blastomeres after equatorial rotation; (C) dorsal view, (D) ventral view, 4 blastomeres after anterior shifting of the c blastomere; (E,O) dorsal view, (F,P) ventral view, 8 blastomeres; (G) dorsal view, (H) ventral view, 8 blastomeres after anterior shifting of the d3.1 blastomere; (I,Q) dorsal view, (J,R) ventral view, 16 blastomeres; (K,S) dorsal view, (L,T) ventral view, 30 or 32 blastomeres. Primary ectoderm, medium stipples; secondary ectoderm, without stipples; mesectoderm, fine stipples; and entoderm, coarse stipples. Anterior-posterior axis, up and down respectively. [(A-L) Redrawn from Teuchert (1968); (M-T) Redrawn from Sacks, (1955).]

radial cleavage pattern. Despite rotation, there is no indication of spirally canted spindles in subsequent cleavages.

Fourth and fifth cleavages are apparently of mixed orientation in both groups but serve to continue the trends initiated in second and third cleavages. Bilateral symmetry is firmly established and presumptive cell layers are clearly identifiable (Fig. 4I-L, Q-T). A blastocoel is also formed, but is reduced quickly in size during the process of gastrulation.

While subsequent embryology in the two groups is parallel in many respects, the blastomeres participating in the formation of germ layers have radically different origins (see Table I). Teuchert's suggestion, that a reversal of orientation may have occurred between the two groups, would solve the problem of entodermal origin since both would then be rooted in the d series. Unfortunately this postulation raises even more serious problems, since it would also necessitate a very unlikely reversal of anterior and posterior ends from one group to the other!

Gastrulation begins with the invagination of two entoblast daughter cells d4.1, 4.2 in macrodasyids and a5.3, 5.4 in parthenogenic chaetonotids, and is followed by epibolic overgrowth of presumptive mesodermal cells by the secondary ectoderm. In macrodasyids an elongate blastoporal furrow develops initially (Fig. 5A). This furrow slowly closes from posterior to anterior, forming a V-shaped stomodeal opening (Fig. 5B) of ever decreasing size until only the mouth opening remains. At no time in Teuchert's study was a proctodeal invagination detected. In parthenogenic chaetonotids a blastopore develops anteriorly. This stomodeal opening is followed by a secondary invagination of posterior cells forming an inverted-V-shaped proctodeal opening (Fig. 5C). The blastocoel is largely occluded at this time by differentiating pharyngeal, intestinal and mesodermal tissues, resulting in stereogastrulae for both groups of gastrotrichs. One need not, however, agree with Teuchert that the body cavity which later develops is a schizocoel. Spaces which she considers to be histogenic artifacts in her preparations are probably remnants of the blastocoel lying between distinct tissue layers that are simply appressed tightly against one another.

The first growth phase includes continued tissue differentiation, elongation and initial flexion of the body. Flexion occurs by means of a transverse fold located in the posterior pharyngeal region and is completed during the second growth phase, which is characterized by continued elongation and organogenic development.

Rates of embryonic development vary by species, temperature, and probably with a host of other environmental parameters. Teuchert (1968) gives time intervals for macrodasyids from spawning to

TABLE I
BLASTOMERIC ORIGINS OF BASIC GERM LAYERS IN EACH OF TWO GROUPS OF GASTROTRICHS[a]

Germ layer	Origin: macrodasyids		Origin: parthenogenic chaetonotids	
Primary ectoderm	a5 None c5.1–5.8	b5 None d5 None	a5 None c5.6,5.8	b5.1–5.4 d5.6,5.8
Secondary ectoderm	a5.3,5.4,5.7,5.8 c5 None	b5.3,5.4,5.7,5.8 d5.7,5.8	a5.5,5.6,5.8 c5.2,5.4,5.5,5.7	b5.5,5.6,5.8 d5.2,5.4,5.5,5.7
Mesectoderm	a5.1,5.2,5.5,5.6	b5.1,5.2,5.5,5.6	a5.1,5.2,5.7	b5.7
	c5 None	d5.5,5.6	c5.1,5.3	d5.1,5.3
Entoderm	a5 None c5 None	b5 None d5.1–5.4	a5.3,5.4 c5 None	b5 None d5 None

[a] After Sacks (1955) and Teuchert (1968).

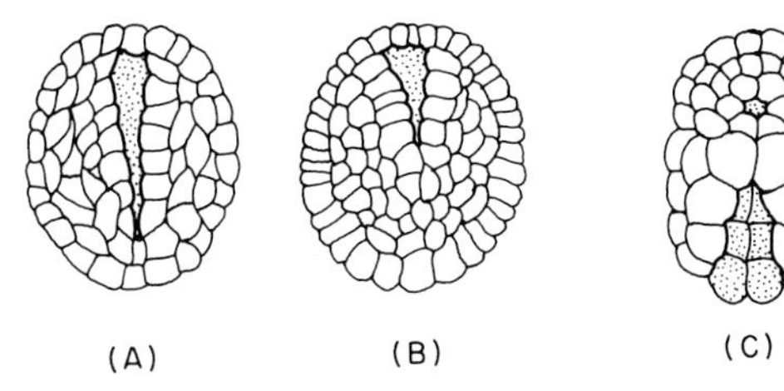

FIG. 5. Stages in the later embryology of *Turbanella cornuta* (A,B) and *Lepidodermella squammata* (C). (A) ventral view, early blastoporal furrow, (B) ventral view, later, partially closed blastoporal furrow, (C) ventral view, stomodeal (blastoporal) opening anteriorly and proctodeal invagination posteriorly. Anterior-posterior axis, up and down respectively. [(A,B) Redrawn from Teuchert (1968), (C) Redrawn from Sacks, (1955).]

hatching at 14-18 ⁰/oo salinity, as shown in Table II. By way of contrast, Swedmark (1955) listed 10 days for the development of *Macrodasys affinis* at 14°C and approximately 30 ⁰/oo salinity. For the freshwater parthenogenic chaetonotid *Lepidodermella squammata*, Brunson (1949) gives a mean of 1.7 days between spawning and hatching at 20°C, while Sacks (1964) lists a mean of 1.1 day for embryonic development at 22°C. Interestingly, both Brunson and Sacks indicate that sexual maturity is reached within 24 hours after hatching in *L. squammata*. Brunson gives the posthatching growth rate as 4 μm/hour at 20°C, while Sacks lists it as 3 μm/hour at 22°C.

10.4.2 Postembryonic Development

Development is direct, the juvenile which results being more or less a miniature of the adult. At the time of hatching all of the adult features necessary to the success of the juvenile are present. Those associated with coordination, locomotion, and feeding are particularly well-developed. Other features, such as size of contractile elements and protective cuticular elaborations, and number of locomotor, sensory, adhesive, and osmoregulatory units, can be expected to increase with body size, but are already functional at time of hatching.

Growth to adulthood includes eidostic changes, those involving body shape and proportions (Hummon, 1971), as well as the addition of metric and meristic increments. An analysis of metric and eidostic changes associated with an ontogenetic growth series was completed for *Chaetonotus testiculophorus* by Hummon (1966). Information on meristic and eidostic changes between juveniles and adult stages is included in Teuchert (1968) for *Turbanella cornuta*. A juvenile, 150 μm long, may bear <22 bundles of ventral locomotor cilia, at

TABLE II
RATES OF EMBRYONIC DEVELOPMENT[a]

Species	Temperature, °C		
	12-14	17-18	20-22
Turbanella cornuta	9-10 Days	4 Days	3 Days
Cephalodasys maximus		7-8 Days	
Dactylopodola baltica		7-8 Days	
Macrodasys caudatus		9-10 Days	

[a] After Teuchert (1968).

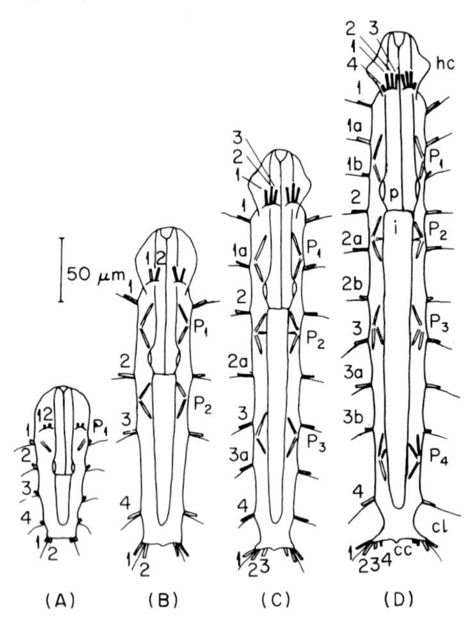

FIG. 6. Eidostic and meristic development of *Turbanella cornuta*. (A) embryo (120 μm long), (B) juvenile (240 μm long), (C) subadult (330 μm long), (D) young adult (400 μm long). Primary adhesive tubules: 1-4 lateral, 1, 2 anterior and posterior. Secondary adhesive tubules: 1a-3a lateral, 3 anterior and posterior. Tertiary adhesive tubules: 1b-3b lateral, 4 anterior and posterior. p, Pharynx; i, intestine; hc, lateral head cone; cl, caudal lobe; cc, medial caudal cone. P1-P4, protonephridial groups. [Redrawn from Teuchert (1968).]

4-6 cilia (approximately 8 μm long) per bundle and 16 adhesive tubules, whereas an adult, 700 μm long, may bear <160 bundles of cilia at 10-14 cilia (25-30 μm long) per bundle and <210 adhesive tubules (Fig. 6). Growth is centered mainly in elongation of the intestinal region and accessory structures, such as head cones and caudal lobes.

TABLE III
LIFE TABLE FOR LEPIDODERMELLA SQUAMMATA[a,b]

x	l_x	d_x	Q_x	L_x	T_x	e_x	m_x	$l_x m_x$
embryo	1000	0	0	1000	11332	11.33	0	–
0-1	1000	0	0	1000	10232	10.23	0.04	0.04
1-2	1000	0	0	1000	9232	9.23	1.22	1.22
2-3	1000	13	13	993.5	8232	8.23	1.29	1.28
3-4	987	13	13	980.5	7238.5	7.33	0.82	0.80
4-5	974	39	40	954.5	6258	6.44	0.33	0.32
5-6	935	79	84	895.5	5303.5	5.68	0.07	0.06
6-8	856	198	232	757	4408	5.10	0	—
8-10	658	224	340	546	2894	4.40	0	—
10-12	434	118	272	375	1802	4.20	0	—
12-14	316	79	250	276.5	1052	3.33	0	—
14-16	237	132	558	171	499	2.10	0	—
16-18	105	79	772	65.5	157	1.50	0	—
18-20	26	26	1000	13	26	1.00	0	—
	0	Σ1000						$R_o = \Sigma 3.72$

[a] Based on data from Sacks (1964).

[b] Key to symbols:

x=age class in days

l_x=survivorship (number alive at beginning of period)

d_x=mortality (number dying during period)

Q_x=age specific mortality rate (death/1000 living at beginning of period)

L_x=number alive at midperiod $[\frac{1}{2}(l_x+l_{x+1})]$

T_x=gross number of days to be lived (Σ(d)L_x to age class x, where (d) represents the number of days in the period)

e_x=mean age expectation in days at beginning of period (T_x/l_x)

m_x=age specific fecundity (birth rate/individual living at end of period)

R_o=net reproductive rate/individual lifetime

Indeed, usually there is sufficient consistency of features so that Renaud-Mornant (1968) was able to describe adequately a new species from juvenile to subadult specimens. The one system which is not developed at hatching, of course, is the reproductive system.

Though no data are available regarding quantitative life histories of macrodasyids, I have prepared a life table (Table III) from information given by Sacks (1964) for the parthenogenic chaetonotid, *Lepidodermella squammata*. The table is based on animals cultured at 22°C, and data are grouped by one day intervals during the reproductive phase of life and two day intervals thereafter. Summarizing its results, the life expectancy (e_x) of an individual at spawning is 11.33 days. At hatching it is 10.23 days, and after having achieved its net reproductive rate (R_0) of 3.73 ova its life expectancy is 5.10 days. The maximum life span of an individual was 19 days. The postreproductive phase is probably absent in most members of natural populations, having been sacrificed to additional mortality, the causes of which are difficult to quantify. This sort of information however is most useful and, if available for natural populations of parthenogenic and hermaphroditic chaetonotids and macrodasyids, would contribute greatly to our knowledge of the reproductive biology of the marine Gastrotricha.

10.5 References

Beatty, R.A. (1957). "Parthenogenesis and Polyploidy in Mammalian Development." Cambridge Univ. Press, London and New York.

Beauchamp, P. de (1930). Le développement des Gastrotriches (Note préliminaire). *Bull. Soc. Zool. Fr.* **54**, 549-558.

Brunson, R.B. (1949). The life history and ecology of two North American gastrotrichs. *Trans. Amer. Microsc. Soc.* **68**, 1-20.

Colinvaux, P.A. (1964). The environment of the Bering land bridge. *Ecol. Monographs* **34**, 297-329.

Darlington, C.D. (1958). "Evolution of Genetic Systems," 2nd rev. ed. Oliver and Boyd, Edinburgh.

Frey, D.G. (1964). Remains of animals in quaternary lake and bog sediments and their interpretation. *Arch. Hydrobiol. (Beih. Ergeb. Limnologie)* **2**, 1-114.

Gerlach, S. (1953). Gastrotrichen aus dem Küstengrundwasser des Mittelmeeres. *Zool. Anz.* **150**, 203-211.

Ghiselin, M.T. (1969). The evolution of hermaphroditism among animals. *Quart. Rev. Biol.* **44**, 189-208.

Goldberg, R.J. (1949). Notes on the biology of a common gastrotrich of the Chicago area *Lepidoderma squamatum* (DuJardin). *Trans. Illinois Acad. Sci.* **42**, 152-155.

Hennig, W. (1966). "Phylogenetic Systematics". Univ. Illinois Press, Urbana, Illinois.

d'Hondt, J.-L. (1965). Coup d'oeil sur les Gastrotriches Macrodasyoides du Bassin d'-Arcachon. *Actes Soc. linnéenne Bordeaux (Ser. A)* **102** (16), 1-16.

d'Hondt, J.-L. (1968). Contribution à la connaissance des Gastrotriches intercotidaux du Golfe de Gascogne. *Cahiers Biol. Mar.* **9**, 387-404.

Hummon, W.D. (1966). Morphology, life history, and significance of the marine gastrotrich, *Chaetonotus testiculophorus* n. sp. *Trans. Amer. Microsc. Soc.* **85**, 450-457.

Hummon, W.D. (1969a). *Musellifer sublitoralis*, a new genus and species of Gastrotricha from the San Juan Archipelago, Washington. *Trans. Amer. Microsc. Soc.* **88**, 282-286.

Hummon, W.D. (1969b). Distributional Ecology of Marine Interstitial Gastrotricha from Woods Hole, Massachusetts, with Taxonomic Comments on Previously Described Species. Ph.D. Thesis, Univ. of Massachusetts.

Hummon, W. D. (1971). The marine and brackish water Gastrotricha in perspective. *In* "Proceedings of the First International Conference on Meiofauna" (N. C. Hulings, ed.), Smithsonian Contrib. Zool. No. 76, 21-23.

Hutchinson, G.E. (1957). Concluding remarks. *Cold Spring Harbor Symp. Quant. Biol.* **22**, 415-427.

Kofoid, C.A. (1894). On some laws of cleavage in *Limax*. A preliminary notice. *Proc. Amer. Acad. Arts Sci.* **29**, 180-203.

Mayr, E. (1957). Difficulties and importance of the biological species concept. *In* "The Species Problem," pp. 371-388. Amer. Ass. Advan. Sci., Washington, D.C.

Mayr, E. (1963). "Animal Species and Evolution." Harvard Univ. Press, Cambridge, Massachusetts.

Mayr, E. (1969). "Principles of Systematic Zoology." McGraw-Hill, New York.

Metschnikoff, E. (1864). Über einige wenig bekannte niedere Thierformen. *Z. Wiss. Zool.* **15**, 450-463; taf. XXXV.

Packard, C.E. (1936). Observations on the Gastrotricha indigenous to New Hampshire. *Trans. Amer. Microsc. Soc.* **55**, 422-427.

Papi, F. (1957). Tre nuovi Gastrotrichi mediterranei. *Pubbl. Staz. Zool. Napoli* **30**, 177-182.

Pennak, R.W. (1963). Ecological affinities and origins of free-living acelomate freshwater invertebrates. *In* "The Lower Metazoa" (E.C. Dougherty, ed.), pp. 435-451. Univ. California Press, Berkeley, California.

Remane, A. (1926). Zur Frage der Summereier der Gastrotrichen. *Zool. Anz.* **69**, 54-56.

Remane, A. (1927). Beiträge zur Systematik der Süsswasser-gastrotrichen. *Zool. Jahrb. (Abt. Systematik)* **53**, 269-320.

Remane, A. (1936). Gastrotricha. *In* "Klassen und Ordnungen des Tierreichs" (H.G. Bronn, ed.) Band 4, Abt. II, Buch 1, Teil 2, Liefrungen 1-2, pp. 1-242. Akad. Verlagsges., Leipzig.

Remane, A. (1943). *Turbanella ambronensis* nov. spec., ein neues Gastrotrich aus der Otoplannenzone der Nordsee. *Zool. Anz.* **141**, 237-240.

Renaud-Debyser, J. (1964). Note sur la faune interstitielle du Bassin d'Arcachon et description d'un Gastrotriche nouveau. *Cahiers Biol. Mar.* **5**, 111-123.

Renaud-Mornant, J. (1967). *Heterolepidoderma foliatum* n. sp. (Gastrotricha, Chaetonotidae), des faciès saumâtres du Bassin d'Arcachon. *Bull. Soc. Zool. Fr.* **92**, 161-166.

Renaud-Mornant, J. (1968). Présence du genre *Polymerurus* en milieu marin, description de deux espèces nouvelles (Gastrotricha, Chaetonotoidea). *Pubbl. Staz. Zool. Napoli* **36**, 141-151.

Riemann, F. (1966). Die interstitielle Fauna im Elbe-Aestuar Verbreitung und Systematik. *Arch. Hydrobiol.* **31** *(Suppl.)*, 1-279.

Robbins, C.E. (1963). Studies on the Taxonomy and Distribution of the Gastrotricha of

Illinois. Ph.D. Thesis, Univ. Illinois.
Robbins, C.E. (1967). The uniparental species concept in the Gastrotricha. Unpublished manuscript, presented at the American Society of Zoologists symposium on the species concept among uniparental animals. *Amer. Zool.* **7**, 177.
Sacks, M. (1955). Observations on the embryology of an aquatic gastrotrich, *Lepidodermella squammata* (Dujardin, 1841). *J. Morphol.* **96**, 473-484, pl. 1-5.
Sacks, M. (1964). Life history of an aquatic gastrotrich. *Trans. Amer. Microsc. Soc.* **83**, 358-362.
Schmidt, P., and Teuchert, G. (1969). Quantitative Untersuchungen zur Ökologie der Gastrotrichen im Gezeiten-Sandstrand der Insel Sylt. *Mar. Biol.* **4**, 4-23.
Simpson, G.G. (1961). "Principles of Animal Taxonomy." Columbia Univ. Press, New York.
Sonneborn, T.M. (1957). Breeding systems, reproductive methods, and species problems in Protozoa. *In* "The Species Problem" (E. Mayr, ed.), pp. 151-324. Amer. Ass. Advan. Sci., Washington, D.C.
Suomalainen, E. (1950). Parthenogenesis in animals. *In* "Advances in Genetics" (M. Demerec, ed.), Vol. 3, pp. 193-253. Academic Press, New York.
Swedmark, B. (1955). Développment d'un Gastrotriche Macrodasyoide, *Macrodasys affinis* Remane. *C. R. Acad. Sci. Paris* **240**, 1812-1814.
Swedmark, B. (1958). On the biology of sexual reproduction of the interstitial fauna of marine sand. *Proc. Int. Congr. Zool., 15th Congr. (London)* pp. 327-329.
Teuchert, G. (1968). Zur Fortpflanzung und Entwicklung der Macrodasyoidea (Gastrotricha). *Z. Morphol. Tiere* **63**, 343-418.
White, M.J.D. (1954). The evolution of parthenogenesis. *In* ""Animal Cytology and Evolution," 2nd ed., pp. 339-365. Cambridge Univ. Press, London and New York.
White, M.J.D. (1961). "The Chromosomes," 3rd ed. Methuen, London.
Wilke, U. (1954). Mediterrane Gastrotrichen. *Zool. Jahrb. (Abt. Systematik)* **82**, 497-550.
Zelinka, C. (1889). Die Gastrotrichen. Eine Monographsiche Darstellung ihrer Anatomie, Biologie, und Systematik. Z. *Wiss. Zool.* **49**, 209-384; tafel XI-XV.

CHAPTER 11

KINORHYNCHA

Robert P. Higgins

11.1 Introduction

The Kinorhyncha (= Echinoderida) are a phylum of microscopic marine invertebrates grouped with the "Aschelminthes" by many authors. Kinorhynchs are segmented pseudocoelomates which closely resemble gastrotrichs. Although members of this phylum are widely distributed and often abundant in marine sediments, they are poorly known. Most of what is known about the Kinorhyncha has been recorded in a monograph by Zelinka (1928). The most recent publications, which may be used for further review, have been written by myself (Higgins, 1969a,b; 1971). The preponderance of literature, aside from the monograph, has dealt with systematics. Only a little is known about the reproduction of the nearly 100 described species.

The phylum Kinorhyncha is divided into two distinct orders, the Cyclorhagida and the Homalorhagida. Ten genera have been described. Nearly all of the information on reproduction of kinorhynchs is based on studies of a single genus from each of the two orders: *Echinoderes* in the Cyclorhagida, and *Pycnophyes* in the Homalorhagida.

11.2 Asexual Reproduction

There is no evidence for asexual reproduction in the Kinorhyncha.

11.3 Sexual Reproduction

11.3.1 Sexual Dimorphism

Kinorhynchs are dioecious. Some taxa exhibit external sexual dimorphic characters; however, most do not. Where such characters are missing, sex can be distinguished only by mature gonads. Primary external dimorphism is developed best in the homalorhagids. Gonopores are located ventrally near the anterolateral margin of each terminal sternal plate (Fig. 1 and 5). Homalorhagid males normally have two penis spines associated with each gonopore. Since copulation has never been observed, it is difficult to determine the exact function of these spines. It seems most likely that they function as do the copulatory spicules of the nematodes. Males may possess a series of genital bristles along the anterior border of the gonopore. The size and cuticular morphology of the gonopore may vary with both sex and species. In the case of *Neocentrophyes* (Higgins, 1969b), gonopores and penis spines are located laterally as in the cyclorhagid kinorhynchs. This characteristic reflects the assumed primitively intermediate phylogenetic position of this recently discovered genus.

In the cyclorhagid kinorhynchs such as *Echinoderes*, gonopores are visible at the lateral margin of the junction of the last two terminal segments (Fig. 4 and 10). In other cyclorhagids the gonopores are extremely difficult to observe. With the exception of *Echinoderes* no external primary sex characters have been noted. The sex of most cyclorhagids must be determined on the basis of the mature gonads. Three penis spines are present near the male gonopore and genital bristles are few in number when compared with the males of homalorhagids.

Secondary sexual dimorphic characters in the homalorhagids include a pair of adhesive tubes in the adult males of the two most common homalorhagid genera—*Pycnophyes* and *Trachydemus*—but not in *Neocentrophyes* or any juvenile stages of the former two genera. Adhesive tubes (Fig. 5) are located anteromedially on the sternal plates of the fourth segment (= second trunk segment). Zelinka (1928) assumed that these tubes function in copulation. In addition to the adhesive tubes, other secondary sex characteristics of homalo-

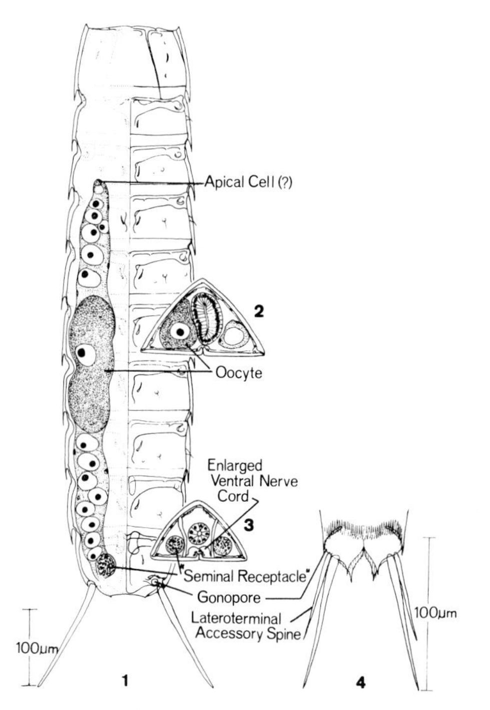

FIGS. 1-4. *Pycnophyes,* adult female. Fig. 1. Left half showing exposed ovary: right half showing ventral surface. Fig. 2-3. Cross sections at the levels indicated. Fig. 4. *Echinoderes,* adult female, ventral aspect of terminal segment.

rhagids include minor differences in the cuticular structure of the terminal segment.

Cyclorhagid secondary sexual dimorphic characters are limited to prominent lateroterminal accessory spines located anterior to the lateroterminal spines of *Echinoderes* females (Fig. 4). Since adhesive tubes can be found in both sexes of some cyclorhagid species and may be partially or entirely missing in others, this character is not helpful in the identification of a male or female.

11.3.2 Sex Determination and Hermaphroditism

No information exists on sex determination, and hermaphroditism has not been observed, nor is it probable. There is little information on sex ratios, but my own casual observations suggest the usual 1:1 ratio.

11.3.3 Anatomy of the Reproductive System and Gametogenesis

Cylindrical gonads, encased in a thin squamous epithelium, are suspended in the pseudocoelom lateral to the intestine and the seg-

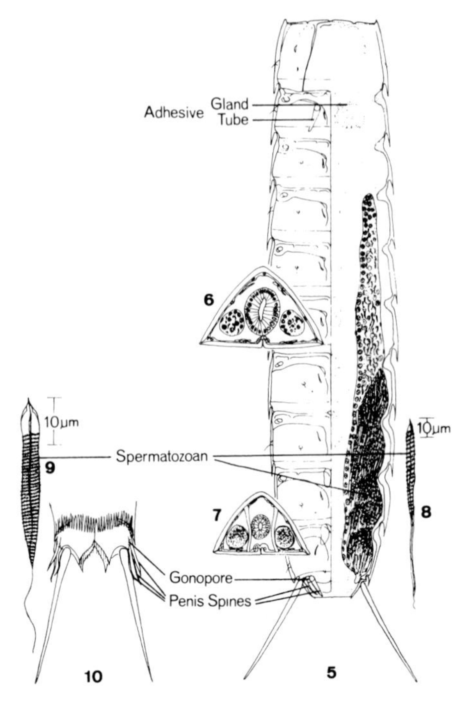

FIGS. 5-10. *Pycnophyes*, adult male. Fig. 5. Right half showing exposed testis: left half showing ventral surface. Fig. 6-7. Cross sections at the levels indicated. Fig. 8. Spermatozoan from *Pycnophyes* [after Zelinka (1928)]. Fig. 9. Spermatozoan from *Echinoderes* [after Zelinka (1928)]. Fig. 10. *Echinoderes*, adult male, ventral aspect of terminal segment.

mentally arranged dorsoventral muscles of the trunk segments. Mature testes (Fig. 5) are slightly larger posteriorly because of the dense accumulation of spermatozoa. Mature ovaries (Fig. 1) commonly develop a large oocyte in the middle. This gives the ovary a slightly distorted appearance from its otherwise uniform outline. The wall of the posterior portion of both testes and ovaries is surrounded by a thin reticulum of muscle fibers similar to those found surrounding the intestine. According to Zelinka (1928) both germ and somatic cells within the ovary are the product of amitotic divisions of a single apical cell. No mitotic divisions have been observed and the subject remains an interesting challenge for research.

Although it is difficult to identify an apical cell or any other distinct cell in the anterior regions of the testis, there is usually a series of rather small, densely stained nuclei with prominent nucleoli. Boykin (1965), in his study of the morphology of *Trachydemus ilyocryptus*, suggests that these are spermatogonia. The area posterior to the spermatogonia is made up of spermatocytes and spermatids.

On the basis of their size, primary and secondary spermatocytes may be recognized. Spermatids have a distorted, somewhat twisted nucleus which changes considerably during spermiogenesis.

The spermatozoa of the Kinorhyncha are arranged in the posterior $\frac{1}{2}$-$\frac{2}{3}$ of the testis when they reach maturity. In *Echinoderes* the spermatozoa are huge, often as much as $\frac{1}{4}$ the length of the adult's body (Fig. 9). In the homalorhagids the spermatozoa are smaller, up to $\frac{1}{10}$ the adult body length as a general rule (Fig. 8). The anterior portion of the typical spermatozoan is sausage-shaped for up to $\frac{2}{3}$ the length of the cell. Although no clearly distinguishable acrosome may be seen, the tip of the cell is hyaline and pointed. Following the minute acrosomal region is a large, dense zone of mitochondria. An axial filament, sometimes extending beyond a sheath in the tail, is present in most of the spermatozoa described by Zelinka (1928).

Oocytes are scattered uniformly throughout the ovary in its early stages of development. As the ovary matures, a single oocyte, rarely two, develops centrally. Smaller oocytes seem to disappear during the growth of this single oocyte, and Zelinka (1928) has suggested that smaller oocytes are nutritive cells. The mature oocyte is richly supplied with granules which are probably yolk. The germinal vesicle is usually spherical and may reach 50 μm in diameter. Its prominent nucleolus may be as large as 10 μm in diameter. No chromatin is evident.

The posterior section of the thin ovary wall becomes a poorly defined oviduct consisting of only three rows of cells. Near the gonopore the oviduct develops a diverticulum called the seminal receptacle because of its general location and the presence of many densely stained bodies which Zelinka identified as sperm (Fig. 1 and 3). Several authors, including myself, are not convinced that Zelinka was correct. Remane (1936) was the first to note that the so-called spermatozoa within the seminal receptacle did not resemble the spermatozoa within the testis of the same species in any manner. One of Remane's suggestions is that this structure might be a gland and the refractile bodies within it might be its secretions. Whereas I have reached no conclusions on the function of this area of the ovary, Boykin (1965) has suggested that this may be the region which supplies oocytes to the anterior portion of the ovary. Clearly more work needs to be done to clarify the function of the "seminal receptacle."

11.3.4 Origin of the Gonads

Gonads begin their development from tissue in the terminal segment of the early juvenile stages. Each gonad develops anteriorly as a solid mass of undifferentiated tissue. Cytodifferentiation begins in the late juvenile stages. No comprehensive studies have been made on this subject.

11.3.5 Reproductive Behavior, Mating, and Oviposition

Many hours of careful observation by several investigators, including myself, have failed to reveal any notable sexual behavior on the part of kinorhynchs. It is assumed that copulation takes place, but the act has not been observed. In certain species in temperate climates, a greater abundance of sexually mature individuals is said to occur in the late winter and early spring. Kozloff (1972), however, has found that two species of *Echinoderes* are reproductive at all seasons. In tropical and semitropical climates I have noted no particular pattern of maturity although extensive year-round studies have not been made.

The first report of oviposition was by Nyholm (1947b) who observed what he thought were homalorhagid eggs laid in the channels excavated by the animals in the uppermost layer of detritus. Nyholm also noted similar bodies enclosed in a ball of mucus attached to the terminal segment of the female. In some instances, they were found deposited in the old cuticle during the molt of "paedogenic" forms of the "Hyalophyes" stage (referring to the late juvenile stage of *Pycnophyes*). Indeed, female homalorhagids, particularly *Pycnophyes*, often are found with a mucous secretion at the terminal segment, but I have not yet observed eggs in this material. It is possible that this mucus is used to attach the eggs to sand grains or detritus, as in the case of many meiobenthic animals. It is also worth noting an interesting correlation between this mucous substance and the distinct enlargement of the two ventral nerve cords in the terminal segments found only in *Pycnophyes* females (Fig. 3). Although no research has been directed to this correlation it should receive some attention by future investigators.

In 1965, I reported finding three female *Pycnophyes frequens*, each with what appeared to be an extruded egg (Higgins, 1965). These dark brown spherical bodies measured approximately 80 μm in diameter and had slightly sculptured walls 5-6 μm thick. These specimens were found in preserved material collected along the coast of Maine in 1962. Six years later, while studying this same homalorhagid species in the region of Woods Hole, Massachusetts, I again found several females each with this same spherical body attached terminally. The living animals were carefully observed over a period of several days but nothing conclusive was noted and no further examples were found.

Eggs of a species of *Echinoderes* (close to *E. dujardini*) have recently been described by Kozloff (1972). These are about 75×65 μm

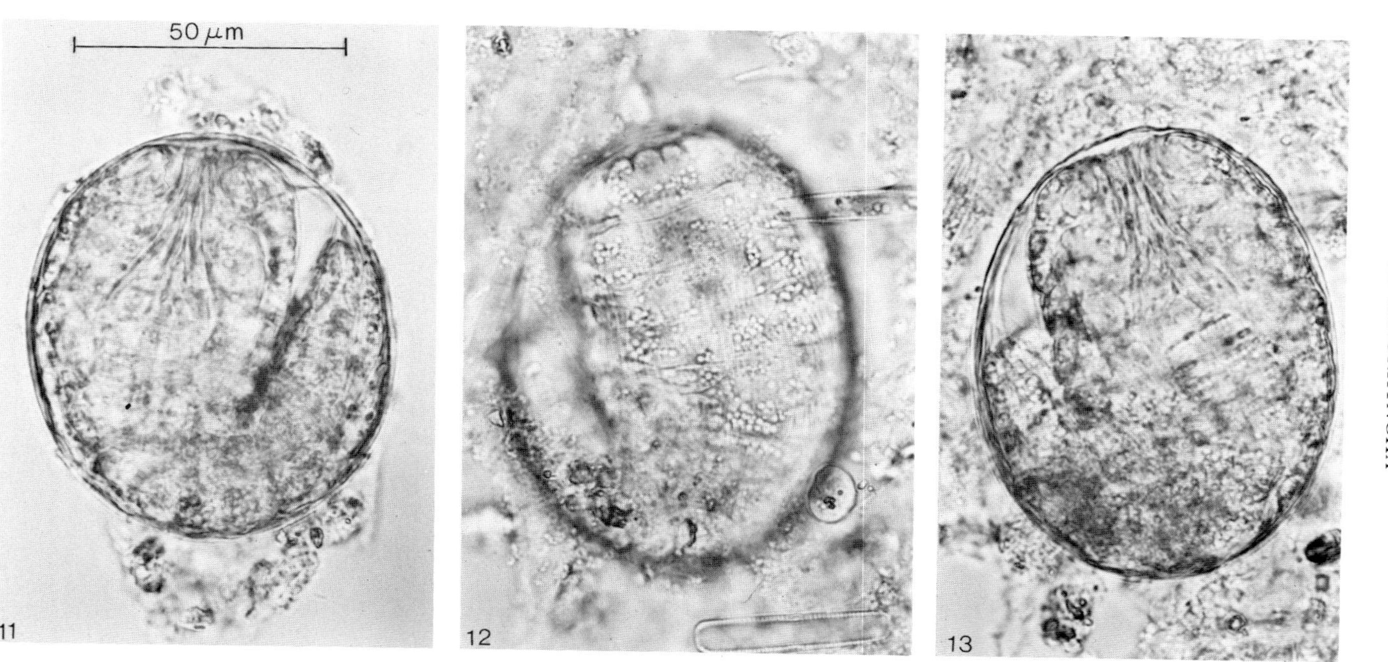

FIGS. 11-13. Juvenile *Echinoderes* immediately prior to hatching shown in different optical sections. [Courtesy of Dr. E.N. Kozloff.]

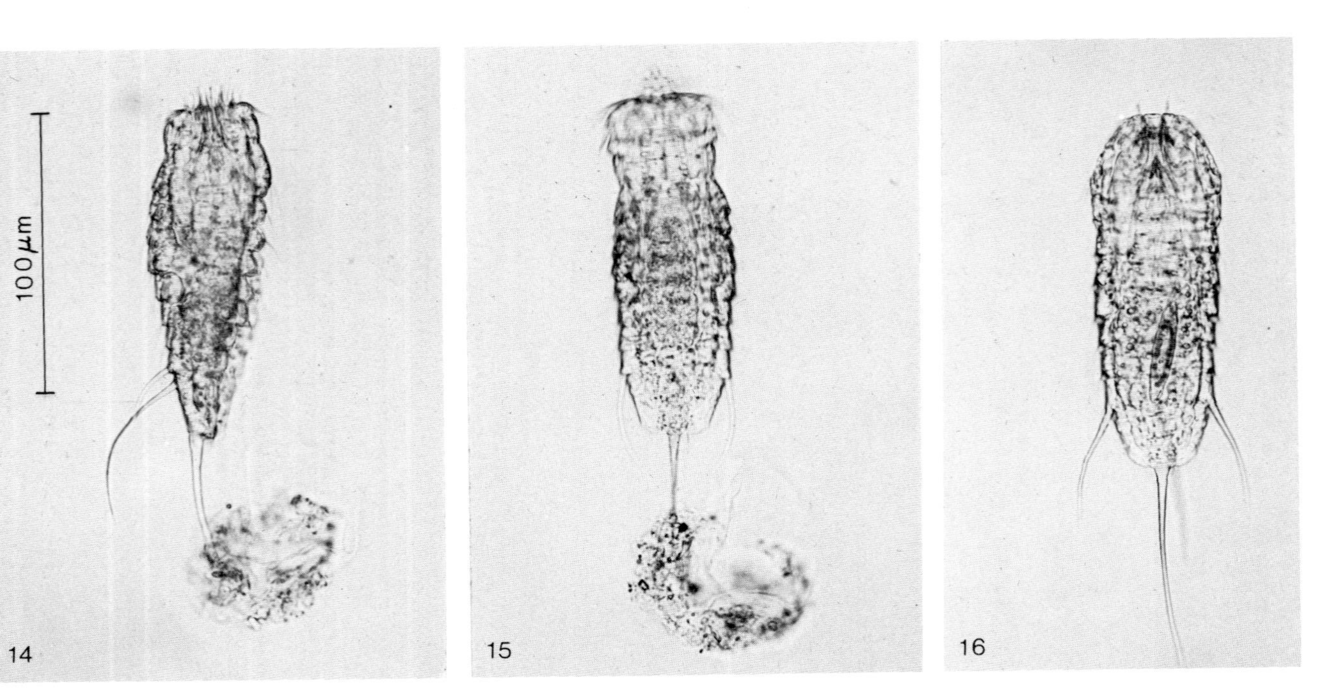

FIGS. 14-16. Sequence of the first "Hapaloderes" stage of *Echinoderes* emerging from egg envelope. Fig. 14. Lateral view from right side. Fig. 15. Ventral view of same specimen. Fig. 16. Ventral view of same specimen, after ingesting a diatom. [Courtesy of Dr. E. N. Kozloff.]

or 80 × 70 μm and have thin and completely transparent envelopes (Fig. 11-13). They are usually so heavily obscured by adherent detritus and microorganisms that it is difficult to recognize them in a search through sediment in which the kinorhynchs are abundant.

11.4 Development

The early development of kinorhynchs — that is, what may be properly called embryology — has not been described. Kozloff (1972), however, has studied the development of *Echinoderes* from the time the young kinorhynch assumes a wormlike form to the point that it hatches. He showed that the early wormlike stage shows no external signs of segmentation and no spines. By the time segmentation becomes evident, the head is rather well developed and scalids have begun to appear. Just before hatching (Fig. 11-13) the juvenile kinorhynch has 11 of the 13 segments characteristic of the adult, and the pattern of spines is like that of 11-segment juveniles found in nature. At the time of hatching (Fig. 14-16) the young kinorhynch straightens and simultaneously everts its head, thus tearing open the envelope within which it is enclosed.

With respect to *Echinoderes*, at least, Kozloff's work contradicts the reports of Nyholm (1947a,b) that there are larval stages, without a head but with long spines, to which segments are added gradually.

According to my observations on *Echinoderes bookhouti* Higgins, 1964 there are at least six stages in its life history (Fig. 17), each derived from a molt. As now confirmed by Kozloff (1972), the earliest stage is 11-segmented. The first molt establishes a 12-segmented juvenile. By the end of the third molt a thirteenth segment is clearly visible. During this process the median terminal spine, characteristic of a series of the first three 'Hapaloderes" stages, shortens relative to the body length. In the case of *E. bookhouti*, whose adult has middorsal spines on segments 6-10, one finds middorsal spines on each of the "Hapaloderes" segments 6 through the midterminal spine of the last segment. At the third molt, with the establishment of the thirteenth segment, the midterminal spine is absent, and midterminal spines persist on segments 6-12. The *Echinoderes* juveniles lacking a midterminal spine are referred to as "Habroderes" stages. At least two such stages are present in *E. bookhouti*. As molting continues the posteriormost middorsal spine will be lost until the adult condition of middorsal spines on segments 6-10 is achieved. During this progressional elimination of middorsal spines, new spines tend not to enlarge in proportion to the increase in body length. Lateral spines may follow this same pattern of development. The more neotenous taxa retain the juvenile pattern of spines on all or nearly all posterior segments, and some have spines on all trunk segments.

A similar series of juvenile stages has been noted by Zelinka (1928) and Nyholm (1947a). In the case of the latter author, he described a series of juvenile stages of *Pycnophyes flaveolatus* having fewer than 11 segments. The first of these Nyholm refers to as a "Leptodemus-like" stage, one without terminal spines as in the juvenile stages of *Trachydemus* which are properly called "Leptodemus" stages. Nyholm's smallest juvenile possessed about seven segments and measured 150 μm in length. I have observed no homalorhagids of less than 11 segments even at the stage closely approximating the size of Nyholm's 7-segmented stage. In the case of *Pycnophyes beaufortensis* Higgins, 1964 (Fig. 18) there is a series of juvenile stages which is not too dissimilar to the series noted in *Echinoderes*. The posthatching stages begin with an 11-segmented "Centrophyes." With each successive molt, a new segment is visible. A series of middorsal spines is present although much less distinctive than the homologous series in *Echinoderes*. The third "Centrophyes" stage molts and the midterminal spine is lost. This establishes a series of

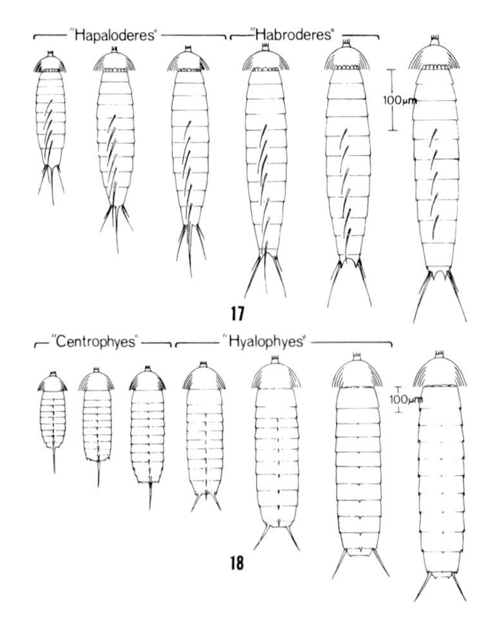

FIGS. 17-18. Fig. 17. Posthatching developmental stages of *Echinoderes bookhouti*. ("Hapaloderes," "Habroderes" and adult) Fig. 18, Posthatching developmental stages of *Pycnophyes beaufortensis* ("Centrophyes," "Hyalophyes" and adult).

three "Hyalophyes" stages, each with 13 segments, hyaline but otherwise resembling the adult. As in the case of the "Habroderes" series, successive molts of the "Hyalophyes" stages will diminish the relative proportions of the middorsal spines until they remain as spinose processes in the adult. Both Nyholm (1947a) and I have noted that the adult may be established without a final molt. In such cases the adult external morphology may vary slightly from that normally characteristic of the species.

In *Trachydemus langi* Higgins, 1964 there is a series of minute spinose processes which diminish in prominence after each molt. A series of six "Leptodemus" stages begins with an eleven-segmented individual, approximately 150 μm in length. Otherwise, as in the adult, the juveniles of *Trachydemus* exhibit no spines on the trunk region.

As noted by Zelinka (1928) and Remane (1936) there are no larval organs which change to adult organs, only a series of juvenile stages in those species so far studied. Juveniles are recognized by their thin, transparent and highly elastic cuticle. The cuticle is weakly divided into segments. Segmental borders are most difficult to define in the posteriormost region of the trunk. Along the middorsal and lateroventral lines the posterior edges of the segments extend posteriorly to form spinose processes in the homalorhagid taxa whereas in the cyclorhagids spines may occur in these places and variably persist in the adult stage. Adhesive tubes, gonopores, and all external primary and secondary sex characters are absent in the juvenile kinorhynch stages. Adults generally have a well-developed chitinous exoskeleton. The segments of the adult trunk are often reinforced along the anterior margin and have an elaborate ventrolateral ball and socket articulation. Adhesive tubes are present in many taxa and some have distinct gonopores and penis spines. The more recognizable adult characters are missing or highly modified in those genera considered neotenous such as *Cateria*, *Neocentrophyes*, and others (Higgins, 1968).

During molting, the entire exoskeleton is lost. This includes the cuticularized foregut and hindgut as in the arthropods. Before molting the hypodermis withdraws from the old cuticle and begins forming a new, very thin cuticle. This process begins in the terminal segment and proceeds anteriorly. The cuticular portion of the head loosens and as soon as this is complete the neck ruptures at its junction with the scalid zone. Within the old cuticle the animal appears to have diminished in size. Its gut appears "inflated" and pulsates during the process of emergence. New scalids form and assist the juvenile's escape from the old cuticle, a process which, from laboratory observations, may require several days.

Acknowledgments

I am grateful for the generous assistance of Dr. E. N. Kozloff, Friday Harbor Laboratories, University of Washington, who made it possible to incorporate his significant studies of *Echinoderes* development into this manuscript.

11.5 References

Boykin, J.D. (1965). The Anatomy of *Trachydemus ilyocryptus* (Kinorhyncha). Ph.D. Thesis. Univ. Washington, 115 pp.

Higgins, R.P. (1964). Three new kinorhynchs from the North Carolina coast. *Bull. Mar. Sci. Gulf Carib.* **14**, 479-493.

Higgins, R.P. (1965). The homalorhagid Kinorhyncha of northeastern U.S. coastal waters. *Trans. Amer. Microsc. Soc.* **84**, 65-72.

Higgins, R.P. (1968). Taxonomy and postembryonic development of the Cryptorhagae, a new suborder for the mesopsammic kinorhynch genus *Cateria*. *Trans. Amer. Microsc. Soc.* **87**, 21-39.

Higgins, R.P. (1969a). Indian Ocean Kinorhyncha: 1, *Condyloderes* and *Sphenoderes*, new cyclorhagid genera. *Smithson. Contrib. Zool.* **14,** 1-13.

Higgins, R.P. (1969b). Indian Ocean Kinorhyncha: 2, Neocentrophyidae, a new homalorhagid family. *Bull. Biol. Soc. Wash.* **82,** 113-128.

Higgins, R.P. (1971). A historical overview of kinorhynch research. *In* "Proceedings of the First International Conference on Meiofauna" (N.C. Hulings, ed.), *Smithson. Contrib. Zool.* No. **76,** 25-31.

Kozloff, E.N. (1972). Some aspects of development in *Echinoderes* (Kinorhyncha). *Trans. Amer. Microsc. Soc.* **91**, 119-130.

Nyholm, K. G. (1947a). Contributions to the knowledge of the postembryonic development in the Echinoderida Cyclorhagae. *Zool. Bidr. Uppsala* **25**, 423-428.

Nyholm, K.G. (1947b). Studies in the Echinoderida. *Arkiv. Zool.* **39A** (14), 1-36.

Remane, A. (1936). Gastrotricha und Kinorhyncha. *In* "Klassen und Ordnungen des Tierreichs" (H.G. Bronn, ed.), Bd. 4, Abt. 2, Buch 1, Teil 2, 1-385, Akad. Verlagsges., Leipzig.

Zelinka, C. (1928). "Monographie der Echinodera." Wilhelm Englemann, Leipzig, 396 pp., 27 pl.

AUTHOR INDEX

Numbers in italics refer to the pages on which the complete references are listed.

C

G

H

I

J

R

S

T

Z

SUBJECT INDEX

S

T

V

Z

TAXONOMIC INDEX

D

E

I

K

L

M

N

O

P

R

Z